International Conference on Novel Functional Materials (ICNFM2018)

MATEC Web of Conferences Volume 238 (2018)

Anshan, China
16 - 18 September 2018

Editor:

Lu Han

ISBN: 978-1-5108-7563-0

Printed from e-media with permission by:

Curran Associates, Inc.
57 Morehouse Lane
Red Hook, NY 12571

Some format issues inherent in the e-media version may also appear in this print version.

No changes have been made to the content of these proceedings. There may be changes to pagination and minor adjustments for aesthetics.

Printed by Curran Associates, Inc. (2019)

For additional information, please contact EDP Sciences – Web of Conferences at the address below.

EDP Sciences – Web of Conferences
17, Avenue du Hoggar
Parc d'Activité de Courtabœuf
BP 112
F-91944 Les Ulis Cedex A
France

Phone: +33 (0) 1 69 18 75 75
Fax: +33 (0) 1 69 28 84 91

contact-edps@webofconferences.org

Additional copies of this publication are available from:

Curran Associates, Inc.
57 Morehouse Lane
Red Hook, NY 12571 USA
Phone: 845-758-0400
Fax: 845-758-2633
Email: curran@proceedings.com
Web: www.proceedings.com

TABLE OF CONTENTS

ICNFM2018 Committee

Conference Chairs

Prof. Zhang Zhiqiang, President of University of Science and Technology LiaoNing (China)
Prof. Ju Dongying, Vice President of Saitama University of Technology (Japan)

Steering Committee

Prof. An Baigang, University of Science and Technology LiaoNing, China
Prof. Luo Xudong, University of Science and Technology LiaoNing, China
Prof. Zhao Hongyang, University of Science and Technology LiaoNing, China
Prof. Li Xianchun, University of Science and Technology LiaoNing, China
Prof. He Zhongliang, University of Science and Technology LiaoNing, China
Prof. Zhou Yanwen, University of Science and Technology LiaoNing, China
Prof. Hu Zhizhi, University of Science and Technology LiaoNing, China

Program Chairs

Prof. Hasebe,Yasushi, Saitama University of Technology, Japan
Prof. Zhang Run, Australian Institute of Bioengineering and Nanotechnology (AIBN), Australian
Prof. Hideki Okamoto, Okayama University, Japan
Prof. Stephen C. Bergmeier, Ohio State University, USA

International Program Committee

Prof. Ramesh Singh, University of Malaya, Malaysia
Prof. Mohd Hamdi Bin Abd Shukor, University of Malaya, Malaysia
Prof. Ahmed El-Shafie, University of Malaya,Malaysia
Prof. Stefan Talu，Technical University of Cluj-Napoca, Romania
Dr. Beddiaf ZAIDI，University of Batna, Algeria
Dr. ANAND VYAS，Hong Kong Polytechnic University, Hong Kong
Prof. RAJIV KUMAR, Goswami Ganesh Dutt Sanatan Dharam College，

Preface

Dear Distinguished Authors and Guests,

ICNFM2018 aims to bring together leading academic scientists, researchers and research scholars to exchange and share their experiences and research results on all aspects of advanced functional materials. It also provides a premier interdisciplinary platform for researchers, practitioners and educators to present and discuss the most recent innovations, trends, and concerns as well as practical challenges encountered and solutions adopted in the fields of advanced functional materials.

The proceeding records the fully refereed papers presented at the conference. The main conference themes and tracks are Advanced Materials Design and Mechanics and other related topics. Hopefully, all participants and other interested readers benefit scientifically from the proceedings and also find it stimulating in the process.

We would like to thank the program chairs, organization staff and the members of the program committees for their work. On behalf of the organizing committee, I would like to thank all the editors for their great support to ICNFM2018. Without their excellent editorial work, ICNFM2018 will not be published so timely and successfully. We are grateful to all those who have contributed to the success of ICNFM2018.

Finally we wish all the authors and attendees of ICNFM2018 a unique, rewarding and enjoyable memory at ICNFM2018 in Anshan, China.

Editor of ICNFM2018

Statement of Peer review

In submitting conference proceedings to *Web of Conferences*, the editors of the proceedings certify to the Publisher that

1. They adhere to its **Policy on Publishing Integrity** in order to safeguard good scientific practice in publishing.
2. All articles have been subjected to peer review administered by the proceedings editors.
3. Reviews have been conducted by expert referees, who have been requested to provide unbiased and constructive comments aimed, whenever possible, at improving the work.
4. Proceedings editors have taken all reasonable steps to ensure the quality of the materials they publish and their decision to accept or reject a paper for publication has been based only on the merits of the work and the relevance to the journal.

Title, date and place of the conference

International conference on novel functional materials (ICNFM2018)
Sep. 16-18, 2018 in Anshan, China.

Proceedings editor(s):

Lu Han, University of Science and Technology LiaoNing, China

Date and editor's signature

November 13. 2018

MATEC Web of Conferences 238, 01001 (2018) https://doi.org/10.1051/matecconf/201823801001
ICNFM2018

Space Charge Accumulation in Silicone Rubber Influenced by Poole-Frenkel Effect

Wenbin Kang[1], Shaoxing Meng[1], Haozhe Cui[2], Yuwei Li[2], Rui Mi[2], Chenyu Yan[2], Shijun Li[2], Daomin Min[2,a]

[1]China Electric Power Research Institute (Wuhan Branch), 430074 Wuhan, Hubei, China
[2]State Key Laboratory of Electrical Insulation and Power Equipment, School of Electrical Engineering, Xi'an Jiaotong University, 710049 Xi'an, Shaanxi, China

Abstract. With the rapid increase of electrical energy consumption in metropolises, more and more power cables are utilized in power grid or urban power network to transmit and distribute electrical energy. Silicone rubber is widely used as polymeric insulating materials of power cable accessories due to their excellent electrical and thermal performances. However, under high dc electric field space charges can accumulate inside the silicone rubber, which will distort the electric field in the bulk of the material, influencing the reliability and safety of operation. A bipolar charge injection and transport model is adopted to investigate the accumulation of space charges and distortion of electric field in silicone rubber. It is found that when charge injection rate is higher than the charge migration rate, space charges will accumulate in the material and the accumulation increases with an increase in electric field. The influence of the Poole-Frenkel effect is then studied. It is found that stronger Poole-Frenkel effect can enhance the nonlinearity of effective carrier mobility of silicone rubber, reducing the accumulation of space charges. This indicates that tuning the nonlinearity of effective carrier mobility by nano-doping can suppress space charges, improving the reliability of power cable accessories.

1 Introduction

Due to the rapid development of the economy, a surging demand for electricity is rising in China. By the end of 2017, State Grid Corporation of China has 1.38 TW of power generating capacity in its service area, growing by 8.5% compared to the previous year and accounting for 78% of the total installed generating capacity in China [1]. In addition, the electrical energy consumptions in metropolises are very high, which are 106G kWh and 152G kWh in Beijing and Shanghai, respectively [2]. The rapid increase of electrical energy consumption in metropolises requires more power cables to be utilized in power grid or urban power network to transmit and distribute electrical energy [3].

Recent years, high voltage direct current (HVDC) power cables are widely used in power grid. Accordingly, the quantity demanded for power cable accessories which are used to interconnect two power cables or a power cable and a power apparatus increases rapidly. Power cable accessories can mitigate the electric field distortion at the cutting point of shielding layer, avoiding the occurrence of electrical breakdown or surface flashover on power cables. However, power cable accessories operate in a harsh environment consisting of distorted electric field, high temperature, and mechanical stress, which can result in fault with high probability. Considering the complexity of cable structure and distribution of dc electric field, the problem is more severely for dc power cable accessories. Statistical results show that more than 70% failures in power cable transmission lines are caused by the faults of power cable accessories [3, 4]. Therefore, as an important component of dc power cable transmission system, the operation safety of power cable accessories is a key issue for the reliability of power system.

For HVDC power cable, the electric field is distributed according to dc conductivity in polymeric insulating materials. Since the dc conductivity of polymeric insulating materials depends strongly on electric field and temperature. The electric field is nonuniform in polymeric insulating materials. Silicone rubber is widely used as polymeric insulating materials in power cable accessories due to their excellent electrical and thermal performances. It is of interest to investigate the space charge formation and accumulation properties in silicone rubber under high dc electric field via a bipolar charge injection and transport model.

2 Bipolar charge injection and transport model

Figure 1 illustrates the schematic of bipolar charge injection and transport model in a polymeric material [5, 6]. A polymeric material with a thickness of d is clamped by two electrodes, and a voltage is applied to the material. Electrons and holes are injected into the conduction band

a Corresponding author: forrestmin@xjtu.deu.cn

E_C and valence band E_V of the polymeric material. The injected electrons and holes will migrate toward anode and cathode under the applied electric field. There are a lot of traps in polymeric materials formed by physical disorders and chemical impurities. These traps can capture mobile electrons and mobile holes during their migration. Trapped charges can be activated to mobile states by thermal energy or Poole-Frenkel effect which means the potential barrier can be reduced by an electric field to assist the detrapping of charges [7-9]. When mobile electrons and mobile holes encounter each other in the material, they will recombine via Shockley-Read-Hall mechanism [10]. While mobile electrons and trapped holes, as well as mobile holes and trapped electrons will recombine via Langevin recombination mechanism [11]. Electrons and holes that are not recombine inside the material will flow out to its electrodes.

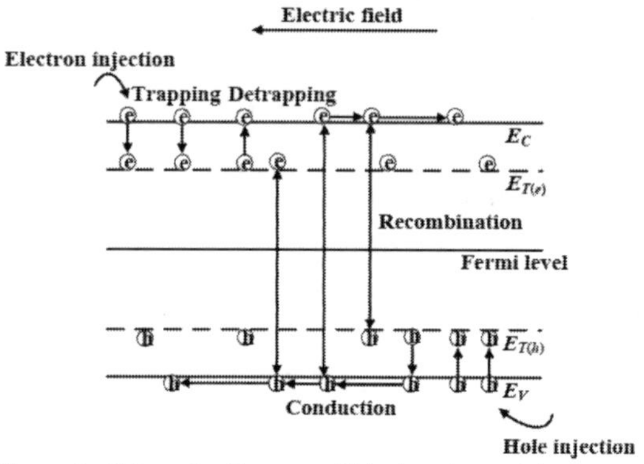

Figure 1. Schematic diagram of bipolar charge injection and transport model in a polymeric material. Here E_C and E_V are the energies of conduction band and valence band respectively, and $E_{T(e)}$ and $E_{T(h)}$ are the energies of deep traps for electrons and holes respectively in the forbidden band.

2.1 Injections of positive and negative charges from electrodes into silicone rubber

The injected current densities of electrons and holes from anode and cathode, $j_{in(e)}$ and $j_{in(h)}$, are determined by electric field, temperature, and effective potential barrier, which are described by following equations [12],

$$j_{in(e)} = 4\psi^2 N_C e \mu_C F \exp\left(-E_{T(e)}/k_B T\right) \exp f^{1/2} \quad (1)$$

$$j_{in(h)} = 4\psi^2 N_V e \mu_V F \exp\left(-E_{T(h)}/k_B T\right) \exp f^{1/2} \quad (2)$$

where N_C and N_V are the densities of states in conduction band and valence band in $1/m^3$ respectively, e is the elementary charge in C, μ_C and μ_V are carrier mobilities in conduction band and valence band in $m^2/(Vs)$ respectively, F is the electric field in V/m, $E_{T(e)}$ and $E_{T(h)}$ are the energies of deep traps for electrons and holes in eV, k_B is the Boltzmann constant, and T is the sample temperature in K. Furthermore, f is the reduced electric field and ψ is the field-dependent parameter. The carrier mobilities in conduction band and valence band are assumed to be 1×10^{-5} $m^2/(Vs)$.

The reduced electric field is determined by Coulomb radius, electric field, and temperature, which is $f=eFr_c/k_BT$. The coulomb radius r_c is equal to $e^2/4\pi\varepsilon_0\varepsilon_r k_B T$. Here ε_0 is the permittivity of vacuum in F/m and ε_r is the dielectric constant of silicone rubber. Then, the field-dependent parameter $\psi(f)$ can be obtained by the following equation $\psi(f)=f^{-1}+f^{-1/2}-f^{-1}(1+2f^{1/2})^{1/2}$.

2.2 Self-consistent drift-diffusion equations

A set of self-consistent drift-diffusion equations are utilized to calculate the transport and accumulation of space charges and the distribution of electric field inside silicone rubber, and the conduction current through the material. The self-consistent drift-diffusion equations consist of charge conduction equation, charge continuity equation, and Poisson's equation [5, 6].

$$j_{c(e)}(x,t) = q_{free(e)}(x,t)\mu_{0(e)}F(x,t) \quad (3)$$

$$j_{c(h)}(x,t) = q_{free(h)}(x,t)\mu_{0(h)}F(x,t) \quad (4)$$

$$\frac{\partial q_{free(e)}(x,t)}{\partial t} + \frac{\partial q_{trap(e)}(x,t)}{\partial t} + \frac{\partial j_{c(e)}(x,t)}{\partial t} = S_e \quad (5)$$

$$\frac{\partial q_{free(h)}(x,t)}{\partial t} + \frac{\partial q_{trap(h)}(x,t)}{\partial t} + \frac{\partial j_{c(h)}(x,t)}{\partial t} = S_h \quad (6)$$

$$\frac{\partial^2 \phi(x,t)}{\partial x^2} = -\frac{q_{net}(x,t)}{\varepsilon_0 \varepsilon_r} \quad (7)$$

where $j_{c(e)}$ and $j_{c(h)}$ are the conduction current densities of electrons and holes in conduction band and valence band in A/m^2 respectively, $q_{free(e)}$ and $q_{free(h)}$ are the carrier densities of electrons and holes in C/m^3 respectively, $q_{trap(e)}$ and $q_{trap(h)}$ are the densities of trapped electrons and trapped holes in C/m^3 respectively, $\mu_{0(e)}$ and $\mu_{0(h)}$ are the carrier mobilities of electrons and holes in silicone rubber in $m^2/(Vs)$ respectively, S_e and S_h are the charge trapping-detrapping-recombination terms for electrons and holes respectively, ϕ is the potential in the material in V, and q_{net} is the net charge density, which is $q_{net}=q_{free(e)}+q_{trap(e)}+q_{free(h)}+q_{trap(h)}$. Furthermore, x is the position in the material and t is the elapse time after the application of voltage on the material.

2.2 Charge trapping-detrapping-recombination dynamic equations

The trapping-detrapping-recombination dynamics for mobile electrons $S_{e\mu}$, trapped electrons S_{et}, mobile holes $S_{h\mu}$, and trapped holes S_{ht} are described by the following equations [5, 6].

$$S_{e\mu} = -P_{tr(e)} q_{free(e)} \left(1 - q_{trap(e)}/eN_{T(e)}\right) + P_{de(e)} q_{trap(e)} - R_{e\mu,h\mu} q_{free(e)} q_{free(h)} - R_{e\mu,ht} q_{free(e)} q_{free(h)} \quad (8)$$

$$S_{et} = P_{tr(e)} q_{free(e)} \left(1 - q_{trap(e)}/eN_{T(e)}\right) - P_{de(e)} q_{trap(e)} - R_{et,h\mu} q_{trap(e)} q_{free(h)} \quad (9)$$

$$S_{h\mu} = -P_{tr(h)} q_{free(h)} \left(1 - q_{trap(h)}/eN_{T(h)}\right) + P_{de(h)} q_{trap(h)} - R_{e\mu,h\mu} q_{free(e)} q_{free(h)} - R_{et,h\mu} q_{trap(e)} q_{free(h)} \quad (10)$$

$$S_{ht} = P_{tr(h)} q_{free(h)} \left(1 - q_{trap(h)}/eN_{T(h)}\right) - P_{de(h)} q_{trap(h)} - R_{e\mu,ht} q_{free(e)} q_{trap(h)} \quad (11)$$

where $P_{tr(e)}$ and $P_{tr(h)}$ are the trapping coefficients for mobile electrons and holes in s^{-1} respectively and $P_{tr(e,h)}=eN_{T(e,h)}/\varepsilon_0\varepsilon_r$. $N_{T(e)}$ and $N_{T(h)}$ are the densities of deep traps for electrons and holes in $1/m^3$ respectively, $R_{e\mu,h\mu}$ is the recombination coefficient between mobile electrons and mobile holes, $R_{e\mu,h\mu}=(\mu_{0(e)}+\mu_{0(h)})/\varepsilon_0\varepsilon_r$. $R_{e\mu,ht}$ and $R_{et,h\mu}$ are the trap-assisted recombination coefficients between mobile electrons and trapped holes and between trapped electrons and mobile holes respectively, $R_{e\mu,ht}=\mu_{0(e)}/\varepsilon_0\varepsilon_r$ and $R_{et,h\mu}=\mu_{0(h)}/\varepsilon_0\varepsilon_r$.

The charge detrapping coefficients P_{de} is described by the following equation [5, 6].

$$P_{de} = \upsilon_0 \exp\left(-\frac{E_T}{k_B T}\right) G(F) \tag{12}$$

where υ_0 is the attempt-to-escape frequency in s^{-1}, which is assumed to be 1×10^{12} s^{-1}, and $G(F)$ is the Poole-Frenkel effect improved by Ieda et al [7].

$$G(F)=\frac{4\gamma}{\alpha^2 F}\sinh\left(\frac{\alpha^2 F}{4\gamma}\right), \alpha^2 F \le 2\gamma \tag{13}$$

$$G(F)=\frac{1}{\alpha^2 F}\left[\left(\alpha F^{1/2}-1\right)\exp\left(\alpha F^{1/2}-\gamma\right)+\exp(\gamma)\right] -\frac{2\gamma}{\alpha^2 F}\exp\left(-\frac{\alpha^2 F}{4\gamma}\right), \alpha^2 F > 2\gamma \tag{14}$$

where α is a coefficient relating to the potential barrier lowering caused by Pool-Frenkel effect, $\alpha=2(e^3/4\pi\varepsilon_0\varepsilon_r)^{1/2}/(mk_BT)$, and m is constant changing from 1 to 2, while γ is a constant which is 3/2 in the present work.

2.2 Charge transport parameters for silicone rubber

The dielectric constant measured by novocontrol broadband dielectric relaxation spectroscopy is about 3.5 at 50 Hz at room temperature. The carrier mobility controlled by shallow traps is 7.0×10^{-14} $m^2/(Vs)$ for both electrons and holes [13]. The densities of deep traps for electrons and holes are 6.2×10^{20} $1/m^3$ according to surface potential decay experiments, while the energies of deep traps for electrons and holes are 0.90 eV [13]. The potential barriers are assumed to be the same as deep traps for the charge injections from cathode and anode into silicone rubber.

3 Simulation results and discussion

3.1. Distributions of space charges and electric fields

Figure 2 shows space charge distributions in silicone rubber at various times under an applied voltage of 50 kV at constant m of 2.0. After the application of voltage, electrons and holes will be injected into the polymeric material by equations (1) and (2) respectively. Since the carrier mobility of the silicone rubber is very low and deep traps in the material can capture mobile charge carriers, injected charges can accumulate in the bulk, which is called homo space charges. As shown in Figure 2, negative charges are accumulated in the area near the cathode, while positive charges are accumulated in the area near the anode. As electrons and holes move toward anode and cathode under electric field respectively, the magnitude and area of space charges increase with time. The magnitudes of positive space charges are 7.27 C/m^3 and 53.68 C/m^3 at t of 10 s and 1000 s, respectively. The widths of space charge accumulation are 0.027 mm and 0.265 mm at t of 10 s and 1000 s, respectively.

Figure 3 shows the distributions of electric fields in silicone rubber at various times under an applied voltage of 50 kV at constant m of 2.0. As homo space charges accumulating in silicone rubber, electric fields near the interfaces between the material and its electrodes decreases gradually, while electric fields in the middle of the material increase with time. The electric fields at the interface between the silicone rubber and cathode are reduced from 5.00×10^7 V/m to 3.64×10^7 V/m and 7.93×10^6 V/m at 50 s and 1000 s, respectively. The electric fields in the middle of the material are 5.05×10^7 V/m and 5.44×10^6 V/m at 50 s and 1000 s, respectively.

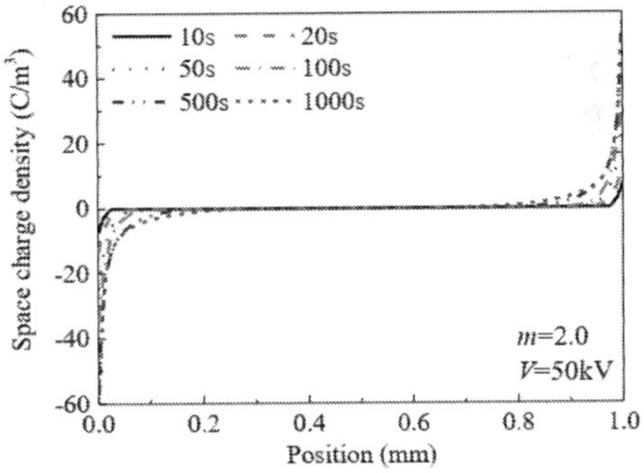

Figure 2. Distributions of space charges in silicone rubber at various times at m of 2.0.

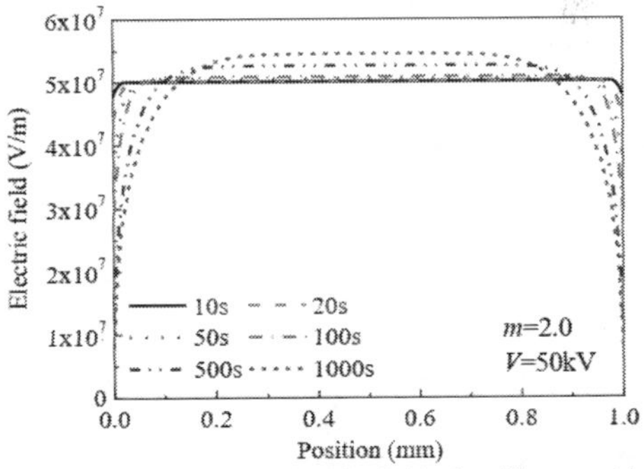

Figure 3. Distributions of electric fields in silicone rubber at various times at m of 2.0.

Figure 4 demonstrates space charge distributions at steady state in silicone rubber at various electric fields at m of 2.0. The higher electric field can result in more electrons and holes being injected into the polymeric material. The magnitudes of positive space charges are 30.92 C/m^3 and 61.06 C/m^3 at electric fields of 1×10^7V/m and 9×10^7V/m, respectively. The area accumulated space charges is also increased by increasing the electric field. The widths of space charge accumulation are 0.026 mm and 0.447 mm at electric fields of 1×10^{7}V/m and 9×10^{7}V/m, respectively.

Figure 5 shows the electric field distributions at steady state in silicone rubber at various electric fields at *m* of 2.0. More homo space charges are injected and accumulated inside the polymeric material can cause severer distortion of electric field. The maximum electric fields in the middle of silicone rubber are 1.01×10^{7} V/m and 1.01×10^{8} V/m at electric fields of 1×10^{7}V/m and 9×10^{7}V/m, respectively.

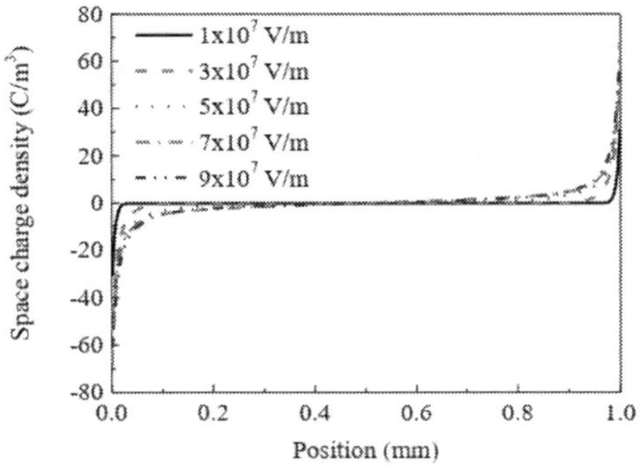

Figure 4. Space charge distributions of silicone rubber at various electric fields at *m* of 2.0.

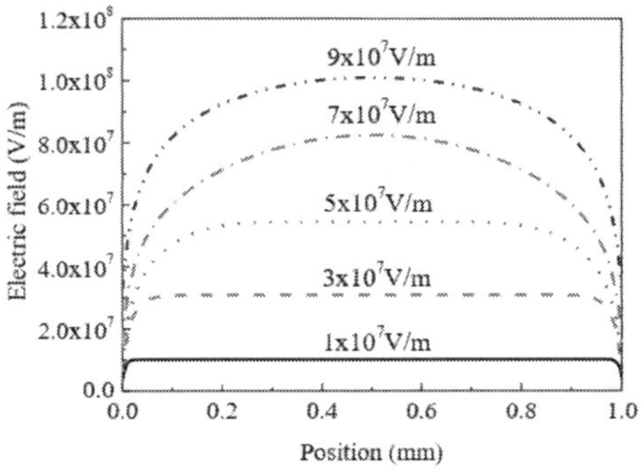

Figure 5. Electric field distributions of silicone rubber at various electric fields at *m* of 2.0.

3.2 Electric field dependent current density and space accumulation

Under the application of electric field, carriers will transport through the material, forming current which can be detected from the external circuit by a pico-ampere meter. Figure 6 demonstrates current density flowed through silicone rubber as a function of electric field at various *m*. The potential barrier width of trapped charges can be modified by the Poole-Frenkel effect. Smaller *m* means stronger Poole-Frenkel effect, which can facilitate the migration of charge carriers. It is observable that the current density is higher at smaller *m*.

The simulation results indicate that changing the constant *m* by incorporating nanoparticles into polymeric materials can tune their nonlinear conductivity. It can maintain high resistivity at relatively low electric fields. At high electric fields, it then can facilitate the migration of charge carriers so that reduce the accumulation of space charges, which is shown in Figure 7. It indicates that tuning the nonlinearity of effective carrier mobility can suppress the accumulation of space charges, improving the reliability of power cable accessories.

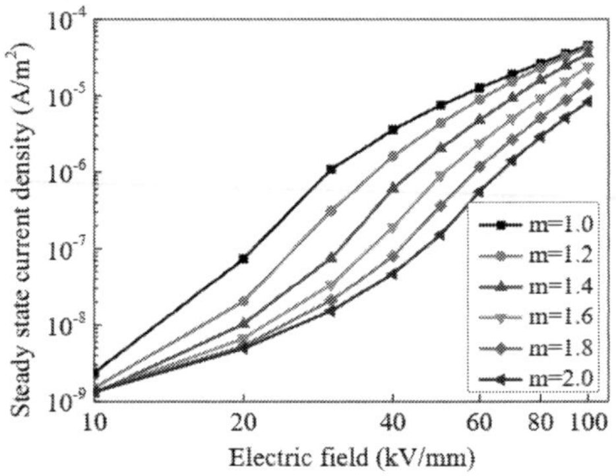

Figure 6. Current density as a function of electric field in silicone rubber at various *m* from 1.0 to 2.0.

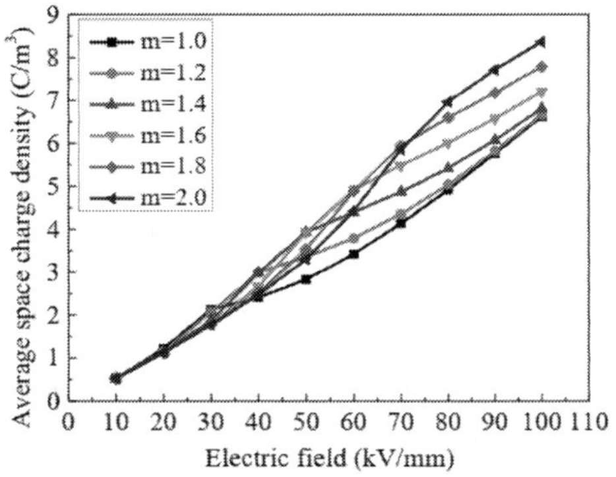

Figure 7. Average space charge accumulation as a function of electric field in silicone rubber at various *m* from 1.0 to 2.0.

4 Conclusion

A model consisting of bipolar charge injection and transport as well as charge trapping-detrapping-recombination dynamics with Poole-Frenkel effect is adopted to investigate the accumulation of space charges and distortion of electric field in silicone rubber. It was found space charges will accumulate in the material when charge injection rate is higher than the charge migration rate, and the accumulation increases with the increase in electric field. Then, we studied the influence of Poole-Frenkel effect on space charge accumulation and external current density. It was found that stronger Poole-Frenkel effect with smaller *m* can enhance the nonlinearity of effective carrier mobility of silicone rubber, reducing the accumulation of space charges, which can improve the operational safety of power cable accessories.

References

This work is supported by the State Grid Corporation of China Science Projects "Study on the Influence of Harmonic Voltage on Cable Joint and Capacitive

Bushing" (EPRIGYKJ(2015)4268) and "Study on computational method of electric field in dc insulation system" (no. 522722160001).

References

[1]http://www.sgcc.com.cn/html/sgcc_main/col2017041241/column_2017041241_1.shtml and https://chinaenergy portal.org /2017-electric-power-industry-summary-statistics/.
[2]http://shoudian.bjx.com.cn/html/20180129/877269.shtml and https://baijiahao.baidu.com/s?id=1594623958914905 384&wfr=spider&for=pc.
[3] Y. Zhou, S. M. Peng, J. Hu and J. L. He, IEEE Trans. Dielectr. Electr. Insul., **24**, 1308-1318, (2017)
[4] Y. J. Fei, Y. X. Zhang and Y. X. Zhou, Adv. Tech. Electr. Eng. Ener., **33**, 30-34, (2014)
[5] D. M. Min, C. Y. Yan, R. Mi, H. Z. Cui, Y. W. Li, W. W. Wang, M. Frechette and S. T. Li, IEEE Nanotechnology Magazine, **2**, 15-22, (2018)
[6] D. M. Min, S. T. Li and Y. Ohki, IEEE Trans. Dielectr. Electr. Insul., **23**, 507-516, (2016)
[7] M. Ieda, G. Sawa and S. Kato, J. Appl. Phys., **42**, 3737-3740, (1971)
[8] J. L. Hartke, J. Appl. Phys., **39**, 4871-4873, (1968)
[9] J. G. Simmons, Phys. Rev., **155**, 657-660, (1967)
[10] W. Shockley and W. T. Read, Phys. Rev., **87**, 835-842, (1952)
[11] J. J. M. van der Holst, F. W. A. van Oost, R. Coehoorn and P. A. Bobbert, Phys. Rev. B, **80**, 235202(8), (2009)
[12] J. C. Scott and G. G. Malliaras, Chem. Phys. Lett., **299**, 115-119, (1999)
[13] D. M. Min, C. Y. Yan, Y. Huang, S. T. Li and Y. Ohki, Polymers, **9**, 1-15, (2017)

Photofunctional hybrid ionogel prepared by doping the terbium complex covalently modified graphene within a stationary ionic liquid matrix

Lijia Liu[1], Meiyu, Wang[1], Qiuping Li[1,a]

[1]*School of Materials and Chemical Engineering, Ningbo University of Technology, No.201, Fenghua Road, Jiangbei District, Ningbo, Zhejiang 315211, P. R. China.*

Abstract. Ionogel is an emerging hybrid material which is usually constructed by confined an ionic liquid into some special matrices, and it is expected to have great potential applications in the field of electrolyte materials, optical-electric materials and so on. Here, a terbium complex covalently modified graphene was introduced into the 1-butyl-3-methylimidazolium bis[(trifluoromethyl)sulfonyl]imide ionic liquid based ionogel matrix. The resulted material is a kind of novel photoluminescent hybrid material which has a transparent appearance and shows strong green under the UV light. The photoluminescent properties of the ionogel has been measured in details, revealing a broad excitation spectra, line-like emission spectra and long luminescence lifetimes which make it great potential applications in the preparation of optical-electric devices, together with the intrinsic conductive property of the embedded ionic liquid.

1 Introduction

Lanthanide complexes functionalized hybrid materials have gathered lots of attentions from the material scientists over the past decades. It is mainly due to the combination of lanthanide complexes with various matrices would have promised a novel multi-functional materials that could be used in some emerging research areas such as biomedical imaging, optical-electrical devices or photoluminescent makers [1]. Duing to effective intramolecular energy transfer (the so-called “antenna effect”), the lanthanide complexes are characterized by long lifetime, narrow-width emission bands in the visible and near infrared regions, large Stokes shift, and it can be easily combined with the other host matrices through the ligand-derived conjugated molecular [2]. However, the poor thermal-stability, chemical-stability and low mechanical strength of traditional lanthanide complexes has limited its practical applications [3-4]. Thus, in order to improve the photo-stability, the lanthanide complexes have been introduced into various organic-inorganic systems to build the so-called lanthanide organic-inorganic hybrid materials which had been proven to have the potential applications for new developments [5-6]. The combination of lanthanide complexes with the other organic and inorganic components could integrate the merits of different matrices together, for instance, the processability, thermal and chemical stability, conductivity, bio-compatibility and even the second luminescence centers. Besides, most of the photosensitive ligands can be easily modified by some couple agent and then linked to the corresponding supported matrix, which also makes the construction of lanthanide organic-inorganic hybrid materials feasible and variable. From this bases, the lanthanide complexes have been covalently grafted to various silicon-oxygen matrices, polymer matrices, carbon matrices etc. and formed numerous luminescent hybrid materials [7-11].

Recently, the nanoparticle composited gel materials which are built by incorporating a nanoparticle into the polymer networks derived gel systems have arouse the general attentions [12]. This combination may result in a synergistic property enhancement of each component, and even holds the promise of providing a multi-functional hybrid materials with applications in diverse fields such as nano-medicine, optoelectronic devices, drug delivery and so on. Especially the new developed ionogel has been put on lots of focus. Ionogel is an emerging hybrid material which is prepared by the confinement of ionic liquid in the well-designed three-dimension network usually comes from the cross-linking silica-oxygen, polymer or metal-organic matrices [13]. Because of the inherited ionic conductivity performance that comes from the stationary ionic liquid, it has been expected to have potential application in the fields such as electrochemical sensing, supercapacitor, luminescent hybrid materials and so on. It has been proved that the introduction of lanthanide complexes into an ionogel matrix would provide a new opportunity to design and synthesis novel hybrid materials, which is hoped to have the outstanding luminescent properties of lanthanide complex and the good ionic conductivity comes from ionic liquids [14-16].

Inspired by the above discussion, we have designed and synthesized a novel ionogel by doped the terbium

[a] Corresponding author: liqiuping@nbut.edu.cn

complex covalently bonded graphene to a poly(methyl methacrylate) (PMMA) confined ionic liquid hybrid matrix. The photoluminescent properties of the hybrid ionogel had been measured in details.

2 Experimental sections

2.1 Materials

2,2′-Bipyridyl-4,4′-dicarboxylic acid (96%, BPC), N,N′-Dicyclohexylcarbodiimide (99%, DCC), 4-Dimethylaminopyridine (99%, DMAP), 1,6-Hexamethyl-enediamine (99%, HMEDA), azodiisobutyronitrile (98%, AIBN), Methyl methacrylate (99.5%, MMA) and Trifluo-roacetylacetone (98%, TFA) were provided by Aladdin reagents Co. LTD. Graphene oxide (GO) was purchased from Shenzhen Turing evolutionary technology Co. LTD. 1-butyl-3methylimidazoliumbis[(trifluoromethyl)sulfonyl] -imide (99%, [BMI][BTSI]) was provided by Shanghai Cheng Jie chemical Co. LTD (China). The $TbCl_3 \cdot 6H_2O$ are prepared by dissolving Tb_4O_7 oxides in concentrated hydrochloric acid in our lab. While the other reagents are analytical pure and provide by China National Medicines Group.

2.2 Characterizations

Infrared (IR) spectra were obtained on Nicolet iS50. Luminescent excitation and emission spectra were recorded on Hitachi F-4600 instrument. Luminescent lifetime was obtained using the Edinburgh Instrument FLS920.

2.3 Preparation of terbium complex modified graphene

The $Tb(TFA)_3(H_2O)_2$ was prepared according to a modified method reported in ref. [17], typically as follows: 3 mmol TFA was dissolved in 15 mL boiling ethanol, and then a solution of 1 N sodium hydroxide was added, followed by the addition of 1 mmol $TbCl_3 \cdot 6H_2O$ in 10 mL of ethanol in an hour. The mixture was refluxed and stirred for 2 h, and then stirred until cooled to room temperature. The NaCl precipitates was removed by filter and the residual solution was collected as the $Tb(TFA)_3(H_2O)_2$ solution which was used directory in future step without further purification.

The BPC modified graphene was synthesized from the GO as follows: Firstly, 100 mg GO was dispersed in 100 mL anhydrous solvent of DMF and dichloromethane (DCM) (v:v = 1:1) by means of supersonic vehicle, then, 1.5 mmol DCC and 1 mmol DMAP were added to active it for 1 h in ice bath. After that, 1 mmol HMEDA was added into the solution and stirred for 24 h at 40 °C. Subsequently, the amino functionalized graphene was collected by filtration with a microporous nylon66 membrane and washed with hot DMF repeatedly (Graphene-NH_2). Meanwhile, 1 mmol BPC was dissolved in anhydrous solvent of DMF and DCC, then, 1.5 mmol DCC and 1 mmol DMAP were added to active it for 1 h in ice bath. In the following, the previously obtained Graphene-NH_2 was added into the mixture and treated with supersonic dispersion for 10 min, the resulting solution was then stirred for 24 h at 40 °C. Finally, the mixture was filtered by the nylon66 membrane and washed with hot DMF to obtain the Graphene-BPC.

The terbium complex functionalized graphene was synthesized as follows: 50 mg Graphene-BPC was then dispersed into the previously obtained $Tb(TFA)_3(H_2O)_2$ solution in ethanol and treated with ultrasound for 30 min, after the ligand exchange procedure in 48 h, the terbium complex functionalized graphene was obtained and collected by filtration with nylon66 membrane, which was denoted as Graphene-Tb.

2.4 Preparation of Graphene-Tb doped ionogel composite

Typically, 2 mg Graphene-Tb was evenly dispersed into 0.5 mL [BMI][BTSI] ionic liquid with the help of ultrasonic, followed by an addition of 0.5 mL MMA monomer and 25 mg AIBN. The mixture was kept in the ultrasonic bath and then raised the temperature to 75 °C to initiate the polymerization of MMA monomer until a transparent bulky-gel was obtained, the resulted composite was denoted as Graphene-Tb-Ionogel.

3 Results and discussion

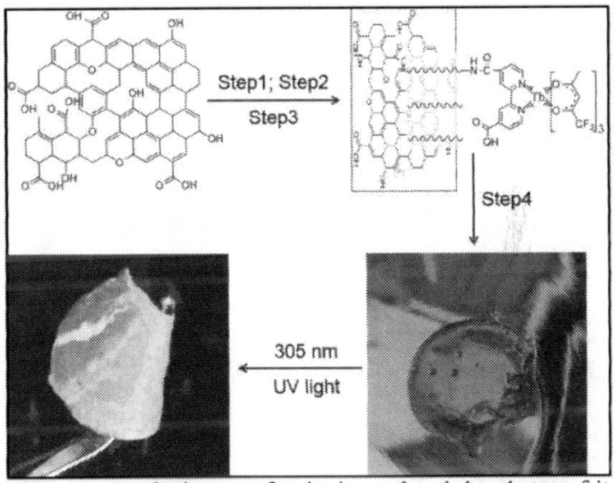

Figure 1. Synthetic route for the ionogel and the photos of it under daylight (right) and 305 nm UV light (left). Step1: DCC, DMAP and HMEDA; Step2: DCC, DMAP and BPC; Step3: $Tb(TFA)_3(H_2O)_2$; Step4: MMA, ([BMI][BTSI]) and AIBN.

In this paper, we had introduced the terbium complex into an PMMA confined ionogel system through the supported graphene matrix. The detailed synthetic pathways to obtain the ionogel and the digital photos of it under daylight and UV lamp are shown in the Figure 1. The ionogel reported here is a transparent, flexible, viscous material, and shows strong green in appearance under the 305 nm UV light. Such material has well processability and excellent photoluminescent properties, the embed ionic liquid also ensure the ionogel a certain conductivity. It is a novel optical/electric dual-functional material which belongs to the field of nanoparticle composited gel materials. The construction procedures of the ionogel had been confirmed by the FTIR spectroscopy (Figure 2). Here, the spectrum A shows a series typical adsorption peaks of graphene oxide. As show in the Figure 2B, the adsorption band around 1595 cm^{-1} and 1545 cm^{-1} are assigned to the -C=O stretching vibrations of the chelate ring of the lanthanide complex moieties and the ring breathing vibration of pyridine moieties, indicating that the surface modification of GO with terbium complex is successful. The spectrum C belongs to the Graphene-Tb-Ionogel, two obvious band at 1729 cm^{-1} and 1576 cm^{-1} are associated with to the -COO- stretching vibration of PMMA and the ring stretching of imidazolium moieties of the embed ionic liquid ([BMI][BTSI]), respectively.

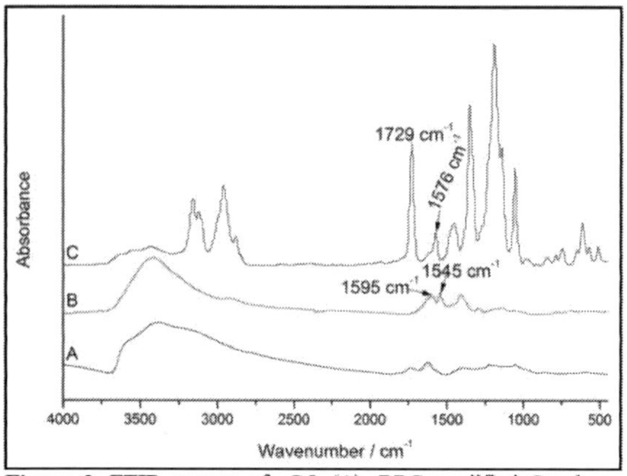

Figure 2. FTIR spectra of GO (A), BPC modified Graphene (B : Graphene-BPC) and the ionogel (C: Graphene-Tb-Ionogel).

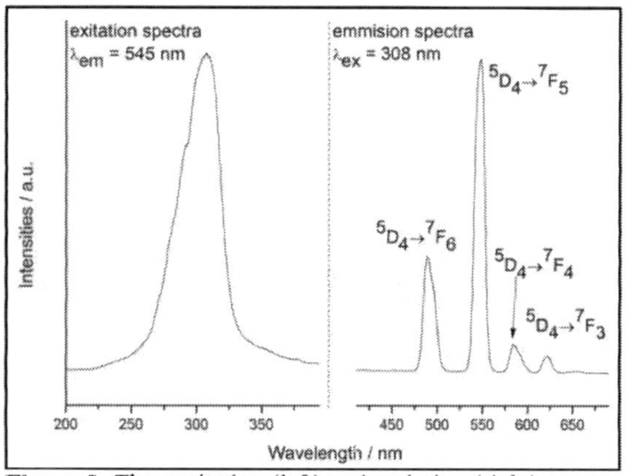

Figure 3. The excitation (left) and emission (right) spectra of the Graphene-Tb-Ionogel.

The excitation and emission spectrum of the ionogel were recorded in room temperature on the Hitachi F-4600 spectrophotometer and shown in Figure 3. The excitation spectrum was measured by monitoring the characteristic emission of terbium ion at 547 nm, which has a broad adsorption peak centered at about 308 nm, suggesting that the combination of TFA and the ionogel system can sensitize the transition of terbium ions effectively. Then, the emission spectra of ionogel was obtained by using the most appropriate wavelength (308 nm) as the excitation wavelength based on its excitation spectra. Subsequently, the corresponding emission spectrum was obtained and found to have four peaks between 450 and 650 nm, which were associated with to the $^5D_4{\rightarrow}^7F_6$ (489 nm), $^5D_4{\rightarrow}^7F_5$ (547 nm) $^5D_4{\rightarrow}^7F_4$ (585 nm) and $^5D_4{\rightarrow}^7F_3$ (622 nm) respectively. The most intense peak appearing at 547 nm can be ascribed to the magnetic dipole transition of terbium ions, which results in the bright-green luminescence, and that is why the sample glowing strong green luminescence when it's exposed to the UV lights (see Figure 1). It's clearly that the emission spectrum is a typical ligand-based emission, meanwhile, there is not apparent f-f transition peaks in the excitation spectrum, indicating that efficient intramolecular energy transfer took place between the TFA ligands and the Tb^{3+} ions, and this makes the ionogel an efficient light conversion material. According to the Dexter's [18] theory, the suitability of the energy level between organic ligand and the corresponding lanthanide ions is critical for energy transfer. Thus, according to the above analysis, the ligand TFA used in our system can transfer energy efficiently to the $^5D_4$ energy level (21000 cm^{-1}) of the Tb^{3+} ions, indicating that the ligand used in our ionogel system and the terbium ions match very well.

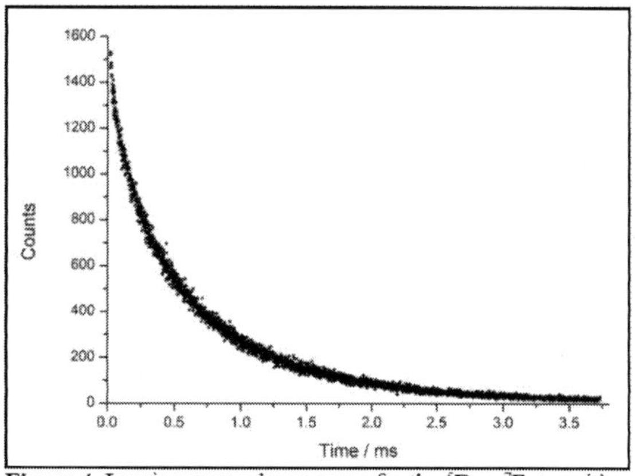

Figure 4. Luminescence decay curve for the $^5D_4{\rightarrow}^7F_5$ transition (545 nm) of the Graphene-Tb-Ionogel.

In order to further explore the photoluminescent properties of the terbium functionalized ionogel, it's photoluminescent decay curve was measured by detecting the $^5D_4$ transition of Tb^{3+} on the FLS 920 fluorescence spectrometer. As shown in the Figure 4, the decay curve can be fitted to an exponential formation [$\ln(S_t/S_0) = -k_1t = -t/\tau$], indicating that the Tb^{3+} ions within the ionogel has similar coordination environment. The photoluminescent lifetime value of Graphene-Tb-Ionogel was determined to be 0.694 ms. The above experimental results indicate that ionogel matrix could serve as an excellent host for lanthanide complexes when constructing the luminescent lanthanide hybrid materials.

4 Conclusions

In summary, a novel terbium complex modified graphene doped ionogel was prepared and characterized. This was achieved by firstly modified the GO with terbium complex via a three-step procedure: the GO was modified with HMEDA and BPC subsequently through carboxyl amino coupling reaction with the help of DCC and DMAP, and then introduced the terbium complex onto its surfaces through a ligand exchange procedures. Second, the terbium complex functionalized graphene was doped into a stationary ionic liquid system, which was confined by the PMMA networks. The resulted material is a kind of composite gel of ionic liquid, polymer, terbium complex and graphene, which has been proven to have outstanding photoluminescent properties. In addition, the methods used for modifying graphene and preparing ionogel are very simple and convenient, it can be easily extended to other nanoparticle composited gel system. Furthermore, the ionogel reported here has viscous, flexible and conductive proprieties, which brings it opportunity in the application of optical-electronic device.

Acknowledgements

This research was funded by the Natural Science Foundation of Zhejiang Province (LQ14B010001), the Natural Science Foundation of Ningbo (2016A610105), the Undergraduate Innovation and Entrepreneurship Training Program (201611058007).

References

1. B. Yan, *Photophysical Applications of Photofunctional Rare-Earth Hybrid Materials* (Springer Singapore, 2017)
2. J. Akr, L. R. Hill, A. L. Thompson, S. Faulkner, T. J. Sørensen, *Dalton Trans.* **47**, 4794(2018)
3. N. Sabbatini, M. Guardigli, J. M. Lehn, *Coord. Chem. Rev.* **123**, 201(1993)
4. J. Kido, Y. Okamoto, *Chem. Rev.* **102**, 2357(2002)
5. L. D. Carlos, R. A. Ferreira, V. d. Z. Bermudez, S. J. Ribeiro, *Adv. Mater.* **21**, 509(2009)
6. J. Feng, H. J. Zhang, *Chem. Soc. Rev.* **42**, 387(2013)
7. X. L. Zheng, M. Y. Wang, Q. P. Li, *Materials* **11**, 677(2018)
8. C. Liu, B. Yan, *Eur. J. Inorg. Chem.* **2**, 279(2015)
9. W. J. Zhang, Y. Chen, S. P. Zhang, S. K. Tang, X. Q. Chen, G. Y. Cui, *J. Inorg. Organomet. Polym. Mater.* **28**, 1294(2018)
10. Y. Wang, K. Čépe, R. Zbořil, *J. Mater. Chem. C* **4**, 7253(2016)
11. K. Binnemans, *Chem. Rev.* **109**, 4283(2009)
12. P. Thoniyot, M. J. Tan, A. A. Karim, D. J. Young, X. J. Loh, *Adv. Science* **2**, 1400010(2015)
13. J. L. Bideau, L. Viau, A. Vioux, *Chem. Soc. Rev.* **40**, 907(2011)
14. Z. P. Fan, Y. G. Wang, Z. X. Xue, L. Zhang, Y. H. Chen, S. M. Zhang, *J. Sol-Gel Sci. Technol.* **72**, 328(2014)
15. H. R. Li, Y. G. Wang, T. R. Wang, Z. Q. Li, *Ionic Liquids and Rare Earth Soft Luminescent Materials* (Springer Berlin Heidelberg, 2016)
16. B. Dai, D. G. Deng, H. Yu, L. Lei, Y. Y. Li, C. X. Li, S. Q. Xu, *Rsc Adv.* **6**, 72149(2016)
17. L. R. Melby, N. J. Rose, E. Abramson, J. C. Caris, *J. Am. Chem. Soc.* **86**, 5117(1964)
18. D. L. Dexter, *J. Chem. Phys.* **21**, 836(1953)

Research on super capacitance properties of modified activated carbon in sodium sulfate electrolyte

Wang Lin[1,2,a], Wang Chun[1], LiuYing[3], Yan Heng[1], Zhang Xiangbo[1]and Xu Hongfeng[2]
[1]*School of Materials and Metallurgy, University of Science and Technology Liaoning, Anshan114051,China*
[2]*School of Materials Science and Engineering, Dalian Jiaotong University, Dalian,116028,China*
[3]*Tangshan vocational college of science and technology,Tangshan,063000,China*

Abstract: In order to improve the specific capacity and energy density of activated carbon electrode materials, activated carbon was modified by surface treatment using hydrogen peroxide. The modified carbons were characterized the changes of pore structure and functional groups before and after modification by N_2 adsorption-desorption isotherms and FTIR spectroscopy. The three electrode system was assembled with modified activated carbon as electrode materials and Na_2SO_4 as the electrolyte. The influence of surface modification on the performance of EDLCs was studied by galvanostatic charge-discharge, cyclic voltammetry, and alternating current impedance. The results show that the surface chemistry of the modified activated carbon material has changed, the specific surface area has changed and the pore size has not changed much with the introduction of new oxygen-containing functional groups. The specific capacity of the activated carbon material modified by 5h is 75F/g, which is 86.1% higher than that before the modification. It can be seen that the change of functional group after the modification of activated carbon can help to increase the specific capacity.

Introduction

Supercapacitors have received wide attention as an efficient, environmentally friendly, and renewable energy device. The traditional capacitor has low power density, short cycle life, slow charge and discharge speed, which makes the application market shrink compared with supercapacitors. Electrolyte and electrode materials are important components of supercapacitors and critical to the performance of capacitors[1-3]. As an important electrode material, activated carbon material has the advantages of larger specific surface, excellent thermal, electrical conductivity, good chemical corrosion resistance, small thermal expansion coefficient, low cost , easy availability and easy industrialization[4-8]. The pore structure and surface chemistry of carbon materials have an important influence on the formation of electric double layers[9-13].

In this paper, the activated carbon was modified with 30% H_2O_2. The modified activated carbon was used as the electrode material of the supercapacitor. Physical characterization was carried out by Fourier transform infrared spectroscopy and N_2 adsorption-desorption method. Activated carbon was used as the electrode material. The cyclic voltammetry, galvanostatic charge and discharge characteristics of activated carbon with different modification time were measured in 1mol/L

[a] Corresponding author: E-mail:wanglin6051@163.com

Na_2SO_4 electrolyte, the specific capacity of the activated carbon was calculated and the effects of different modification conditions on the capacitance properties of carbon materials were investigated.

1 Experimental

1.1 Materials and reagents

activated carbon (AC), PTFE emulsion (PTFE), acetylene black, foam nickel, 30% H_2O_2, Na_2SO_4, absolute ethanol, etc. All reagents are analytical grade.

1.2 Activated carbon modification

The activated carbon for the study was prepared by SUNRISE POWER. A carbon (AC-0.5-AC-6) was prepared through oxidation of AC (1 g) with 50 mL of 35 wt% H_2O_2 solution under stirring for 0.5h-6h at room temperature. Recorded as AC-0.5, AC-1,AC-2,AC-3, AC-4,AC-5 and AC-6, untreated activated carbon is recorded as AC-0.The resultant material was extensively rinsed with distilled water. Then they were dried in a vacuum drying oven.

1.3 Performance characterization

1.3.1Fourier transform infrared spectrometer (FTIR)

FTIR was carried out by KBr tableting method. The appropriate KBr and activated carbon before and after modification were mixed, grinded and tableted. The TENSOR 27 Fourier infrared spectrometer was used to analyze the changes in surface functional groups of carbon materials, with wavelength range of 250~4000cm^{-1}.

1.3.2 Nitrogen absorption and desorption experiment

The nitrogen (N_2) isotherm adsorption/desorption test is one of the methods for testing parameters such as specific surface area and pore size distribution of materials. The specific surface area and pore distribution of the material were tested using a Micrometeritics ASAP 2000 specific surface area tester, vacuum pretreatment at 393K, absorption/desorption at liquid nitrogen temperature (-196℃), and data was recorded and analysed by computer.

1.3.3 Electrochemical characterization

Electrodes (1cm diameter) were pressed from a mixture of the studied carbons (80 wt%), PTFE(10 wt%), carbon black (10 wt%). Three-electrode with Hg/Hg_2SO_4 as a reference electrode comprised the studied carbon and Pt electrode as working and counter electrodes, respectively. Cyclic voltammetry and galvanostatic charge/discharge cycling were measured by a CHI660D electrochemical workstation in 1mol/L Na_2SO_4 electrolyte solution. The specific capacitance (Cs) can be calculated according to the following equation:

$$Cs = It/mU \qquad (1)$$

where in I (A), t (s), m (g) and U(V) represent the discharge current, discharge time, mass of active material and window potential, respectively.

2 Results and discussion

2.1 Infrared spectroscopy

In order to investigate the Changes in oxygen-containing functional groups on the surface of activated carbon. Fourier transform infrared (FTIR) spectroscopy was conducted. The result is shown in Figure 1.

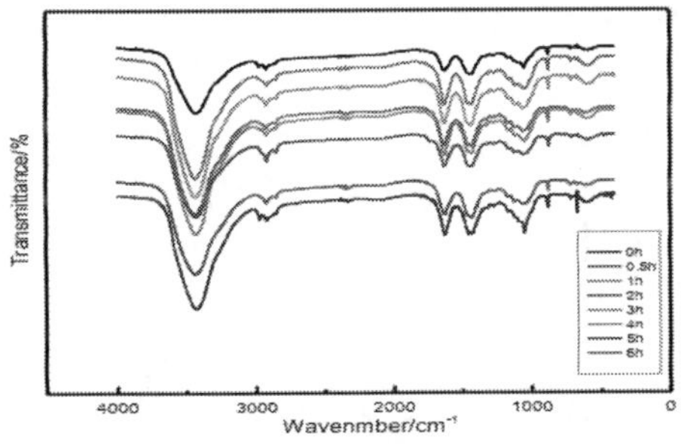

Figure 1. FTIR spectra of activated carbons before and after modification

Figure 1 is a FTIR spectrum of a carbon material before and after surface modification. It can be seen from the figure that after H_2O_2 modification treatment, the -COO-anti-extension peak of AC-0 at 1570 cm^{-1}

 https://doi.org/10.1051/matecconf/201823801003

disappears after modification, and the -OH stretching vibration peak of the carbon material modified by H_2O_2 is obviously enhanced at 3430 cm^{-1}, especially the AC enhancement with the modification time of 5h is most obvious. The C-O stretching vibration and O-H out-of-plane bending occurred at 1264cm^{-1} and 669cm^{-1} respectively, and the carbonyl stretching vibration peak appeared at 1640cm^{-1}, indicating that the modified CO content may increase, and the surface of the carbon material was introduced with -OH [14-18]. It can be seen that the surface modification has a great influence on the surface chemistry of the carbon material. Through the modification of hydrogen peroxide, some functional groups and groups are introduced on the surface of the activated carbon, which is beneficial to improve the capacitance performance of the activated carbon electrode.

2.2 N_2 adsorption-desorption isotherm

Fig.2 is nitrogen adsorption-desorption isotherms of activated carbons. It can be seen from the figure that the activated carbon which was modified that the adsorption amount of N_2 was changed at different extents, but the shape of the adsorption curve and the desorption curve did not change substantially. All adsorption isotherms are classic V-type isotherms. The adsorption capacity in the low pressure is very low and slowly increases. The relative pressure at 0.9 is a significant lag in the desorption curve. According to the trend of the adsorption isotherm and the hysteresis loop, it can be seen that all activated carbon is a mesoporous or microporous material.

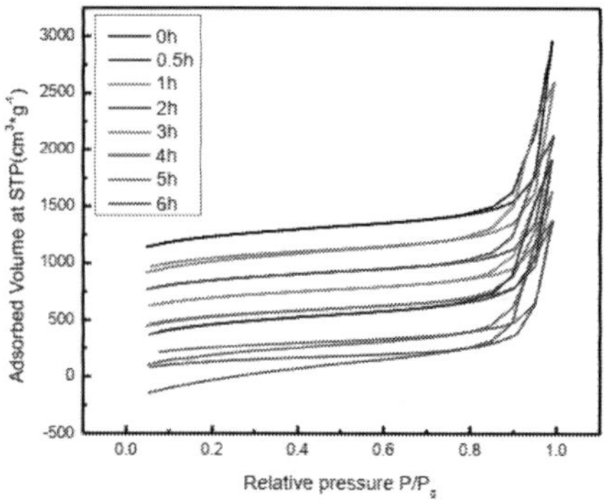

Figure 2. Nitrogen adsorption-desorption isotherms of activated carbons

The changes in specific surface area and pore structure parameters of carbon materials before and after modification are displayd in table 1. The modification treatment did not change in the pore structure of the micropores and mesopores of the activated carbon, and the sample still retained the characteristics of the microporous material. However, the specific surface area of the carbon material are decreased at different modification times. The total pore volume and micropore volume of carbon materials are reduced.The average pore size of the modified carbon material is decreased slightly. The results of different modification time modification treatments, except for a certain decrease in specific surface area, other pore structure parameters did not change much.

Table1. Specific surface area and pore diameter of active carbon

S_{BET} (m^2/g)	D (nm)
1593.241	3.345
1513.394	3.335
1458.420	3.365
1313.478	3.366
1296.835	3.354
1261.308	3.376
1143.590	3.360
1069.074	3.036

2.3 Electrochemical properties

To compare the electrochemical properties of carbon materials before and after modification. CV and galvanostatic charge-discharge measurements were performed. CV curves of AC-5 obtained at different scan rates and CV curves obtained at a scan rate of 20 mV/ s for different treatment time are shown in Fig. 3 and Fig.4.

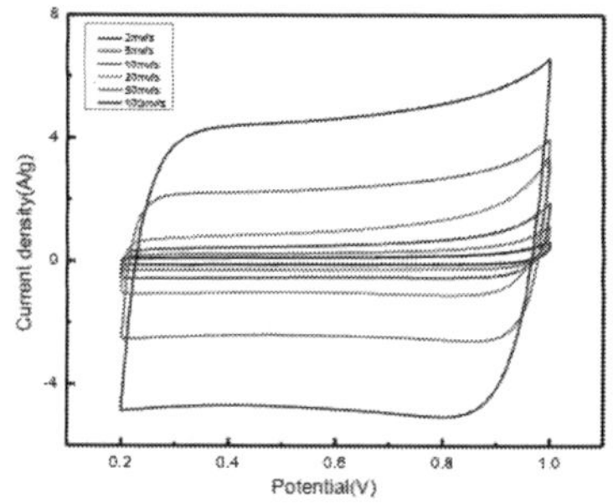

Figure 3. AC-5 CV curves obtained at different scan rates

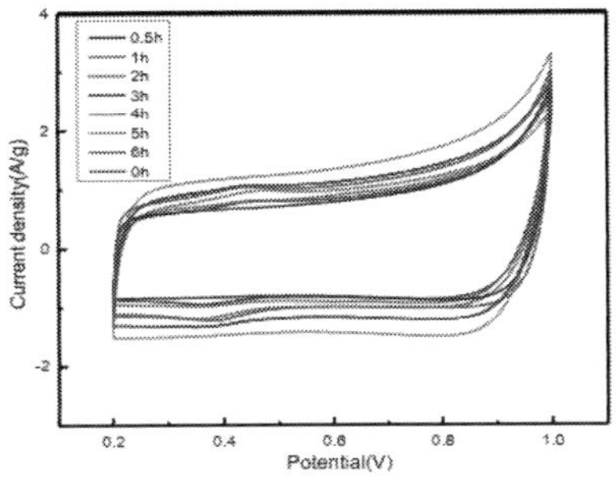

Figure 4. AC-5 CV curves obtained at a scan rate of 20 mV/ s for different treatment time

It can be seen from Fig.3 and Fig.4 that the cyclic voltammetry curves all show an approximately rectangular shape and do not contain obvious redox peaks, indicating that electrostatic physical adsorption is the main energy storage method for electrode materials. However, due to the existence of a large number of micropores in the activated carbon, the electrolyte infiltration and the positive and negative ions penetrate the pores in different ways, resulting in different equivalent series resistance (ESR), and the cyclic voltammetry deviates from the ideal rectangular shape to some extent. As the scanning speed increases, the response current increases proportionally, indicating that the electrolyte ions migrate faster and have less resistance in the activated carbon tunnel. The cyclic voltammetry curve is basically rectangular, indicating good reversibility.

The galvanostatic charge/discharge characteristics of AC-0-AC-6 in 1mol/L Na_2SO_4 were recorded in Fig.5.

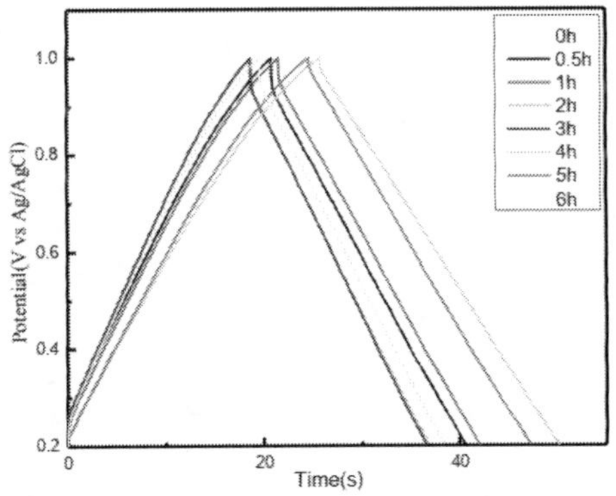

Figure 5. The charge-discharge curves obtained at 2A/g for sample AC-0-AC-6 in 1mol/L Na_2SO_4

It can be seen from Fig.5 that the charge-discharge curve has a relatively symmetrical isosceles triangle shape, indicating that the capacitor has good capacitance characteristics, and at the moment of charging and discharging, the potential drop is small, indicating that the internal resistance of the activated carbon electrode is small. Accroding to the charge-discharge curve, the specific capacitance value of the activated carbon electrode can be calculated. Fig.6 shows the specific capacitance of different samples calculated from their discharge processes at different current densities.

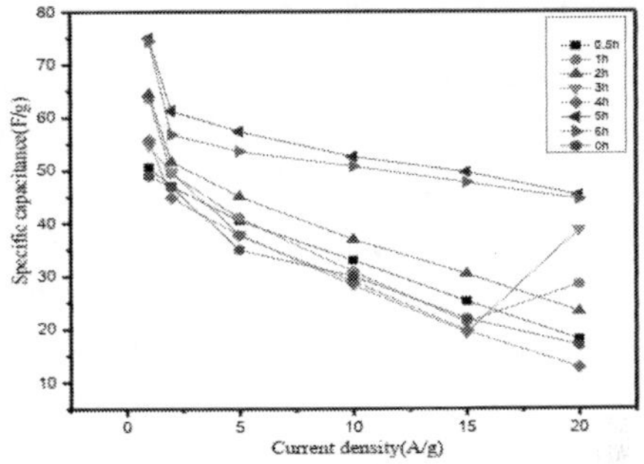

Figure 6. The specific capacitance of AC-0-AC-6 at different current densities

It can be seen from Fig.6 that the variation rules of activated carbon at different modification times are not much different. Before the modification time 5h, the specific capacity gradually increased, and reached the maximum at the modification time of 5h. When the modification time is 6h, the specific capacity does not continue to increase, it is slightly smaller than the modification time of 5h, indicating that the electrochemical performance of the activated carbon is the best at 5h.

3 Conclusion

(1)The modification treatment leads to large changes in the surface chemistry of the carbon, and the original -COO- content is reduced. Functional group =CO and –OH on the carbon surface are increased. the specific capacity of the carbon materials are increased, thereby improving the electrochemical performance of the electrode.

(2)By comparing the relationship between capacitance and structure of activated carbon materials with different modification times, due to the pore structure, surface chemistry and wettability of carbon materials, with the increasing of the specific surface area, the specific capacitance is not increasing always. There is no direct correlation between the specific capacity of activated carbon materials and the change in specific surface area.

(3)It have been shown by both galvanostatic charge/discharge and cyclic voltammetry experiment that the surface functional groups introduced by the modification of carbon materials can effectively increase their specific capacity. By comparing the activated carbon electrode materials with different modification time, the specific capacity of activated carbon with the modification time of 5h is the highest. When the current density is 1A/g, the specific capacity is the largest, which is 75F/g compared to the modification time of 0h (specific capacity 40.3 F/g) increased by 86.1%, and it was found that the modification time was not as long as possible.

Acknowledgement

This work was supported by Provincial College Students Innovation and Entrepreneurship Training Program (NO. 201710146000029) , Youth fund by University of Science and Technology Liaoning (NO.2015QN27) and National College Students Innovation and Entrepreneurship Training Program(NO. 201810146004)

Reference

[1] J.P. Cheng, Q.L. Shou, J.S. Wu, F. Liu, V.P. Dravid, X.B. Zhang.J. Electroanal. Chem. **698** (2013)
[2]Yu A P, Davies, Chen Z W. Wiley-VCH,(2011)
[3] LIU Ya-Fei, HU Zhong-Hua*,XU Kun, ZHENG Xiang-Wei, GAO Qiang. Acta Phys. -Chim. Sin.(2008)
[4] Meier J C, Katsounaros I, Galeano C, etal. Energy Environ Sci.(2012)
[5] K. Fic, G. Lota, M. Meller and E. Frackowiak, Energy Environ. Sci.**5**, 5842–5850 (2012)
[6] Frackowiak E. Phys. Chem. (2007)
[7]Kotz R，Carlen.ElectrochimicaAct,**45**,2483(2000)
[8]Qu D Y. J. Power Sources, **109**, 403-411(2002)
[9]Burke A. J.Power sources，**91**,37-50(2000)
[10]Sharma P. J.hydrogen energy:，**34**,4889-4899(2009)
[11]Zhang Y. IJHE,**34**,4889-4899(2009)
[12]Hamon M A, Hu H, Bhowmik P, etal. Chem.Phys.Lett., **347**, 8-12(2001)
[13] Rosca I D, Watari F, Uo M, etal. Carbon, **43**(2005)
[14]Qu D, Shi H. J.Power Sources **74**,99-107(1998)
[15]Bichat H P，RAY.E.Carbon,**48**,4351-4361(2010)
[16]He X. J PHYS CHEM SOLIDS, **70**,738-744(2009)
[17]J.P.Zheng. J.Power Sources，**62**(1996)
[18]R.N.Reddy. J.Power Sources，**132**(2004)

Magnetoelectric effect in a Terfenol-D and electret capacitor system

Tian-Xiang Gao, Yan-Wei Zhou, Su-Su Wang, Yun Zhou[a]

College of Science, China Jiliang University, Hangzhou 310018, China

Abstract. A capacitor system based on a Terfenol-D rod and an electret film has been proposed. We used the traditional dynamic measurement to investigate the magnetoelectric effect of the system at the frequency of 50 kHz. The results show that the ME coefficient of the system is significantly affected by the DC bias magnetic field. The maximum tunability of the magnetoelectric effect based on the DC bias magnetic field is more than 10%. The induced AC voltage of effective values increases almost lineally with increasing AC signal magnetic field. The detected magnetoelectric effect of the system exhibits hysteresis, and the magnetoelectric hysteresis loops are almost unchanged with the test period. These findings provide a type of capacitor for potential applications in magnetic detections and magnetoelectric sensors.

1 Introduction

In previous decades, materials with magnetoelectric (ME) effect have drawn significant interest for their wide applicability in ME multifunctional devices [1-3]. Because the ME effect of single-phase ME material is weak, many researches focus on complex composites and combined configurations [4-5]. In these composites and configurations, most of them are made of magnetostrictive materials and piezoelectric materials [6-7]. Electrets have the property of possessing a permanent electric field at their surface [8], which make them be widely used in energy harvesters, air filters, microphones, and so on [9-11]. However, Electrets-based ME configurations are seldom reported.

Taking the above-mentioned factors into consideration, the aim of this study is to investigate the ME effect of a capacitor system based on magnetostrictive materials and electrets. In this system, the magnetostrictive material can stretch out and draw back under an applied magnetic field, which change the distance between the two plates of the capacitor. Then an output voltage between the two plates can be measured. Thus a ME response is observed. This ME response of this capacitor system is quite different from that of ME materials. Therefore, it is necessary to investigate the ME effect of this system, which also can provide useful information for practical applications of electrets.

2 Experimental details

Figure 1 illustrates the geometry and working principle of the capacitor system. The system consists of a magnetostrictive Terfenol-D rod, a dielectric elastic latex plate, an electret film plate, and a copper plate as shown in Figure 1a. Both the electret film and the Terfenol-D rod were commercially obtained with the same diameter of 8 mm. The length of the Terfenol-D rod is about 30 mm, and the thicknesses of the latex pate and electret film are approximately 0.15 mm and 0.02 mm, respectively. The electret film was attached to the copper plate using conductive glue. The saturation magnetostriction of Terfenol-D, the surface charge density of the electret film, and the relative dielectric constant of the latex plate were reported to be approximately 1300 ppm, 62 μC/m^2, and 8 respectively [12]. As can be seen in Figure 1a, the Terfenol-D and copper plate were very closely spaced-filled with the latex plate and electret film, and faced one another. The Terfenol-D, latex pate, electret film and copper plate formed a parallel plate capacitor system.

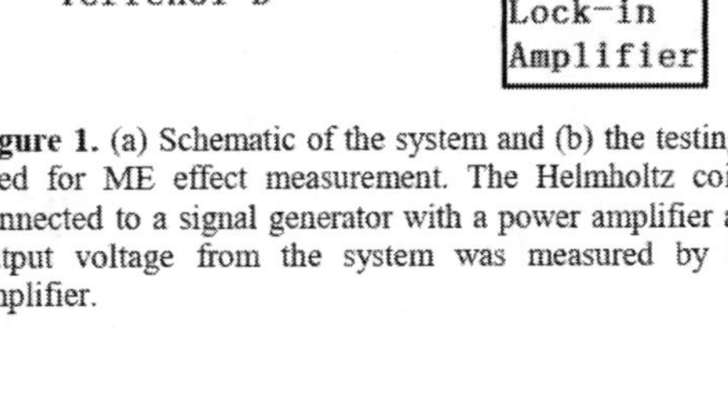

Figure 1. (a) Schematic of the system and (b) the testing setup used for ME effect measurement. The Helmholtz coils was connected to a signal generator with a power amplifier and the output voltage from the system was measured by lock-in amplifier.

[a] Corresponding author: zhouyun@cjlu.edu.cn

The system was placed in a plastic capsule and two leads from the capacitor were connected to a lock-in amplifier as mentioned before [12].The ME effect measurement was performed using the dynamic method, as shown in Figure 1b. The ME coefficient (α) was determined by measuring the output voltage under a small applied sinusoidal AC magnetic field of 50 kHz (generated by Helmholtz coils, 0.45 Oe) superimposed onto the DC magnetic bias (generated by an electromagnet) [12-13]. All measurements were performed at room temperature.

3 Results and discussion

When a magnetic field applies to the capacitor system, the length of the magnetostrictive Terfenol-D changes, the distance between the parallel plates (*d*) is changing accordingly. Therefore the capacitance of the capacitor changes. Figure.2 shows the capacitance (*C*) as functions of DC bias field *H* at the signal magnetic field (about 0.45 Oe) frequency of 50 kHz. Figure 2 shows that the capacitance significantly affected by the DC magnetic field. *C* increases with increasing *H*. Some previous studies indicate that the magnetostriction of Terfenol-D is affected by the DC magnetic field [14]. When *H* is applied to the Terfenol-D, its length will increase with increasing *H*, and the distance between the parallel plates *d* decreases. Therefore, the increase of *C* with increasing *H* is obtained based on equation below

$$C=\varepsilon A/d \tag{1}$$

Where ε is the dielectric constant of the capacitor, *A* is the plate area.

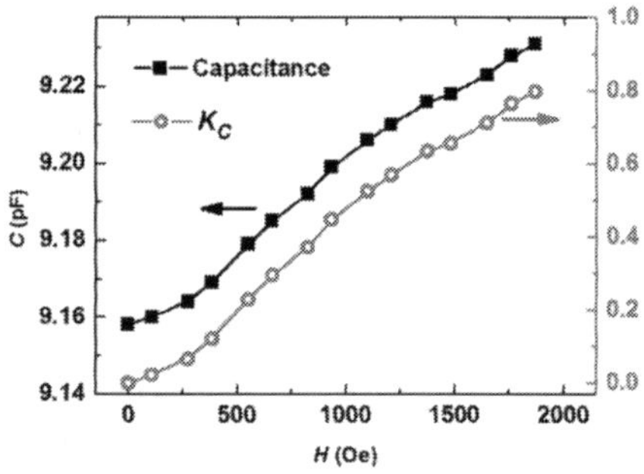

Figure 2. The capacitance (*C*) and the magnetocapacitance effect of the capacitance (*MC*) as functions of DC bias magnetic field (*H*) at the frequency of 50kHz for the system.

Based on the strong field dependence of *C*, it is worthwhile to investigate the magnetocapacitance effect (*MC*) with varying *H*. *MC* can be expressed as

$$MC=100\% \times(\alpha(\mathrm{H})-\alpha(0))/\alpha(0) \tag{2}$$

Where *C*(*H*) and *C*(0) are the capacitance under an applied DC magnetic field and zero field at the frequency of 50 kHz, respectively. *MC* increases with increasing *H* as shown in Figure 2. both the *C* and the *MC* with *H* show the same changing trend as magnetostriction of Terfenol-D does[14]. The result coincide with the equation (1). The maximum value of K_M is about 0.8% at *H*=1760 Oe. This means that hat this system can be used in magnetic detectors and sensors.

In the parallel plate capacitor system, the electret film maintains a fixed charge. Therefore, a voltage across the parallel plate is obtained. As a magnetic field is applied to the Terfenol-D rod, it moves back and forth, changing the distance between the two plates of the capacitor system, which, in turn, creates an induced voltage difference [12]. When an AC signal magnetic field is superimposed on the DC magnetic field *H*, an AC signal will be induced and can be measured by an lock-in-amplifier . The ME effect coefficient can be expressed as:

$$\alpha_M=\delta U/\delta H \tag{3}$$

Where δH is the AC signal magnetic field, δU is the induced voltage.

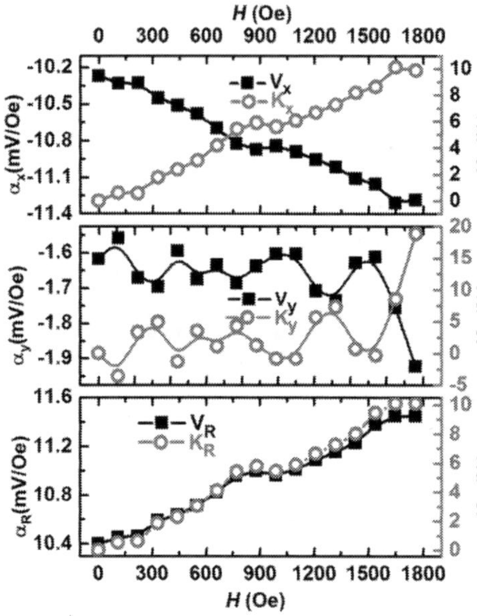

Figure 3. The real part, image part and effective values of the ME effect coefficient (α_x, α_y and α_R) and their tunabilities (K_x, K_y and K_R) as functions of DC bias magnetic field (*H*) at the frequencie of 50 kHz for the system.

Figure 3 shows that the real part, image part and effective values of the ME effect coefficient (α_x, α_y and α_R) as functions of DC bias magnetic field (*H*) at the frequency of 50 kHz for the system. The result shows that both α_x and α_R are significantly affected by the DC magnetic field. α_x decreases almost linearly with increasing *H*, while α_R increases almost linearly with increasing *H*. However, the changes of α_y with *H* is not evident. α_R can be expressed as

$$\alpha_R{}^2=\alpha_x{}^2+\alpha_y{}^2 \tag{4}$$

So the positive value of α_R can be observed as shown in Figure 3.

Based on the strong H-dependence of the ME effect coefficient, it is worthwhile to investigate its tunability (K) with varying H. K can be expressed as

$$K = 100\% \times |(\alpha(H) - \alpha(0)) / \alpha(0)| \quad (4)$$

where $\alpha(H)$ and $\alpha(0)$ are the ME effect coefficient under an applied DC magnetic field and zero field, respectively. The tunabilities of α_x and α_R (K_x, K_R) increase almost linearly with increasing H as shown in Figure 3. The highest value of K_x, K_R are more than 10% at H=1760 Oe, which indicates that this capacitor system may be used in magnetic detectors and sensors.

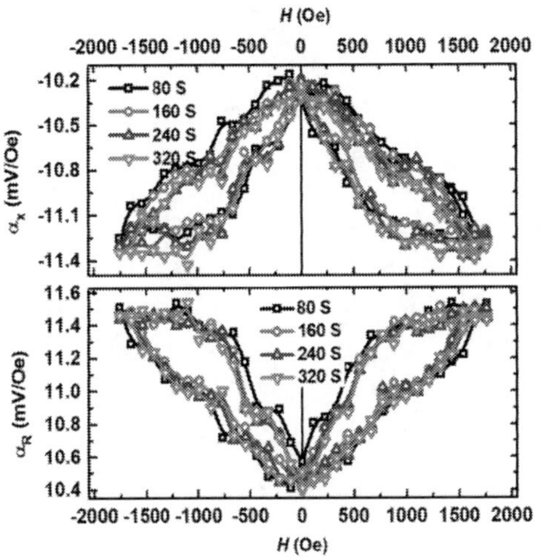

Figure 4. Test period dependence of the real part and effective values of the ME hysteresis effect coefficient (α_x, α_R) loops at the applied signal field frequency of 50 kHz.

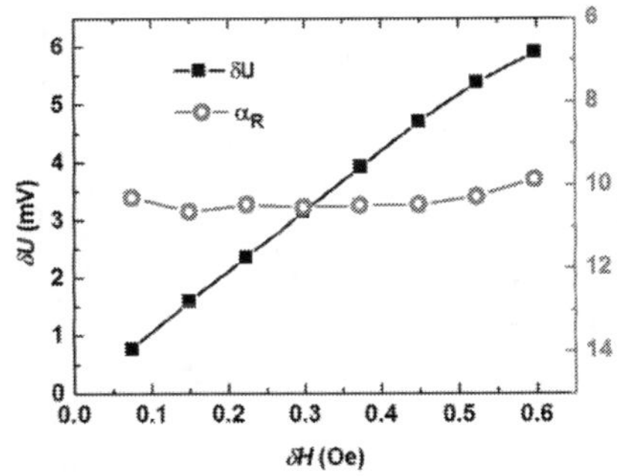

Figure 5. Induced AC voltage of effective values (δU) and the effective values of the ME effect coefficient (α_R) as a function of applied AC signal magnetic field (δH) at the frequency of 50 kHz.

Figure 4 shows the test period dependence of the real part and effective values of the ME hysteresis effect coefficient (α_x, α_R) loops at the applied signal field frequency of 50 kHz. Both α_x and α_R vs. H have been measured by varying H from zero to a positive maximum (1760 Oe), then from the positive maximum to a negative maximum (-1760 Oe), and finally, from the negative maximum to a positive maximum. This is a test period (T). As H increases from zero to a positive maximum, α_x decreases to a minimum value, while α_R increases to a maximum value. As H decreases from positive maximum to zero, both α_x and α_R do not reach their previous values, thereby exhibiting hysteresis. As H is reversed from zero to the negative maximum, the same behavior is observed. The hysteresis behavior indicates that the ME effect of the system needs time to respond to the varying H. To see the ME effect with test period (T), we measured the ME effect of the system with T at 80, 160, 240 and 320 s, as shown in Figure 4. The result shows that both the real part and effective values of the ME hysteresis loops are almost unchanged with T, which is much different from that of the ME composites [13].

To explore the correlation of the effective value of the induced AC voltage (δU) with the applied AC magnetic signal (δH), the values of δU with the applied δH at frequency of 50 kHz was measured, as shown in Figure 5. The result indicate that with the increase of δH, δU increases almost linearly, which means that this system can be used in AC magnetic detectors and sensors. The effective value of the ME effect coefficient (α_R) of 50 kHz is also obtained by $\delta U/\delta H$ as shown in Figure 5. α_R is almost unchanged with the increase of δU. However, as δU is too big, the α_R is slightly bigger than the others. The reason for the phenomenon is that the magnetostriction of Terfenol-D almost increase linearly with the increasing magnetic field at a low value range [14]. The signal magnetic field (δH) is very small and less than 0.5 Oe, so the changing distance between the two plates of the capacitor (δd) increases linearly with increasing δH. Accordingly, δU increases linearly with increasing signal magnetic field as shown in Figure 5.This means that the AC magnetic signal should not be big in the ME effect measurement for this kind of system. Inaddition, this means that this capacitor system can be used in AC magnetic detectors and sensors.

4 Conclusions

In summary, we prepared a capacitor system composed of a Terfenol-D rod and an electret film, and studied the ME effect coefficient using dynamic measurement. A significant ME effect with high tunability at the frequency of 50 kHz was observed. The results indicate that both the real part and effective values of the ME effect coefficient are strongly dependent on the bias magnetic field. The bias-field-dependent ME hysteresis of the system is almost unchanged with the test periods. The induced effective value of AC voltage increases almost linearly with the increasing AC magnetic signal. The results indicate that this type of capacitor system has potential practical uses in energy converter, magnetic detection and sensors.

Acknowledgment

This work is supported by the National Natural Science Foundation of China (NSFC No. 11547048).

References

1. Fiebig M, J. Phys. D: Appl. Phys. **38,** R123 (2005)
2. Fusil S, Garcia V, Barthélémy A and Bibes M, Annu. Rev. Mater. Res. **44**,91 (2014)
3. Gonçalves R, Larrea A, Zheng T, Higgins M J, Sebastian V, Lanceros-Mendez S and Martins P, Eur. Polym. J. **84**, 685(2016)
4. Palneedi H, Maurya D, Kim G Y, Priya S, Kang S J L, Kim K H, Choi S Y and Ryu J, Appl. Phys. Lett. **107**, 012904 (2015)
5. Tian Y, Shen S, Cong J, Yan L, Wang S and Sun Y, J. Am. Chem. Soc. **138**, 782 (2016)
6. Chen A, Zhao Y, Li P, Zhang X, Peng R, Huang H, Zou L, Zheng X, Zhang S, Miao P, Lu Y, Cai J, Nan C-W, Adv. Mater. **28**, 363 (2016)
7. Ummer R P, Raneesh B, Thevenot C, Rouxel D, Thomas S and Kalarikkal N, Rsc Adv. **6**, 28069 (2016)
8. McCarty L S, Whitesides G M, Angew, Chem. Int. Edit. **47**, 2188 (2008)
9. Chen G, Li Y, Xiao H and Zhu X, J. Mater. Chem. A **8**,5(2017)
10. Chen G, Xu Y, Chen X, XIao H, **25**, 797(2018)
11. Shin K, Kim J, Lee H, Seo D and Moon W, J. Acoust. Soc. Am. **140**, 3033 (2016)
12. Zhou Y, Chen G, Sun Y, Sun Y., Zhou H, Cao D, Xu L, Jiao Z, Sensor. Actuat. A-Phys. **270**,46(2018)
13. Zhou Y, Zhou H M, Ye Y X and Jiao Z W, Appl. Phys. A **122**,142 (2016)
14. Zheng X J, Liu X E, J. Appl. Phys. **97**, 053901 (2005)

The preparation, characterization and application of glycol aqueous base graphene oxide nanofluid

Li Yihuai[1,2,3], Wu Zihua[1,3,a], Xie Huaqing[1,3], Tang Dingcong[1], Wang Yuanyuan[1,3], Li Zhen[2]

[1] *Research Center of Resource Recycling Science and Engineering, Shanghai Polytechnic University, Shanghai, 201209, China*

[2] *School of Environmental and Chemical Engineering, Shanghai University, Shanghai 200444, China*

[3] *School of Environmental and Materials Engineering, College of Engineering, Shanghai Polytechnic University, Shanghai 201209, China*

Abstracts: Enhancement of the heat transfer of the cold side is one of the approaches to improve the performance of thermoelectric generator systems (TEGs). In order to investigate viability and further performance of the TEG for waste heat recovery in industry area, a small low-temperature waste heat thermoelectric generator setup has been constructed with graphene oxide (GO) nanofluids as coolants in the paper. The results showed excellent stability of GO nanofluids through the preparation of two-step method to be applied in the TEGs .The highest open output voltage of TEG system were obtained by 0.15% weight fraction of GO nanofluids as coolant when the temperature difference (△T) was designed at 95K and the hot side temperature was fixed at 373K in the TEGs. Compared with conventional glycol-water base fluid as coolant, the highest open voltage enhancement ratio has reached 65.26% in the TEGs.

1 Introduction

Thermoelectric generator (TEG) is a type of solid-state devices that can directly convert thermal energy into electricity by the Seebeck effect of semiconductor materials. Although its efficiency is a little low (<7%), it has the noteworthy advantages of simple structures, the absence of moving parts, silence, reliability, environmental friendliness, and so on. Nowadays, TEGs have been used in many fields such as automotive waste heat, solar panel, and geothermal energy power generation, etc[1-6].

As with any heat engine operating on a thermodynamic cycle, the system of a TEG includes heat source, power system, cooling system and data acquisition system. The efficiency of the TEGs partly depends on the temperature difference between the hot and cold sides of the TEGs, which is determined by the hot source and cooling method. Usually, heat source of the TEGs are from waste heats, such as automobile exhaust, industrial waste heat, and so on[7-10]. In[11] investigated the performance of TEG heated by exhausted gas from an actual engine. Madan[12] chose dispenser printer as hot source and designed TEG devices. The conditions of the hot side of the TEG cannot change in the practical application as it is expected. So, when the temperature of the heat source was limited, the thermal transfer capabilities of the working fluid on the cold side of the TEGs directly have an effect on the output performance of the TEGs. It is a better way to improve the heat transfer at the cold side of TEG for maximizing thermal efficiency as high as possible. Until now, cooling system commonly adopts the air natural cooling or takes off residual heat by base fluids, such as water, ethylene glycol and so on. These fluids are inexpensive, availability and high specific heat capacity. However, these conventional common fluids have very low thermal conductivities[13] which limit their further development.

Nanofluid was found to exhibit better thermo-physical properties, such as higher thermal conductivity and higher viscosity[14-19].These are important factors because thermal conductivity and viscosity would directly contribute to high output performance of the TEGs. In order to improve the thermal conductivity of these conventional liquids, many researchers have focus on this aspect. Deasy[20] simulated the design of a passive liquid cooling system for TEGs and experimentally verified the simulation results. Nguven[21] experimentally studied the behavior and heat transfer enhancement of Al_2O_3 nanofluids for cooling of microprocessors or other electronic components. Zhou[22]

[a] Corresponding author: yhli@sspu.edu.cn

utilized water loaded with CuO nanoparticles to optimize based heat exchange schemes. The results showed the conversion efficiency of CuO nanofluid-cooled TEGs was 14% higher than that of pure water-cooled.

The preparation technology and performance of nanofluids have become more and more comprehensive and thorough in recent years. Graphene, found by Novoselov and co-workers[23], is a flat monolayer of sp2-bonded carbon atoms packed into a honeycomb lattice. It has attracted much attention due to two-dimensional structure, unique physical and chemical properties such as high thermal conductivity[24]. It is reported that a single layer of graphene have a thermal conductivity of 5300W/m•K by experimental measurements[25]. Chen[26] experimentally obtained that a maximum thermal conductivity of suspended single-layer graphene was 2450.55W/m•K. Besides, graphene is an ultralight material because of its low density which is to the benefit of nanofluids preparation. However, it is found that graphene is difficult to be dispersed in polar solvents directly due to the hydrophobic property of graphene in the process of the experiment. Yu[27] for the first time selected the hydrophilic graphene oxide as the additive in a nanofluid and the results showed that the thermal conductivity enhancement ratios of GO nanosheets nanofluids were up to 76.80% compared to the corresponding base fluid. The team further developed a facile technique to produce ethylene glycol based nanofluids containing GO nanosheets and the thermal conductivity enhancement ratios were up to 86% [24]. Ijam[28] prepared GO nanosheets /glycol-water-based nanofluids which kept stable for more than 5 months, the results showed that the improvement in thermal conductivity was up to11.7%. And then, hydrophilic graphene nanofluids have been explored for practical applications instead of traditional fluids, such as in solar collectors[29-31], miniature loop heat pipe[32-35] , and so on. It is proved that GO and homologous series nanofluids are effective media of heat transfer. In the study, GO and its stable nanofluids were prepared by the chemical and two-step method, respectively. An experimental scale test system of TEG was constructed and GO nanofluids as coolant were used instead of base fluid in the cooling system. The effect of GO nanofluids on the output open voltage of the TEG system at different concentration and different temperature were studied. The results could provide a fundamental understanding of the optimization of large-scale TEG systems.

2 Experimental

Graphite (code no.GF-1S2) was from by Shanghai Shanshan Technology limited Company. Other chemical reagents (H_2SO_4, H_3PO_4, $KMnO_4$, H_2O_2) were analytical reagent and from Sinopharm Chemical Reagent Co. Ltd.. All reagents were used without any further treatment. The synthesis of GO was adopted from Huang et al.[36] based on modified Hummer's method.

The prepared GO were dispersed in the mixture with ethylene glycol and deionized water (V/V=1:1) without adding any surfactant solution. The mixture is homogenized by ultrasonication (KQ2200E Sonicator, KunShan Ultrasonic Instruments Co.Ltd., China) for about 1 hour to suspend the GO in mixed reagent. A larger fraction of 0.5wt.% GO nanofluid was prepared according to the above two-step process and it was diluted to prepare 0.05wt.%, 0.10wt.%, 0.15wt.% and 0.20wt.% GO nanofluids by severely magnetic stirring and ultrasonication. X-ray diffraction (XRD) patterns of the selected samples were recorded by the reflection scan with nickel-filtered Cu Kα radiation (D8, Bruker-AXS, Germany);The Raman spectra were gained with a FT-Raman microscope using a laser with a wavelength of 532nm (Senterra, Bruker, Germany); The thermal conductivity of samples was measured with a thermal conductivity analyzer (C-Therm TCi, C-Therm Technologies Ltd., Canada); Viscosity of GO nanofluids was measured as a function of temperatures using viscometers (DV2THA, Brookfield, USA).

The TEG test system and measurement uncertainties were described in detail in our previous paper [37]. The schematic of the experimental system is shown in Fig.1. The system was made up of a pump (Masterflex L/P, 77962-20, USA), a thermostatic bath (DCW-2008, Ning Bo Scientz Biotechnology Co., Ltd., China), TEG, a heating plate (IKA, C-MAG-HP4, Germany), a heat sink and a data acquisition unit. The interfaces between the TEG, a heating plate and a heat sink were covered thermal grease to reduce the thermal contact resistance. The detailed parameters of TEG are shown in table 1. GO nanofluids as coolant were poured into the liquid cabin firstly and then were brought into the whole TEG system by pump. The open output voltage (V) and the cold surface temperature of TEG were collected by the data acquisition system, for example, the open output voltage of TEG was detected by the electrical leads line 2 and 3, the inlet temperature (T_{in}), the outlet temperature (T_{out}) and the designed condenser's temperature (T_w) were obtained by the thermocouples line 4, 5 and 1, respectively. When the temperature at the hot side of TEGs is fixed, the voltage of TEGs depended on the performance of the cold side of TEGs. All the experimental data were got by repeated more than twice.

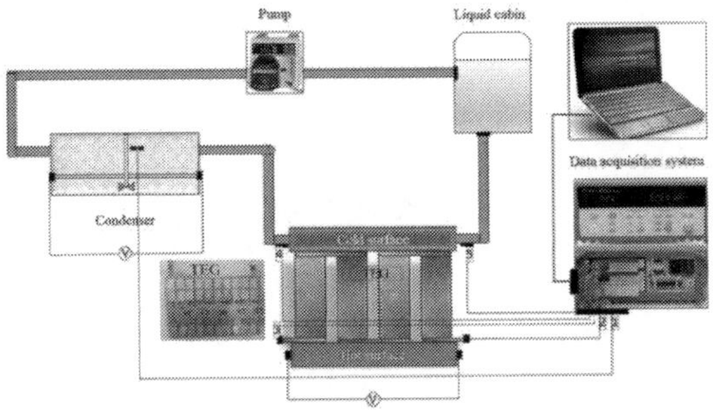

Fig.1 Schematic of the experimental setup

Table 1 Specifications of the TEG

Specifications	Value
Type of material	Bi_2Te_3
Maximum current (A)	7
Maximum voltage (V)	3.5
Maximum power (W)	7.6
△Tmax (°C)	270
Dimensions(L×W×H) (mm^3)	50*50*3.5
Number of couples	49

3 Results and discussions

3.1 Micro-structure characterization

Fig.2 shows the XRD results of graphene oxide and the pristine graphite. The sharp diffraction peak observed at 2 θ =26.501° associated with (002) reflection of graphite and interlayer distance of 0.336nm calculated by Bragg' s equation. After oxidation of graphite, a strong peak at 11.46 o associated with (001) reflection of GO with an interlayer distance of 0.77nm , the stacking height and the number of layers are 5nm and about 7, respectively, calculated by Scherrer equation. The increase in interlayer distance is due to the included functional groups between the graphene sheets after the oxidation[28] .

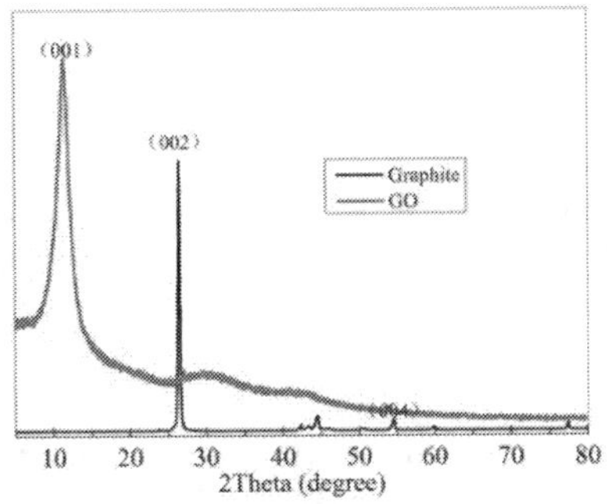

Fig.2 XRD patterns for graphite and GO

Raman spectroscopy is a widely used tool for the characterization of the structure of graphitic materials. From Fig.3, the main feature in Raman spectra of graphitic materials are so-called G and D peaks, which lie at around 1560 and 1360 cm-1 for visible excitation[38]. The G peak is caused by the Raman active E2g phonon of C sp2 atoms. The D peak refers to the sp3 hybridization of carbon atoms and structure defect and its relative signal strength (compared to the G peak) depends strongly on the amount of disorder in the carbon material[39]. In graphite, the ratio of D peak to G peak intensity (ID/IG) is 0.28 indicating that the graphite is ordered structure with low defects; In graphene oxide, the G peaks is shifted and broadened to 1590cm-1 and the D peaks is shifted to 1540cm-1, the ratio of ID/IG is 1.04 which is higher than that of graphite. This is due to severe oxidation of graphite by strong acid during preparation and the more defects caused. Besides, since this ratio is greater than 1, it is plausible to say that some oxygen groups on the surface of graphene were introduced during the preparation[40]. As described in the reference[41], there is no doubt that the high thermal conductivity is diminished by defects and this defects have direct influence on the heat transport, but this defects are very beneficial for the stability of the GO nanofluids without surfactant addition.

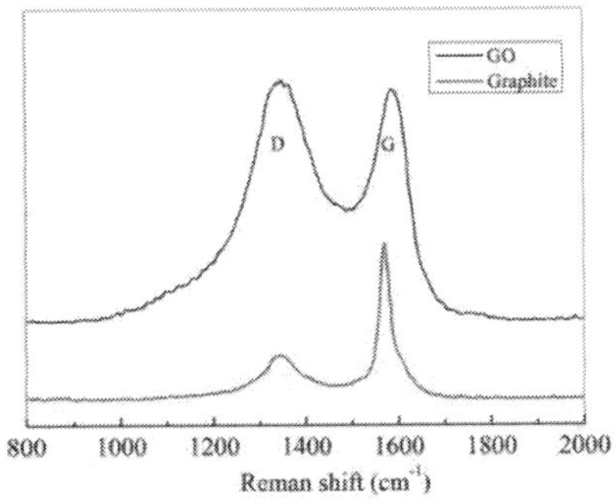

Fig.3 Raman spectra for graphite and GO.

3.2 Effect of GO nanofluids on the voltage of the TEG

The effect of GO nanofluids on the heat exchange of cold side of the TEGs was studied. Fig.4 shows the distribution of T_{ave} ($T_{ave}=(T_{in}+T_{out})/2$) and T_w versus time. The results indicated that the temperature curves were nearly a straight line when time was over 900s and reaches a steady state of the system. All the experimental data were collected when the starting time of system was over 900s. Besides, it was found that the fluctuation of the temperatures with 0.15wt.% GO nanofluid as the coolant are larger than those of the base fluid as the coolant. The result may be caused by Brownian movement of nanoparticles dispersed in the base fluid.

Fig.5 presents the effect of the GO nanofluids on the voltage of the TEG at different temperature and flow rate. The hot side temperature was fixed at 373K, The designed temperature difference ($\triangle T$) was obtained through varying the cold side fluid temperature. The highest voltage were 0.432 and 0.412V with 0.15wt.% GO nanofluids as coolant when $\triangle T$ was at 95 and 90K, respectively. The voltage improvement ratio is up to 4.85% when $\triangle T$ was increased by 5K at the test system of the TEG. The change of voltage of the TEG at different concentration and same temperature difference can be observed from Fig.5 (a). When $\triangle T$=95K, the voltage of the TEG increases with increasing the particles fraction when it is lower than 0.15wt.%. While the fraction is higher than 0.15wt.%, the voltage of the TEG decrease with increasing solid loading. The same tendency occurred at 90K which can be seen from Fig.5 (b). The enhancement percentage was calculated based on the expression below:

$$\text{Enhancement ratio}=((V-V_0)/V_0)\times 100\%$$

where V and V_0 are the open voltage of the tested TEG device with nanofluid and the base fluid as the coolant, respectively. The highest enhancement of 60%, 65% were achieved at $\triangle T$= 90 and 95K with 0.15wt.% GO nanofluid rather than 0.20 wt.% GO nanofluid , respectively. This behavior may be due to more GO added in the base liquid would increase the thermal conductivity as well as the viscosity

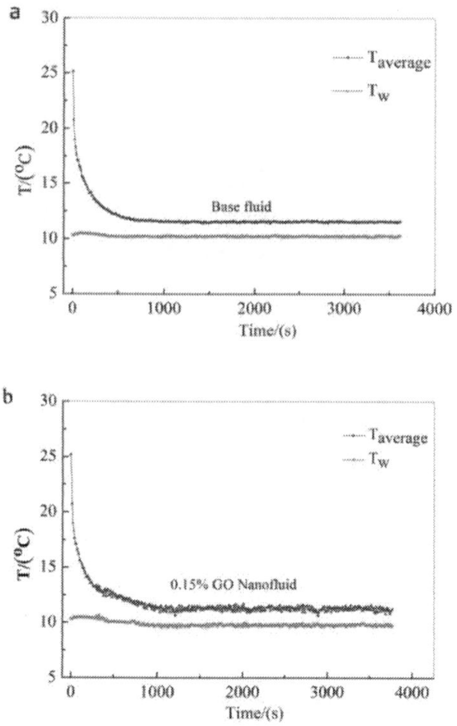

Fig.4 Temperature distributions of $T_{average}$ and T_w with base fluid and 0.15wt.% GO nanofluids

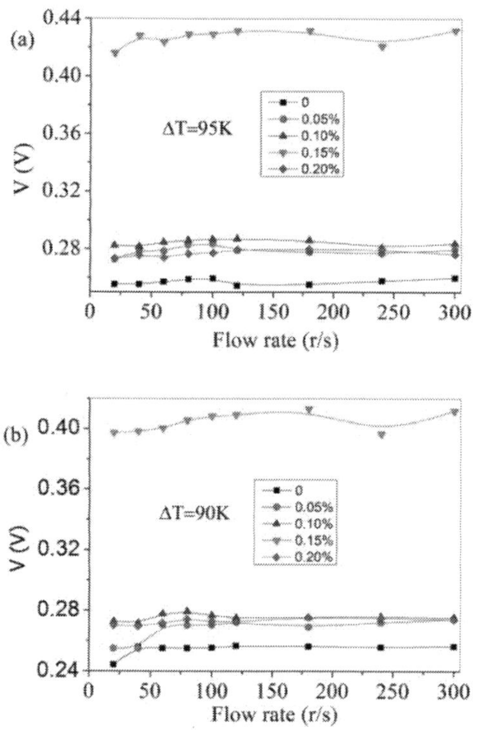

Fig.5 Effect of the GO nanofluid on the voltage of TEGs at different conditions

4 Conclusions

In the present experimental study, GO have been prepared by using Modified Hummer's method and the GO dispersing glycol-water based nanofluids were further prepared by a two-step method. The interlayer spacing of GO increases due to the formation of functional groups between the graphite layers and more defects in the structure of GO after oxidation. These functional groups and defects are the main reasons of the stability of GO nanofluids without any surfactant addition. To thoroughly investigate the heat transfer enhancement and behavior of GO nanofluid as coolant of the TEGs. The open output voltage of TEG can be increased nearly 60% and 65% compared with glycol-water mixed base fluid at the weight fraction of 0.15% when the temperature differences($\triangle$T) are 90 and 95K, respectively. So, in a word, using GO nanofluids is an effective approach to enhance the heat transfer performance of the coolant in the TEGs.

Acknowledgements

This work was supported by the National Natural Science Foundation of China (Grant Nos. 51676117& 51476095), the Natural Science Foundation of Shanghai (Grant No. 14ZR1417000), the Program for Professor of Special Appointment (Young Eastern Scholar, Grant No. QD2015052) at Shanghai Institutions of Higher Learning, the Key Subject of Shanghai Polytechnic University (Material Science and Engineering, Grant No.XXKZD1601) and Gaoyuan Discipline of Shanghai – Environmental Science and Engineering (Resource Recycling Science and Engineering).

References

1. T. Wang, W. Luan, W. Wang, S.-T. Tu, Applied Energy, 136 (2014) 860-865.
2. W. He, S. Wang, X. Zhang, Y. Li, C. Lu, Energy, 91 (2015) 1-9.
3. S. Islam, I. Dincer, B.S. Yilbas, Energy, 93, Part 1 (2015) 1246-1258.
4. M.F. Remeli, L. Kiatbodin, B. Singh, K. Verojporn, A. Date, A. Akbarzadeh, Energy Procedia, 75 (2015) 645-650.
5. T.H. Kwan, X. Wu, Applied Energy, 165 (2016) 297-307.
6. G. Li, G. Zhang, W. He, J. Ji, S. Lv, X. Chen, H. Chen, Energy Conversion and Management, 112 (2016) 191-198.
7. L. Wang, W. Cai, H. Zhao, C. Lin, J. Yan, Applied Thermal Engineering, 94 (2016) 314-323.
8. İ. Temizer, C. İlkılıç, Renewable and Sustainable Energy Reviews, 63 (2016) 141-151.
9. J.-H. Meng, X.-D. Wang, W.-H. Chen, Energy Conversion and Management, 120 (2016) 71-80.
10. X. Wang, B. Li, Y. Yan, S. Liu, J. Li, Applied Thermal Engineering, 102 (2016) 176-183.
11. B.d. In, H.i. Kim, J.w. Son, K.h. Lee, International Journal of Heat and Mass Transfer, 86 (2015) 667-680.
12. D. Madan, Z. Wang, P.K. Wright, J.W. Evans, Applied Energy, 156 (2015) 587-592.
13. M. Afrand, A.A. Nadooshan, M. Hassani, H. Yarmand, M. Dahari, International Communications in Heat and Mass Transfer, 77 (2016) 49-53.
14. C. Selvam, E.C. Muhammed Irshad, D.M. Lal, S. Harish, Experimental Thermal and Fluid Science, 77 (2016) 188-196.
15. A.M. Aly, S.E. Ahmed, Journal of Nanofluids, 5 (2016) 110-119.
16. M.A. Akhavan-Behabadi, F. Hekmatipour, B. Sajadi, Experimental Thermal and Fluid Science, 78 (2016) 10-17.
17. M.A. Akhavan-Behabadi, M. Shahidi, M.R. Aligoodarz, M. Fakoor-Pakdaman, Applied Thermal Engineering, 106 (2016) 916-924.
18. Q. Li, A. Shirazi, C. Zheng, G. Rosengarten, J.A. Scott, R.A. Taylor, Energy, 96 (2016) 253-267.
19. M. Hosseinzadeh, S.Z. Heris, A. Beheshti, M. Shanbedi, Journal of Thermal Analysis and Calorimetry, 124 (2016) 827-838.
20. M.J. Deasy, N. Baudin, S.M. O'Shaughnessy, A.J. Robinson, Applied Energy, 205 (2017) 499-510.
21. C.T. Nguyen, G. Roy, C. Gauthier, N. Galanis, Applied Thermal Engineering, 27 (2007) 1501-1506.
22. S. Zhou, B.G. Sammakia, B. White, P. Borgesen, C. Chen, International Journal of Heat and Mass Transfer, 81 (2015) 639-645.
23. A.K. Geim, K.S. Novoselov, Nat Mater, 6 (2007) 183-191.
24. W. Yu, H. Xie, X. Wang, X. Wang, Physics Letters A, 375 (2011) 1323-1328.
25. A.A. Balandin, S. Ghosh, W. Bao, I. Calizo, D. Teweldebrhan, F. Miao, C.N. Lau, Nano Letters, 8 (2008) 902-907.
26. L. Chen, H. Xie, W. Yu, B. Wang, Z. Wu, International Journal of Thermal Sciences, 94 (2015) 221-227.
27. W. Yu, H. Xie, W. Chen, Journal of Applied Physics, 107 (2010).
28. A. Ijam, A. Moradi Golsheikh, R. Saidur, P. Ganesan, Journal of Materials Science, 49 (2014) 5934-5944.
29. M. Vakili, S.M. Hosseinalipour, S. Delfani, S. Khosrojerdi, Solar Energy Materials and Solar Cells, 152 (2016) 187-191.
30. S. Mesgari, R.A. Taylor, N.E. Hjerrild, F. Crisostomo, Q. Li, J. Scott, Solar Energy Materials and Solar Cells, 157 (2016) 652-659.
31. A.K. Hussein, Renewable and Sustainable Energy Reviews, 62 (2016) 767-792.

32. T. Tharayil, L.G. Asirvatham, V. Ravindran, S. Wongwises, International Journal of Heat and Mass Transfer, 93 (2016) 957-968.
33. E. Sadeghinezhad, M. Mehrali, M.A. Rosen, A.R. Akhiani, S. Tahan Latibari, M. Mehrali, H.S.C. Metselaar, Applied Thermal Engineering, 100 (2016) 775-787.
34. M. Mehrali, E. Sadeghinezhad, R. Azizian, A.R. Akhiani, S.T. Latibari, M. Mehrali, H.S.C. Metselaar, Energy Conversion and Management, 118 (2016) 459-473.
35. K.M. Kim, I.C. Bang, International Journal of Thermal Sciences, 100 (2016) 346-356.
36. N.M. Huang, H.N. Lim, C.H. Chia, M.A. Yarmo, M.R. Muhamad, International journal of nanomedicine, 6 (2011) 3443-3448.
37. Y. Li, Z. Wu, H. Xie, J. Xing, J. Mao, Y. Wang, Z. Li, Science China Technological Sciences, 60 (2017) 1168-1174.
38. K.N. Kudin, B. Ozbas, H.C. Schniepp, R.K. Prud'homme, I.A. Aksay, R. Car, Nano Letters, 8 (2008) 36-41.
39. W.S. Sarsam, A. Amiri, M.N.M. Zubir, H. Yarmand, S.N. Kazi, A. Badarudin, Colloids and Surfaces a-Physicochemical and Engineering Aspects, 500 (2016) 17-31.
40. M.B. Moghaddam, E.K. Goharshadi, M.H. Entezari, P. Nancarrow, Chemical Engineering Journal, 231 (2013) 365-372.
41. H. Xie, W. Yu, Y. Li, L. Chen, Nanoscale Research Letters, 6 (2011) 124.

Preparation of ultrafine iron phosphate micro-powder from phosphate slag

Fangjing Sun[1,2], Yi Zhang[1], Jiawei Zhang[1], Xixi Yan[1], Xiaoyu Liu[1,2], Lijun Wang[1,2,a]

[1]*School of Environmental and Materials Engineering, Shanghai Polytechnic University, Shanghai 201209, P. R. China;*
[2]*Research Center of Resource Recycling Science and Engineering, Shanghai Polytechnic University, Shanghai 200444, China.*

Abstract: In this experiment, ultrafine iron phosphate micro-powder was prepared by hydrothermal method which used phosphate slag as an iron source. The effects of reaction temperature, surfactants type and amount on its particle size were explored. The samples were characterized by using Malvern Laser Particle Size Analyzer (MS2000), X-Ray Diffractometer (XRD), Scanning Electron Microscope (SEM) and Energy Dispersive X-Ray Spectroscopy (EDX). The results showed that at 160 °C, 1 wt%CTAB, monoclinic iron phosphate micro-powder was obtained with an average particle size about 0.4 μm which also has a good dispersion in aqueous solution.

1 Introduction

With the improvement of people's living standards, transportation such as cars, ships and airplanes have gradually entered our lives. And the requirements for people incorrosion resistance and service life of their metal parts have become highest than ever before[1]. In order to prevent the metal surface being corroded and improve its wear resistance, phosphating technology has been widely used in the steel industry [2]. Phosphorization is one of the key processes in the metal industry before painting[3-5]. Its function is to improve the adhesion and wear resistance of the coating effectively. However, along with the phosphating process, the production of phosphate slag will be inevitable[6-8]. The production of phosphate slag not only reduces the quality of the workpiece, but also affects the phosphating process, and if it is directly discharged, it will cause serious pollution to the soil, water and air[9-13]. More importantly, the phosphating slag contains a large amount of Fe^{3+}, Zn^{2+}, PO_4^{3-}, etc., which are all recyclable materials[14-16]. On the one hand it can be used to reconfigure the phosphating solution, on the other hand, it can be used to prepare iron phosphate[17-19]. In recent years, China's environmental problems have aroused great concerns from the government. The people's awareness of environmental protection has greatly deepened. The treatment and disposal of phosphate slag has become one of the hotspots of research in recent years.

At present, some researches have been done on the utilization of phosphate slag both internally and externally[17-18]. The commonly used methods are: physical method (adsorption, reverse osmosis, electrodialysis and contact filtration method), chemical precipitation method, biological method, etc[19-24]. However, based on the treatment technology of phosphate slag is not perfect, many companies still use simple accumulation, incineration or landfill method[2]. This subject has studied the composition and properties of phosphate slag and the synthetic process of recycling and reuse. It has a deep understanding of it and has mastered the technology of phosphate slag to purify iron phosphate[5,8,25-26]. I have adopted a new method to purify and prepare iron phosphate micro-powder with smaller particle size. The purified and modified iron phosphate micro-powder has high dispersibility and can be used as catalyst[15,25], adsorbent and slow release fertilizer[26]. It can also be used as a precursor material for the preparation of lithium iron phosphate cathode[27]. Through the purification and utilization of iron phosphate in the phosphate slag, thousands of tons of iron phosphate can be recovered every year, and the economic value created can reach millions, at the same time it also solves a major problem of environmental pollution.

In summary, the subsequent research and application of the purified iron phosphate has theoretical research value and practical application value for promoting the phosphate slag resource treatment.

2 Materials and Methods

2.1 Experimental equipment and Chemicals

2.1.1 Experimental equipment

[a] Corresponding author: ljwang@sspu.edu.cn

Experimental Equipment: GSH type reaction kettle (Taixing City Construction Company); HJ-6 multi-head magnetic stirrer; SHB-III circulating water multi-purpose vacuum pump and DNF-6021 vacuum drying oven.

Characterization and performance test Equipment: X-Ray Diffractometer Diffractometer (XRD) (D8-Advance, Bruker Corporation, Germany) (λ = 0.154056 nm); Scanning Electron Microscope (SEM) (S-4800, Hitachi Corporation, Japan); Malvern 2000 Laser Particle Size Instrument (MS2000); American Micromeritics ASAP2020M Specific Surface Analyzer (BET); Japan Shimadzu 720 X-ray fluorescence spectrometer (EDX).

2.1.2 Chemicals

phosphate slag (Zhejiang Haiyan Metal Products Co., Ltd.); phosphoric acid (analytical grade, content ≥85.0 wt%); Cationic surfactant (CTAB), Analytically pure; Anionic surfactant (SDS), Analytically pure; Amphoteric surface activity Agent (hexamethylenetetramine), Analytically pure; Anhydrous ethanol, Analytically pure and Deionized water.

2.2 Experimental method

(1) Weigh a certain amount of purified phosphating slag in a beaker, add concentrated phosphoric acid (85.0 wt%), cationic surfactant (CTAB) and deionized water in a certain proportion and stir evenly under a magnetic stirrer;
(2) Re-transfer the reaction vessel, hydrothermal reaction for 12 h, vacuum filtration, drying at 80 °C to obtain ultra-fine iron phosphate ($FePO_4$ $2H_2O$);
(3) Calcined at 250 °C for 6 h to obtain a pale yellow anhydrous iron phosphate sample.

It was found by the Malvern 2000 laser particle size analyzer that the obtained powder sample had a particle size of about 0.4 μm.

3 Results and discussion

3.1 Effect of reaction temperature on the particle size of iron phosphate micropowder

The cationic surfactant (CTAB) was selected as the dispersant. The effect of hydrothermal temperature on the particle size of iron phosphate was studied under the condition of controlling the CTAB amount to 0.4g and the feed ratio being 4:2:1:1. It can be seen from Fig.1 that the particle size of iron phosphate decreases first and then increases when the temperature rises. At 160 °C, the particle size is the smallest, which the average particle size is about 0.4 μm. The reason for this trend may be: before 160 °C, as the temperature increases, the nucleation rate of the crystal grains also increases, and the crystal nucleus is too late to grow. At this time the particle size is getting smaller and smaller, along with the temperature increasingly, the number of crystal nuclei increases sharply, the molecular motion is severe. However, the probability of collision between crystal grains and crystal grains increases, so that agglomeration occurs, and the particle size of the iron phosphate particles becomes larger.

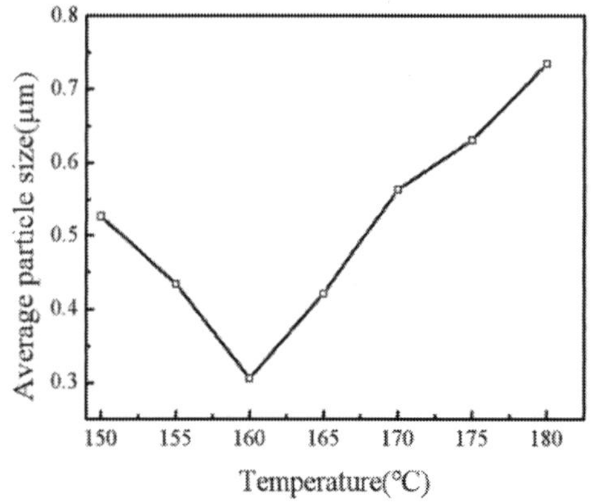

Figure 1. Effect of reaction temperature on particle size of iron phosphate micro-powder

3.2 Effect of reaction ratio on the particle size of iron phosphate

During the whole reaction process, the reaction ratio of purified iron phosphate and phosphoric acid, absolute ethanol and deionized water also has a great influence on the particle size. The reaction temperature is controlled at 160 °C for 12 h. Under this condition, the influence of the feed ratio of the raw materials on the particle size was explored. It can be seen from Table 1 that the particles of the particles were purified when the iron $FePO_4$: H_3PO_4:C_2H_6O: H_2O=4:2:1:1 was purified and the particle size is the smallest. At the same time, it can be seen that the content of phosphoric acid and deionized water have significant influences on the particle size. When no water is added, the particle size is large, this probably because the deionized water provides a liquid environment for its nucleation and together with it achieved a dispersion effect.

Table 1 Effect of reaction ratio on the particle size of iron phosphate

	$FePO_4 \cdot 2H_2O$: H_3PO_4: C_2H_6O: H_2O				
Reaction ratio	4:1:2:0	4:1:2:1	4:2:1:0	4:2:1:1	4:2:2:0
Average particle size (μm)	0.705	0.697	0.581	0.407	0.671

3.3 Effect of the types of surfactant on the particle size of iron phosphate micropowder

Surfactants are mainly classified into three major categories: cationic surfactant, anionic surfactant, and zwitterionic surfactant. This process controls the hydrothermal reaction temperature at 160 °C for 12 h, and three types of surfactant is used as dispersant to regulate the particle size of iron phosphate. The most commonly used surfactant is: Cetyltrimethyl Ammonium Bromide (CTAB), Sodium Dodecyl Sulfate (SDS) and Dodecyl Sulfopropyl Betaine (DSB). As shown in Figure 2 below, the three surfactants have different effects on the particle size of iron phosphate under different surfactant dosages. The results show that the cationic surfactant can effectively control the growth of the nucleation of the iron phosphate micropowder and can disperse the fine powder of iron phosphate well, while the anion and zwitterion have a weaker regulation effect on the iron phosphate particle size. The reason may be that the phosphate anion moiety binds to the cationic surfactant to prevent the growth of the iron phosphate particles.

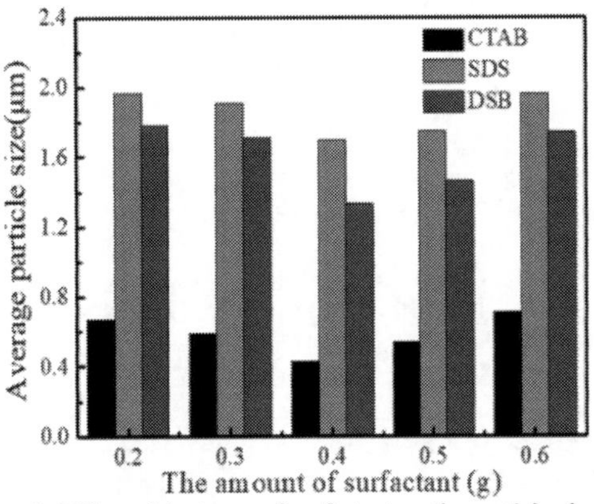

Figure 2. Effect of the types of surfactant on the particle size of iron phosphate micro-powder

3.4 Effect of the amount of cationic surfactant CTAB on the particle size of iron phosphate micropowder

Through the above comparison, we have used cationic surfactant (CTAB) as the preferred surfactant, and the amount of surfactant is one of the factors affecting the particle size. When other reaction conditions are not changed, the influence of the amount of CTAB on its particle size can be investigated. As shown in Fig.2, it can be visually seen that as the amount of CTAB increases, the particle size of the obtained sample first decreases and then increases. When the amount of CTAB is 0.4 g, that is 1 wt% of the amount of iron phosphate, the particle size of the sample is minimized. The reason why the particle size is first reduced and then increased may be that when the amount of CTAB is small, a thin film can be formed on the surface of the sample to hinder the formation of other chemical bonds, and the growth of crystal grains is limited, and the particle size is very small. When the amount of CTAB is too large, the surfactant coated on the surface of the sample accumulates, and the van der Waals force between the CTAB molecules causes them to stick together, causing the entire particle to become larger.

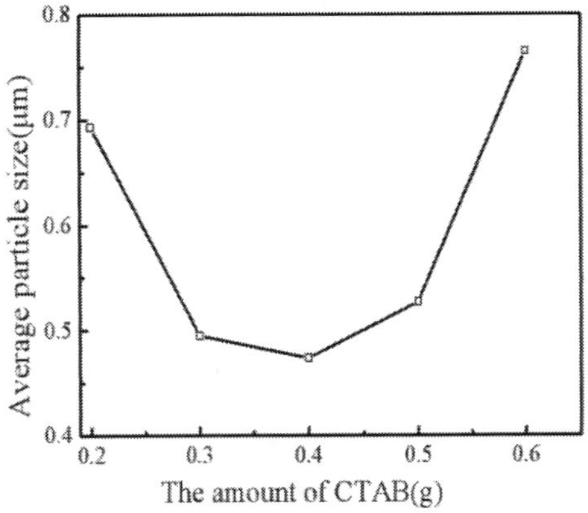

Figure 3. Effect of CTAB amount on particle size of iron phosphate micro-powder

3.5 Characterization

3.5.1 SEM of sample

According to the above experimental investigation, the optimal purification conditions for the preparation of iron phosphate micropowder were selected. As shown in Fig. 3, the hydrothermal temperature was 160 °C, and the reaction ratio was: after the purification, iron phosphate: phosphoric acid: ethanol: water = 4:2:1:1, the reaction time is 12 h, the SEM image of the sample prepared under the condition that the amount of surfactant (CTAB) is 1 wt% It can be seen that the particle size of the sample is relatively uniform, relatively scattered. The average particle size is about 0.4 μm.

Figure 4. Hydrothermally prepared sample SEM image

3.5.2 XRD of sample

Fig.5 is a comparison of the hydrothermally prepared sample and the standard iron phosphate XRD. The prepared sample spectrum is consistent with the standard spectrum, and the peaks of the three strong peaks are sharp The XRD test results indicates that the micropowder belongs to Monoclinic crystal compared to the standard PDF card 33-0666.

https://doi.org/10.1051/matecconf/201823802002

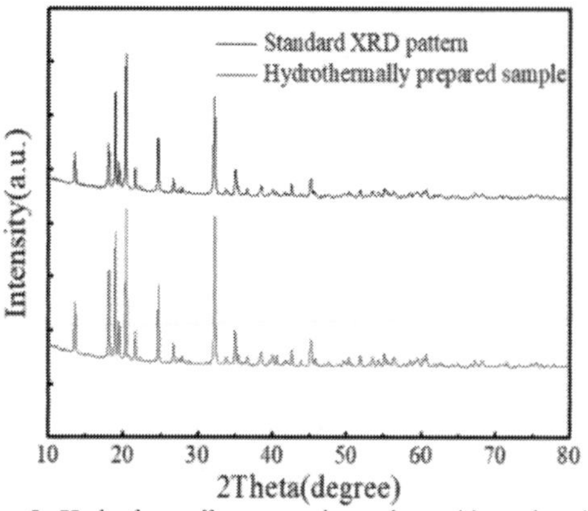

Figure 5. Hydrothermally prepared samples and iron phosphate standard XRD spectra

4 Summary

In this paper, the solid waste phosphating slag is purified by a simple process, in order to prepare micron-sized iron phosphate powder. The XRD test results show that the prepared iron phosphate belongs to the monoclinic system, which is a regular-sized, flat-surfaced micron-sized sheet with a thickness.The thickness thereof is about 200 nm, and the average particle diameter is about 0.4 μm.

In this experiment, the particle size is controlled by changing the reaction temperature, the type and amount of surfactant, and after a large number of experiments, relatively preferred conditions are obtained: purified iron phosphate: phosphoric acid: absolute ethanol: deionized water = 4:2:1:1, the hydrothermal temperature is 160 °C, the reaction time is 12 h, and the amount of the surfactant (CTAB) is 1wt% of the amount of iron phosphate.

This experiment provides new ideas and new methods for the resource treatment of phosphating slag. The prepared iron phosphate has a good application in the fields of agriculture, ceramics, environmentally friendly coatings and surface passivation.Because of its unique adsorption performance, ion exchange performance and electrochemical performance, it has also become an increasingly important application in the fields of industrial wastewater and lithium battery electrode materials.

Acknowledgments

This work was supported by Engineering Key Subject of Shanghai Polytechnic University (No. XXKZD1601), Shanghai Polytechnic University Graduate Program Foundation (EGD16YJS012) and Gaoyuan Discipline of Shanghai-Environmental Science and Engineering (Resource Recycling Science and Engineering). The authors acknowledge the use of equipment, and the scientific and technical assistance of Research Center of Resource Recycling Science and Engineering, Shanghai Polytechnic University.

References

[1] X Y Wu, J C Huang, Q C Lang, et al. Progress and prospects of research on phosphating slag resource[J].Recy. Circ.**8**(8): 38-41(2015).

[2] X Y Xiong, T Q Xiong, Q Q Song, et al. Metal surface treatment phosphati slag comprehensive utilization technology[J]. Meta. Pow. **2**: 43-49(2015).

[3] Y Lu, Y T Song, Y L Liu. Study on preparation and properties of phosphate corrosion resistant coatings[J]. Elect. Rev, **30**(118): 19-22.(2016).

[4] H Q Fan, S Y Li. Research progress on the status and resource utilization of solid waste phosphate slag[J]. Mode. Chem. **27**(s1): 97-99, 101(2007).

[5] X Q Liu, S N Zhang, L J Wang, et al. Effect of calcination temperature on preparation of LiFePO4/C positive electrode material by purifying phosphate Slag[J]. Journal S. P. U.**34**(1): 26-30(2017).

[6] F P Ma, X W Chen, J Tian. Production technology of tricalcium phosphate by using phosphating waste residue[J]. Inor S. I. **47**(9): 54-55(2015).

[7] J T, X W Chen, R P Jiang. Zinc phosphating solution waste slag comprehensive utilization technology[J]. Mode C. I. **35**(6): 155-156(2015).

[8] D F Wu, S N Zhang, L J Wang, et al.Preparation of new catalyst and photocatalytic reaction for waste phosphate slag hydrothermal reaction[J]. Journal S. P. U.**33**(4): 283-288(2016).

[9] H Y Ye, D M Zheng, H Q Li, et a.Study on preparation of ultrafine dihydrate ferric phosphate[J].Inor S. I. **44**(4): 59-61(2012).

[10] P Z Zhao, H B Liu, H H Zheng, et al. Facile synthesis of $FePO_4{\bullet}2H_2O$ submicrometer-discs. Mate L. **123**:128-130(2014).

[11] H J Song, Y L Sun, X H Jia. Hydrothermal synthesis of iron phosphate microspheres constructed by mesoporous polyhedral nanocrystals. Mater C. **107**: 182-188(2015).

[12] G C Li, S Zhang, H G Peng, et al. Growth and shape control of orthorhombic $Fe_5(PO_4)_4(OH)_3{\bullet}2H_2O$ single crystalline dendrites. Jour C. G. **312**(20): 2967-2971(2010).

[13] Y M Zhu , S Z Tang, H H Shi, et al . Synthesis of $FePO_4{\bullet}xH_2O$ for fabricating submicrometer structured $LiFePO_4/C$ by a co-precipitation method. Cerams Interl, **40**(2): 2685-2690(2014).

[14] T B Zhang, X B Cheng, Q Zhang, et al. Construction of a cathode using amorphous $FePO_4$ nanoparticles for a high-power/energy-density lithium-ion battery with long-term stability. Jour P. S. **324**: 52-60(2016).

[15] F K Behbahani, F MGolchin. A new catalyst for the synthesis of 2-substituted perimidines catalyzed by $FePO_4$. Jour T.U. Sci. **11**(1): 85-89(2017).

[16] X Wang, J Zhuang, Q Peng, et al. A general strategy for nanocrystal synthesis, Nature,**437**: 121–124 (2005).

[17] L S Zhong, J S Hu, H P Liang, et al. Self-assembled 3D flowerlike iron oxide nanostructures and their application in water treatment, Adv. Mater. **18**: 2426–2431(2006).

[18] C Chen, W Chen, J Lu, et al.Transition-metal phosphate colloidal spheres, Angew. Chem. Int. Ed. **48**: 4816–4819(2009).

[19] H Okawa, J P Yabuki, Y H Kawamura, et al. Synthesis of $FePO_4$ cathode material for lithium ion batteries by a sonochemical method, Mater. Res. Bull. **43**: 1203–1208(2008).

[20] A Mamoru. Oxidation activity of iron phosphate and its characters, Catal. Today **85**: 193–198(2003) .

[21] K Kandori, T Kuwae, T Ishikawa. Control on size and adsorptive properties of spherical ferric phosphate particles, J. Colloid Interface Sci. **300**: 225–231 (2006).

[22] Z C Shi, A Attia, W L Ye, et al. Synthesis of $FePO_4$ by direct solid state reaction at ambient temperature and its charge–discharge properties, Electrochim. Acta **53**: 2665–2673 (2008).

[23] F Cao, D X Li. Biotemplate synthesis of monodispersed iron phosphate hollow microspheres, Bioinsp. Biomim. **5**: 16005–16010 (2010).

[24] L M Wei, Y Zhang, L J Wang, et al. Synthesis of nitrogen-doped carbon
nanotubes-$FePO_4$ composite from phosphate residue and its application as effective Fenton-like catalyst for dye degradation. Jour E. Sci,1-9(2018).

[25] Y Zhang, Z F Ying, L J Wang, et al. Modified iron phosphate/polyvinyl alcohol composite film for controlled-release fertilisers[J]. RSC Adv.**8**: 18146-18152(2018).

[26] R Cai, Y P Du, W Y Zhang, et al. Synthesis of porous amorphous $FePO_4$ nanotubes and their lithium storage properties, Chem. Eur. J. **19**:1568–1572(2013) .

Synthesis of iron phosphate-SAPO-34 composite and its application as effective absorbent for wastewater treatment

Ying Cao[1,2], Lianmei Wei[1], Xianzhen Song[1], Xixi Yan[1], Xiaoyu Liu[1] and Lijun Wang[1,2,a]

[1]*Facultyof Environmental and Materials Engineering, Shanghai Polytechnic University, Shanghai 201209, P.R. China*
[2]*Research Center of Resource Recycling Science and Engineering, Shanghai Polytechnic University, Shanghai 201209, P.R. China*

Abstract. High-purity $FePO_4$ was purified from iron-based phosphating slag as raw material, and $FePO_4$@SAPO-34 was synthesized by hydrothermal crystallization method under the action of templating agent-diethylamine. The synth esized $FePO_4$@SAPO-34 samples were characterized by x-ray diffraction (XRD), scanning electron microscopy (SE M) and Fourier transform infrared spectroscopy (FT-IR). The effects of different crystallization time on the morpholo gy and crystallization of $FePO_4$@SAPO-34 crystals were investigated. The removal of heavy metal ion wastewater by low-cost $FePO_4$@SAPO-34 was investigated. The experimental results show that when the reaction time is 180 °C an d the reaction time is 72h, the crystallization of $FePO_4$@SAPO-34 is the best. When the dosage is 0.6g, the removal e fficiency of heavy metal ions is the highest.

1 Introduction

Phosphating is a commonly used surface treatment technology. Phosphating slag is a precipitated substance that is produced on the surface of a metal phosphating process. Its main components are metal ions such as phosphate, iron and zinc [1]. In order to effectively control the pollution of phosphating slag, researchers have carried out a lot of research work and made some progress. In this study, the iron phosphate is purified by phosphating slag and has various crystal structures. It is provided with unique catalytic properties, ion exchange properties, etc. Research reports show that iron phosphate is negatively charged (zeta = -30 ~ -40mV) in wastewater, and Its electronegative properties produce electrostatic adsorption on positively charged ions [2].

The molecular sieve has a uniform pore structure. According to the molecular size, it can selectively adsorb. Thus, it is mostly used for adsorbents and catalysts [3]. In 1982, SAPO-34 was first synthesized as a SAPOS series molecular sieve by Wilson et al [4]. It has CHA structure and medium strength acid center. SAPO-34 exhibits excellent performance in adsorption and catalysis [5]. Because SAPO-34 has a neutral framework structure, it is difficult to provide the active center required for many reactions [6]. Therefore, it is very important to modify the material. In summary, this study prepared a new catalytic adsorption material of $FePO_4$@SAPO-34. New materials were applied to remove heavy metal ions from wastewater.

2 Materials and methods

2.1 Materials

Phosphate residue was supplied by an auto components company in Zhejiang, China. $Pb(NO_3)_2$ (analytical regent(AR),⩾99.9%), pseudo boehmite (Al_2O_3,⩾67.5%), phosphoric acid (AR, ⩾85%) and diethylamine (AR, 99%) were obtained from Sinopharm Chemical Reagent Co. Ltd., of Shanghai, China. Silica solution (SiO_2, ⩾30%) was purchased from Qingdao Haiyang Chemical Co.,Ltd[4]. Other chemicals used in the present work were of ananlytical grade and used without further purification. Deionized water was produced by an ultrapure water system (UPT-I-60L, Super Pure Technology Corporation, China).

2.2 Adsorbent preparation

A certain amount of phosphating slag, water and phosphoric acid were uniformly mixed in an appropriate ratio, and placed in an oven at 80℃ for 4 h, and the sample was suction filtered and dried overnight; after three times of reaction, a phosphating slag was obtained to purify the sample iron phosphate. The experiment used hydrothermal synthesis of $FePO_4$@SAPO-34. Firstly, the silicon source, the phosphorus source, the aluminum source, the $FePO_4$ and the appropriate amount of water are uniformly mixed according to a certain ratio, and then

[a] Corresponding author: ljwang@sspu.edu.cn

the templating agent diethylamine is added, and after pretreatment, it is transferred to a polytetrafluoroethylene reactor, and heated to 180 °C, and is self-generated[5]. Crystallize under pressure for different times, after suction filtration, washing and drying. Then, it was baked in a muffle furnace at 550 °C for 6 h to obtain a modified new material.

2.3 Heavy metal ion removal experiment

The experiment of removal of heavy metal ions in simulated wastewater was carried out with a material dosage of 0.6 g, lead-containing ($Pb(NO_3)_2$) wastewater concentration of 100 mg/L. Adesignated intervals, samples of the solution were collected and immediately filter to monitor the adsorption efficiency by inductively coupled plasma-atomic emission spectrometry (ICP-AES) (ICAP 6000, Thermo Corporation, USA). All adsorption experiments were stirred at 350 r/min. After adsorption experiment, the $FePO_4$@SAPO-34 adsorbent was collected with strong acid and then reused.

2.4 Characterization methods

The morphology of the samples was characterized by sca-nning electron microscopy (SEM) (S-4800, Hitachi Corp-oration,Japan) with the accelerating voltage at 10 kV and current at 10 mA. X-ray diffraction (XRD) (D8-Advance, Bruker Corporation, Germany) measure ments were un-dertaken to record the phase composition and crystalline structure of the powder samples, operating at 40 kV and 40 mA with a monochromatized Cu Kα radiation (λ = 0.15418 nm) source. The chemical bonds on the surface of the samples were examined by Fourier transform infrared spectroscopy (FTIR) (Vertex70, Bruker Corporation, Germany). KBr was chosen as a diluent. The surface area was measured based on the Brunauer Emmett Teller (BET) (ASAP-2020M, Micromeritics Corporation, USA) model over a relative pressure range of 0.05–1.

3 Results and discussion

3.1 XRD characterization

Figure 1 is an XRD pattern of a sample synthesized at different crystallization times. Crystallization plays a role in transforming metastable materials into stable crystals. So crystallization time is one of the most important factors that affect the synthesis of zeolite. Fig1. shows five peaks at 9.48°, 16.2°, 20.7°, 25°and 31°, corresponding to (101), (211), (104), (401) of SAPO-34 (JCPDS), respectively [6-7]. With the extension of time, the peak intensity gradually increased. When t=72 h, the peak intensity was the strongest, but when the time was extended to 84 h, the peak intensity weakened and the impurity peak began to appear. The crystallinity of SAPO-34 was set to 100%. The relative crystallinity of $FePO_4$@SAPO-34 is calculated as follows [7-8]:

$$X_i(\%) = X_s \frac{\sum I_i}{\sum I_s} \quad (1)$$

Where, X_i is relative crystallinity of the sample; X_S is equivalent to 100%; $\sum I_i$ is the sum of the five characteristic peak intensities of the sample; $\sum I_s$ is the sum of the five characteristic peak intensities of the standard sample.

From Fig. 3, the relative crystallinity of $FePO_4$@SAPO-34 increased from 40% to 98%, indicating that the crystallinity of the sample increased with the crystallization time. But when the time is too long, it could lead to the formation of other crystals. Too short or too long crystallization time was not conducive to the synthesis of $FePO_4$@SAPO-34. The crystallization time was the most favourable for the synthesized $FePO_4$@SAPO-34 at 72 h.

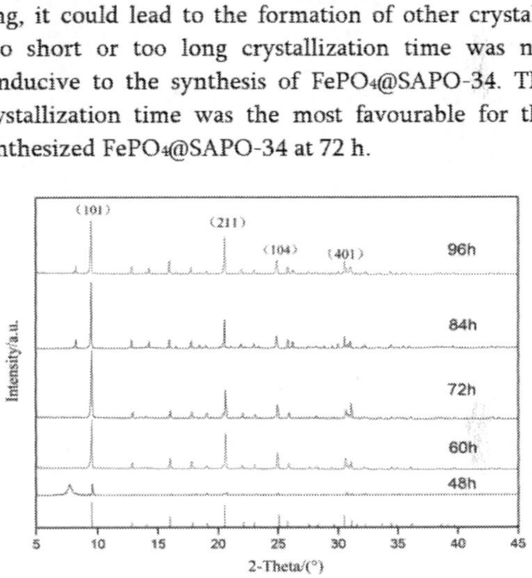

Figure 1. –X-ray diffraction (XRD) patterns of $FePO_4$@SAPO-34 at different crystallization times

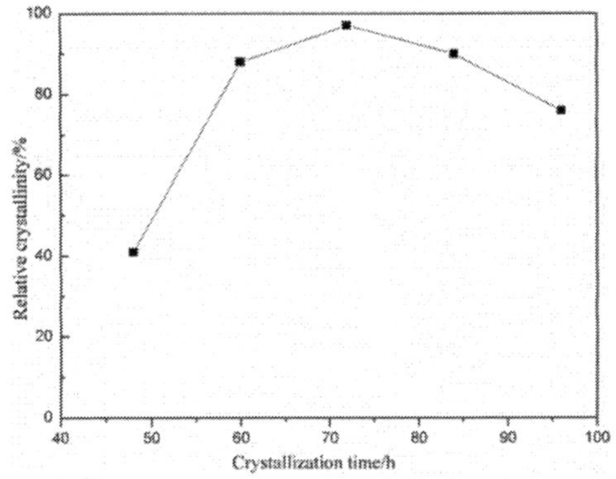

Figure 2. Crystallinity of $FePO_4$@SAPO-34 at different crystallization times

https://doi.org/10.1051/matecconf/201823802003

3.2 SEM characterization

The scanning electron microscopy (SEM) images showed, as the time increased from 48 h to 72 h, the shape of the material became more and more complete. At 180 ℃, t=72 h, $FePO_4$@SAPO-34 had a typical cubic small square structure of chabazite-based molecular sieves, indicating that the crystal growth was perfect [9-10]. When the time was gradually increased to 84-96 h, the morphology of $FePO_4$@SAPO-34 particles was destroyed, and some amorphous substances appeared in the synthesized product.

Figure 3. Scanning electron microscopy (SEM) images of $FePO_4$@SAPO-34 at different crystallization times

3.3 FT-IR and BET characterization

FT-IR analysis was further carried out to verify $FePO_4$@SAPO-34 in the reaction process. FT-IR spectrum was sensitive to the skeletal structure of the molecular sieve, and could better detect the molecular sieve crystallite structure. SAPO-34 has a periodic arrangement of PO^{2+}, AlO_4 and SiO_4 tetrahedral framework structures. As shown in Fig 4, the spectrum of $FePO_4$@SAPO-34 was basically consistent with the spectrum of SAPO-34. Peak at 2320 cm^{-1} could be ascribed to the bending vibration of CO_2, while those at 730cm^{-1} , 640cm^{-1} , 575cm^{-1} and 530cm^{-1} could be assigned to the stretching vibration of OPO (O-Al-O), the double six-membered ring vibration peak and PO_4. It may be due to moisture in the air during the test. When $FePO_4$ was doped, since the electron energy of iron was smaller than the electron energy of silicon, the bonding strength of the sample was enhanced, and the corresponding frequency was increased[11-13].

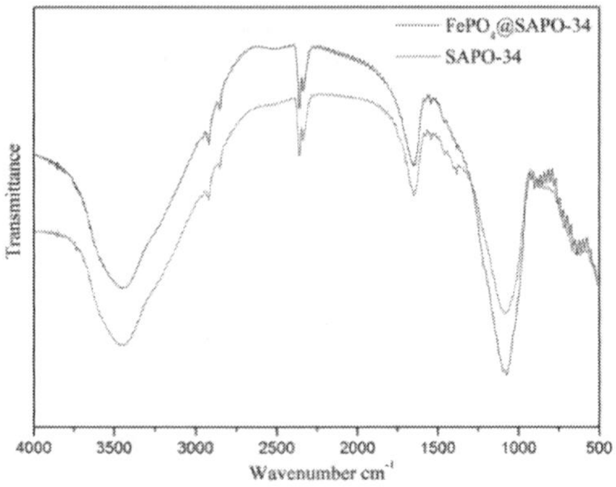

Figure 4. FT-IR image of $FePO_4$@SAPO-34 and SAPO-34 samples

From Table 1 ,the specific surface of SAPO-34 was 391.2 m^2.g-1, and the specific surface of $FePO_4$@SAPO-34 was 362.8 m^2.g-1. It had little difference and large specific surface area.

Table 1. BET of $FePO_4$@SAPO-34 and SAPO-34

Sampal	$FePO_4$@SAPO-34	SAPO-34
BET surface area/(m^2.g^{-1})	362.8	391.2

3.4 Adsorption of pb^{2+} experiments

The removal effect of Pb^{2+} by SAPO-34 and $FePO_4$@SAPO-34 materials has been shown in Fig.5. It can be seen that SAPO-34 has better absorbing capacity than $FePO_4$@SAPO-34 in the first 45min, which was ascribed to the large specific surface area for pure SAPO-34 with a uniform pore structure[13-15]. However, the removal efficiency of two materials has been changed. The removal efficiency of $FePO_4$@SAPO-34 exceeded to that of SAPO-34 due to the electrostatic effect of FePO4 in aqueous solution, further attracting cations which has been reported in other papers. Thus, the $FePO_4$@SAPO-34 showed excellent adsorption performance in pollution removal.

On this basis, this experiment studied the effect of $FePO_4$@SAPO-34 dosage on Pb^{2+} adsorption. As shown in Figure 6, as the amount of material increases, it can be seen that, the amount of Pb^{2+} adsorption increases. The increase was larger and then tended to be flat. The adsorption efficiency is calculated as follows:

$$\eta = \frac{C_0 - C_i}{C_0} \times 100\% \quad (2)$$

Where, C_0 is original concentration; C_i is concentration during dynamic adsorption equilibrium. η is adsorption efficiency. When the addition amount was 0.2 g, the adsorption rate was only 67.3%. When the amount added was increased to

0.6 g, the removal rate of Pb^{2+} was 89%. When the amount added was 1 g, the adsorption rate was 92%. As the number of exchanged active particles increases, the amount of adsorption of metal ions becomes large[14-16]. Therefore, the amount of adsorbent has a great influence on the adsorption effect.

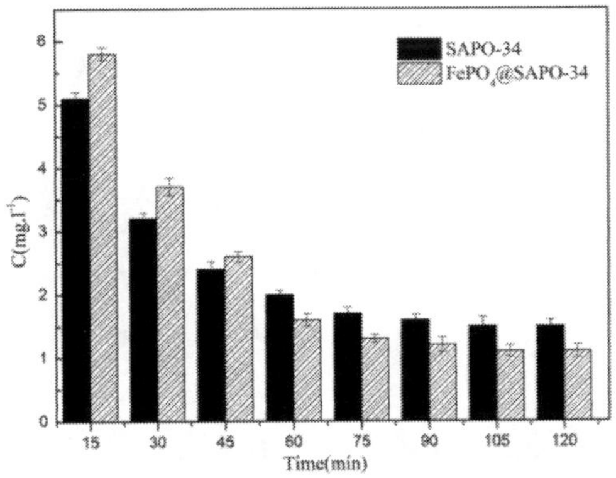

Figure 5. Pb^{2+} removal effect image of $FePO_4$@SAPO-34

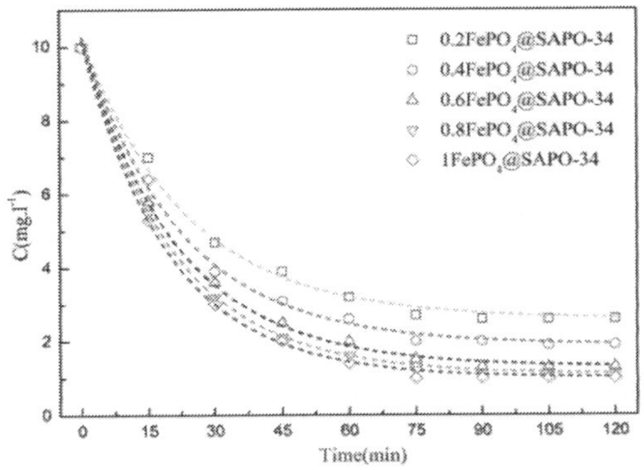

Figure 6. Adsorption effect diagram for different adsorbent dosages

Conclusions

In this study, an effective phosphate residue recovery method was proposed and be used to fabricate a $FePO_4$@SAPO-34 adsorbent. Experiments were conducted to investigate the effect of different crystallization times on the crystal form of the adsorbent. The results showed that the synthesized sample crystal form was more complete at 72 h. Experiments have found that new materials can effectively remove heavy metal ions from wastewater. Compared with SAPO-34, the new synthetic material passes through the electro-negative properties of $FePO_4$, so it electrostatically adsorbs positively charged ions. And FePO4 provides a metal active center for SAPO-34. The synergy between SAPO-34 and $FePO_4$@SAPO-34 improves the adsorption performance of new materials.

Acknowledgement

This work was supported by Shanghai Polytechnic University Graduate Program Foundation (No. EGD18YJ0005), Material Science and Engineering Key Subject of Shanghai Polytechnic University (No. XXKZD1601), and Gaoyuan Discipline of Shanghai-Environmental Science and Engineering (Resource Recycling Science and Engineering).

References

1. L. Wei, Y. Zhang, S. Chen, L. Zhu, X. Liu, L. Kong, L. Wang, J. Environ. Sci. 1–11 (2018)
2. Q. Sun, Z. Xie, J. Yu, National Science Review, 1–17 (2017)
3. L J. Wang, L L. Xie, H. Yuan, Q H. Li, Q Z. Li, ACA CHIMICA SINICA. 65, 170-172 (2007)
4. L J. Wang, H T. Zhao, Z X. Hao, H. Yuan, L L. Xie, Z. Tian, S W. Chen, J F. Wang, Q H. Li, ACTA CHIMICA SINICA. 65, 1929-1932 (2005)
5. L J. Wang, H T. Zhao, Z X. Hao, L L. Xu, L L. Xie, Z. Tian, H. Yuan, N C. Bing, Q H. Li, ACTA CHIMICA SINICA. 66, 1317-1321 (2008)
6. W P. Zhu, Y P. Xue, Y. Li, S X. Tian, G. Yue, Modern Chemical Industry. 30 (2010)
7. L. Long, S. Xiao, Z Y. Li, S H. Qi, Y N. Men, W B. ao, Y N. Liu. BULLETIN OF THE CHINESE CERAMIC SOCIETY. 34 (2015)
8. A H. Xing, Q Y. Feng, X F. Zhang, J D. Jiang, INDUSTRIAL CATALYSIS. 24 (2016)
9. M. Han, Dalian University of Technology. (2011)
10. S. Lin, Jilin University. (2011)
11. Y. Wang, Taiyuan University of Technology. (2008)
12. S Andonova, S Tamm, C Montreuil, et al. Applied Catalysis B Environmental, 180 (2016)
13. G. Feng, P. Cheng, W. Yan, M. Boronat, X. Li, J. Su, J. Wang, Y. Li, A. Corma, R. Xu, J. Yu, 351 (2016)
14. M. Han, Dalian University of Technology. 2011
15. W J. Wang, C Z. Gen, Environmental Science and Technology. 27 (2014)
16. Q. Sun, Z. Xie, J. Yu, National Science Review, 1–17 (2017)

MATEC Web of Conferences **238**, 02004 (2018)
ICNFM2018
https://doi.org/10.1051/matecconf/201823802004

Microwave Synthesis of AlFeCuCrNi /TiB_2 High-Entropy Alloy Matrix Composites

Lijuan Lan[a],Tianjiao Pu,Yingying Gu,Chengyan Zhu and Heguo Zhu

College of Materials Science and Engineering, Nanjing University of Science and Technology. Nanjing, 210094; P. R. China

Abstract. The Al-Fe-Cu-Cr-Ni-Ti-B system was microwaved to generate high entropy alloy matrix composites reinforced by TiB_2 particles. The micro structure and reaction process of the composites were observed and investigated by modern analysis methods, including X-ray diffraction (XRD), scanning electron microscopy (SEM), X-ray energy dispersive spectroscopy (EDS) and differential scanning calorimeter (DSC) analysis. The results show that AlFeCuCrNi /TiB_2 composites can be prepared by microwave heating method. The matrix structure was FCC, and the reinforcement TiB_2 showed regular geometric morphology in the matrix and evenly distributed in the matrix when the volume fraction of the reinforcement is 10%. When the volume fraction of the reinforcement increased to 15%, TiB_2 partially aggregates in the matrix, and the system activation energy was 195.69 kJ/mol.

1. Introduction

Conventional alloys usually have one or two elements as the main elements. To meet industrial requirements, researchers add appropriate amounts of other elements to improve the properties of the alloy [1]. In the 1990s, scholar Ye Junwei et al [2] from Taiwan first proposed the concept of multi-principal high-entropy alloy, which attracted attention from domestic and foreign scholars. The consist of multi-principal high-entropy alloy is at least five elements, each of them takes up a content, ranging from 5% to 35%.This kind of alloy's properties are determined by a variety of principal elements [3]. It is found that the microstructure of high-entropy alloys is not so complex as predicted by traditional alloy design concepts. Due to the high entropy effect, the structure of high-entropy alloys is mainly composed of a single BCC, FCC or HCP solid solution phase, and also shows high strength, hardness, excellent wear resistance, corrosion resistance, low temperature resistance, high temperature oxidation resistance and temper softening resistance [4-6].

Mixed entropy [7] is an important thermodynamic feature of high-entropy alloys distinguished from traditional alloys. Gibbs Free Energy Law:

$$\Delta G=\Delta H - T\Delta S \quad (1)$$

(ΔG is the change of free energy, H is the mixed enthalpy, T is the thermodynamic temperature, and S is the mixed entropy). Although the high-entropy alloy has high mixed entropy, the free energy of the system can be reduced, and the intermetallic space is largely suppressed. The formation of the compound makes it easier for the alloy to form a solid solution rather than an intermetallic compound when solidified [8].

At present, domestic and foreign scholars are increasingly mature in the study of high-entropy alloys, and there is less progress in high-entropy alloy-based

[a]*Corresponding author: lan_lj@163.com;*

https://doi.org/10.1051/matecconf/201823802004

composites. In this paper, the TiB_2 enhanced AlFeCuCrNi high-entropy alloy matrix composites prepared by microwave sintering were studied. The prepared samples were analyzed by X-ray diffraction (XRD) and scanning electron microscopy (SEM). The microstructures and distribution of different TiB_2 volume fractions were studied. The DSC curve at different heating rates was also tested, and the reaction kinetics was studied to determine the activation energy of the reaction.

2. Experimental

Ni powder (purity 99.8%), Fe powder (purity 99.4%), Cu powder (purity 99.5%), Al powder (purity 99.7%), Cr (purity 99.7%), Ti powder (purity 99.5%), B powder (purity 99.5%) were arranged according to a certain ratio. In the AlFeCuCrNi /TiB_2 high-entropy alloy matrix composite, the molar ratio is

Al:Fe:Cu:Cr:Ni=0.5:1:1.5:1:1, and the volume fraction of TiB_2 were respectively 10% and 15%. The sample is 30g per serving. The prepared powder was dried in a dry box. The powder mix was ball-milled in a stainless steel vacuum jar for 4 h and compacted under a pressure of 180 MPa into cylindrical specimens. Then, put the crucible with the pressure block into the microwave oven, closed the valve, vacuum, opened the control panel, set the heating and cooling program for microwave radiation, raised it to 1100 °C, and after cooling for 5 min, cool down to 100 °C with the furnace, took out the sample. The sample was ground, polished and mechanically polished and investigated by XRD and SEM equipped with an energy dispersive spectroscopy (EDS).Another dry powder of TiB_2 volume fraction of 15% was pressed into a thin sample, and 5-10 mg was placed in a differential scanning calorimeter to measure the DSC curve at different heating rates.

3. Results and discussion

3.1 Thermodynamic analysis of the reaction system

In the Al-Fe-Cu-Cr-Ni-Ti-B system, the main reactions that may occur are as follows:

$$Ni+Al \rightarrow NiAl$$

$$\Delta G^{\theta}_{NiAl,T} = -118407 + 4.10T \quad (2)$$

$$3Ni+Al \rightarrow Ni_3Al$$

$$\Delta G^{\theta}_{Ni_3Al,T} = -153134 + 4.14T \quad (3)$$

$$2Ni + 3Al \rightarrow Ni_2Al_3$$

$$\Delta G^{\theta}_{Ni_2Al_3,T} = -280420 + 8.15T \quad (4)$$

$$Ni+3Al \rightarrow NiAl_3$$

$$\Delta G^{\theta}_{NiAl_3,T} = -150624 + 4.17T \quad (5)$$

$$Ti+2B \rightarrow TiB_2$$

$$\Delta G^{\theta}_{TiB_2,T} = -281500 + 7.9T \quad (6)$$

From (1-1) to (1-5), it is clear that each equation is negative in the temperature range of the experiment. That is, from the thermodynamic point of view, each reaction may spontaneously proceed. Figure 1 is a graph showing the change of Gibbs free energy with temperature for NiAl, Ni_3Al, Ni_2Al_3, $NiAl_3$ and TiB_2. The figure shows that the Gibbs free energy of TiB_2 is the lowest, that is, the reaction is most likely to occur,so we can indicate that the reinforcement can be formed in the system.

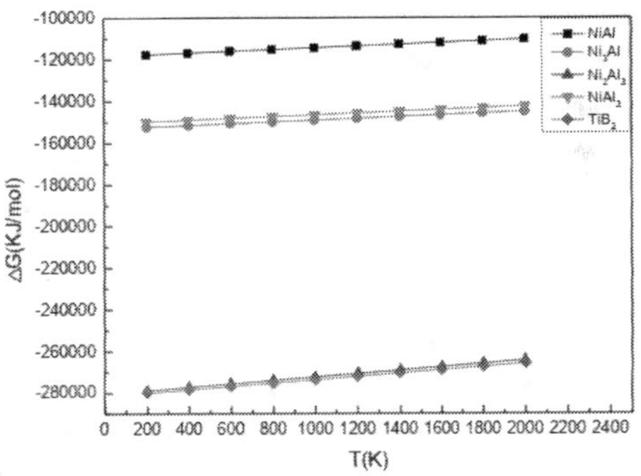

Fig.1 the change of Gibbs free energy with temperature for NiAl, Ni_3Al, Ni_2Al_3, $NiAl_3$ and TiB_2

3.2 SEM at different volume fractions and their corresponding XRD

Figures 2 and 3 respectively show SEM and XRD images with volume fractions of 10% and 15%. It can be seen from Fig. 2 that the bulk particles are distributed on the substrate with a scale of about 1 um. The corresponding XRD indicates that the reaction product of the system

https://doi.org/10.1051/matecconf/201823802004

mainly has two phase structures, including the matrix phase and TiB_2. The matrix phase's structure is FCC. It can be seen from Fig. 3 that the TiB_2 particles are segregated and the dispersion is decreased, but the gap between the particles is reduced. It can be explained that the volume fraction is increased, and the heat of the thermal explosion reaction increases after the action of the microwave, which is favorable for diffusion and reduces the gap.

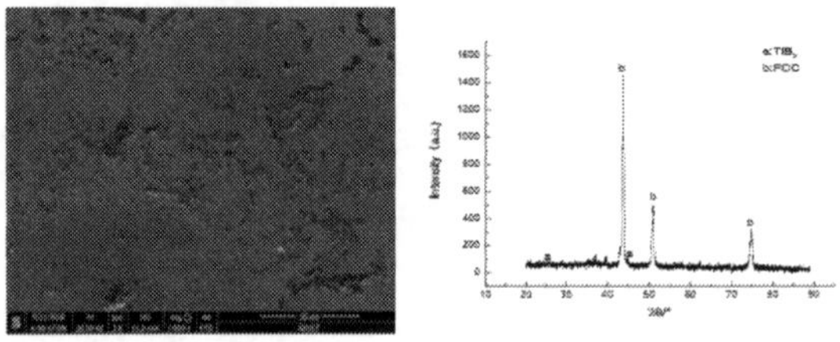

Fig.2 SEM and XRD images of the composites with volume fractions of 10%

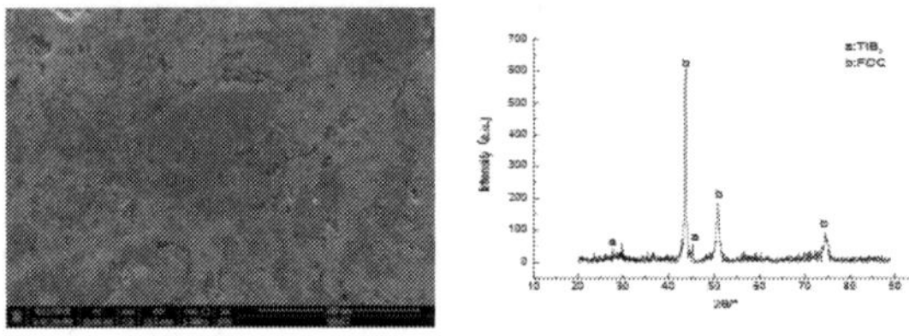

Fig.3 SEM and XRD images of the composites with volume fractions of 15%

3.3 DSC analysis at different heating rates (volume fraction 15%)

Fig. 4 is a DSC graph respectively showing heating rates of 10 K/min, 15 K/min, 20 K/min, and 25 K/min, when the volume fraction of the reinforcement is 15%. The figure indicates that there are obvious exothermic peaks at 800-900K under the four heating rates. With the increase of heating rate, the exothermic peak continuously moves back, the peak temperature is 835K, 842K, 844K, 859K, which is caused by the hysteresis of heat transfer. The higher the heating rate, the particle surface gets the energy of the reaction easier when the heating rate increases. Also, the transfer of heat takes time, which causes a temperature difference between the inside and outside of the particle, and the diffusion of the reactant takes time. Therefore, when the surface of the sample reaches the reaction temperature, the heat accumulated in the sample is not enough to support the reaction due to the large temperature gradient. When the heat is transferred to the inside, the temperature of the furnace continues to rise, that is, the reaction temperature rises.

Calculated according to the kinetic model proposed by Kissinger [9]:

$$\frac{d(\ln\frac{\beta}{T_m^2})}{d(\frac{1}{T_m})} = -\frac{E_a}{R}$$

Where: β-temperature rate, K/min; E-phase change activation energy, kJ / mol; R-gas constant, J / mol·K; T_m-peak temperature of the exothermic peak, K; C-constant. For the $\ln\beta/T^2 \sim 1/T$ plot, fit the line, as shown in Figure 5. Since the slope of the line is -Ea/R, the Ea value is obtained. By measuring the DSC curve at four heating rates, the reaction activation energy is 195.69 kJ/mol.

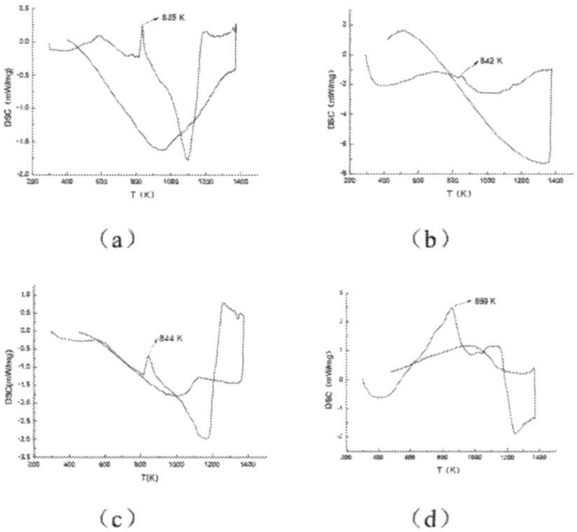

Fig.4 DSC curves of the powder mixture generated at different heating rates

(a) 10K/min (b) 15K/min (c) 20K/min (d) 25K/min

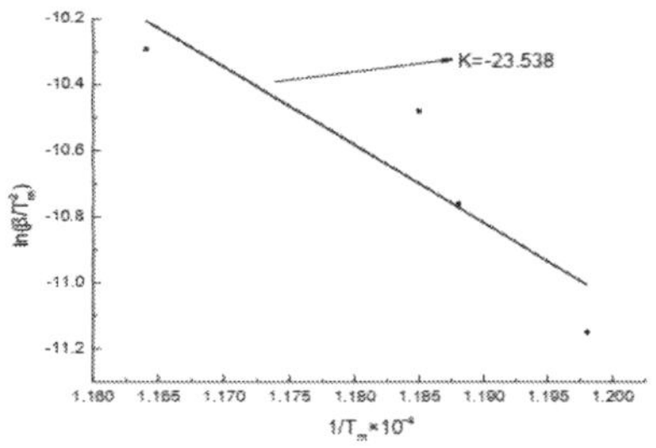

Fig.5 Plots of the $\ln(\beta / T_m^2)$ - $1/T_m$ of the for the Al-Fe-Cu-Cr-Ni-Ti-B system

4. Conclusions

(1) AlFeCuCrNi /TiB_2 high-entropy alloy matrix composites can be prepared by microwave heating method. The matrix structure was FCC, and the reinforcement TiB_2 showed regular geometric morphology and dispersed in the matrix.

(2) When the volume fraction of the reinforcement is 10%, TiB_2 is more evenly distributed in the matrix. When the volume fraction of the reinforcement increases to 15%, TiB_2 partially aggregates in the matrix, and the system activation energy was 195.69 kJ/mol.

Acknowledgements

This work was supported by the National Natural Science Foundations of China (No.51571118 and 51371098); National undergraduate scientific research training project of China (201710288071); Natural Science Foundations of Jiangsu Province, China (No. BK20141308).

References

[1] Greer A L. Nature. **366**. 303(1993).
[2] Yeh J W, Chen S K, Lin S J, et al. Advanced Engineering Materials .**6**.299(2004).
[3] Yeh J W, Chen S K. Metallurgical and Materials Transation. **A35**. 2533(2004).
[4] Zhang Y, Zuo T T, Cheng Y, et al. Scientific Reports. **3**. 1355(2013).
[5] Santodonato L J, Zhang Y. Nat Commun. **6**(2015).
[6] Lim X. Nature. **533**. 306(2016).
[7] Zhang D, Zhang N. Journal of Beijing Union University(Natural Sciences).**21**.4(2007).
[8] Li Z L , Sun H F, Gao P, et al. Hot Working Technology.**39**.62(2010).
[9] Zhang LC, Xu J, Eckert J. Journal of Applied Physical.**100**.(2006)

Study on the Preparation of TiO_2-SiO_2 Composites from Rice Husk Ash

Yu Liang, Xiaoyu Ma*, Suping Cui, Zhihong Wang, Yali Wang
College of Materials Science and Engineering, Beijing University of Technology, China

Abstract: In this paper, TiO_2-SiO_2 composites with different SiO_2 contents were prepared by coprecipitation method using rice husk ash as raw material. The effects of the concentration of alkaline leaching solution and reaction temperature on the extraction rate of SiO_2 from rice husk ash were investigated. The optimum formation conditions of TiO_2-SiO_2 composites were explored, and the microstructure and properties of the composites were analyzed. The experimental results showed that the extraction rate of SiO_2 from rice husk ash reaches the maximum when the concentration of alkaline leaching solution is 1.5 mol/L and the reaction temperature is 90℃ . When the pH value is 7, the TiO_2-SiO_2 composites can achieve the optimum formation condition. TiO_2-SiO_2 composites are composed of approximately spherical particles with obvious pore structure of the surface and piling up in fluffy state. The Si atoms enter the crystal structure of TiO_2 with Ti-O-Si bonds connected to Ti atoms, which inhibits the growth of TiO_2 grains and leads to the increase of the transition temperature of TiO_2 from anatase to rutile.

1 Introduction

In recent years, nanoscale TiO_2 has been widely used in the fields of seawater desalination, photocatalysis and solar cells for its high weather ability and excellent optical and mechanical properties[1-3].However, the pure TiO_2 has many defects such as higher price, smaller specific surface area and weaker acid, leading thatthe development demand of many fields is limited[4]. Therefore, in order to overcome these defects and improve the physical and chemical properties of TiO_2, it is usually combined with SiO_2, Fe_2O_3or CeO_2 to form composite oxide materials with low density and high porosity[5-6]. Among them, SiO_2 has high bonding strength, large specific surface area and good thermal stability[7],in addition, TiO_2-SiO_2 composites after adding SiO_2 can have excellent properties of TiO_2 and SiO_2 at the same time[8]. Therefore, TiO_2-SiO_2 composites have become the focus of research. At present, most laboratories apply sol-gel or coprecipitation to synthesis TiO_2-SiO_2 composites that were made from titanium butoxide and tetraethyl orthosilicate[9-10].However, the hydrolysis rate of ester organic compounds is faster, the experimental conditions are demanding, and the price of themare relatively high, so the industrial production of TiO_2-SiO_2 composites are limited in some extent. It is found that there are a lot of amorphous SiO_2 in rice husk ash from agricultural waste[11],and the alkaline solution of SiO_2 is easy to react with $Ti(SO_4)_2$ solution[12]. Accordingly, TiO_2-SiO_2 composites can be prepared by cheap rice husk ash and $Ti(SO_4)_2$ as precursors, which can not only rationalize the utilization of agricultural waste, but also protect the environment and create higher commercial value[13].

In this paper, cheap and readily available rice husk ash and $Ti(SO_4)_2$ were used as silicon source and titanium source respectively to prepare TiO_2-SiO_2 composites, which were applied to coprecipitation method to adjust the pH value by the ammonia titration. The

Corresponding author: Xiaoyu Ma. Email: maxiaoyu@bjut.edu.cn microstructure and properties of TiO_2-SiO_2 composites were studied by means of BET, SEM, FT-IR and XRD.

2 Experiment

2.1 Preparation of TiO_2-SiO_2 Composites

First, the rice husk ash and alkaline leaching solution are heated and mixed in a flask, and the sodium silicate solution is obtained through filtration. Then add appropriate concentration of $Ti(SO_4)_2$ solution slowly and regulate pH value through ammonia water in it. Last, TiO_2-SiO_2 composites with different SiO_2 contents were obtained after filtration, drying and calcining in muffle furnace (FR-1236). The prepared samples were represented by TiO_2-SiO_2(x), where x was the content of SiO_2 in TiO_2-SiO_2 composites.

2.2 Characterization of TiO_2-SiO_2 Composites

Micromeritics's TriStar 3020 high speed automatic specific surface area and pore analyzer was used for the specific surface area and pore structure analysis (BET) of the TiO_2-SiO_2 composites. The specific surface area is calculated by Brunauer-Emmett-Teller and the pore size distribution is obtained by measuring the N_2 desorption isotherm and calculating it by Barrett-Joyner-Halenda (BJH); The crystal structure and morphology of TiO_2-SiO_2 composites were analyzed by XRD-7000 X-ray diffraction analyzer produced by Shimadzu Corporation. Cu is the target. Ka is the radiation source. The tube voltage is 40KV and the scanning range is 10~70 degrees; The FTIR-8400S infrared spectrometer was used for analyzing the TiO_2-SiO_2 composites by Fourier infrared spectroscopy (FT-IR). The sample was dried, ground and pressed test and the test range was 400-1400cm^{-1}; The QUANTA 600F scanning electron microscope (SEM) of FEI company was used for analyzing the microstructure of the surface of TiO_2-SiO_2 composites.

3 Discussion of results

3.1 The influence factors on the extraction of SiO_2 from rice husk ash

The composition of rice husk ash showed that the content of SiO_2 in rice husk ash reached 37.5% and the rests are carbon and a small amount of metal[8]. Its component analysis results are shown in Table 1

3.1.1 The effect of concentration of alkaline leaching solution on extraction rate of rice hull ash

The 20.0g rice husk ash where the mass of SiO_2 is about 7.5g from the table 1 was reacted with the 200ml alkaline leaching solution whose OH^- concentration is 0.5~4mol/L respectively. The relationship between the extraction rate of rice husk ash and OH^- concentration is shown in Fig. 1. With the increase of OH^- concentration, the extraction rate of rice husk ash is becoming larger and larger. When the concentration of OH^- is 1.5mol/L, the reaction of SiO_2 in rice husk ash and alkaline leaching solution are gradually complete, and the extraction rate of rice husk ash reaches the maximum. As the concentration of OH^- in the alkaline leaching solution continues to increase, the excess OH^- may react with the metal in rice husk ash, the residual sediment increases and the extraction rate of rice husk ash begins to decreases. So the optimum concentration of OH^- in the alkaline leaching solution is 1.5 mol/L. Fitting the curve before the extraction rate of rice husk ash reaching the maximum value, a straight line equation can be obtained, $y=41.26667+37.4x$. Verifying the extraction rate of SiO_2 in rice husk ash with alkaline leaching solution that the concentration of is 1.2 mol/L is 86%, which is approximated to 86.15% that is calculated. The more OH^- in alkaline leaching solution, the more reaction amount to SiO_2. Accordingly, there is a linear relationship between the extraction rate of SiO_2 before reaching the maximum value and the concentration of OH^- in alkaline leaching solution.

Table 1 the main components of rice husk ash

Component	SiO_2	K_2O	P_2O_5	CaO	MgO	SO_3	Al_2O_3	C
Content (%)	37.5	1.8	1.2	0.6	0.5	0.4	0.2	57.8

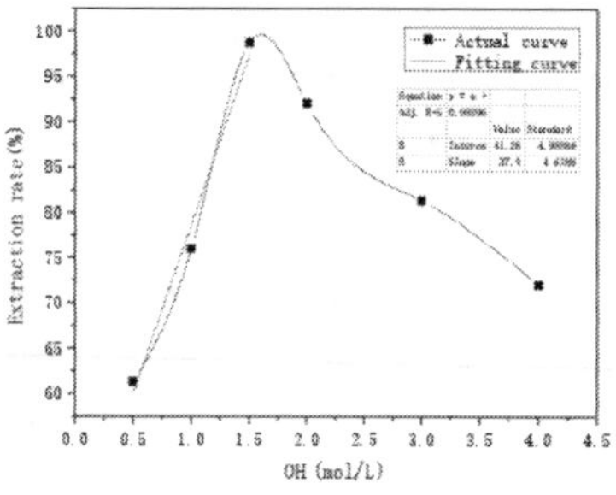

Fig. 1 The relationship between the extraction rate of rice husk ash and OH^- concentration in alkaline leaching solution

3.1.2 The effect of reaction temperature on extraction rate of rice hull ash

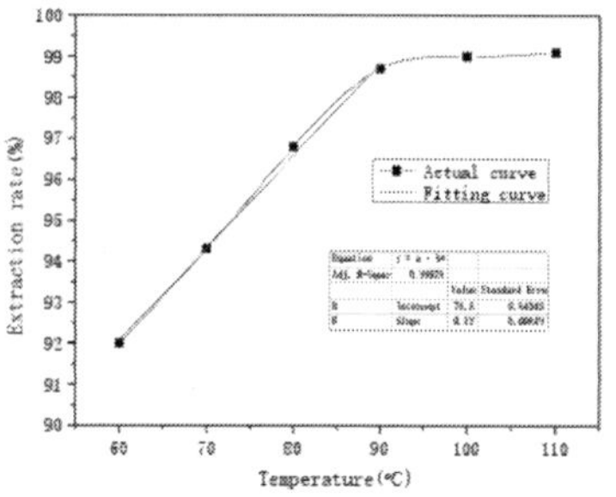

Fig. 2 The relationship between the extraction rate of rice husk ash and reaction temperature

In order to fully extract SiO_2 from rice husk ash, the relationship between the extraction rate of rice husk ash and the reaction temperature was investigated, as shown in Fig. 2. With the increase of reaction temperature, the extraction rate of rice husk ash increases gradually. When the temperature rises to 90℃ , the extraction rate of rice husk ash reaches 98.7%. As the temperature continues to raise, the extraction rate of rice husk ash tends to be gentle. The fitting analysis of the curve before 90℃ can be seen that the extraction rate of rice husk ash is linear with temperature. The equation is $y=78+0.226x$. Verifying the extraction rate of SiO_2 in rice husk ash at 60℃ and 75℃ are 92.65% and 95% respectively, which are approximated to 92.69% and 94.95%, both of which are calculated. Therefore, when the reaction temperature is less than 90℃ , the extraction rate of rice husk ash increases linearly with the increase of reaction temperature. It can also be proved that the reaction between rice husk ash and alkaline leaching solution accords with the first order reaction of kinetic equation.

3.2 The characterization of TiO_2-SiO_2 composites materials

3.2.1 The effect of pH value on the generation of TiO_2 and SiO_2

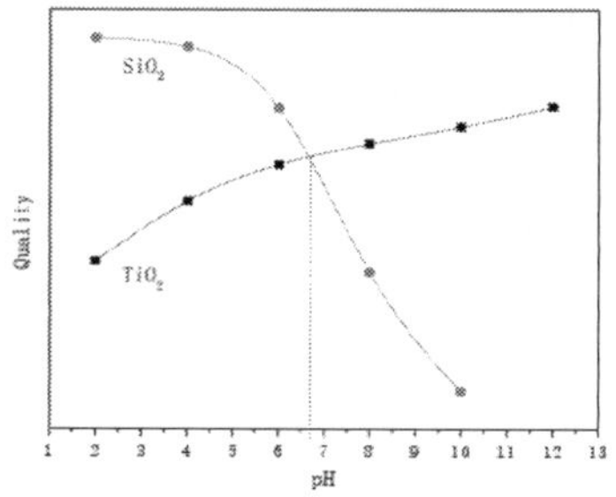

Fig. 3 The effect of pH value on the generation of TiO_2 and SiO_2

The pH value is a key factor affecting the generation of TiO_2 and SiO_2. By adjusting the pH of the solution, it can not only determine the ratio of TiO_2 to SiO_2 in TiO_2-SiO_2 composites, but also obtain the optimal generation condition of TiO_2-SiO_2 composites. As shown in Fig. 3, when pH is 10, white silica sol begins to form in sodium silicate solution. With the decrease of pH, more and more white colloid forms in the solution, because of the H^+ first reacting with SiO_3^{2-} in the solution[13],which indicates that the quality of SiO_2 is bigger and bigger. When the pH of solution is 2, the mass of SiO_2 reaches the maximum value, which indicates that SiO_2 is more suitable for generation in acid solution. When the pH value is 2, white titanic acid begins to be produced in $Ti(SO_4)_2$ solution. As the pH increases, more and more white precipitate generates in $Ti(SO_4)_2$ solution, which indicates that the amount of TiO_2 is increasing. Considering the optimum generation amount of two oxides, It can be determined that the optimum pH value of TiO_2-SiO_2 composites is 7.

3.2.2 Analysis of specific surface area and pore structure of TiO_2-SiO_2 composites

 https://doi.org/10.1051/matecconf/201823802005

Table. 2 Specific surface area and pore distribution of TiO_2-SiO_2 composites with different SiO_2 content

Sample	S_{BET} (m^2/g)	Pore volume (cm^3/g)	Pore diameter (nm)
TiO_2	145.7	0.33	9.7
TiO_2-SiO_2(0.2)	253.6	0.62	7.9
TiO_2-SiO_2(0.5)	355.5	0.65	7.6
TiO_2-SiO_2(0.8)	413.9	0.64	7.4
SiO_2	463.9	0.70	6.5

In order to investigate the effect of SiO_2 content on the surface structure of TiO_2-SiO_2 composites, the specific surface area and pore structure of the composites were analyzed in this paper, as shown in Table 2.

It can be seen from the table that the specific surface area and pore volume of pure TiO_2 are the smallest. With the increase of SiO_2 content, the specific surface area of TiO_2-SiO_2 composites is larger and larger, and the pore volume also increases gradually. In the TiO_2-SiO_2(0.8), the content of SiO_2 is high and Si atoms affect the formation of Ti atoms, resulting in the complete collapse of the framework of TiO_2. Therefore, the pore volume decreases. When the content of SiO_2 is 100%, the specific surface area and pore volume reach the maximum value. In the TiO_2-SiO_2 composites, the pores of SiO_2 particles are occupied by the TiO_2 powders, so the specific surface area and pore volume decrease rapidly when the content of TiO_2 powders are high[14].

3.2.3 Analysis of SEM test results

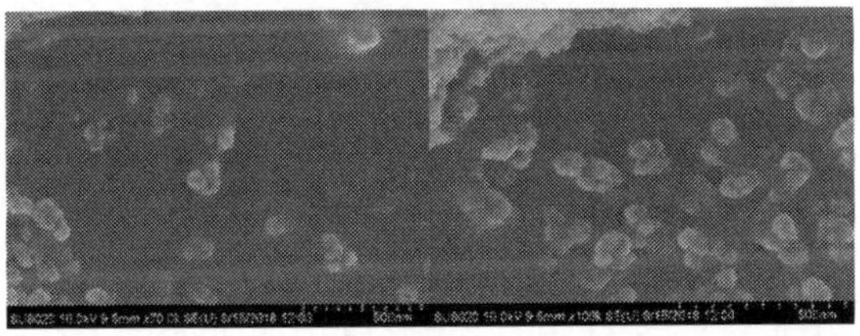

(a) (b)

Fig. 4 Microstructural diagrams of TiO_2-SiO_2(0.5) composites
(a) Under the condition of 70,000 times; (b) Under the condition of 100,000 times

Fig. 4 are SEM diagrams of TiO_2-SiO_2(0.5) composites under the condition of 70,000 times and 100,000 times. It can be seen from the diagram that the TiO_2-SiO_2 composites are made up of nearly spherical particles, and the particle size is about 40~50nm. The distribution of TiO_2-SiO_2 composites is more dispersed, mostly consisting of three or four spherical particles with different sizes. It shows that SiO_2 and TiO_2 are combined with stable chemical bonds.

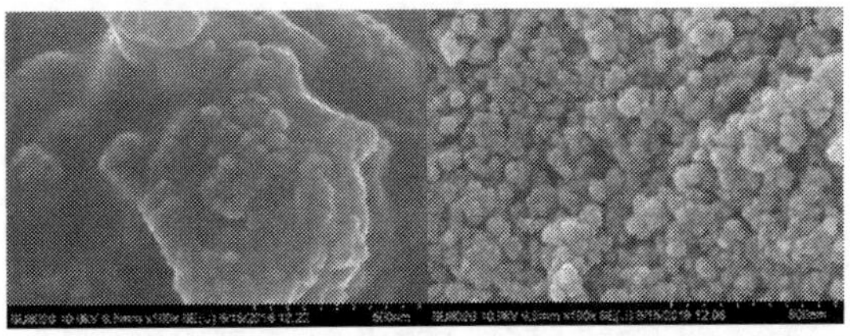

(a) (b)

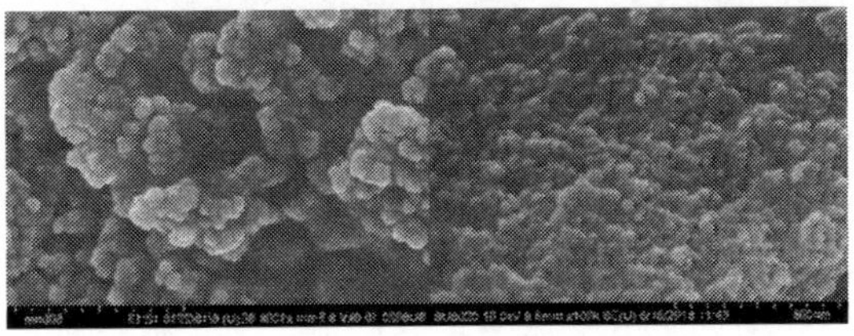

(c) (d)

Fig. 5 SEM diagrams of TiO_2-SiO_2 composites with different SiO_2 content
(a) Pure TiO_2; (b) TiO_2-SiO_2(0.2);
(c) TiO_2-SiO_2(0.5); (d) TiO_2-SiO_2(0.8)

Fig. 5 are SEM diagrams of TiO_2-SiO_2 composites with different SiO_2 content at the magnification of 100 thousand times. As is shown in the Fig. 5(a), pure TiO_2 particles are large, the surface is smooth and the internal pore structure is not obvious, so the specific surface area and pore volume are relatively low. With the addition of SiO_2, the particles on the surface of TiO_2-SiO_2 composites decrease and uniform gradually, the surface smoothness reduces, and the pore structure becomes more and more obvious. When the content of SiO_2 reaches 80%, the surface of the composites exhibit melting shape, which exists a large number of pore structures, and the fine and uniform particles are fluffy in the surface. Therefore, the specific surface area and pore volume are relatively higher, which are in agreement with the analysis results of Table 2.

3.2.4 FT-IR analysis of TiO_2-SiO_2 Composites

In order to determine the form of SiO_2 and TiO_2 in

TiO_2-SiO_2 composites, FT-IR tests were performed on three oxides of SiO_2, TiO_2 and TiO_2-SiO_2(0.5) respectively, as shown in Fig. 6.

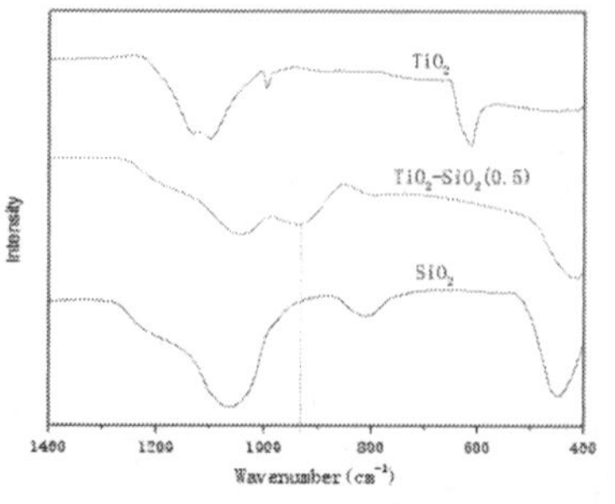

Fig. 6 the FT-IR diagram of three oxides

As can be seen from Fig. 6, the three oxides have different infrared absorption peaks at different positions, of which the absorption peak at 610cm^{-1} belongs to the stretching vibration of the Ti-O-Ti bond, and the absorption peak at 445cm^{-1}, 810cm^{-1}and 1066cm^{-1} belong to the stretching vibration of the Si-O-Si bond[15]. It has been proved that the infrared absorption peak of 920~970cm^{-1} is the characteristic absorption peak of the Ti-O-Si bond[16]. The infrared characteristic absorption peak of Ti-O-Si bond can be detected clearly in TiO_2-SiO_2 composites, while the infrared characteristic absorption peak of Ti-O-Ti bond disappears. It proves that the combination of TiO_2 and SiO_2 in the form of chemical bonds, rather than simple physical mixing, which is in line with the analysis results of SEM. During the formation of TiO_2-SiO_2composites, Si atoms enter the network framework of TiO_2 in the form of substituted cations and connect with Ti atoms in the form of Ti-O-Si bond. As the Ti atomic radius (0.064nm) is larger than the radius of the Si atom (0.041nm), and the length of Ti-O bond (0.179nm) is longer than the length of Si-O bond (0.159nm), which cause the growth of the TiO_2 crystal being inhibited and the grain size becomes smaller[17]. Therefore, The Ti atom transforms from octahedral coordination to tetrahedral coordination gradually[18].

3.2.5 Analysis of XRD test results

In order to study the effect of SiO_2 content and calcination temperature on the crystal structure of TiO_2, different kinds of samples were tested by XRD respectively. The results are shown in Fig. 7.

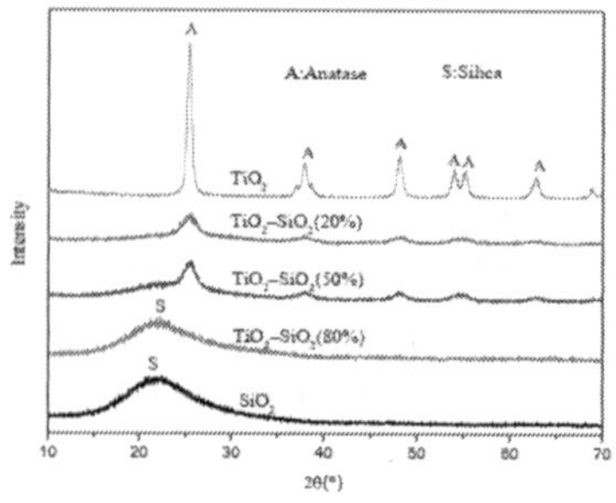

(a)

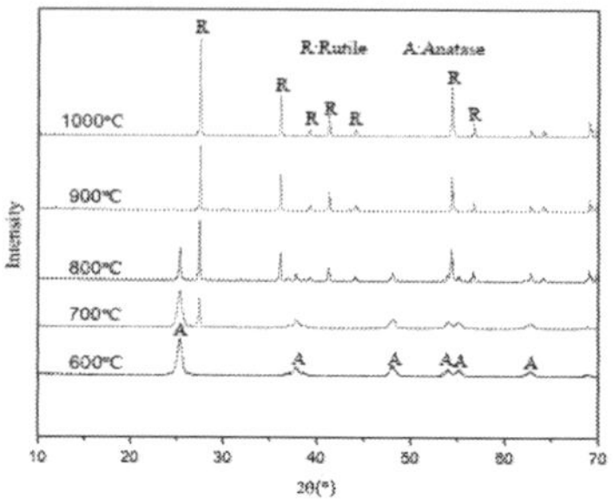

(b)

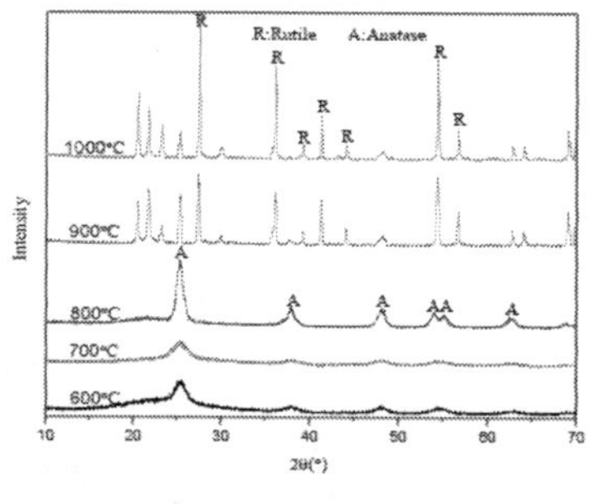

(c)

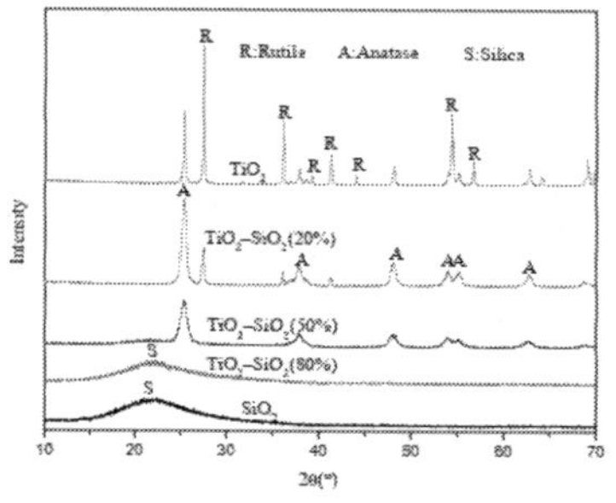

(d)

Fig. 7 XRD diagrams under different conditions

(a) TiO_2-SiO_2(x) at 600℃; (b) Pure TiO_2 at 600~1000℃
(c) TiO_2-SiO_2(0.5) at 600~1000℃; (d) TiO_2-SiO_2(x) at 800℃

XRD test chart of TiO_2-SiO_2 composites with different SiO_2 contents at 600℃ are shown in Fig. 7(a). When the content of SiO_2 is less than 20%, the diffraction peak of anataseTiO_2 exists only in TiO_2-SiO_2 composites, and the peak value is highest and the width is narrowest. With the addition of SiO_2, the characteristic peaks of anataseTiO_2 gradually generalize and the peaks value gradually decrease, but the characteristic peak of SiO_2 gradually becomes obvious. When the content of SiO_2 is higher than 50%, the characteristic peaks of anataseTiO_2 disappear, and a steamed bread shaped peak is found near 22.5 degrees, indicating that the TiO_2-SiO_2composites was mainly in the form of amorphous SiO_2[19]. Therefore, as SiO_2 content increases, the crystallinity of TiO_2 in TiO_2-SiO_2 composites decreases gradually.

Fig. 7(b) and Fig. 7(c) are XRD test results of pure TiO_2 and TiO_2-SiO_2(0.5) composites at 600~1000 temperature. As can be seen from Fig. 7(b), when the temperature is 700℃ , the pure TiO_2 begins to appear weak rutile characteristic diffraction peaks, indicating that part of the anataseTiO_2 changes to the rutile structure gradually. When the temperature rises to 900℃ , the characteristic diffraction peak of anatase is disappeared, and TiO_2 transforms into rutile structure completely. As is shown in Fig. 7(c), compared with pure TiO_2, the phase transformation temperature of TiO_2-SiO_2(0.5) composites is higher. When the temperature rises to 900℃ , the rutile characteristic diffraction peak appears. Therefore, the addition of SiO_2 increases the transition temperature of TiO_2 from anatase to rutile and also improves its thermal stability.

Fig. 7(d) are the XRD test result of TiO_2-SiO_2 composites with different SiO_2 contents at 800℃ . Pure TiO_2 has the strongest rutile characteristic peak, the width of which is the narrowest. With the increase of SiO_2 content, the diffraction peaks of rutile decrease gradually and become wider and wider. When the content of SiO_2 is higher than 50%, the characteristic diffraction peaks of rutile disappear and the characteristic diffraction peaks of anatase become generalized. The characteristic diffraction peaks of TiO_2 in TiO_2-SiO_2 composites disappeared gradually with the increase of SiO_2 content. Some studies have shown that the transformation of TiO_2 from anatase to rutile is caused by lattice stress and cell volume[20].The smaller the cell volume is, the higher the transition temperature is. Accordingly, with the increase of SiO_2 content, it can be speculated that the grain size of TiO_2-SiO_2 composites decreases gradually, which is consistent with the results of SEM and FT-IR.

4 Conclusion

1. Using rice husk ash as silicon source and $Ti(SO_4)_2$ as titanium source, the TiO_2-SiO_2 composites with different SiO_2 content can be prepared by coprecipitation method. When the concentration of OH^- in alkaline leaching solution is 1.5 mol/L and the reaction temperature is 90℃ , the extraction of SiO_2 from rice husk ash reaches the maximum value. SiO_2 and TiO_2 can reach the optimal amount of generation by adjusting pH to 7 with ammonia water.

2. Most of the TiO_2-SiO_2 composites are composed of 3~4 approximately spherical particles, which accumulate in a fluffy state. The surface of the composites are rough and the pore structure is obvious. With the increase of SiO_2 content, the specific surface area and pore volume increased gradually.

3. During the formation of TiO_2-SiO_2 composites, Si atoms enter the crystal structure of TiO_2 in the form of substituted cations and connect to Ti atoms in the form of Ti-O-Si bond, which inhibits the crystal growth of TiO_2, makes the grain size of TiO_2 smaller, and eventually leads to the transition temperature of TiO_2 in TiO_2-SiO_2 composites from anatase to rutile increases. Besides, as theSiO_2 content increases, the crystal transition temperature increases gradually.

Acknowledgement

This study has been supported by the The National Natural Science Foundation of China (51502008); National Key R&D Plan (2017YFB0310802).

Reference

[1] K. Hashimoto, H. Irie, A. Fujishima. J. JPN J APPL PHYS,. **44**, 8269-8285 (2014).

[2] R. Wang, K. Hashimoto, A. Fujishima, et al. J. Adv. Mater, **10**, 135-138 (2010).

[3] P. Zhang, F Y. Lv. J. ENERGY, **82**, 1068-1087 (2015).

[4] S. Cao, N. Yao, K L. Yeung. J. J SOL-GEL SCI TECHN, **46**, 323-333 (2008).

[5] W. Dong, Y. Sun, C W. Lee, et al. J. J AM CHEM SOC, **129**, 13894-13904 (2007).

[6] L. Pinho, M J. Mosquera. J. APPL CATAL B-ENVIRON, **134-135**, 205-221 (2013).

[7] M A L. Vargas, M. Casanova, A. Trovarelli, et al. J. APPL CATAL B-ENVIRON, **75**, 303-311 (2007).

[8] Y. Liu. D. JILIN U, (2013).

[9] J. Krzakroś, J. Filipiak, C. Pezowicz, et al. J. Acta Bioeng Biomech, , **11**, 21 (2009).

[10] C. Hu, Y. Tang, Z. Jiang, et al. J. APPL CATAL A-GEN, **253**, 389-396 (2003).

[11] D. Ouyang, K. Chen. J. Journal of Chinese Electron Microscopy Society, **22**, 390-394 (2003).

[12]Y. Cheng. D. JILIN U, (2017).

[13] Y. Liu. D. JILIN U,(2015).

[14] F Y. Sun, J H. Lin, K F. Ren, et al. J. B CHIN CERAM SOC, **35**, 870-874 (2016).

[15] X. Gao, I E. Wachs. J. CATAL TODAY, **51**, 233-254 (1999).

[16] S. Zhu, D. Zhang, X. Zhang, et al. J. MICROPOR MESOPOR MAT, **126**, 20-25 (2009).

[17] C F. Song, K L. Meng, P. Yang, et al. J. THIN SOLID FILMS, **413**, 155-159 (2002).

[18] J B. Miller, E I. Ko. J. J CATAL, **159**, 58–68 (1999).

[19] P C. Arroyomartínez, N A. Sánchezflores, M E, Villafuertecastrejón, et al. J. J.braz.chem.soc, **29**, 1 (2018).

[20] L Y. Wang, Y P. Sun, B S. Xu. J. J MATER SCI, **43**, 1979-1986 (2008).

Study on reaction mechanisms of iron matrix compositesreinforced with in situ Al_2O_3 particles reacted byAl-Fe_2O_3-Fe system

Chengyan Zhu, Jie Jiang, Heguo Zhu *

College of Materials Science and Engineering, Nanjing University of Science and Technology. Nanjing, 210094; P. R. China

Abstract.Iron matrix composites reinforced with Al_2O_3 particles were fabricated in situ through exothermic dispersive (XD) reaction from a powder mix of pure elements Fe, Al and Fe_2O_3. The fabrication and reaction mechanisms were investigated by using X-ray diffraction (XRD), scanning electron microscopy (SEM), X-ray energy dispersive spectroscopy (EDS) and differential scanning calorimetry (DSC) analysis. The reaction between Al and Fe_2O_3 was found to occur in three steps. Their activation energies are 203.8 kJ/mol, 1100.9 kJ/mol and 380 kJ/mol, respectively. DSC analysis shows that the reaction peak shifts to a higher temperature with an increase in the heating rate. When the heating rate is 10 K/min, the rates of the three reactions exhibit a similar trend, i.e., the reaction rate is very slow at the start, and then it increases rapidly, finally it becomes slow again. Their reaction times are 456, 672 and 650 s, respectively.

1. INTRODUCTION

Iron matrix composites (MMCs) exhibit many attractive features, such as good thermal and electrical conductivities, high strength and fracture toughness and excellent wear resistance. *In situ*iron matrix composites are new additions, whose reinforcements are formed through one or more *in situ* reactions among raw materials. They have attracted considerable attention in recent years owing to clean reinforcement/matrix interface, high thermal stability, good isotropic properties and low fabrication costs [1,2].

Alumina ceramic (Al_2O_3) particles possess outstanding physical and mechanical properties, for example, high modulus and hardness, low density, high melting temperature (2323K), excellent oxidation resistance and thermal stability in metal melt. They are considered to be one of the best reinforcements in the fabrication of metal matrix composites [3]. Recently, iron matrix composites reinforced with TiC particles were synthesized, and their mechanical properties and wear resistance were studied [4]. In contrast, reports on Al_2O_3 particles reinforced Fe-based composites are very limited. Georgioua *et al.* [5] prepared a nanostructured cermet coating consisting of alumina particles dispersed in a Fe–Cu–Al matrix by supersonic spraying, and identified a strong effect of processing parameters and chemical composition of the powders on the structural and mechanical characteristics of the resulting (Fe–Cu–Al) + Al_2O_3 cermet. Accordingly, a 'structure–property' relationship was established for this type of nanostructured cermet coatings. In another study, Al_2O_3 particles were mixed with Fe_3Al powders and then sintered at 1373K, and resultant iron based composites have fine grain size and high hardness [6]. Babakhani *et al.* [7] fabricated iron–alumina composite foams through a combination of combustion synthesis (CS) and spark plasma sintering (SPS). They found that applying pulsed electric current to the powder mix led to the formation of Al_2O_3 through a highly exothermic reaction between Al and Fe_2O_3, i.e., $XFe + 2Al + Fe_2O_3 = Al_2O_3 + (X+2)Fe$. They also demonstrated that the time to reach the maximum combustion temperature increased with increasing X.

In this paper, *in situ* Fe-based composites reinforced with Al_2O_3 particles were prepared, for the first time, from Fe-Fe_2O_3-Al powder mixtures through exothermic dispersive (XD) synthesis. The reaction process was explored and clarified by using differential scanning calorimetry (DSC), X-ray diffraction (XRD), scanning electron microscopy (SEM) and energy dispersive spectroscopy (EDS), assisted by thermodynamic analysis.

2. EXPERIMENTS PROCEDURES

2.1. Materials

Pure Fe (40-50 μm), Fe_2O_3 (30-50 μm) and Al powders (3-50 μm) were used as raw materials in fabricating Fe-Al_2O_3 composites. To produce the composites having an Al_2O_3 volume fraction of 40%, the powder mixture of Fe, Fe_2O_3 and Al were prepared with their volume fractions derived from the reaction equation $2Al + Fe_2O_3 \rightarrow Al_2O_3 + 2Fe$. The weight ratio of Al, Fe_2O_3 and Fe in the mixed powder is thus 10.48: 31.06: 58.50. The

[a] Corresponding author: zhg1200@sina.com

https://doi.org/10.1051/matecconf/201823802006

powder mix was ball-milled in a stainless steel vacuum jar for 2 h and compacted under a pressure of 180 MPa into cylindrical specimens with a diameter of 30 mm and a length of 5 mm. The cylindrical specimens were heated to a temperature above 1473 K in a vacuum furnace and then held at this temperature for 10 min before cooling down to room temperature inside the furnace. The composite samples were mechanically polished and investigated by using XRD (Rigaku D/MAX2400) and SEM (Quanta 2000)equipped with an energy dispersive spectroscopy (EDS).

2.2. Thermodynamic analysis

The layout of STA449C DSC thermal analyzer is similar to that of XD device shown in our previous paper[8]. When the temperature in the vacuum furnace of the DSC reaches critical values, chemical reactions take place in the powder mixture, in a manner similar to what occurs in the XD device. The mass of the sample placed in the DSC crucible is 5~15 mg cut from the cylindrical specimen, the environment and condition for the sample being tested in the DSC crucible are similar to those used in the XD furnace. Therefore, the DSC curve can be used to determine the temperatures at which the reactions between the Fe, Fe_2O_3 and Al take place during the heating and cooling stage.

In order to obtain the DSC curves, the sample was heated in argon atmosphere in the furnace of STA449Cthermal analyzer, wherein the temperature was increased from the ambient (298 K) to 1473 K at heating rates of 10, 20and 30 K/min, respectively. Subsequently, the activation energies of the reactions occurring in the Al-Fe_2O_3-Fe system during the heating process were calculated and discussed.

3. RESULTS AND DISCUSSION

3.1 Activation energy

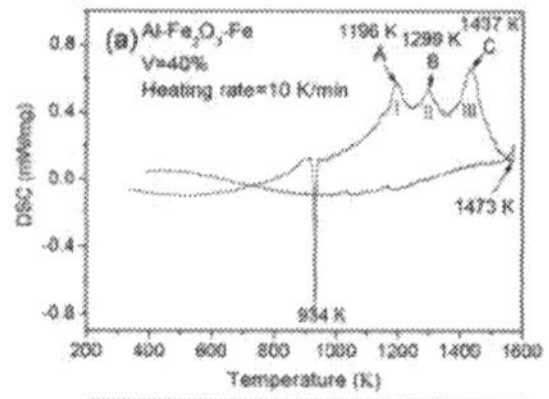

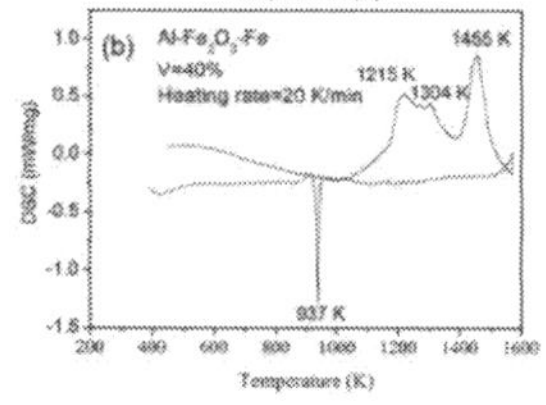

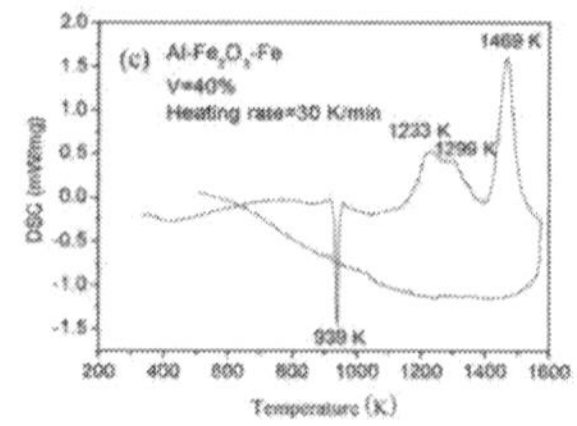

Fig.1DSC curves of the powder mixture generated at different heating rates: (a) 10 K/min, (b) 20 K/min and (c) 30 K/min.

Activation energy (E) is the minimum amount of energy required for a chemical reaction to take place. It can be calculated by Kissinger equation [9]:

$$\frac{d\left(\ln\frac{\beta}{T_m^2}\right)}{d\left(\frac{1}{T_m}\right)} = -\frac{E}{R} \quad (1)$$

where T_m is the temperature corresponding to the exothermic peak, β is the heating rate, and R is air constant (8.314 J/mol). As shown in Fig. 1a, three exothermic peaks appear at temperatures of 1196 (point A), 1299 (point B) and 1437 K (point C), respectively, during the heating process, meaning that the entire reaction of the Al-Fe_2O_3-Fe system are composed of three steps, designated here as step I, step II and step III. No peaks took place during the cooling stage, indicating that the reaction products are stable.

With the increase of the heating rate, all the exothermic reaction peaks shift towards higher temperatures (Fig. 1b,c).The reason is that when the heating rate increases, the rate of thermal energy input increases and the time available for reaction decreases. Transport of reactants (through diffusion), wetting, and final chemical reaction are all time-dependent processes. However, there is no sufficient time to improve the reaction conditions with increasing heating rates. Therefore, some of region would react at a higher temperature, resulting in the shifting of exothermic peaks to the higher temperature.According to the dependence of peak temperature on heating rate, the relation of $\ln\left(\beta / T_m^2\right) \sim 1/T_m$ for each individual reaction is plotted in Fig. 2. By fitting data points using a straight line and setting its slope equal to $-\frac{E}{R}$, the activation energies of the three reactions are calculated and found to be 203.8, 1110.5 and 380 kJ/mol, respectively. Consequently, step Iis the easiest, while the stepII is the most difficult.

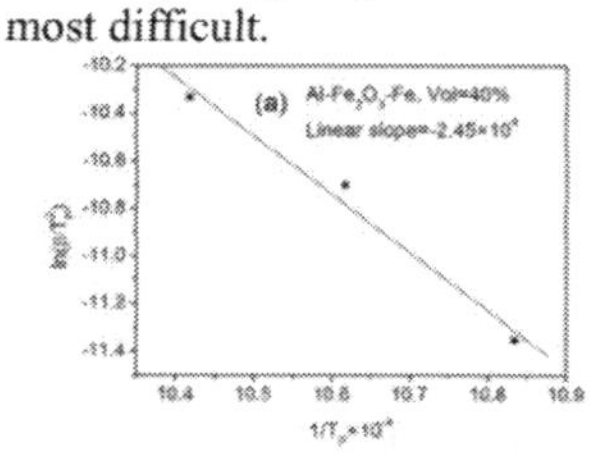

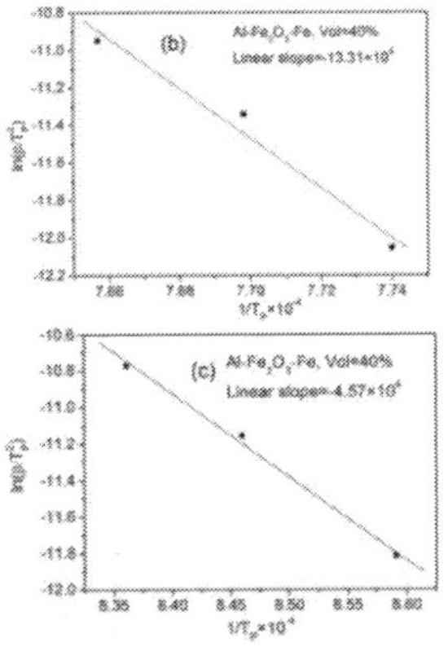

Fig.2 Plots of the $\ln(\beta/T_m^2)$ - $1/T_m$ of the three step reactions for the Al-Fe_2O_3-Fe system: (a) step I; (b) step II; (c) step III.

To understand the time-dependent behavior of reactions, the rate (α) of each of three steps was examined and compared (Fig. 3), by analyzing the DSC results obtained at the heating rate of 10 K/min [10]. All the three reactions exhibit a similar trend; that is, they start slowly, and then acceleraterapidly before becoming slow again [11]. Three steps took 456, 672 and 650 s, respectively. It can be seen from Fig. 3 that the reaction time of step II is the longest and its slope is also the smallest. Notably, the activation energy of step II reaction is also the highest among the three steps.

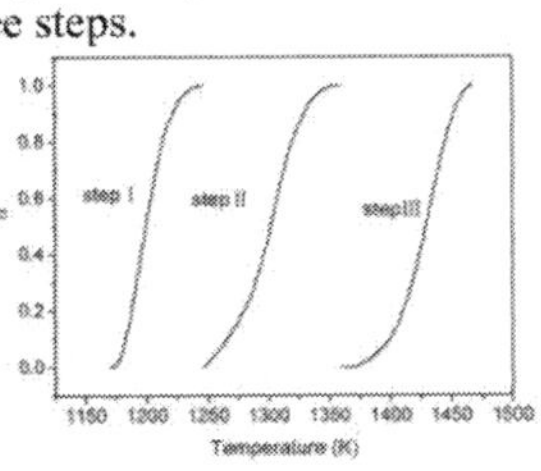

Fig.3 Converting fraction (α) of the three steps reactions versus temperature at a heating rate of 10 K/min

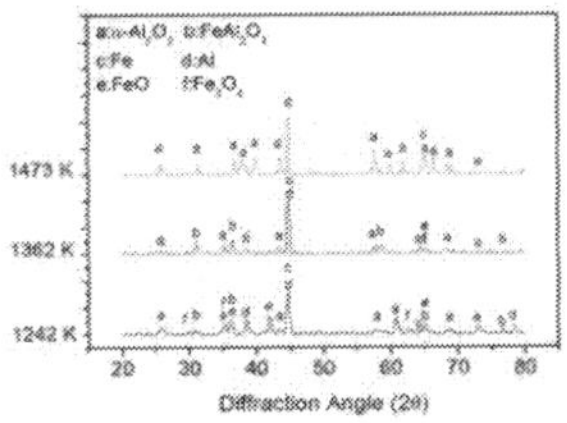

Fig.4 XRD diffraction patterns of sintered products in Al-Fe_2O_3-Fe system

3.2 Intermediate reactions

In order to identify the nature of the transitional phases in the reaction, three green compacts were heated at the rate of 10 K/min up to temperatures at which the reactions occur, i.e., 1242, 1362 and 1473 K, respectively. Their corresponding XRD patterns are shown in Fig. 4. The composite resulting from the first reaction is found to comprise Al, Al_2O_3(s), FeO, Fe_3O_4 and $FeAl_2O_4$. The SEM microphotograph of the composite is shown in Fig. 5a. EDS data of an intermediate phase FeO (labeled as A in Fig. 5a) is shown in Fig.5b.Accordingly, the reaction for step I can be expressed as:

$4Al(l) + 12Fe_2O_3(s) = Al_2O_3(s) + 5FeO(s) + 6Fe_3O_4(s) + FeAl_2O_4(s)$ (2)

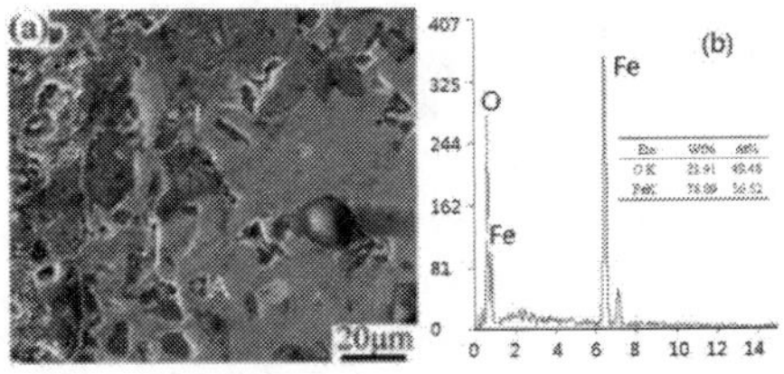

Fig.5SEM microphotograph (a) and EDS data (b) of the composites at heating temperature of 1242 K.

The variation of standard Gibbs free energy ΔG_T^0 of Equation 2 with temperature can be expressed as [12,13]:

$$\Delta G_T^0 = -1815365 - 176.9T \quad (3)$$

Note that ΔG_T^0 isless than zero for step I reaction, meaning it can take place spontaneously.

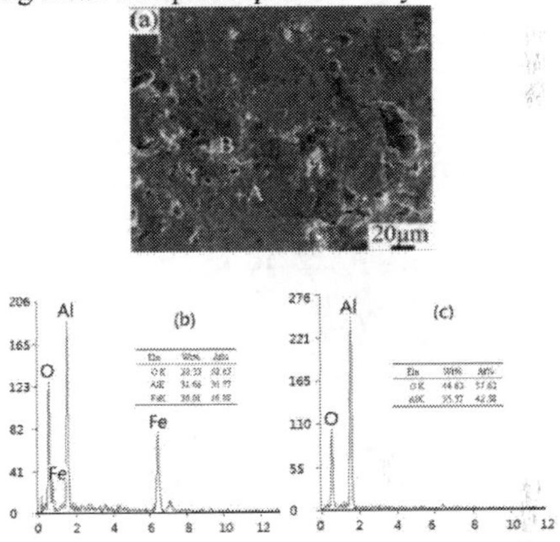

Fig.6SEM microphotograph (a) and EDS datum of the intermediate phase $FeAl_2O_4$(b) and stable phase Al_2O_3(c) from the composites at heating temperature of 1362 K.

When the heating was terminated at 1362 K, the resulting composite consists of Al, Al_2O_3, Fe and $FeAl_2O_4$(Fig.4). Its SEM microphotograph is shown in Fig. 6a. EDS data of the intermediate phases $FeAl_2O_4$ and Al_2O_3 (marked as A and B in Fig. 6a) are shown in Fig. 6b and 6c, respectively.Hence, the stepII reaction can be written as:

$4FeO(s) + 3Fe_3O_4(s) + 10Al(l) = 4Al_2O_3(s) + 12Fe(s) + FeAl_2O_4(s)$ (4)

It shows that the intermediate phases FeO and Fe_3O_4 produced from step I reacted with Al to form the phases α-Al_2O_3 and $FeAl_2O_4$. The variation of standard Gibbs free energy ΔG_T^0 of Equation 4 as a function of temperature can be expressed as:

$$\Delta G_T^0 = -4233380 + 384.8T \quad (5)$$

Note that ΔG_T^0 is less than zero for step II reaction, meaning it can also take place spontaneously.

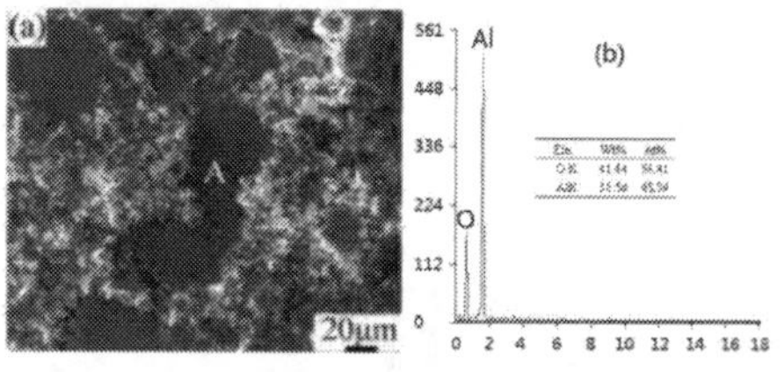

Fig.7SEM microphotograph (a) and EDS data (b) of the composites at heating temperature of 1473 K.

When the heating was terminated at 1473 K, the phases of the resultant composite consist of Al_2O_3 and Fe (as seen in Fig.4). The SEM microphotograph of the composite is shown in Fig. 7a. EDS data of the stable phase Al_2O_3 is labeled as A in Fig.7b. Hence, the stepIII reaction can be written as:

$$3FeAl_2O_4(s) + 2Al(l) = 4Al_2O_3 + 3Fe \quad (6)$$

The intermediate phase $FeAl_2O_4$ produced in previous steps reacted with Al in step III to form the phase α-Al_2O_3. The variation of standard Gibbs free energy ΔG_T^0 of Equation 6 can be calculated according to:

$$\Delta G_T^0 = -774269 + 89.8T \quad (7)$$

Obviously, ΔG_T^0 is less than 0 for step III reaction, which means it can also take place spontaneously.

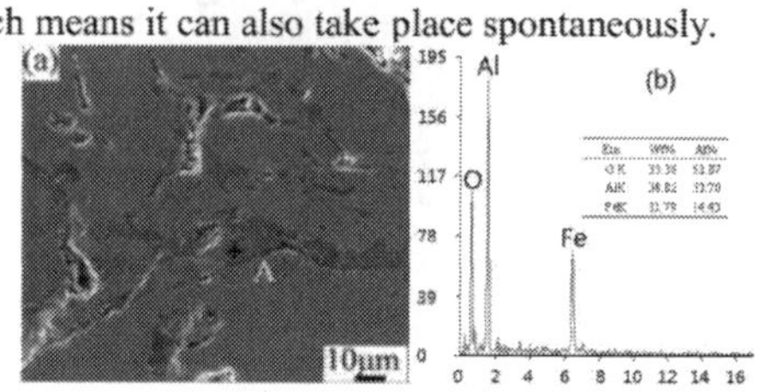

Fig.8SEM microphotograph (a) and EDS data (b) of the composites with reinforcement volume fraction of 10% at heating temperature of 1473 K.

Based upon the above analysis, when the starting materials with a composition of 10.48 wt% Al: 31.06 wt% Fe_2O_3: 58.50 wt% Fe are heated from room temperature to 1473 K, three step reactions occur, giving rise to three transitional phases, i.e., FeO, Fe_3O_4 and$FeAl_2O_4$. Therefore, the heating temperature is an important factor that affects the combustion results. On the other hand, with a decrease in reinforcement volume fraction, the sum of the exothermic heat would be reduced, leading to a decline in combustion temperature. For example, when the reinforcement volume fraction is 10 vol%, the intermediate phase $FeAl_2O_4$remains in the final composite (Fig. 8).

4. CONCLUSIONS

In situ Fe-Al_2O_3 composites were fabricated from Al-Fe_2O_3-Fe system by using XD method. Reaction mechanisms and process were investigated. The following conclusions can be drawn:

1) During the heating process, three exothermic peaks can be observed in the DSC curves of the Al-Fe_2O_3-Fe system. There are no characteristic peaks during cooling process. With the increase of the heating rate, all the peaks moved to higher temperatures.

2) Three intermediate reactions are identified:
Step I: $4Al(l) + 12Fe_2O_3(s) = Al_2O_3(s) + 5FeO(s) + 6Fe_3O_4(s) + FeAl_2O_4(s)$,
Step II: $4FeO(s)+3Fe_3O_4(s)+12Al(l)=4Al_2O_3(s)+12Fe(s)+FeAl_2O_4(s)$, and
Step III: $3FeAl_2O_4(s) + 2Al(l) = 4Al_2O_3 + 3Fe$.

3) The activation energies of the three reactions are calculated, which are 203.8, 1100.9 and 380 kJ/mol, respectively.

4) At a heating rate of 10 K/min, the times it took for the three reactions to complete are 456, 672 and 650 s, respectively. Their corresponding kinetic curves are similar, i.e., the reaction rate is very slow at the initial stage, and then it increases rapidly and, finally, it becomes slow again. Among the three reactions, the step II is the longest and also is the most difficult one.

ACKNOWLEDGEMENTS

This work was supported by the National Natural Science Foundations of China (No.51571118 and 51371098) and Natural Science Foundations of Jiangsu Province, China (No. BK20141308).

REFERENCES

[1]Yudong Sui, Mojin Zhou, Journal of Alloys and Compounds, 2018, 741:1169-1174.
[2]Bo Song, Zhiwei Wang, Qian Yan, Yuanjie Zhang, Yusheng Shi. Materials Science and Engineering: A, 2017,707: 478-487
[3] Viseslava Rajkovic, Dusan Bozic, Aleksandar Devecerski, Milan T. Jovanovic.Materials Characterization, 2012, 67: 129 – 137.
[4] Wen F, Lin T, Liu XQ. Processing Engineering, 2012, 27: 1738-1743.
[5] Georgioua EP, Achanta S, Dostac S, Fernandez J, Matteazzi P, Kusinski J, Piticescu RR, Celis JP.Applied Surface Science, 2013, 275: 142-147.
[6] Yaping Bai, Jiandong Xing, Haoliang Wu, Zhen Liu, Yimin Gao, Shengqiang Ma. Materials Design, 2012, 39: 211-219.
[7] Babakhani A, Zahabi E, Yavari Mehrabani H. Journal of Alloys Compound, 2012, 514: 20-24.
[8]ZhuHG, Dong K, Wang H, Li JL, Xie ZH. Powder Technology, 2013, 246: 456–461.
[9] Zhang LC, XuJ, Eckert J.Journalof Applied Physical,100: 033514.
[10] Zhu HG, Min J, Li JL, Ai YL, Ge LQ, Wang HZ. Composite Science and Technology, 2010, 70 : 2183-2189.
[11] Zhu HG, Jiang YL, Song JZ, Li JL, Paul Munroe, Xie ZH.Journal of Materials Science, 2013, 48: 929-935.
[12] Ye DR, Hu JH. Hand book of thermodynamics data for in organic substance. Beijing: Metallurgy Industry Press, 2002: 27-72.
[13] Knacke O. Thermochemical properties of inorganic substances. Berlin: Springer-Verlag, 1991: 30-150.

Preparation of water-soluble asphalt carbon dots and hyperchromic effect to Cr (VI)

Lijuan Gao [a], Jia Sui , Xiaoqin Qiao, Chaoshuai Hu and Xuefei Zhao[b,*]
Engineering Research Center of Advanced Coal Coking and Efficient Use of Coal Resources, University of Science and Technology Liaoning, Anshan 114051, China

Abstract Water-soluble asphalt carbon dots（W-CDs）was prepared from medium - temperature coal tar pitch as raw material, via mixed acid oxidation，alkali solubilization and acid precipitation method. It was observed that as-prepared W-CDs were irregular particles with good water dispersibility and the particle size was controlled by the pH of the acid precipitation solution by a polarizing microscope . Infrared spectroscopy confirmed that the surface of the W-CDs was functionalized by hydrophilic and color-promoting functional groups, which had good water solubility; the thermal weight loss proved to be good thermal stability below 200°C; the fluorescence spectrum proved that the fluorescence emission peak position of the carbon dots appears red shift with increase of excitation wavelength, with fluorescence conversion property, and its emission light intensity decreases, belonging to non-resonant fluorescence type of direct jump fluorescence; UV spectrum proves that the addition of W-CDs can be chromium (VI) CrO_4^{2-} content The detection limit was increased from milligram to microgram; the molar absorptivity of CrO_4^{2-} was increased by 4 times. The experimentally optimized W-CDs-ultraviolet-viscometry method for measuring CrO_4^{2-} conditions: W-CDs to chromium ratio of 8:2, pH of 8, hexamethylenetetramine as a surfactant, and the solution was allowed to stand at room temperature for 45 min.

Introduction

Carbon nanomaterials refer to carbon materials with at least one dimension of less than 100 nm in the dispersed phase.Carbon quantum dots (CQDs) is a new member of the carbon nanomaterial family. Wang [1] reviewed the application fields and detection methods on CQDs. CQDs as a rising star of carbon nanomaterials, by virtue of their in water solubility, chemical inertness, low toxicity, easy functionalization and high fluorescence intensity, photobleaching resistance, adjustable color of light and other optical properties, have displayed tremendous momentum in numerous fields such as biosensing, bioimaging, drug delivery, optoelectronics, photovoltaics and photocatalysis, as well as in metal ions, anions, small organic molecules and biomolecules. Mao[2] used fluorescent CQDs as probes to establish a new method for the determination of Cu^{2+} based on the fluorescence quenching effect of Cu^{2+} on carbon quantum dots. Niu [3] used a bottom-up electrochemical (EC) method to prepare green fluorescent nitrogen-doped carbon quantum dots (N-CQDs), based on the principle of on-off fluorescence (FL), the photoluminescence intensity of N-CQDs is further extended to ON-OFF-ON and ON-OFF-ON-OFF

[a]email:gaolijuan@ustl.edu.cn, [b]email: zhaoxuefei@163.com

by the nature of Fe^{3+} quenching, used for the detection of Fe^{3+} in specific environments. Gao [4] produced an on-off-on alternately controlled CQDs fluorescent nanosensor to detect iron Fe^{3+} in living cells. At the same time, glutathione (GSH) can increase the fluorescence of CDs/Fe^{3+} mixed solution, and effectively distinguish cancer cells from normal cells based on the difference in the content of glutathione between cancer cells and normal cells. Li [5] studied a novel electrochemical sensor for simultaneous determination of Cd^{2+} and Pb^{2+} by anodic stripping voltammetry used by NCQDs-GO. In the hybrid system, both N-CQDs and GO have a large electroactive surface area and rich oxygen-containing functional groups, which providing the active site for the compound effectively adsorbs heavy metal ions through electrostatic interaction, improving sensor sensitivity. Li[6] using CQDs composite TiO_2 under visible light reduced Cr (VI), after 120 min light reaction, the reduction of chromium (VI) rate reached 99. 2%, which is the 1.5 times of the reduction efficiency of TiO $_2$. Jiang [7] synthesized TiO_2 nanocomposite photocatalysts incorporating CQDs by sol-gel method, CQDs have up-conversion fluorescence characteristics, and the visible light response of TiO_2 samples doped with CQDs is significantly improved. Chen[8] constructed a novel sensor based on Pt nanoparticle-CQDs/ionic liquid functionalized graphene oxide (PtNPs-CDs/IL-GO) nanocomposites to detect H_2O_2. Zhang [9] used CQDs to improve the charge transfer and electrocatalytic properties of the $BiVO_4$-SnO_2 interface. Ye [10] demonstrated that CQDs, as an environmentally friendly photosensitizer, can extend the range of light absorption to the entire visible range significantly. Wang[11] synthesized a novel full-spectrum-driven CQDs (CQDs)/Bi_2WO_6 (CBW) hybrid material to provide full spectrum utilization for enhanced photoactive CQDs heterojunctions in contaminant degradation and energy conversion for reference. Zhang[12] synthesized a CQDs/$BiPO_4$ nanocomposite photocatalytic material with enhanced visible light absorption and charge separation, the photocatalytic performance of CQDs/$BiPO_4$ composites is better than that of pure under simulated sunlight or visible light. Hua[13] prepared α-BO / N-CQDs composites by low temperature hydrothermal synthesis in one step. When N-CQDs are combined with α-Bi_2O_3 nanotubes, the absorbance increases. The introduction of N-CQDs can inhibit the photogenerated electron-hole recombination, promote the production of hydroxyl radicals and superoxide radicals, and improve the degradation rate of rhodamine B by the catalyst under simulated sunlight. Wei[14] prepared NCDs/TiO_2 composites using physical mixing and aging methods. Since NCDs can absorb both ultraviolet light and visible light, the range of light absorption of the material is effectively broadened. In addition, the composite material can effectively prevent the recombination of holes and electrons, and the photocatalytic efficiency is greatly improved. As the content of NCDs in the composite increases, the photocatalytic hydrogen production performance also increases. The recombination of carbon quantum dots and catalyst can not only promote the separation of photogenerated electrons and holes, enhance the utilization efficiency of light by the system, improve the photocatalytic performance, but also enhance the stability of the catalyst [15]. The chromium in natural waters is mainly in the valence state of Cr (III) and Cr (VI), of which Cr (VI) is hundreds of times more toxic than Cr (III) [16]. With the continuous development of industry, a large amount of chromium-containing waste is produced in the production process, which directly damages the health and safety of aquatic products and human bodies. Therefore, it is practical to establish a method of detecting Cr (VI) in water with high density and low detection limit. In this paper, Water-soluble asphalt carbon dots（W-CDs）was prepared from medium - temperature coal tar pitch as raw material, via mixed acid oxidation，alkali solubilization and acid precipitation method. Carbon dots have an up-conversion function of absorbing visible light to emit ultraviolet light and suppressing electron-hole recombination properties, make use of that to enhance the color effect of Cr (VI), and to establish a method for measuring the trace amount of CrO_4^{2-} by the W-CDs-UV method, the detection limit of Cr(VI) (CrO_4^{2-}) content is increased from milligram to microgram.

1 Experimental section

1.1 Preparation of the W-CDs

Preparation of water-based asphalt carbon dots（W-CDs）

have obtained under the condition of medium - temperature coal tar pitch (Shanxi huazheng company provides, with softening point 83.6°C) with particle size less than 0.2mm as carbon source, 7:3 (V(sulfuric acid) /V(nitric acid)) mixed acid as oxidant; the ratio of carbon source asphalt to mixed acid 1g:20mL, Constant temperature 38°C for 7.5 h; The following, the contents of the flask were poured into deionized water to terminate the reaction, and place for the night at room temperature; then was filtered using Buchner funnel and washed with water until the filter cake was neutral. The filter cake was transferred to 500mL flask, mixed with 1 mol/L NaOH until the pH of the solution was greater than 12, and stirred at 80°C for 1 h to fully dissolve; and filtered while hot. The filtrate was added with 1 mol/L hydrochloric acid to adjust the pH of the solution to 6, and vacuum filtered, and the filter cake was recorded as W-CDs 1#; the filtrate was further precipitated with 1 mol/L hydrochloric acid to adjust the pH of the solution to 4, and vacuum filtered, and the filter cake was recorded as W-CDs 2#; the filtrate is further precipitated with 1 mol/L hydrochloric acid to adjust the pH of the solution to 2, and vacuum filtered, and the filter cake was recorded as W-CDs 3#.

1.2 Characteristic Instruments

The W-CDs were characterized by Fourier transform infrared spectrometer, UV-visible spectrophotometer, fluorescence spectrophotometer, and thermal weight loss tester.

1.3 Preparation of the W-CDs solution

Accurate weighing 0.2000 g the as-prepared W-CDs Put into a 100mL bottle, dissolved using NaOH aqueous solution(pH 10), and made up to a volumetric flask of 100mL to obtain 2.0 mg/mL W-CDs solution.

1.4 Preparation of CrO_4^{2-} solution

0.5000g Cr_2O_3 (analytically pure, Shenyang Dongxing Reagent Factory) solid powder was dissolved in 250 mL of 1 mol/L hydrochloric acid and reacted for 2 h, and the mixture was filtered to remove a trace amount of white solid to obtain 3 mg/mL CrO_4^{2-} solution.

2 Results and discussion

2.1 Dispersion of the W-CDs

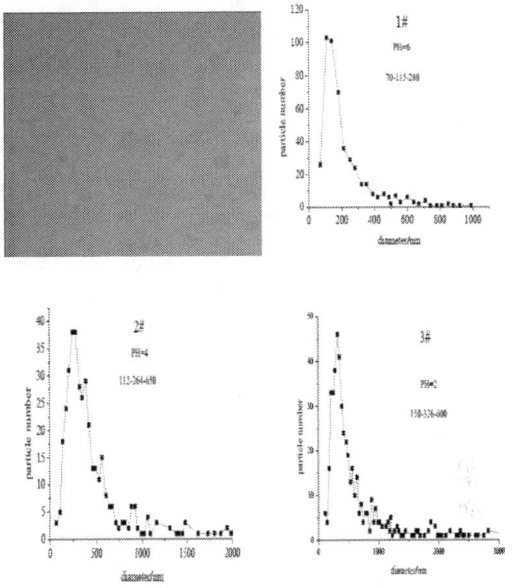

Figure 1. Polarized micrograph and particle size distribution curve of the W-CDs dispersed in water

The morphology of the as-prepared W-CDs were characterized by polarimetric microscope. The polarimetric microscope image (Fig.1) and the corresponding curve of size distribution (Fig.1) illustrate that the W-CDs could be homogeneously dispersed in water solvent, with irregular shape, and hydrodynamic diameters ranging (Fig. 1 (1#)) from 70 to 280nm, average size 115nm. The particle size distribution curve(Fig.1 (1# 2# 3#)) shows that hydrodynamic diameters of W-CDs is related to the pH value of the solution. The particle size of W-CDs decreases with increasing pH .The pH of the solution is used to control the size of W-CDs.

2.2 Characterization of as-prepared W-CDs

To study the apparent characteristic and optical behavior of the W-CDs, FTIR and UV–vis were applied in detail. FTIR spectroscopy was used to investigate the surface functional groups of W-CDs. As shown in Fig.2, compared with the spectrum of the raw material asphalt, the W-CDs display new absorption peak, and the intensity of the absorption peak change greatly. These peaks are attributed to O-H stretching vibration (3610cm^{-1}), C-H stretching vibration (3045cm^{-1}), C=O symmetric stretching vibration (1719cm^{-1}), C=C stretching vibration of aromatic ring skeleton (1593cm^{-1}), N-O symmetry and antisymmetric stretching vibration of the nitro group (1336cm^{-1} and 1520cm^{-1}) , O-S-O symmetric stretching vibration of the sulfonic acid group (1042cm^{-1}). But bending vibration of the alkyl chain of the aromatic ring (735cm^{-1}) and symmetric stretching vibration (1380 cm^{-1}) disappeared. It can be seen that main types of reactions are oxidation, sulfonation, nitrification, after the medium temperature asphalt treated with mixed acid, and reaction occur in the aromatic ring side chain. more oxygen-containing functional groups whit hydrophilic and facilitated color enhancement function, such as nitro group, hydroxyl group, carboxyl group, sulfonic acid group, are introduced into the surface of the as-prepared W-CDs.

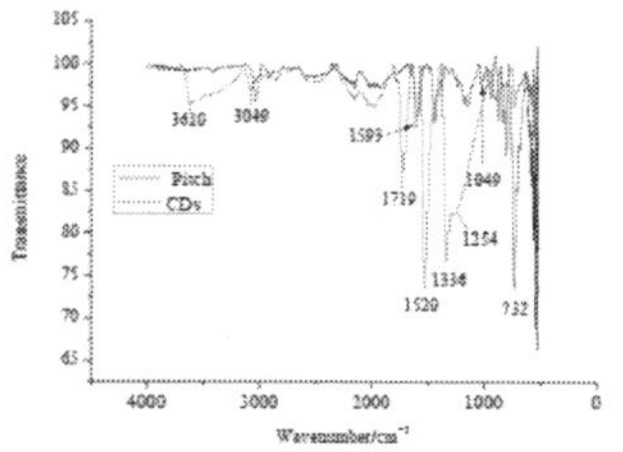

Figure 2. FT-IR spectra of the raw pitch and W-CDs

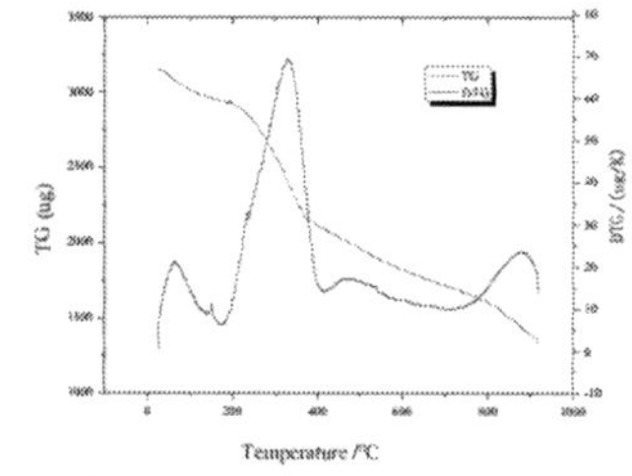

Figure 3. TG/DTG curves of the W-CDs

The TG/DTG curves of the as-prepared W-CDs are presented in Fig.3. As shown in Fig.3, thermal decomposition is clearly revealed. It can be seen from the TG curve that the quality of W-CDs decreases with increasing temperature, and the DTG curve can be roughly divided into three stages according to the image: the first stage, from 36℃ to 150℃, are attributed to the evaporation of some small molecules; the second stage, from 200℃to 400℃, are attributed to the pyrolysis of the asphalt original structure functional group; the third stage: from 700℃to 900℃, are attributed to the removal of hydrophilic group in W-CDs. These results show that the temperature of W-CDs be below 200℃when applied.

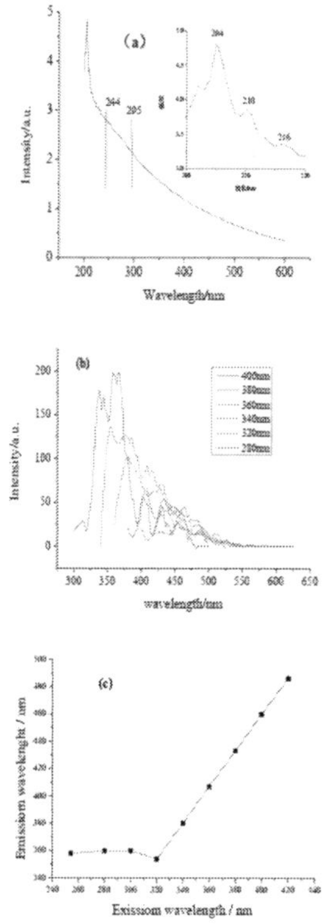

Figure 4. UV-Vis absorption and FL spectra of the W-DCs

As shown in Fig.4(a), the solution of the W-CDs have strong UV-vis absorption peak at 204nm and 210nm, while there is a weak absorption peak at 250 nm , but the absorbance decreases with the increase of absorption wavelength when it is greater than 300nm, and there is no maximum absorption peak.

The photoluminescence (PL) emission spectra of the W-CDs in the excitation wavelength from 280 nm to 400 nm are shown in Fig. 4(b). The position and intensity of the fluorescence emission peak of the W-CDs are related to the excitation wavelength. The emission peak of the W-CDs appears the red-shifted in position and reduced intensity whit the excitation wavelength increases. The excitation avelength, which results in emission spectrum absorption peak is the strongest, is 280 nm, meanwhile the emission wavelength is 360nm.Which proves that W-CDs has fluorescence conversion properties, It belongs to the non-resonant fluorescence type of direct jump fluorescence. In order to investigate the

https://doi.org/10.1051/matecconf/201823802007

photoluminescence characteristics of W-CDs, especially the dependence on excitation light, excitation spectra of W-CDs in excitation wavelength from 254 nm to 420 nm are shown in Fig. 4(c). When the excitation wavelength is increased from 254 nm to 320 nm, it can be seen that the maximum emission wavelength is almost constant, meanwhile the intensity gradually heighten; When the excitation wavelength is increased from 320 nm to 420 nm, it can be seen that the maximum emission wavelength is red-shifted, meanwhile the intensity gradually decreases.

From the theoretical analysis of the surface state, it is considered that the carbon quantum dots are surface-defective luminescence. When the surface electrons of the carbon quantum dots are excited by ultraviolet light, electron-hole pairs are formed, and electrons are instantaneously captured by surface defects during the migration process, thereby generating fluorescence. Meanwhile, a photon with adequate energy can excite an electron in the HOMO level of CQDs, transferring the excited electron to the LUMO level of CQDs. The emission from CQDs shifted to higher wavelength as the intensity increases, which is characteristic of surfacedefect emission derived from CQDs as reported by others. which may arise from the π - π * transition of the nanocarbon particles and from the excitation energy sink of the surface states.

2.3 Effect of the W-CDs on CrO_4^{2-} molar absorption coefficient

Figure 5 (a) compares the UV-Vis absorption spectra of CrO_4^{2-} and the as-prepared W-CDs. As shown in Fig. 5(a), it was found that the maximum absorption peaks appear at 248 nm, 280 nm and 350 nm for the CrO_4^{2-} solution, and the absorption intensity decreases in turn; the maximum absorption peak appears less 250 nm for the as-prepared W-CDs suspension liquid, and no maximum absorption peak appears as wavelength larger 250 nm. The as-prepared W-CDs suspension liquid, at 350 nm, has almost no absorption; but the CrO_4^{2-} solution has a weak maximum absorption. In order to reduce the influence of the absorption of the the W-CDs itself, the measurement wavelength of CrO_4^{2-} is selected to be 350 nm.

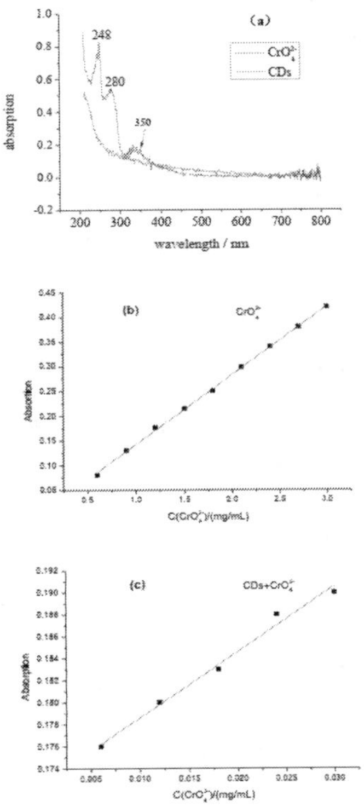

Figure 5. Effect of the as-prepared W-CDs on CrO_4^{2-} absorption coefficient

Figure 5 (b) shows the linear correlation of the absorbance of the CrO_4^{2-} solution at wavelength of 350 nm, and Figure 5 (c) shows the linear correlation of the absorbance at wavelength of 350 nm for a CrO_4^{2-} solution with 6 mg/L the as-prepared W-CDs. It can be seen from Fig. 5 that the pure CrO_4^{2-}solution can measure the linear range of ultraviolet absorption from 0.6 mg/mL to 3.0 mg/mL, and the linear linear regression equation: A=0.00202+0.140 (C/(mg/mL)), correlation coefficient R=0.9995; With the addition of the W-CDs, the CrO_4^{2-} solution can measure the linear range of UV absorption from 0.006 mg/mL to 0.030 mg/mL. The linear regression equation: A=0.1726+0.600 (C/(mg/mL)), correlation coefficient R=0.9940.

According to Beer's law, $A = -\log(I/I_0) = -\lg T = \varepsilon c$; where A: absorbance, I_0, I: intensity of incident light and transmitted light, T: transmittance, ε: molar absorption coefficient of sample, c: sample concentration. It can be seen from Fig. 5 that, added to the W-CDs, a smaller concentration of CrO_4^{2-} can be measured accurately. The measured value of concentration (c) becomes smaller, and the corresponding value of absorbance (A) increases. According to Beer's law, ε increases from 0.140 to 0.600. Generally, ε becomes larger, indicating that the absorption band of the measured substance becomes stronger, which is favorable for the meas urement of the absorbance, and it can be seen that the W-CDs can enhance the CrO_4^{2-} molar absorption coefficient.

2.4 Optimization of conditions for measuring CrO_4^{2-} by W-CDs-UV-vis spectrophotometry

To optimize the conditions for measuring CrO_4^{2-} by W-CDs-UV-visible spectrophotometry, influence factors such as temperature, ratio of W-CDs to CrO_4^{2-}, time, pH, surfactant and other factors were investigated by experiment.

The solution for the experiment were prepared .The CrO_4^{2-} solution (3mg/mL) were mixed with the W-CDs solution (0.020mg/mL) according to different volume ratios, and the absorbance was measured at 360nm. The effect of the volume ratio of the W-CDs to CrO_4^{2-} on the absorbance of the test solution is shown in Fig. 6(a), as the proportion of the W-CDs to CrO_4^{2-} increases, the absorbance increases, and reaches the maximum at V(CDs):V(CrO_4^{2-}) = 8: 2, so the best W-CDs to CrO_4^{2-} is 8:2. Fig. 6(b) shows the effect of time and temperature on the absorbance of the test solution. Under optimized W-CDs to CrO_4^{2-} ratio 8:2, the test solution were heated to 30 ℃, 45 ℃, 60 ℃, 75 ℃, 90 ℃ respectively, and measured the absorbance at 0 min, 15 min, 30 min, 45 min, 60 min, the results are shown in Fig. 6 (b). It can be seen from Fig. 6(b) that the absorbance at different temperatures increases first and then decreases with time, and the maximum value appears at 45 min, so the optimal time is 45 min. The maximum absorbance decreases with the increasing temperature, so the room temperature 30℃ is the optimum temperature.

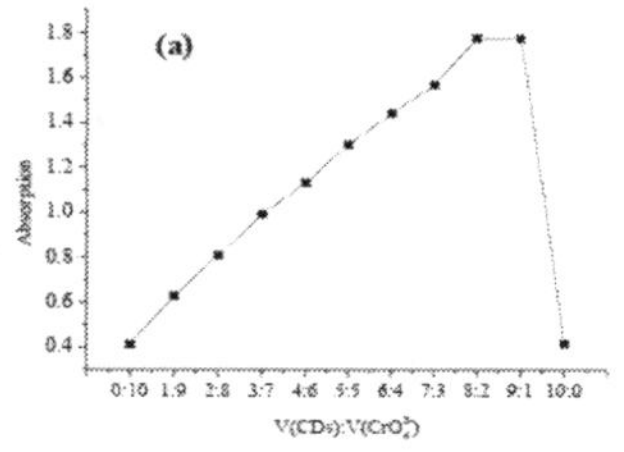

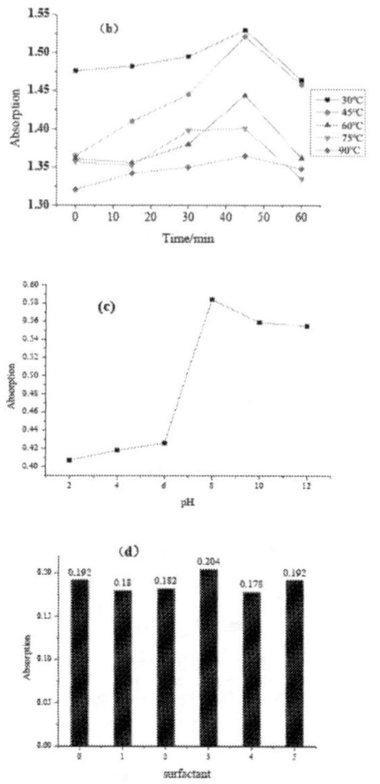

Figure 6. Conditions for measuring CrO_4^{2-} by W-CDs - UV-vis spectrophotometry

Fig. 6 (c) shows the effect of pH on the absorbance of the test solution. Six solutions whit W-CDs to CrO_4^{2-} ratio 8:2, adjusted the pH values to 2, 4, 6, 8, 10, 12 respectively, and measured their absorbance, the results are shown in Fig. 6(c). According to Fig. 6(c), as the pH value of the solution increases, the value of the absorbance increases first gradually and then decreases when pH reaches 8, the absorbance is maximum, so the selected pH value is 8.

Fig.6 (d) presents the effect of the addition of surfactant on the absorbance of the test solution. X-axis indicate selected surfactants (0 (no surfactant), 1(polyacr-ylamide), 2 (sodium hydrogen phthalate), 3(hexameth-ylenetetramine), 4 (adipate), 5 (polyethylene glycol)). Compared with the non-surfactant mixture, the absorbance increases when No. 3 surfactant (hexamethylenetetramine) added, and other surfactants have no significant effect. Therefore, hexamethylenetetramine is selected as the added surfactant.

3 Conclusion

(1) Carbon dots with hydrophilic and color-promoting functional groups on the surface can be prepared by the water-insoluble medium-temperature asphalt as raw material, via mixed acid oxidation, alkali solubilization and acid precipitation method; the particle size of the carbon dots can be controlled by the pH value of the acid precipitation solution, and the particle size distribution range is 70nm to 280nm at the pH value of 6, with a mean value of 115nm, the carbon dots are irregular particles and disperse well in water.

(2) When the particle size of the as-prepared carbon dots is larger than 300nm, the UV absorbance decreases with the increase of the absorption wavelength, and there is no maximum absorption peak; and the fluorescence emission peak red-shifted with the increase of the excitation wavelength, which has fluorescence conversion property; emission intensity reduced, which is a direct linear fluorescence of non-resonant fluorescent type. The addition of as-prepared W-CDs can increase the UV molar absorption coefficient of CrO_4^{2-} by 4 times, and the density of detection increases from mg level to ug level.

(3) Optimization of conditions for measuring CrO_4^{2-} by as-prepared W-CDs-visible spectrophotometry : W-CDs (0.020mg/mL) to CrO_4^{2-}(3mg/mL) volume ratio 8:2, the pH 8, and the hexamethylenetetramine surfactant, the solution is allowed to stand at room temperature for 45 min.

Acknowledgements

This work was supported by the National Natural Science Foundation of China (U1361126), the Specialized Research Fund for the Doctoral Program of Higher Education (20132120110001).

References

[1] R. Wang, K.Q. Lu, Z.R. Tang , Y.J. Xu. J. Mater Chem A, **5**, 3717-3734（2017）

[2] X.J. Mao , H.Z. Zheng, Y.J. Long. Journal of Southwest University (N atural Science Edition), **32(9)**: 40-43（2010）

[3] F.S. Niu , Y.L. Ying , X. Hua , Y.S. Niu , Y.H. Xu , Yi-.T. Long. Carbon, **127**，340-348(2018)

[4] G. Gao, Y.W. Jiang , H.R. Jia , J.J. Yang , F.G. Wu . Carbon, **134**, 232-243(2018)

[5] L.b. Li, D. Liu, A.P. Shi , T.Y. You. Sens. Actuators, B **255** ,1762–1770(2018)

[6] Y.R. Li, Z.M. Liu, Y.C. Wu, J.T. Chen, J.Y. Zhao, F.M. Jin, P. Na. Appl. Catal., B: Envir. **224**, 508 –517(2018)

[7] T.T. Jiang，Y. Shi，J. Ke，K.l. Xu. J. NU. (Nat. Sci.)，**39（1）**，183-142（2018）

[8] D.D. Chen , X.M. Zhuang , J. Zhai , Y.Yyun. Zheng , H. Lu , L.X. Chen . Sens. Actuators, B , **255** , 1500 –1506(2018)

[9] Z.M. Zhang , X. Jiang , J.F. Mei , Y.X. Li , W.H. Han , M.Z. Xie, F.C. Wang , E.Q. Xie. Chem. Eng. J. **331** ,48–53(2018)

[10] K.H. Ye, Z.L. Wang, J.W. Gu, S. Xiao, Y.F. Yuan, Y.Zhu, Y.M. Zhang, W.J. Mai and S.H. Yang. Energy Envir. Sci, **10**, 772—779（2017）

[11] J.J. Wang , L. Tang , G.M. Zeng , Y.C. Deng , H.R. Dong ,Y.N. Liu , L.L. Wang , B. Peng , C. Zhang , F. Chen. Appl. Catal., B: Envir , **222**,115–123(2018)

[12] Z.J. Zhang, J.Y. Xu, H.B. Zeng, N. Zhang. J J. Inorg. Mater, **33 (5)**,582-586（2018）

[13] C.H. Hua，H.C. Ma，X.L. Dong，X.F. Zhang. Chem. J. Chin. Univ, **39(2)**:200-205（2018）

[14] J. Wei, X.D. Li, H.Z. Wang, Q.H. Zhang, Y.G. Li. J. Inorg. Mater , **30(9)**:925-930（2015）

[15] J.Q. Pan, Y.Z. Sheng, J.X. Zhang, J.M. Wei, P. Huang, X. Zhang , B.X. Feng . J. Mater. Chem. A, **2**, 18082–18086（2014）

[16] H.R. Wu，L. Jiang，J. Wu，S.N. Li，R.Cheng，H.Y. MA. Guangzhou Chem. Ind, **46 (5)**:50-52（2018）

Anti-acid Corrosion Property of Concrete Improved by Microstructure Optimizing

L Yu[1], *S X* Zhang[2], *and L* Chen[2]

[1]Research institute of highway ministry of transport, No.8 yard, Xi tu cheng Road, Hai-dian district, Beijing, China.
[2]Beijing Xin-Qiao Technology Development Co., LTD, No.8 yard, Xi tu cheng Road, Hai-dian district, Beijing, China

Abstract. To enhance the performance of anti-acid erosion, silica fume, silicon nitride, organic silicone emulsion agent and fly ash ceramsite are all mixed into concrete as admixture. The acid corrosion environment is simulated by laboratory test. Compressive strength of concrete with different curing ages are tested. Compressive strength and pore distribution of concrete dipping in sulphate and in hydrochloride for 90d are tested. The result shows that concrete with additional materials has better anti-acid corrode performance than the common one. The compressive strength of concrete with additional materials is higher than the common concrete. The pore distribution curve of the concrete with additional materials can keep steady no matter in sulphate or in hydrochloride. This kind of concrete with high anti-acid corrosion durability has been used in the piles construction of the bridges which is surrounded by acid soil and play an important role in enhancing the durability of the bridge.

1 Introduction

A bridge called Lao Hu Shan Chong is located on Mengzi highway in Yunnan province. There was a scrap iron treatment plant before. Now the bridge will be built there. By detection, 12 piles from No.1 to No.3 piers will be in an environment of ions including H^+, Cl^- and SO_4^{2-} . So, making a kind of anti-acid erosion concrete material is necessary.

There are several methods to produce anti-acid erosion concrete including raw materials and mix proportion optimized [1]. For example, put some micro or nano materials into concrete to fill macro pores [2][3]. Hydrophobic materials are also often used to improve the permeability performance of concrete [4]. Internal curing materials can be used to decrease the probability of early age cracking in concrete [5]. All of these can enhance the concrete microstructure durability.

Therefore, these kinds of methods are all used to optimize the concrete microstructure. Two series of concrete has been made. One is the control group with no additions and the other is the treatment group with micro materials, hydrophobic materials and internal curing materials in it. And then put the two groups of specimens into acid liquid. After 90 days, test the microstructure, compressive strength, permeated height of Cl^- ions to verify the anti-acid performance of concrete. Prepare for the pile construction of Lao Hu Shan Chong Bridge in Mengzi city, Yunnan Province in China.

2 Raw materials and mix proportion

2.1 Raw materials

The main binding material is P.O 42.5 Portland cement which is produced by Hua Xin cement plant. The fine aggregate is nature sand whose fineness modulus is 2.8. The coarse aggregate is crushed stone whose diameter is 5-20mm. Polycarboxylic acid water reducer whose water-reducing rate is 20% has been used.

Silica fume is necessary. It is used for filling the micro pores in concrete which is produced by Shang Hai Yao Qian Construction Coating and Decorate Co., LTD.. The properties of silica fume are listed in table 1.

Table 1. The properties of silica fume

Diameter *μm*	Density *kg/m³*	Specific surface area m^2/kg
0.2	3200	15000

The other silica material is nano silicon nitride which is produced by Qin Huang Dao Yi Nuo New Materials Development Co., LTD. It is used for filling the nano pores in concrete. The properties of silicon nitride are listed in table 2.

Table 2. The properties of silica nitride

Diameter nm	Density *kg/m³*	Specific surface area m^2/kg
20	90	50000

* Corresponding author: yuleimabel@163.com

The water-proofing agent of organic silicone emulsion is used in concrete which is produced by Nan Xiong Ding Cheng new material technology Co., LTD. It is used to resistance the permeating of harmful ions. The properties are listed in table 3.

Table 3. Properties of organic silicone emulsion

Style	PH	Effective Ingredient
FS-170	7.0	>95%

Fly ash ceramsite is used for inner curing for concrete. The properties are listed in table 4.

Table 4. Properties of silica nitride fly ash ceramsite

Apparent density kg/m^3	Diameter *mm*	Saturated percent sorption %
1950	2-5	42.3

2.2 Mix proportion

There are two mix proportions. One is the control group which has no adding materials. It will be shown as symbol C in the following paragraphs. The other is the treatment group which includes silica fume, silicon nitride, organic silicone emulsion agent and fly ash ceramsite. This group will be named T in the following paragraphs. Table 5 shows the mix proportions of concretes.

Table 5. Mix proportions of concrete Unit: kg/m3

ID	Cement	Water	Nature sand	5mm-10mm crushed stone	10mm-20mm crushed stone	Silicon nitride	Silica fume	Ceramsite	Organic silicone emulsion agent	Water reducer
C	360	170	817	325	758	0	0	0	0	2.88
T	360	170	721	308	720	2	20	54.5	3.82	2.05

The slumps of C and T are all 180mm±10mm.

3 Test methods

3.1 Specimens prepare

For each mix proportion, making 6 groups of concrete and each group has 3 specimens. Four groups are used for testing the 7d, 28d, 56d and 90d compressive strengths. One group is used for dipping in the acid solution for 90d and then testing the compressive strength. One group is used for testing the depth of penetration of Cl- ion. All of the specimens are 100mm

3.2 Acid solution prepare

Two kinds of solution have been made. One includes hydrochloric acid (HCl) and sodium chloride (NaCl). The other includes sulfuric acid (H_2SO4) and sodium sulfate (Na2SO4). The PH value of these two solutions is all 2. The percentage of salt is all 3%. During the digging period, the PH value of solution should be test every day. If necessary, add additional acid to keep the PH value of the solution staying at 2.

3.3 Microstructure test

After testing the 90d compressive strength, the specimens which had been dipped in the acid solution for 90d will be crushed by a machine. The round shaped mortar particles about 10g will be selected. After washing these particles with ultrasonic cleaner, put them into the air dry oven and dry under the temperature of 105℃±5℃ for 24h. Then put these particles into dryer and cool under nature environment. Test the pore structure by Mercury Injection Apparatus [6].

3.4 Penetration depth test

Split off the specimens after 90d dipping in the acid solution. Use phenolphthalein to titrate the split surface. The colour of split surface where has been eroded by Cl^- will become white. Partition it by 10 sections and read the length of white place.

4 Test results and analysis

4.1 Compressive strength analysis

After curing the specimens for 7d, 28d, 56d and 90d, test the compressive strengths and compare them. Fig. 1 shows the different strength along with different ages.

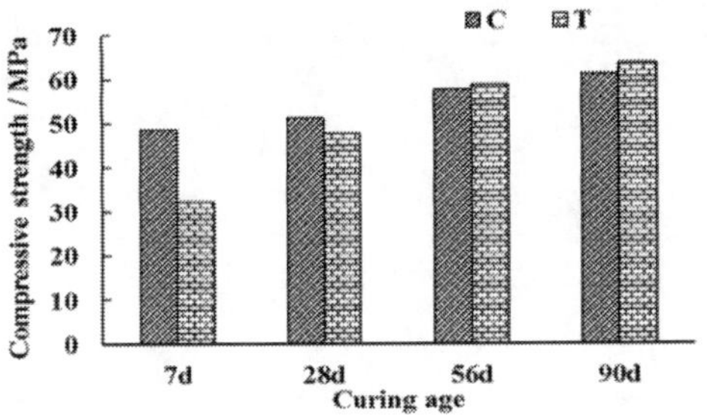

Fig.1. Compressive strength VS curing age

From Fig. 1, the 7d compressive strength of T concrete is lower than C concrete. But 28d, 56d and 90d compressive strengths of T concrete are all higher than C concrete. By increasing of curing age, the strength of T is higher than C. This is because the micro pores have been filled by nano and micro silica materials. In order to analyse the reason, pore structure of C and T concretes have been detect by Mercury Injection Apparatus. Fig. 2 shows the pore distribution of concretes.

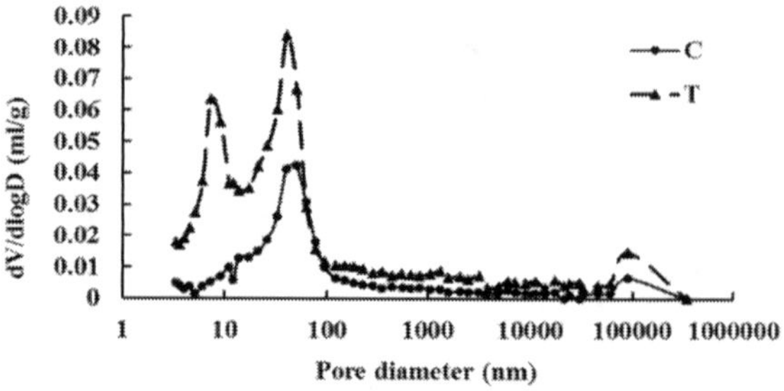

Fig.2. Pore distribution at the curing age of 28d

From Fig. 2, the amount of the pore whose diameter is under 100nm in T group is more than C group. According to the theory of IO•M Butt, pores under 100nm is good to the compressive strength of concrete. So T is better than C due to the filling of silica fume, silicon nitride, organic silicone emulsion agent and fly ash ceramsite.

4.2 Anti-acid performance analysis

Acid is bad for concrete including appearing, compressive strength, permeability. Partial reasons are that acid can destroy the pore structure in concrete.

4.2.1 Appearance of concrete

Fig.3 and Fig. 4 shows the concrete appearance after 90d digging in acid solution.

In the solution of sulfuric acid and sodium sulfate, the concrete surfaces are denudated. In the solution of hydrochloric acid and sodium chloride, the concrete surfaces are as same as before.

Fig.3. Sulfuric acid **Fig.4.** Hydrochloric acid

Scrape the exfoliation on the concretes which have been dipped in sulfuric acid and then dry and weigh it. The exfoliation from C is about 1.9% while the one from T is just 0.6%. So T group is more corrosion resistance.

4.2.2 Compressive strength

To compare two different kinds of concrete, the rate of compressive strength decreasing has been calculated by formula 1.

$$\Delta CS = \frac{CS - CS_0}{CS_0} \times 100\% \quad (1)$$

△CS—The rate of compressive strength decreasing.
CS—Compressive strength of concrete after 90d dipping in the acid solution.

CS_0 — Compressive strength of concrete with the same curing age but without acid solution corroded.

Fig. 5 is about the concretes which are dipped in sulfuric acid solution for ages of 28d, 56d and 90d. C-SO4 is the control group which has no addition materials in it. T-SO4 is the treatment group which has silica fume, silicon nitride, organic silicone emulsion agent and fly ash ceramsite in it.

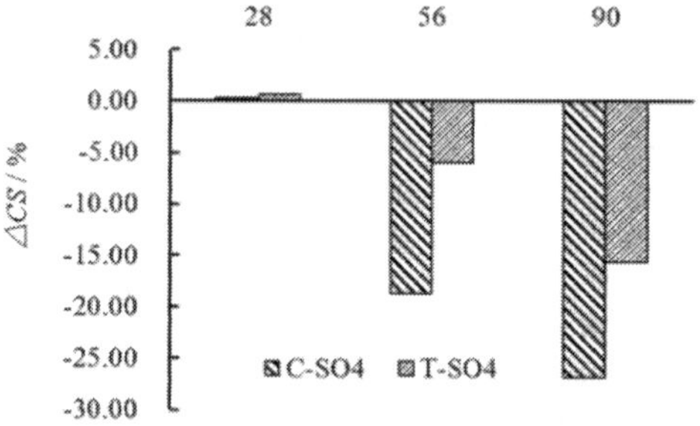

Fig.5. Compressive strength decreasing rate of concrete in sulfuric acid and sodium sulfate solution

From Fig. 5, in the same solution, at the age of 90d, the decreasing of compressive strength of T-SO4 group is less than C-SO4 group. It means T-SO4 group has better anti-acid corrode performance. The data at the age of 28d is unusual. The compressive strength is higher than curing group. This is because the sodium sulfate which can react with the ettringite in the surface layer of concrete. Due to this reason, the pores in the concrete have been filled and the compressive strength is higher than before. But by the time passing, the delay ettringite expands and the strength of concrete decreases.

Fig. 6 is about the concretes which are dipped in hydrochloric acid solution for ages of 28d, 56d and 90d. C-Cl is the control group which has no addition materials in it. T-Cl is the treatment group which has silica fume, silicon nitride, organic silicone emulsion agent and fly ash ceramsite in it.

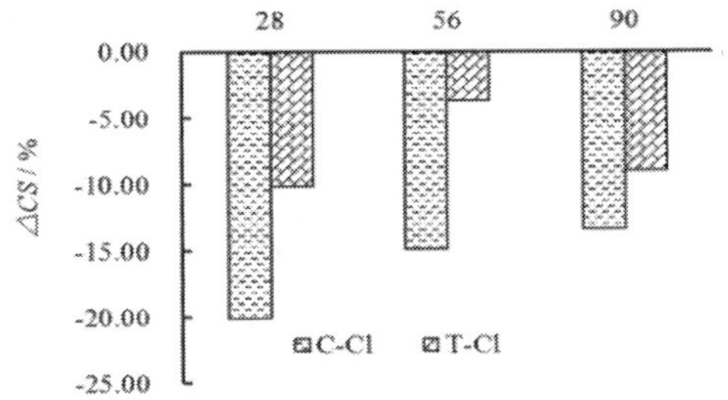

Fig.6. Compressive strength decreasing rate of concrete in hydrochloric acid and sodium chloride solution

From Fig. 6, C-Cl group is worse than T-Cl. That means the concrete which has additional materials has better anti-acid and chlorine salt corrode performance.

4.2.3 Pore distribution

To compare the vary of pore distribution of concrete, Fig. 7 and Fig. 8 are plotted. In Fig. 7, there are two lines. C is the concrete which hasn't been dipped in acid solution but only curing in water. C-SO4 is the concrete which has been dipped in sulfuric acid solution. In Fig. 8, there are also two lines. T is the concrete which hasn't been dipped in acid solution but only curing in water. T-SO4 is the concrete which has been dipped in sulfuric acid solution.

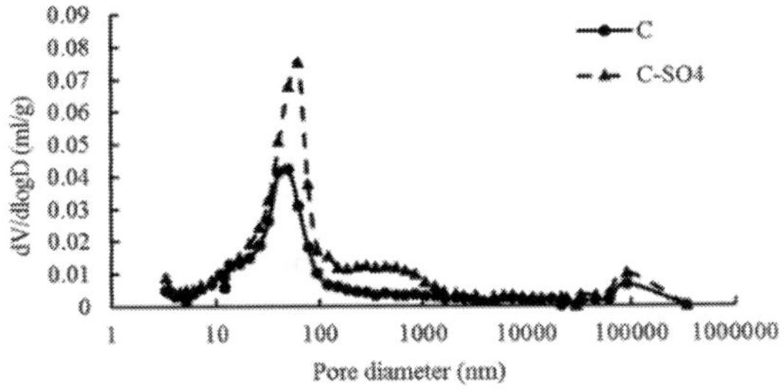

Fig.7. pore distribution of control group concrete under sulfuric acid solution

From Fig. 7, after erosion, the amount of pores whose diameter is among 100nm-1000nm increase. This dimension of pores can make the anti-permeability of concrete decrease. Therefore, the control group of concrete has a worse anti-acid corrode performance.

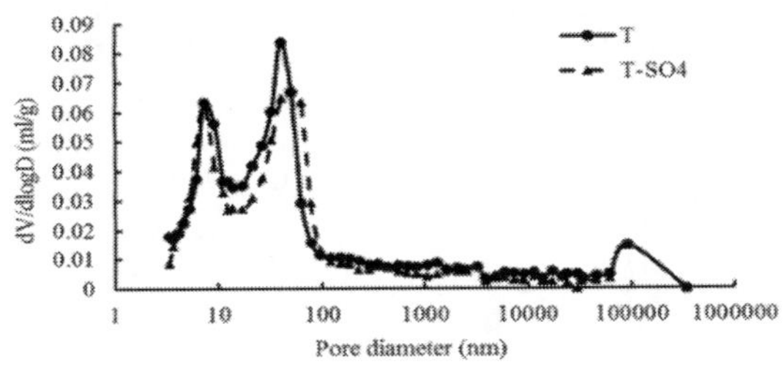

Fig.8. pore distribution of treatment group concrete under sulfuric acid solution

From Fig. 8, after eroded, the pore distribution still keeps as similar as before. Comparing Fig. 7 and Fig. 8, Treatment group behave better than the control group for the erosion of sulfuric acid and sodium sulfate mixture solution. The same situation also happens in the hydrochloric acid and sodium chloride mixture solution. Therefore, from pore distribution figure, it can be concluded that the concrete with additional materials has better anti-acid performance.

5 Conclusions

To enhance the performance of anti-acid erosion, silica fume, silicon nitride, organic silicone emulsion agent and fly ash ceramsite are all mixed into concrete as admixture. By compressive strength test and pore structure test, it can be concluded that the concrete with these admixtures has better anti-acid performance than common concrete without mixing these admixtures. Some conclusions can be shown as follow.

(1) The compressive strength increase along with the curing age increasing. The early age compressive strength of concrete with additional materials is lower than the common concrete. While, after 28d curing, the compressive strength of concrete with additional materials is higher.

(2) No matter in sulphate or in hydrochloride, the compressive strength of concrete with additional materials is higher than the common concrete. The pore distribution curve can keep steady and the pores whose diameter is 100-1000nm don't not increase.

(3) Concrete with silica fume, silicon nitride, organic silicone emulsion agent and fly ash ceramsite has good anti-acid corrode performance.

Acknowledgement

This work was supported by the Basic Research Found for National Public Welfare Scientific Research Institutions [grant numbers 2017-9004].

References

1. F.Yuan Dong, S.Suo Zheng, M.Chen Song. Materials Review A: Summarize Section.,**32**,8 (2018) (In Chinese)
2. L.Yu. *Micro-Properties of Cement Concrete*. (China Building Industry Press, 2017) (In Chinese)
3. J-Guo Sun, Y.Ling GU. Concrete. 10,4(2013) (In Chinese)
4. Y. Yang, H.Ying ZHU, X.Min WANG, L.Huan LU, X.Li XIE. Concrete. **9**,4(2017) (In Chinese)
5. S. Han, M.Zhe An, R. Guo, G. Liu. Journal of Building Materials. **18**,6(2015) (In Chinese)
6. L. Yu. *Study on durability of cement concrete in multi erosion*. (Doctor Degree of Beijing University of Technology, 2013) (In Chinese)

MATEC Web of Conferences **238**, 02009 (2018) https://doi.org/10.1051/matecconf/201823802009
ICNFM2018

Research on the Optimal Combination of Concrete Aggregates Based on Bolomey Equation

Fan Wang[1,*], *Shansuo* Zheng[2], and *Bin* Wang[1]

[1]School of Civil & Architecture Engineering, Xi'an Technological University, Xi'an, 710021, China
[2]School of Civil Engineering, Xi'an University of Architecture and Technology, Xi'an, 710055, China

Abstract. Concrete mixture with high density is one of the important guarantees for its high strength, especially the gradation combination of coarse or fine aggregate. In order to overcome the deficient market supply of sand and gravel for meeting the need of high strength concrete and special concrete, based on the Bolomey equation, combining with the corresponding regulations and grading requirements of the concrete aggregates in the current CHINA standards as follows: Standard for technical requirements and test method of sand and crushed stone (or gravel) for ordinary concrete (JGJ 52), Sand for construction (GB/T 14684), Pebble and crushed stone for construction (GB/T 14685), establish a set of practical numerical calculation methods of the optimal combination on the continuous gradation of concrete aggregates. After concrete-preparation test verification, with the optimal combination of Aggregates, concrete mix proportion is easy to determine with higher guarantee rate, its strength can be raised up to 20% with better durability.

1 Introduction

Large numbers of studies showed that the aggregate gradation hads important impact on the performance, mechanical properties, long-term performance and durability of concrete. Aggregate gradation could be divided into two categories: continuous-graded gradation and single-graded gradation. Among them, single-graded gradation could also be called discontinuous gradation, which refered to the gradation of taking out one or more intermediate gradations in continuous grading. Some research had shown that gap graded aggregate mixed in the concrete of greater workability can cause concrete segregation easily, which is mixed in the concrete mixture with vibratory compaction and lower workability. Classic theory of the aggregate gradation had much as follows: Maximum density proposed by W. B. Fuller and S. E. Thompson[1], J. Bolomey's grading formula[2], the American C. A. G. Weymouth's particle interference theory[3], thick and surrounding stacking theory proposed by Chinese Liu Chongxi[4], the grain composition fractal theory[5-10]. The above theories were theoretical and inconvenient to apply in practical engineering. Therefore, in order to cope with the sand and stone with uneven quality and different graduation in the market, a set of practical numerical calculation methods of the optimal combination on the continuous gradation of concrete aggregate was proposed based on Bolomey equation.

2 Bolomey equation

Bolomey equation [2,11] is one of the continuous grading theories, which was based on the famous particle size distribution computation theory derived from "ideal particles". As for the aggregate particle, the percent passing through the d sieve opening was calculated according to Eq. 1:

$$P_{pd} = \frac{A - B + (100 - A) \times \sqrt{\dfrac{d}{D_{\max}}}}{100 - B} \times 100\% \qquad (1)$$

Where $P_{pd} / \%$ is the percent passing through the d-sieve opening, and $D_{\max} / mm$ is Maximum particle size of aggregates, and B is the percentage of colloidal particles of all solid particles, and A is the value of the empirical coefficient depends on the slump of concrete mix and the type of aggregates, as shown in Table 1.

Table 1. Value of A depends on concrete slump and aggregates.

Type of aggregate particles	slump of concrete (mm)		
	≤50	50~100	≥150
natural	8	10	12
artificial or machine-made	10	12	14

* Corresponding author: ship_wang2955@sina.com.

3 Evaluation of particle size distribution

The particle size distribution or grading of aggregates was evaluated and measured by the sieve analysis test. The results of the sieve analysis could be expressed by the sieve table, the fineness modulus and the sieving graph. Sieve analysis test should be according to Standard for technical requirements and test method of sand and crushed stone (or gravel) for ordinary concrete (JGJ 52), test square hole sieves with the following openings: 80mm, 63mm, 50mm, 40mm, 31.5mm, 25mm, 20mm, 16mm and 10mm, 5mm, 2.5mm and 1.25mm, 630μm, 315μm and 160μm. Among them, nominal diameter less than 5.00 mm of rock particles is known as sand, nominal diameter greater than 5.00 mm of rock particles is known as stone. The degree of fineness modulus was calculated according to Eq. 2:

$$M_f=\left(\sum_{i=1}^{n}\beta_i\right)/100 \quad (2)$$

Where M_f is the fineness modulus, and n is the number of test square holes sieves used, and β_i is the cumulative passing sieve residue of stage sieves.

4 Simplified optimal combination of aggregates

Concrete works are often poured in the way of pumping, and the slump of concrete is required to be 50mm~150mm. In addition, gravel particles usually are used to produce concrete. Therefore, the simplified numerical method of the optimal combination of aggregate particles can be applied according to Bolomey equation.

4.1 Best combination of sand

At present, the natural sand is widely used in engineering, which is preferable to that $A's$ value is 10, and Bolomey equation could be simplified as Eq. 3:

$$P_{pd}=10+\sqrt{1620d}\ (\%) \quad (3)$$

Accordingly, the particle size distribution of the ideal natural sand particles could be calculated. Especially, the cumulative percent passing on sieve size 630μm was 58.1%. According to Standard for technical requirements and test method of sand and crushed stone (or gravel) for ordinary concrete (JGJ 52), the natural sand having ideal particle size distribution belonged to the Ⅱ-area sand used in concrete preferentially. Furthermore, the fineness modulus of ideal natural sand particles obtained based on the minimum void volume was 2.71, could be classified into medium sand.

With the importance of environmental protection and the increasing shortage of resources, the application of artificial or machine-made sand had become more and more. As for the artificial or machine-made sand, it was preferable to that $A's$ value is 12, and Bolomey equation could be simplified as Eq. 4:

$$P_{pd}=12+\sqrt{1548.8d}\ (\%) \quad (4)$$

Accordingly, the particle size distribution of the ideal artificial or machine-made sand particles could be calculated. Particularly, the cumulative percent passing on sieve size 630μm was 56.8%. According to Standard for technical requirements and test method of sand and crushed stone (or gravel) for ordinary concrete (JGJ 52), the artificial or machine-made sand having ideal particle size distribution also belonged to the Ⅱ-area sand used in concrete preferentially. Moreover, the fineness modulus of ideal artificial or machine-made sand particles obtained based on the minimum void volume was 2.65, could be also classified into medium sand.

For two sand of different fineness modulus, the optimum fineness modulus should be close to the ideal sand particle based on the minimum space volume. Through the linear relationship, the combination ratio could be calculated as Eq.5. For three or more types of sand, it also could be obtained by this analogy.

$$S1:S2=\left(\left|M_{fs2}\text{-}M_{fs}\right|\right):\left(\left|M_{fs}\text{-}M_{fs1}\right|\right) \quad (5)$$

Where M_{fs} is the fineness modulus of the ideal sand, the value for natural sand is 2.71, the value for artificial or machine-made sand is 2.65, and M_{fsi} is the fineness modulus of sand, and Si is the mass ratio of sand.

4.2 Best combination of crushed stone

Taking the value of A as 12, the ideal particle size distribution could be calculated. Specially,, the cumulative percent passing of continuous grading crushed stone on each sieve size was consistent basically with the relevant quality requirements of Standard for technical requirements and test method of sand and crushed stone (or gravel) for ordinary concrete (JGJ 52).Taking this one step further, the fineness modulus of ideal crushed stone particles obtained based on the minimum void volume was shown in Table 2.

Table 2. Fineness modulus of ideal crushed stone.

D_{max} /mm	10	16	20	25	31.5	40
M_{fg}	6.0	6.18	6.35	6.59	6.91	7.30

For two crushed stone of different fineness modulus, the optimum fineness modulus should be close to the ideal crushed stone particle based on the minimum space volume. Through the linear relationship, the combination ratio could be calculated as Eq.6. For three or more types of crushed stone, it also could be obtained by this analogy.

$$G1:G2=\left(\left|M_{fg2}\text{-}M_{fg}\right|\right):\left(\left|M_{fg}\text{-}M_{fg1}\right|\right) \quad (6)$$

Where M_{fg} is the fineness modulus of the ideal crushed stone, and M_{fgi} is the fineness modulus of two types of crushed stone respectively, and Gi is the mass ratio of sand.

4.3 Best combination of sand and crushed stone

Similarly, according to the simplified Bolomey equation, based on the minimum void volume mechanism, the fineness modulus of ideal sand and crushed stone particles was calculated and shown in Table 3, Fig. 1 and Fig. 2.

Table 3. Fineness modulus of ideal sand and crushed stone.

D_{max} /mm	natural sand	artificial or machine-made sand
10	3.49	3.42
16	4.07	3.99
20	4.40	4.32
25	4.78	4.69
31.5	5.23	5.14
40	5.74	5.65

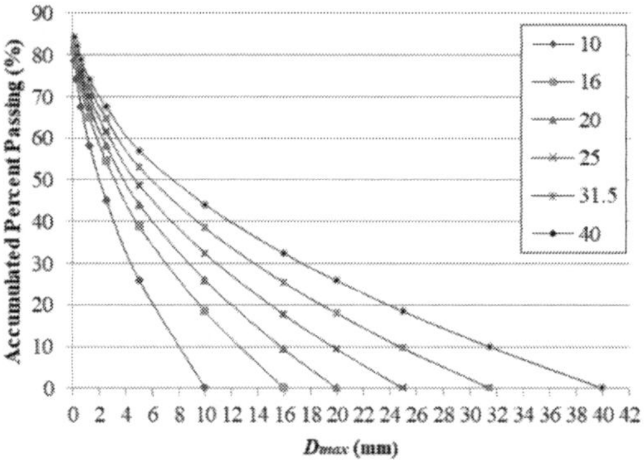

Fig. 1. Sieve analysis results for natural sand and crushed stone.

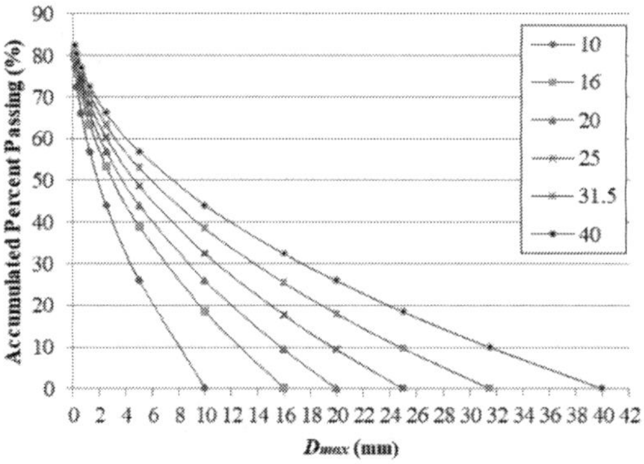

Fig. 2. Sieve analysis results for artificial or machine-made sand and crushed stone

For the optimal combination of sand and crushed stone, whose fineness modulus should be close to the ideal sand and crushed stone. According to the linear relation of the fineness modulus, the combination proportion could be calculated as Eq.7.

$$S:G=\left(\left|M_{fg}\text{-}M_f\right|\right):\left(\left|M_f\text{-}M_{fs}\right|\right) \tag{7}$$

Where S is the mass ratio of sand, and G is the mass ratio of crushed stone, M_{fs} is the fineness modulus of sand with optimal self-combination first, and M_{fg} is the fineness modulus of crushed stone with optimal self-combination first.

Moreover, if the slump of concrete was required to be 50mm~150mm, combining optimally sand with crushed stone for concrete mixture could approach to the combination proportion shown in Table 4.

Table 4. Fineness modulus of ideal crushed stone.

D_{max} /mm	natural sand	artificial or machine-made sand
10	3.22	2.99
16	1.55	1.49
20	1.15	1.11
25	0.87	0.85
31.5	0.67	0.65
40	0.56	0.55

5 Application

Producing the concrete with strength grade of C70, the materials used were as follows: Ordinary Portland Cement of 52.5 grading (P•O 52.5R) according with GB/T 21372 and GB 175 on chemical composition, mineral composition and physical properties, natural river sand with fineness modulus of 2.8 and bulk density of 2.6g/cm², continuous grading crushed stone with fineness modulus of 6.6 and maximum particle size of 31.5mm, non-retarding Polycarboxylates High Performance water-reducing admixture potable water.

The concrete mixtures prepared for use in this study were described in Table 5. The cement content of 460kg/m³, w/c of 0.25 and water-reducer's dosage of 2% remained constant between two mixtures. For the MD-2, $S:G$ is 0.62 through the optimal combination ways in this study.

Table 5. Concrete mixture design.

Mixture designation	Cement content (kg/m³)	Sand (kg/m³)	Crushed stone (kg/m³)	w/c
MD-1	460	855	970	0.25
MD-2	460	700	1130	0.25

Note: In the naming convention MD is "mixture designation".

Nine groups of separate concrete mixtures were cast for each combination and their compressive strength was measured, shown in Fig. 3. The MD-1 mixture exhibited 78.7Mpa of maximum compressive strength at day 28. However, the MD-2 mixture gained 90.6Mpa of minimum compressive strength at day 28, increasing strength about 20% compared to the MD-1 mixture. Obviously, its durability would be better for good compactness.

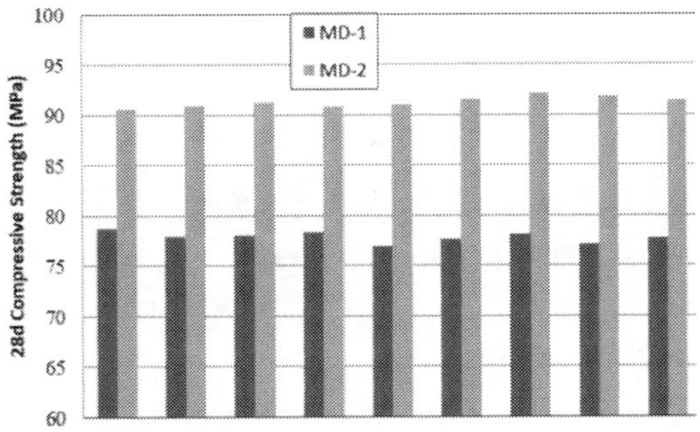

Fig. 3. Compressive strength through 28 days

6 Conclusions

Sand or/and crushed stone of any fineness could be combined into ideal graded particles by formula (5-7), which could not only meet the standard requirements, but also overcome the market supply situation that their quality does not good.

Because of well-graded aggregates in the concrete mixture, the fluctuation range of its test mixture ratio was reduced effectively, and its design mixture ratio was more easily determined.

Making full use of the particle gradation effect of concrete aggregates could effectively improve its strength and durability, could also observe high guarantee rate.

Acknowledge

Assisted by the Project of Shaanxi provincial department of education project (16JK1374).

References

1. W. B. Fuller, S. E. Thompson, JTD, ASCE, **59**, 1:67-143 (1907)
2. Bolomey J., *Bulletin*, ASCE, 16:22-24(1927)
3. C. A. G. Weymouth, Rock Products, 2:26(1933)
4. Chongxi L.,J.H.E., 2:59-66(1964). (In Chinese)
5. Chongxi L., The Changjiang River, 12:34-38(1956). (In Chinese)
6. Feng R., *Academic Dissertation*, D.L.U.T., 2007. (In Chinese)
7. Qun Y., Zhongyin G., Liping C., J.B.M., **9**, 4:419-422(2006). (In Chinese)
8. Qianyuan W., Jingshuang H., Rock and Soil Mechanics, **18**, 3:93-100(1997). (In Chinese)
9. Qianyuan W., Qing Z., J.G.M.E., **13**, 2:109-117(1994). (In Chinese)
10. Qianyuan W., Qing Z., J.I.A.E. of Qingdao, **15**, 2:61-68(1994). (In Chinese)
11. Fan W., *Academic Dissertation*, X.A.U.A.T., 2014. (In Chinese)

Relationship between Anti-frozen Durability and Pore Structure on Concrete in Freeze and Thaw Cycle Environment

Chao Wang[1,*], *Zhao-Lei* Liu[2], *Lei* Yu[2]

[1]Guangdong Road and Bridge Construction Development Co. LTD. No.445, Tian Run Road, Tian-he District, Guangzhou City, Guangdong Province, China.
[2]Beijing Xin-Qiao Technology Development Co. LTD. No.8 yard, Xi Tu Cheng Road, Hai-dian District, Beijing, China

Abstract. To detect which micro structure index can make an effected on the anti-frozen durability on concrete, 3 kinds of concrete with different mix proportion have been made. The dynamic modulus of elasticity and pore structure index were detected and the relativity of them was analysed. The result shows that the dynamic modulus of elasticity decreases after freezing and wary with porosity. Critical pore size becomes larger. The gel pore amount decreases. The capillary and macro pore amount increase. There is a good relativity between frost-resistance performance and pore parameters. By decreasing porosity and the amount of macro pore, increasing gel pore amount, the anti-frozen durability can be enhanced.

1 Introduction

According to China Highway Natural Division [1], two thirds areas of China locate on places where the freeze and thaw situations appear sometimes. The concrete structure in these areas can be affected by freeze and thaw cycle [2]. This can lead to concrete surface peeling off or become unconsolidated [3].

There are some papers to find the mechanism of concrete frost heave and conclude it as the inducement of concrete macro performance decrease[4][5]. But there is rarely study on how to estimate. There is lack of method and result to estimate the frost heave. But, there should be a relationship between macro and micro performance [6].

Therefore, finding the quantization relationship to estimate the degree of macro performance decrease by micro structure change is necessary. So that, the macro performance of concrete can be changed by optimizing the pore structure.

2 Test plan

2.1 Raw materials

The raw materials include the P.O 42.5 cement, fly ash I, nature sand with medium sand, 5-15mm coarse aggregate, tap water, polycarboxylic acid high performance water reducer, rosin air conditioner.

2.2 Mix proportion

The mix proportion of 3 kinds of concrete are shown as table 1.

Table 1. Mix proportion of concrete kg/m^3

ID	Cement	Fly ash	Water	Sand	Stone
1	542	0	220	573	1 065
2	542	0	220	573	1 065
3	461	81	220	573	1 065

The air content of No.1 concrete is the nature air content without any rosin air conditioner. No.2 and No.3 concrete have the same air content which is adjusted by rosin air conditioner. The slumps of these 3 kinds of concrete are all (10±3)cm.

2.3 Specimens

For each mix proportion, 3 specimens which size is 100mmx100mmx400mm has been made for anti-frozen test. The control group specimens for anti-frozen test are also 3 pieces which will be kept in curing room until the treatment group goes to test date.

For micro pore structure test, 3 specimens which size is 100mmx100mmx100mm need to be made. They will be put into the freeze and thaw machine until the recycle is finished. Then crush them and detect its pore structure parameters. There are still 3 cubic specimens which will be kept in the curing room. After recycle finished, they will be crushed as the treatment group and test their pore structure.

2.4 Test methods

Put the harden concretes whose mould have been removed after 24h into the standard curing room whose temperature is 20 ℃ ±2 ℃ and relative humidity is 95%±3%. After 27d curing, put the treatment specimen into fast freeze and thaw test machine. By reference, the fast freeze and thaw cycle test is done as the Standard for Test Methods of Long-term Performance and Durability of Ordinary Concrete [7]. Every 25 times recycle, take the specimen out from freeze-thaw test machine and test the dynamic modulus of elasticity. The control group specimen who is kept in the standard curing room will be test on the same date of treatment group.

After finishing the freeze-thaw recycle, take the cubic specimen out and crush by jaw crusher. Griddle these crushed sand by 4.74mm and 2.36mm square hole sieve. Select 10 gram particle on the 2.36mm sieve. Use ultrasonic washing unit to clean these particle for 5 minute. Put them into oven and dry 24h with a temperature of 105℃±5℃.Take them out and cool at nature environment. And then test the pore structure parameters with Mercury Injection Apparatus.

3 Test result and date analysis

The dynamic modulus of elasticity of concrete have detected after every 25 recycles. Micro pore structure data including porosity, average pore size, critical pore size, fractal dimension, percentage composition of pore with different range of size and pore size distribution are all obtained.

3.1 Analysis of dynamic modulus of elasticity

From Fig. 1 to Fig. 4 shows the changing of dynamic modulus of elasticity for 3 kinds of concrete.

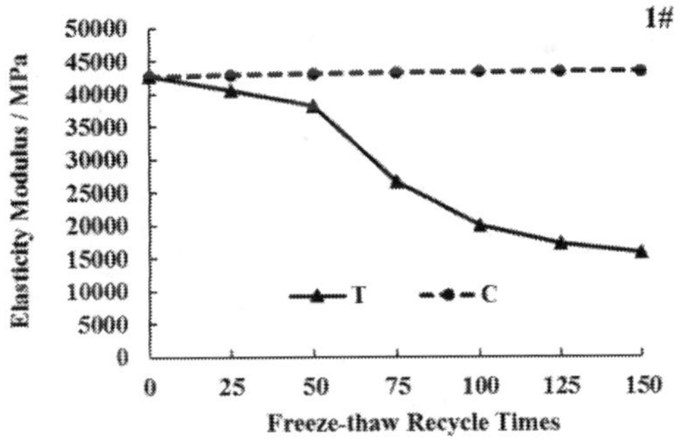

Fig.1. Elasticity modulus VS freeze-thaw recycle times for 1# Concrete

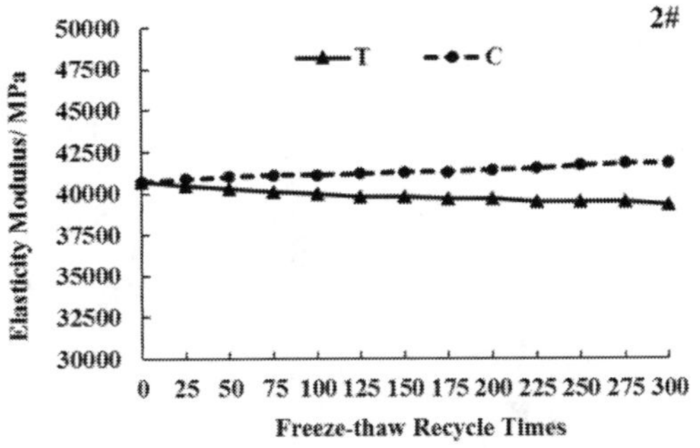

Fig.2. Elasticity modulus VS freeze-thaw recycle times for 2# Concrete

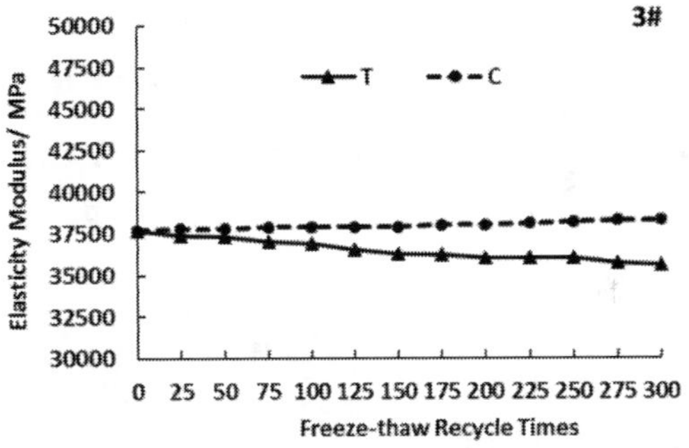

Fig.3. Elasticity modulus VS freeze-thaw recycle times for 3# Concrete

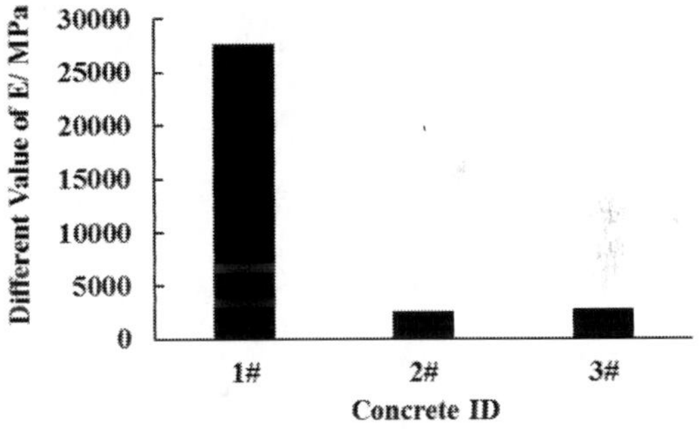

Fig.4. Different value of elasticity modulus at the end of test for 3 kinds of concrete

From Fig. 1, the elasticity modulus of 1# concrete decrease heavily. Only 150 recycles can make 1# concrete destroyed. 2# and 3# concrete don't be destroyed after 300 recycles. So 2# and 3# concrete are better than 1# concrete on anti-frozen durability.

On the second, 2# concrete with air entraining agent whose elasticity modulus is 40.7GPa is larger than 1# concrete without any air entraining agent. 3# concrete with fly ash whose modulus is 37.7GPa is a little lower than 2# concrete. From Fig. 4, the different value of modulus between 2# and 3# concretes are similar. Therefore, the air entraining is the main factor to enhance concrete anti-frozen durability.

3.2 Analysis of pore structure parameters

From Fig. 5 to Fig. 8 shoows the change of pore structure parameters before and after freeze-thaw recycle.

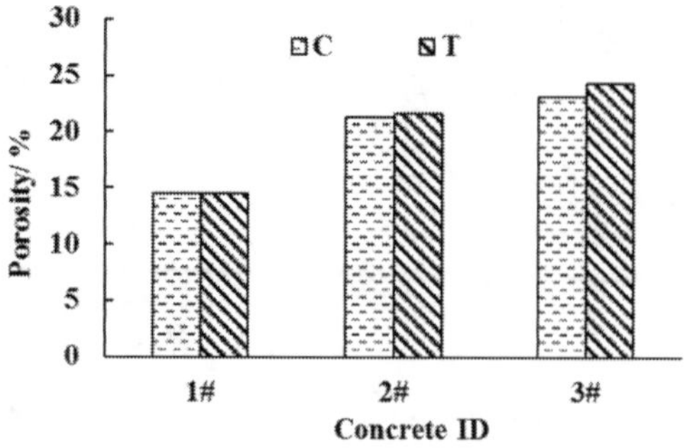

Fig.5. Porosity of different concrete

Porosity is an index to estimate the compactness of concrete. The larger the porosity, the lower the density. Fig. 5 shows that the porosity becomes larger after freeze and thaw recycle. It means freeze and thaw recycle make concrete more porosity.

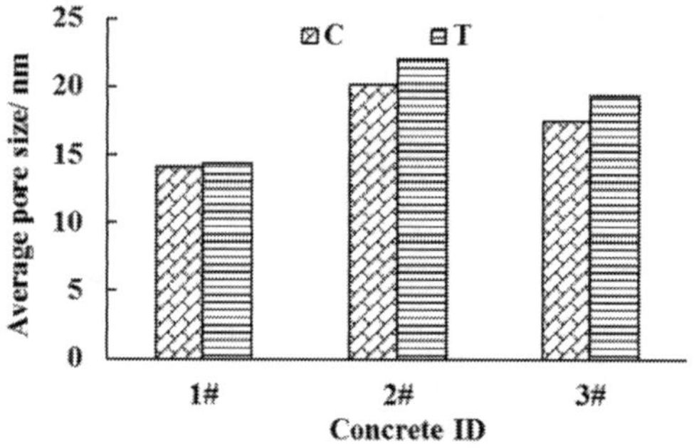

Fig.6. Average pore size of different concrete

Average pore size can estimate the pore size generally. The larger Average pore size, the worse the concrete. From Fig. 6, it can be seen freeze and thaw recycle make concrete worse.

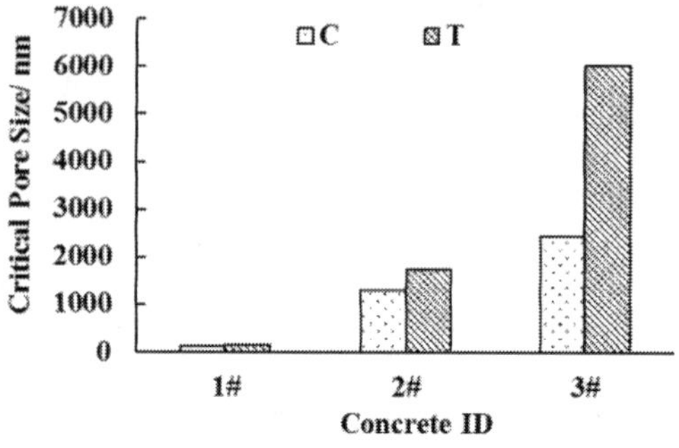

Fig.7. Critical pore size of different concrete

Critical pore can make all range of pore penetrated. The larger the critical pore, the worse the permeability. Fig. 7 shows that freeze and thaw recycle can make critical pore larger. It is harmful to the permeability of concrete.

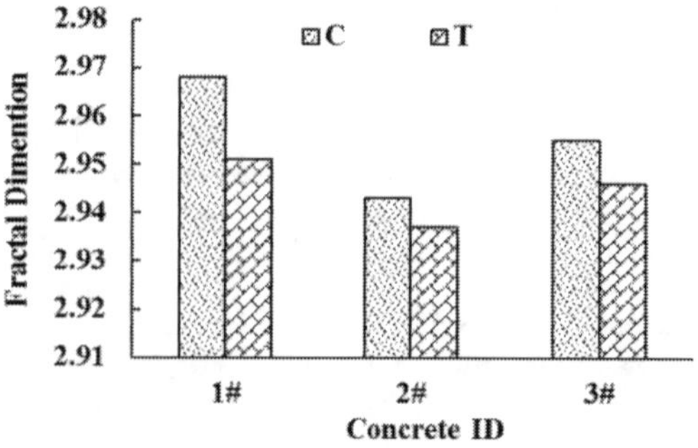

Fig.8. Fractal dimension of different concrete

Fractal dimension can describe the complex of the pore structure of concrete. The larger the fractal dimension, the more complex the outline of pore. Fig. 8 shows that freeze and thaw recycle can make this value smaller. It means that freeze and thaw recycle can make the outline of pore simple.

3.3 Analysis of pore grading

Pore can be divided into 4 grading according to I.O Butte. They include gel pore whose pore size is smaller than 10nm, transitional pore whose pore size is among 10-100nm, capillary pore whose pore size is among 100-1000nm and macro pore whose pore size is larger than 1000nm. Actually, capillary pore and macro pore are harmful to concrete. From Fig. 9 to Fig. 12 shows the difference of these pores before and after freeze and thaw recycle.

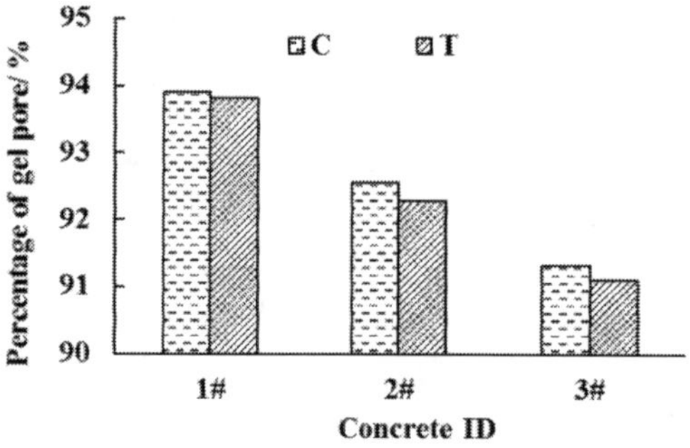

Fig.9. Percentage of gel pore of different concrete

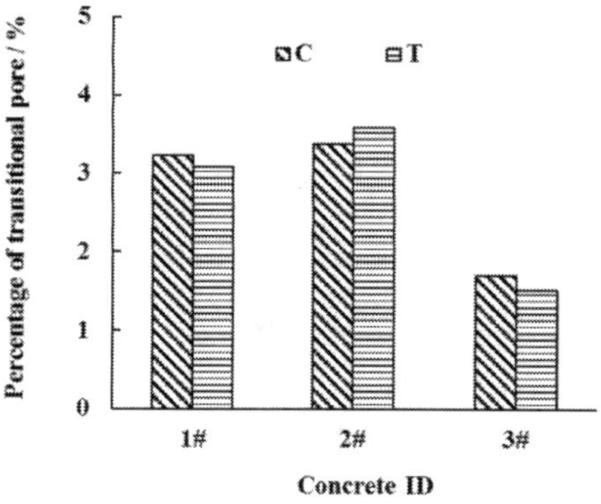

Fig.10. Percentage of transitional pore of different concrete

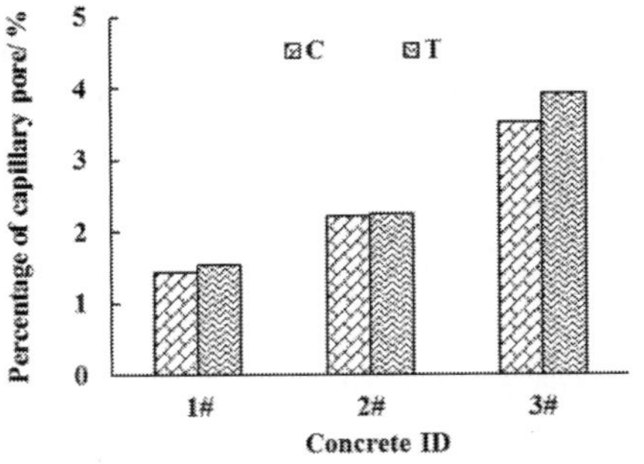

Fig.11. Percentage of capillary pore of different concrete

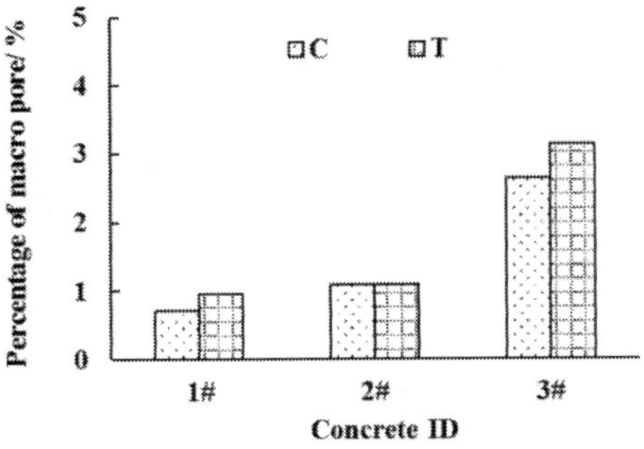

Fig.12. Percentage of macro pore of different concrete

From Fig. 9, Fig. 11 and Fig. 12, freeze and thaw recycle decrease the amount of gel pore but increase the amount of capillary and macro pore. These changing are harmful to concrete.

Fig. 10 shows that the amount of transitional pore of 2# concrete becomes larger after freeze and thaw recycle. The amount of transitional pore of 1# and 3# concrete are fewer than before. It is because that 3# concrete has fly ash in it and is denser than 2# concrete.

Therefore, 2# and 3# concrete are better than 1# because the air entrained and fly ash addition.

3.4 Pore distribution

Fig. 13 is the pore distribution curve. It shows that the curves of 1# concrete move towards the macro pore size direction.

For 2# and 3# concrete, pore distribution curve move towards micro pore size direction.

From pore distribution figure, it can be seen that the amount of pores in 2# and 3# concrete whose size is among 100nm-1000nm become larger than the ones which haven't air entrained.

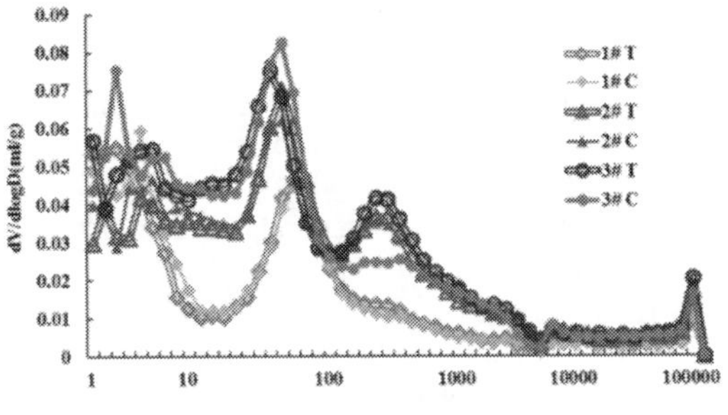

Fig.13. Pore size distribution

3.5 Relationship between Elasticity modulus and pore structure

The trends of changing of elasticity modulus and pore structure indexes are listed as table 2.

Table 2. List of macro and micro indexes after frozen

Index	Changing trend	Relativity with E
Modulus	↓	/
Porosity	↑	negative
Average pore size	↑	negative
Critical pore size	↑	negative
Fraction dimension	↓	positive
Gel pore content	↓	positive
Capillary pore content	↑	negative
Macro pore content	↑	negative

Elasticity modulus can reflect the anti-frozen durability of concrete and can be detected without destroy the specimen.

From table 2, in order to enhance the anti-frozen durability of concrete, decrease the porosity, average pore size, average pore size and critical pore size or you can increase the friction dimension of concrete. What's more, increase the content of gel pore and decrease the capillary and macro pores.

4 Conclusions

To detect which micro structure index can make an effected on the anti-frozen durability on concrete, 3 kinds of concrete with different mix proportion have been made. The dynamic modulus of elasticity and pore structure index were detected and the relativity of them was analysed. The result can be concluded as follow.

(1) Freeze and thaw recycle can make concrete modulus decrease especially for concrete without air entrained. For air entrained concrete, the anti-frozen durability is much better than ones without air entrained.

(2) Porosity and average pore size of concrete become larger after frozen. The concrete will become less dense than before. The critical pore size will become larger so that the anti-permeability becomes worse than before. Fraction dimension become less so that the outline of pore become simple.

(3) Freeze and thaw recycle decrease the percentage of gel pore; increase the percentage of capillary pore and macro pore. The amount of macro pores is more and micro pores is less. The denser the concrete, the less the transition pores after frozen.
(4) After freeze and thaw recycle, pore distribution curve move towards macro pore direction. Air entrained can enlarge the amount of pores among 100nm to 1000nm.
(5) By decreasing the porosity, average pore size and critical pore size, increasing fraction dimension, the anti-frozen durability of concrete can be enhanced. By increasing the content of gel pore and decreasing the content of capillary and macro pore, the anti-frozen durability of concrete can also be enhanced. China Highway Natural Division

Acknowledgement

This work was supported by the Basic Research Found for National Public Welfare Scientific Research Institutions [grant numbers 2017-9004].

References

1. JTJ 003-1986. China Highway Natural Division. 1987-1-1 implement. (In Chinese)
2. W.ping Yang. *Effect of inorganic salts on pore structure and frost resistance of concrete.* (Doctor Degree of Harbin University of Technology,2009) (In Chinese)
3. S.Ping Zhang, M. Deng, M.Hu TANG. Journal of Materials Science and Engineering. **26**,5(2008). (In Chinese)
4. R.Mu, N.Jiao, F.Fen Jiao. Concrete.**7**,4(2016). (In Chinese)
5. Valenza J J, Scherer G W. Cement and Concrete Research. **37**,14(2007): 1007-1021.
6. L. Yu. *Study on durability of cement concrete in multi erosion.* (Doctor Degree of Beijing University of Technology, 2013) (In Chinese)
7. GB/T 50082-2009. *Standard for Test Methods of Long-term Performance and Durability of Ordinary Concrete.* (In Chinese)

Investigation of Chemical Cleaning of Supercritical Superheater Oxide Scale

Hongfeng Li[1], *Qingtang* Xue[1], *Xinhui* Nie[1], and *Yunfei* Xu[2,*]

[1]Guodian Science and Technology Research Institute, Nanjing, Jiangsu 210023, P.R.China
[2] Department of Environment Science and Engineering, North China Electric Power University, No.689, Yonghua North Road, Baoding, Hebei Province, 071000, P.R.China

Abstract. Recent years, as the development of thermal power plant, superheater oxide film caused by over temperature and tube explosion and the turbine blade erosion problem more and more serious. The research and development of chemical cleaning technology of superheater oxide film is one of the most effective ways to solve this problem. The chemical cleaning of superheater is rather difficult than that of boiler, for mainly the reason as below. First, the scale of superheater tube, differences in microstructure, scales dense and need to use higher cleaning medium concentration and longer cleaning time. Second, the materials of superheater complex, involves ferritic steel and austenitic stainless steel such as 12Cr1MoV, T22, T91, TP347 and SUS304. For controlling chemical cleaning process, the acid corrosion sample is employed to slow inhibitor on different metal and alloys. Also, the traditional weight-loss method to evaluate the corrosion rate of materials in chemical cleaning process is discussed to monitor the corrosion process.

1 Background

The thermal power generation enterprises have long been plagued by the problem of superheater oxide scale. When the scale in the superheater tube grows to a certain extent, the thermal expansion coefficient between the scale and the superheater metal matrix is different, and when the unit starts and stops or the loaded changes. During the change process, the scale will fall off due to the change of the main steam temperature, and the falling scale may block the superheater tube and cause local over-temperature burst or turbine blade erosion [1-3]. Chemical cleaning is usually used to improve the heat transfer efficiency of the equipment, slow down the formation and shedding of the superheater scale, reduce the long-term creep damage of the pipeline and the erosion of the turbine[4-5].

The chemical cleaning technology can effectively remove the oxide scale on the surface of the superheater tube, and avoid the superheater tube burst accident caused by the falling off of the scale during the long service period. In the "Guidelines for Chemical Cleaning of Boilers in Thermal Power Plants" (DL/T794, China), the guiding principle of "chemical cleaning" can be carried out for the chemical cleaning of superheaters, "the amount of scale exceeds 400g/m^2, or when the explosion occurs due to oxide scale falling off." There are no specific rules for cleaning conditions.

2 Thickness of oxide scale detection technology

At present, the commonly used oxide scale detection methods include radiation detection, magnetic detection, high-frequency ultrasonic detection, etc. [6-8]. Among them, magnetic detection technology and high-frequency ultrasonic detection methods are more common. (1) The radiation detection method is based on the principle that x-ray or γ-ray is different in intensity attenuation when irradiating the metal substrate and the oxide, similar to the chest piece for medical diagnosis, the scale is intuitive and easy to observe with low cost and operation. Limited to the field environment, and because the scale is an epitaxial film grown on a metal substrate, the resolution is low. (2) The high-frequency ultrasonic detection method is based on the principle of different density and acoustic impedance between the components of the two-phase material at the steel and oxide scale interface. The operation is convenient with highly efficient and the precision can reach 0.01mm, also the thickness of the inner wall of the tube can be measured. T91 tubing commonly used in superheaters can be used for scale testing. (3) The magnetic detection method is based on the different principles of ferromagnetism of austenitic steel and its inner wall oxide scale (mostly different valence oxides of iron). This method can measure the accumulation of scale in the tube with flexible and efficient, but it is only suitable for austenitic stainless steel tubes. The scale non-cutting detection technology of the superheater tube needs further study

In addition to non-destructive testing of superheater technique, the cut-tube inspection is often considered a more accurate method of obtaining oxide scale thickness. For the as discussed method to obtain the scale thickness,

* Corresponding author: lucifertan@163.com

https://doi.org/10.1051/matecconf/201823802011

it is necessary to define "the peeling critical thickness of the scale" and "the peeling average thickness of the scale". The former is the minimum thickness at which scale peeling occurs. When the superheater and the reheater tubes reach the critical thickness of the scale peeling, there is a risk of the scale falling off during the shutdown process of power plant. The latter refers to the statistical average thickness of the scale of the scale-shedding unit. When the superheater and the reheater tube reach the average thickness of the scale peeling, the risk of scale peeling during the shutdown process is high.

By counting the scale peeling of superheater tubes such as 12CrMoV, T23, T91 and TP347, the critical thickness of scale peeling of various materials was obtained, as shown in the following table. For low-chromium ferritic steel of superheater, the inner wall scale peeling tendency is relatively low, and the critical thickness is 0.22 mm; for T91 steel, the inner wall scale peeling tendency is high, and the critical thickness of peeling is 0.16 mm; For the stainless steel, the inner wall scale peeling tendency is greater, and the critical thickness of peeling is 0.07 mm. When the thickness of the oxide of different materials exceeds the critical thickness, the supervision of the peeling of these materials shall be strengthened, and the chemical cleaning shall be considered to remove the scale[9].

Table.1. Critical thickness of scale peeling of various inner walls of superheater steel (mm)

Materials	the peeling critical thickness of the scale	the peeling average thickness of the scale
12CrMoV	0.22	0.30
T23	0.17	0.21
T91	0.16	0.21
TP347	0.07	0.092

3 Research of cleaning process of the over heater and reheater

The glycolic acid-formic acid composite organic acid cleaning process has the advantages of high descaling efficiency, low corrosion rate, low chloride ion content, easy disposal of waste liquid, and is a safe and efficient cleaning medium. According to the characteristics of the scale component, a glycolic acid-formic acid composite organic acid was used as the cleaning medium for the chemical cleaning of superheater oxide scale.

The scope of this chemical cleaning includes: part of the water supply pipeline, the economizer and its header, the water wall and its header, steam drum and down pipe.

The cleaning is carried out by using 2~4% glycolic acid + 1~2% formic acid, adding special organic acid corrosion inhibitor and related additives in the cleaning solution, using citric acid rinsing and hydrogen peroxide passivation. The specific cleaning process parameters are as follows.

The pickling process parameters of the composite acid concentration of 4~6%, the corrosion inhibitor concentration of 0.3~0.5%, the concentration of the cosolvent 0.05~0.1%, cleaning temperature 80~90°C, cleaning time is expected 8~12h, the specific time is from the pickling curve.

The rinsing process parameter rinsing liquid is 0.2~0.5% citric acid, pH 3.5~4.0 (ammonia water adjustment), temperature 50~70°C with rinsing time 2~4h.

Passivation process parameter passivation solution is rinsing liquid + ammonia water + 0.1~0.2% hydrogen peroxide, pH 9.5~10.0 (ammonia water adjustment), temperature 50~55°C, passivation time 2~4h.

After the cleaning, the water-cooled wall was cut and examined. As a result, the deposit on the inner surface of the tube was completely cleaned, and the metal substrate was exposed. The passivation film was intact and evenly dark gray.

In this chemical cleaning, the average corrosion of the corrosion indicator sheet is 16.2g/m2, and the average corrosion rate is 1.01g/(m2·h), which is much smaller than the value specified in DL/T794-2001" Guidelines for Chemical Cleaning of Thermal Power Plant Boilers". Use a two-column format, and set the spacing between the columns at 8 mm. Do not add any page numbers.

4 Monitoring of ferrous ions during chemical cleaning

In boiler chemical cleaning, there are two ways to enter the iron ions in the wash solution. The medium in contact with the inner surface of the boiler metal and its surroundings forms a variety of ferrites under long-term chemical and electrochemical action. Compounds, which are intercalated in metal surface deposits (scales), are dissolved in the cleaning solution along with surface deposits during pickling. The reaction mechanism is as follows[10]:

$$Fe_2O_3+6H^+\rightarrow 2Fe^{3+}+3H2O$$

$$Fe_2O_3FeO+8H^+\rightarrow 2Fe^{3+}+Fe^{2+}+4H_2O$$

$$FeOOH+3H^+\rightarrow Fe^{3+}+2H_2O$$

$$FeO+2H^+\rightarrow Fe^{2+}+H_2O$$

The Fe^{3+} produced by the above reaction passes through the "shield" of the corrosion inhibitor, participates in the depolarizing corrosion of Fe, and produces 1.5 times Fe^{2+} than Fe^{3+}. As the concentration of Fe^{3+} in the cleaning solution continues to increase, this corrosion is also becoming more and more serious.

During the cleaning process, the corrosion resistance of the corrosion inhibitor is limited or insufficient, and other reasons cause the boiler metal to be inevitably corroded to varying degrees. This is mainly due to the depolarizing corrosion of H^+ to Fe:

$$Fe+2H^+\rightarrow Fe^{2+}+H_2$$

In the above two pathways, the iron ions produced by the first cause mainly appear in the high oxidation state of Fe^{3+}, which can also be explained by the formation of metal surface deposits which need to undergo a longer oxidation reaction process, while iron The valence electron configuration 3d64s2 always has a strong tendency to transition to its outer electron 3d5 half-filled stable structure. The iron ion produced by the second reason is generally present in the low oxidation state of Fe^{2+}. This is advantageous for accurately determining the source and quantity of Fe^{2+} in the cleaning solution.

When the boiler is chemically cleaned, the acid concentration, Cu^{2+}, Fe^{3+} and Fe^{2+} concentration are generally monitored during the cleaning process. When the concentration of Fe^{3+} exceeds a certain value, an appropriate amount of reducing agent is added to the cleaning liquid to reduce the concentration thereof, and the corrosion of the boiler metal is slowed down. Usually the Fe^{2+} is not given enough attention. The significance of monitoring Fe^{2+} is only when the cleaning liquid is used. When the concentration and the content of Fe^{2+} are not significantly changed, the pickling can be ended. There is no corresponding indicator monitoring requirement for the change of Fe^{2+} concentration, and emergency measures should be taken when the concentration change of Fe^{2+} is abnormal. This gives people an impression that there seems to be no direct causal relationship between the change in the concentration of Fe^{2+} in the cleaning solution and the success or failure of a boiler cleaning. This is a problem that needs to be deepened. It is well known that in order to investigate the degree of corrosion of boiler metal in cleaning, it is required to carry out the coupon test in synchronization, and to infer the degree of corrosion of the metal in the cleaning according to the weight loss value of the coupon after cleaning. Although this method has certain reference value, it is fundamentally incapable of guiding the safe cleaning of the boiler in a timely and correct manner, especially when there is an abnormality in the process. For small screening tests for corrosion inhibition efficiency before cleaning, sometimes the results of the test are inconsistent with the results of actual scale cleaning. This is because modern engineering technology involves many variables, sometimes due to one of the factors. Accidental omissions can lead to unpredictable results.

During the pickling process, the content of Fe^{2+} is gradually increased. However, if the content of Fe^{2+} continues to rise rapidly, the acid washing should be stopped urgently. There are two reasons for the increase of Fe^{2+} content. First, Fe^{2+} in the cleaning solution comes from the scaly. In the middle, the second is the result of the strong action of H^+, Fe^{3+} and H^++Fe^{3+} on Fe. There are two ways to deal with this situation. One method is that as long as the Fe^{3+} content is not seriously exceeded, it is not necessary to pay attention to Fe^{2+}. One method is to stop cleaning, and the first method is generally used, and the Fe^{3+} content is used as a monitoring indicator for boiler cleaning. It has been found through practice that high content of Fe^{3+} is important, and sometimes low content of Fe^{3+} is detected because it is in a trough after a new round of Fe^{3+} to Fe^{2+} action, so it is easy to be ignored. In short, no matter where the source of corrosion comes from, its reaction products will be expressed in the form of Fe^{2+}, and if it is developed, it will endanger the safe operation of the boiler in the future, leaving a stable situation.

5 Treatment of cleaning waste solution

Citric acid is the most widely used organic acid in equipment cleaning. It is mainly used to clean oxide scale on the surface of equipment with high cost. However, the treatment of citric acid waste liquid is difficult. Domestic citric acid cleaning waste liquid is usually discharged to the coal yard and incinerated with coal burning. This method is simple and easy, but may cause boiler water wall coking and burner corrosion. Therefore, its use has been restricted in recent years[11].

The pollutants in the pickling waste liquid mainly include the residual agent added during the pickling process and the copper and iron dissolved in the system. The treatment process is as follows: (1) Sampling: Sampling and analysing the acidity of the waste liquid, and calculating the amount of limestone required to neutralize the waste liquid. (2) Neutralization and aeration oxidation: the acid washing waste liquid is discharged to the waste liquid pool, the aeration device is aerated, the limestone is slowly added to the pH of the waste liquid is about 12, and the aeration is continued for 24 hours and then allowed to stand for more 12 hours. (3) Circulating filtration: start the waste pump, filter the waste liquid through the filter and discharge it back to the waste liquid pool. (4) Emissions: continue to filter until the indicators meet the requirements.

The main pollutant of the passivation liquid waste liquid is sodium nitrite. Since sodium nitrite will become a carcinogen nitrosamine under certain conditions, it must be converted into harmless substances. The treatment process of the passivation waste liquid is as follows: sodium hypochlorite oxidation, alkali neutralization, aeration, stirring degreasing and COD, standing, activated carbon + multi-media filter in addition to COD and colour, sampling and analysis of pH and TOC to pass and emission. (1) Sampling: Sampling and analysing the concentration of sodium nitrite in the waste liquid, and calculating the amount of sodium hypochlorite required. (2) Add sodium hypochlorite oxidation: Discharge the waste liquid to the waste liquid pool, slowly add sodium hypochlorite, turn on the aeration device to aerate, and mix the waste liquid with sodium hypochlorite. (3) Neutralization, aeration oxidation: slowly add limestone, adjust the pH of the waste liquid to 7.5-9, continue to aerate for 24h and then let stand for 12h. (4) Circulating filtration: start the waste liquid pump, filter the waste liquid through the filter and discharge it back to the waste liquid pool. (5) Emissions: continue to filter until the indicators meet the requirements and then effluent.

Rinse, that is, rinse after washing, is carried out after the end of the passivation process, which contains a small amount of passivation solution, and the treatment process is as follows: (1) Sampling: Sampling and analysing the pH of the waste liquid, and calculating the amount of sulfuric acid (or limestone) required accordingly. (2) Neutralization and aeration oxidation: the passivation waste liquid will be discharged to the waste liquid pool, the aeration device will be aerated, and 50% sulfuric acid (or limestone) will be slowly added until the pH of the waste liquid is between 7.5-9.

Continue to aerate for 24 hours and then let stand for more 12 hours. (3) Circulating filtration: start the waste liquid pump, filter the waste liquid through the filter and discharge it back to the waste liquid pool. (4) Emissions: continue to filter until the indicators meet the requirements and then effluent.

The treated pickling waste can be discharged to the outside after meeting the standard.

6 Conclusion

The problem of superheater scale has long plagued thermal power generation companies. When the scale grows to a certain extent, due to the large difference in thermal expansion coefficient between the scale and the superheater metal material, the scale peeling occurs due to the change of the main steam temperature during the start-stop and load-lifting process of the unit, which greatly increases Risk of superheater squibs and turbine blade erosion caused by scale. The chemical cleaning of the superheater is mainly to slow the formation and shedding of the scale, reduce the long-term creep damage of the pipeline and the erosion of the turbine. Research on chemical cleaning technology for superheater scales has received increasing attention. Choose the appropriate cleaning agent and additives, depending on the material of the superheater. Choosing the right process conditions and evaluating and monitoring the cleaning process will help to obtain safe and reliable superheater cleaning technology.

References

1. J Y Cao, J Chen, Electr. Power, **36** 20 (2003)
2. E. W. Delong. Corros., Houston, NACE, (1985)
3. J. W. Horvath, H. D. Johnson. TAPPI Press : Proceedings of the 1998 TAPPI International Engineering Conference, Miami, 505 (1998)
4. Q F Jiao, B Yao, Electr. Power, **36** 46 (2004)
5. L G Wu, J F Shen,H Z Fan, et al. Electr. Power, **46** 46 (2013)
6. C Y Wang, P Wang, Clean. World, **2** 21 (2011)
7. G Z Qin, Chemical Equipment and Anticorrosion, **6** 29 (2000)
8. C W Yin, H Q Yang, Journal of Xi'an Technological University, **2** 87 (2013)
9. J Y Cao, J X Song. Therm. Power Gener., **09** 29(2009)
10. G H Dai, Northeast Electric Power Technology, **2** 36(1998)
11. X L Zhang, Y G Wang, M X Sun,H Q Li. Corros. Sci. Prot. Technol., **24** 536 (2012)

Preparation and Property of Wireless Charging Magnetic Coupling Component Used Core-Shell Magnetic Nanoparticle

Kunming Qian[1,*], *Yansong* Zhang[1], *Jinghui* Huang[1], *Jie* Hao[1], *Song* Ji[1]

[1]Inner Mongolia Metals Institute, 199 Lingyun Road, Ningbo, 315103, China

Abstract. In order to improve the performance of magnetically coupled material, this paper illustrates the synthesis of Core-shell magnetic nanoparticle exploiting microwave plasma. Result shows that the core w is comprised of pure Fe and shell is comprised of ferric oxide .The radius and thickness of core is about 8nm and 2nm respectively. Obtained nanoparticle shows great dispersity. The result of mathematical model calculation which obtained from Core-shell magnetic particle is consistent with the experiment result.

1 Introduction

An urgent need for wireless charging promote magnetic material and magnetic component[1], because special doping Ferrite Magnetic material has great magnetic property, which is important to Magnetically coupled component. Research shows that magnetic nanoparticle doped Ferrite material has high Magnetic flux density and low high-frequency loss, which are suitable for wireless charge system. However, magnetic nanoparticle shows properties of reunion and oxidation, both do damage to stability. An efficient solution is that coating a dense, anti-oxidated, easy and stable dispersed shell on the surface of nanoparticle. The key points of exploring high-performance magnetic nanoparticle are synthesize Core-shell magnetic nanoparticle and stable dispersion, both are hot topics now[2-10].

This paper illustrates the research on the synthesis of Core-shell magnetic nanoparticle using microwave plasma. We are looking forward to synthesizing magnetic particle with good dispersity and build mathematical model of Core-shell magnetic nanoparticle which can illustrate their property.

2 Experiment

DT4 pure Fe is used as raw material; Low-temperature plasma nanoparticle generator, which take advantage of microwave plasma, is used in the process where nanoparticle is synthesized, so all the processes including production, capture, stable dispersion work under normal temperature, normal pressure, oxygen free and air tightness constantly.

Then, Core-shell magnetic nanoparticleis obtained after in-situ epitaxial growth of anti-oxidatedshell. D/max-2500/PC X ray diffractometer, TECNAIG2 transmission electron microscope, Quanta 250FEG scanning electron microscope is used in physical and chemical analysis.

3 Result and analysis

3.1 Microstructure

Fi.1 is TEM image of Core-shell magnetic nanoparticle, Fig.2 is EDS image.Fig.3 is TEM diffraction pattern of magnetic nanoparticle. Fig.4 is XRD spectrum of magnetic nanoparticle.

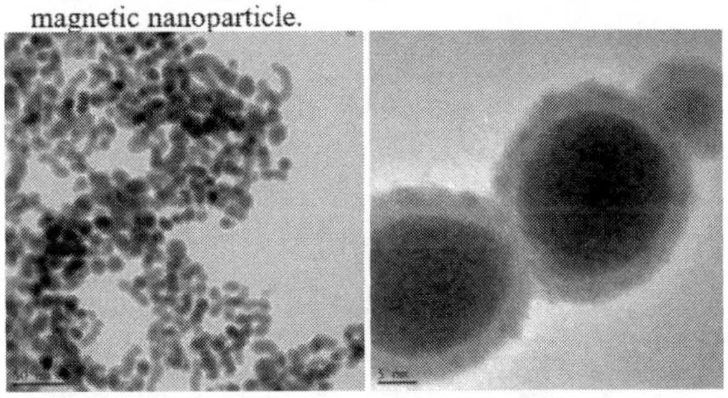

Fig.1. TEM image of Core-shell magnetic nanoparticles

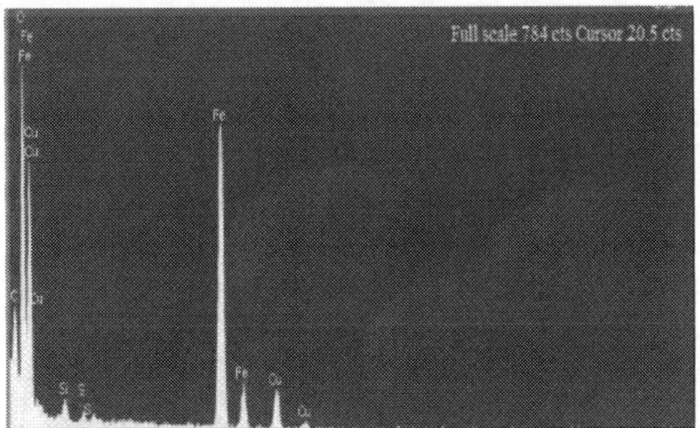

Fig.2. EDS image of Core-shell magnetic nanoparticles

From images we can see, the magnetic nanoparticles obtained from the mentioned process have uniform grain

---
* Corresponding author: qiankm@126.com

size. The nanoparticle, which consists of an inner dense center encapsulated inside a light colored shell, has grain size vary from 5nm to 10nm. After in-situ epitaxial growth process for magnetic nanoparticle, controllable in-situ growth of anti-oxidated shell is available on the surface of magnetic nanoparticle's crystal nucleus. The obtained Core-shell magnetic nanoparticle has great densification. Also, energy spectrum analysis shows that the shell is comprised of Fe3O4 and core is comprised of pure Fe.

Fig.3. TEM diffraction pattern of magnetic nanoparticle

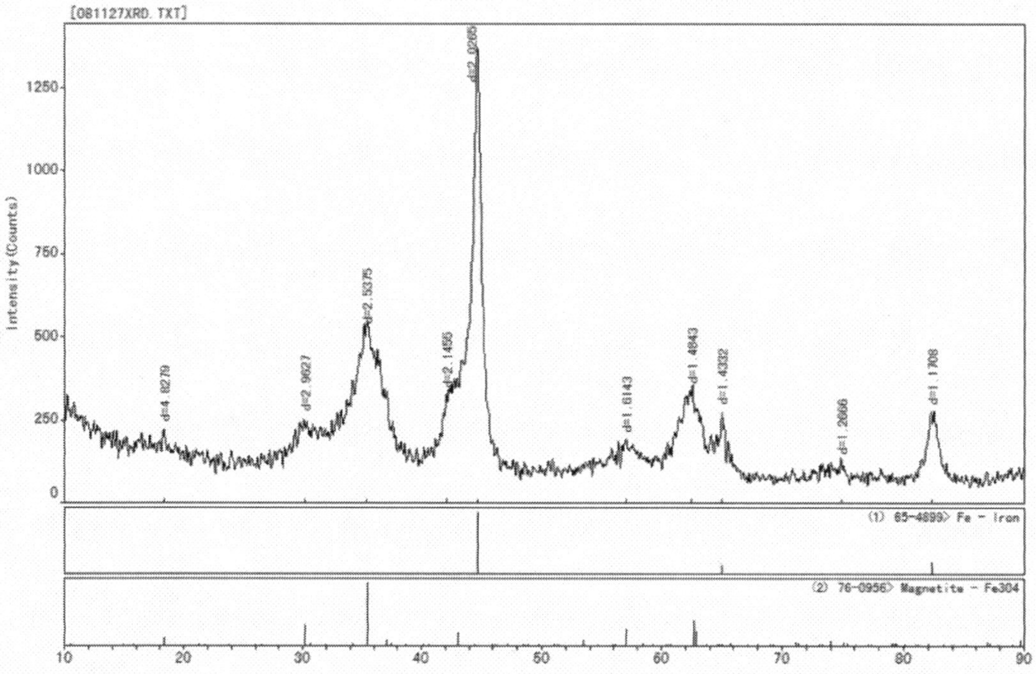

Fig.4. XRD spectrum of magnetic

3.2 Dispersity of Magnetic nanoparticle

Fig.5 is the TEM image of Core-shell magnetic nanoparticle after dispersion, stable dispersion can be seen from the image. Also, the great sphericity of ferric core enable in-situ epitaxial growth of ferric oxide which comprises shell on core and then produce a ceramic-liked dense layer, so Core-shell magnetic nanoparticle synthesized has great mobility, which is important for stable dispersion. The main application of this particle is dopant for soft magnetic component since the great mobility and dispersity is good for dopant doping in main base material for shaping process.

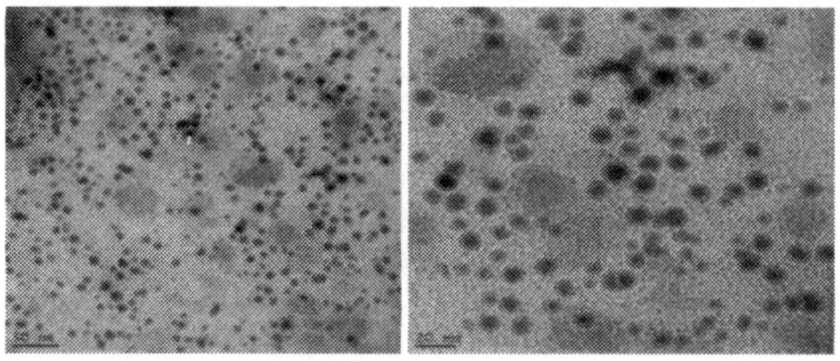

Fig.5. TEM image of Core-shell magnetic nanoparticle after dispersion

4 Characterization and Model building of Core-shell magnetic nanoparticle

Core-shell magnetic nanoparticle has high specific surface area, large surface energy, high surface activity and is unsaturated, so this nanoparticle is totally different with normal particle. Compared with past research which focused on synthesis, structure and application, this paper based on basic concept of magnetization, build and characterize the mathematical model of Core-shell magnetic nanoparticle in order to rich the mechanism research of Core-shell magnetic nanoparticle.

Assume that ρ_l, m_l, V_l is density, mass and volume of Core-shell nanoparticle respectively.

ρ_c, m_c, V_c is density, mass and volume of space between particles respectively.

ρ_d, m_d, V_d is density, mass and volume of surfactant respectively.

ρ_m, m_m, V_m is density, mass and volume of core respectively.

ρ_o, m_o, V_o is density, mass and volume of magnetic shell respectively.

r is radius of core.

δ_1 is the thickness of shell structure.

δ_2 is the thickness of surfactant.

n-V_l is the number of magnetic particle per volume.

n^* is density of particles.

M_s is the saturation magnetization of magnetic particle.

I ischaracteristic saturation magnetization for material.208emu/g for Fe.

Density and radius are already known, so the relationship of M_S, n, r can be obtained, according to the basic concept of magnetization:

$$M_S = \frac{I \times m_m}{m_l} = \frac{I \times \rho_m \times V_m}{V_l \times \rho_l} \quad (1)$$

$$m_m + m_0 + m_c + m_d = m_l \quad (2)$$

$$m_m = n \times \frac{4}{3}\pi r^3 \times \rho_m \quad (3)$$

$$m_o = n \times \frac{4}{3}\pi\left[(r+\delta_1)^3 - r^3\right] \times \rho_o \quad (4)$$

$$m_d = n \times \frac{4}{3}\pi\left[(r+\delta_1+\delta_2)^3 - (r+\delta_1)^3\right] \times \rho_d \quad (5)$$

$$m_c = V_c \times \rho_c \quad (6)$$

$$V_m = n \times \frac{4}{3}\pi r^3 \quad (7)$$

$$V_d = n \times \frac{4}{3}\pi\left[(r+\delta)^3 - r^3\right] \quad (8)$$

$$\rho_l = \frac{m_m + m_o + m_d + m_c}{V_l} \quad (9)$$

$$n^* = \frac{n}{V_l} \quad (10)$$

$$r + \delta_1 = R \quad (11)$$

From (1) to (11) :

$$M_s = \frac{I \times \rho_m \times n \times 4\pi r^3}{3\rho_l V_l} = \frac{I \times \rho_m \times n \times 4\pi r^3}{3\rho_l V_l} \quad (12)$$

$$\rho_l = \{n\frac{4}{3}\pi r^3 \rho_m + n\frac{4}{3}\pi[R^3 - r^3]\rho_o + n\frac{4}{3}\rho_d\pi[(R+\delta_2)^3 - R^3] + \rho_c[V_l - n\frac{4}{3}\pi(R+\delta_2)^3]\}/V_l \quad (13)$$

$$n = \frac{3M_s\rho_l V_l}{4\pi I\rho_m r^3} \quad (14)$$

Put (12) into (13):

$$\rho_l = \rho_c + \frac{3M_s\rho_l\{\frac{4}{3}\pi r^3\rho_m + \frac{4}{3}\pi\left[R^3 - r^3\right]\rho_o + \frac{4}{3}\rho_d\pi\left[(R+\delta_2)^3 - R^3\right] - \frac{4}{3}\pi\rho_c(R+\delta_2)^3\}}{4\pi I\rho_m r^3} \quad (15)$$

$$r = \left[\rho_l M_s \frac{(R+\delta_2)^3(\rho_d - \rho_c) + R^3(\rho_o - \rho_d)}{\rho_m I(\rho_l - \rho_c) + M_s\rho_l(\rho_o - \rho_m)}\right]^{\frac{1}{3}} \quad (16)$$

$$\delta_1 = R - \left[\rho_l M_s \frac{(R+\delta_2)^3(\rho_d - \rho_c) + R^3(\rho_o - \rho_d)}{\rho_m I(\rho_l - \rho_c) + M_s\rho_l(\rho_o - \rho_m)}\right]^{\frac{1}{3}} \quad (17)$$

$$n^* = \frac{n}{V_l} = \frac{3M_s\rho_l}{4\pi I\rho_m r^3} \quad (18)$$

MS is known by vibrating sample magnetometer. Other known parameters are showed in Table 1:

Table 1. Known parameters of Core-shell magnetic nanoparticle

ρ_l(g/cm^3)	ρ_c(g/cm^3)	ρ_d(g/cm^3)	ρ_m(g/cm^3)	ρ_0 (g/cm^3)
1.6	0.69	0.91	7.86	5.24
δ_2(cm)	M_s(emu/g)	*I* (emu/g)	V_1 (mm^3)	R/cm
2×10^{-7}	50	208	0.06	10×10^{-7}

Calculated results :

r = 7.6 nm

δ_1 = 2.3 nm

n^* = 2.5×10^{16} /mm^3

The result is consistent with what is obtained from TEM image.

5 Conclusions

(1) Synthesis Core-shell magnetic nanoparticle using microwave plasma.

(2) Obtained magnetic nanoparticle shows great dispersity.

(3) The calculation result of mathematic model is consistent with what is obtained from experiment.

Acknowledgement

We appreciate the support from Ningbo international cooperation project(2016D10002).

References

1. Xie W Y, Chen W, Magnetic component in Wireless Power Transmission.J. Magnetic component and power source
2. Yang L M, Wang D W etc, Synthesis of ferric nanoparticle by plasma. J. Nuclear Fusion and Plasma Physics
3. Wei Z Q, Zhu L, etc, Synthesis and characterization of Carbon-Encapsulated Fe Nanoparticles by DC Carbon Arc Plasma, Chinese journal of vacuum science and technology. J
4. Liang X, Wang Z J, Liu C J, Size-controlle dsynthesis of colloidalgold nanoparticles at room temperature under the influence of glow discharge. J. Nanoscale.Res.Lett， 2010, **5**:124-129
5. Chiang W H, Richmonds C,Sankaran R M,

Continuous-flow atmospheric- pressure microplasmas: a versatile source for metal nanoparticlesynthesis in thegas or liquid phase plasma sources. J. Sci.Techn, 2010, **30**:34011
6. Kaneko T, BabaK, Hatakeyama R, Gas-liquid interfacial plasmas: basic properties and applications to nano-material synthesis. J. Plasma Phys. Contr. Fusion, 2009, **51**:24011
7. WeiZ H, Liu C J, Synthesis ofmonodispersegold nanoparticles in ionicliquid by applyingroom temperature plasma. J. Material Letters, 2011, **65**:353-355
8. Sankaran R Mohan, Microplasma synthesis of nano-particles. R. ICPIG, 2009, **29**
9. Ouyang H W, Meng X J, Huang S C, etc, Technological progresses of fabricating nano-iron and nano-iron oxide powders. J. Materials Science and Engineering of Power Metallurgy, 2008, **13(6)**:315-322
10. Dieter Vollath, Plasma synthesis ofnanopowders. J. Nanoparticle Research, 2008, **10**:39-57
11. Ping Liu, Yuming Wang, Study on twin stacking faults in ultrafine nickel. J. Materials and Design, 2000, **21(3)**:155-157
12. Granqvist C G, Buhrman R A, Ultrafine metal particles. J. Appl Phys, 1976, **47(5)**:2200-2219

Interface Detection of Concrete and Slurry in Cast-in Place Pile

Zhaolei Liu[1], *Lei* Yu[2,*], and Shuxiong Zhang[2]

[1] Research institute of highway ministry of transport, No.8 yard, Xi tu cheng Road, Hai-dian district, Beijing, China.
[2] Beijing Xin-Qiao Technology Development Co., LTD, No.8 yard, Xi tu cheng Road, Hai-dian district, Beijing, China

Abstract. To find the interface of concrete and slurry in a cast-in place pile, the numerical simulation method and Laboratory tests are performed. Pressures along a pipe were obtained. Differential pressures of fixed distance were calculated. Differential pressure curves were plotted. The result shows that pressure along a pipe is differ. Differential pressure remains constant in certain mediums but changes across different mediums. The interface of concrete and slurry can be found from differential pressure plots. If transducers whose number is at least two can be tied on a fixed place of pipe, differential pressure will be obtained by pulling pipe out gradually, and the interface of the concrete and slurry will be found easily. This is useful for keeping pipe under concrete and ensuring pipe quality.

1 Introduction

The cast-in place pile is part of underground construction. Pipe is used for pouring concrete into a pile hole [1]. Every pouring reaches a depth of 6 meters, at which point the pipe must be lifted. To make pouring continuous, the pipe must be kept in concrete at a depth of at least 2 meters [2]. Therefore, pipe depth control is necessary. If a pipe is pulled out of concrete, the pile will be separated into two parts [3]. Currently, people always use a drop-hammer to detect the top of the concrete [4]. This method is manual and inefficient because pouring progress has to be stopped. To find the interface of concrete and slurry in cast-in place pile in a timely way, differential pressure was measured by numerical simulation and laboratory tests.

2 Material and methods

2.1 Numerical simulation

ICEM CFD 16.0 is used to build the geometric model and mesh generation. Fluent 16.0 is used to build the fluid numerical calculation model. Tecplot 360 EX 2015 R1 software is used for numerical calculation result reduction.

2.1.1 Geometric model abstracted

During pouring, the pipe will be lifted when its bottom is 6 meters beneath the concrete. This is a cyclic process. In certain construction projects, sizes such as the diameter of pile holes, pipe diameters and pipe wall thicknesses are fixed. Investigation shows that the diameter of pile hole is usually 1.2 meters. Pipe diameters are usually 30 centimeter. Pipe wall thicknesses are 8 millimeter. The geometric model of a pouring system can be abstracted as the following figure.

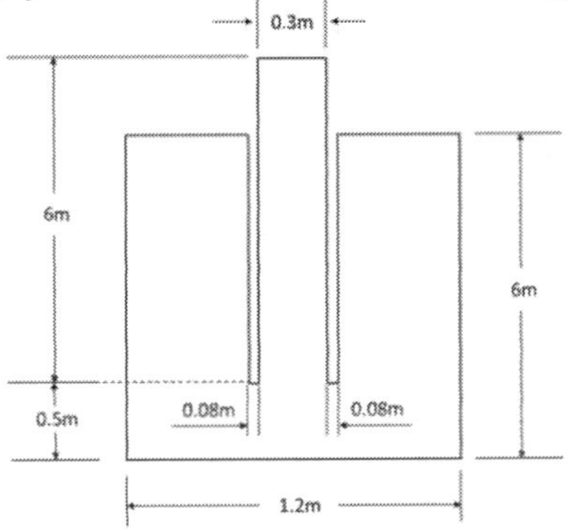

Fig. 1. Sketch map of geometric model of cast-in place pile system

2.1.2 Mesh generation

Mesh generation steps are shown as the following figures.

Step 1 Geometric graph drawing
The point on left bottom is (0, 0, 0). Define coordinates as shown in Table 1. Fig. 2 is a sketch map of a geometric graph.

Table 1. Coordinates list for every point

Point	1	2	3	4
Coordinate	0,0,0	1.2,0,0	1.2,6,0	0.758,6,0
Point	5	6	7	8
Coordinate	0.758,0.5,0	0.75,0.5,0	0.75,6.5,0	0.45,6.5,0
Point	9	10	11	12
Coordinate	0.45,0.5,0	0.442,0.5,0	0.442,6,0	0,6,0

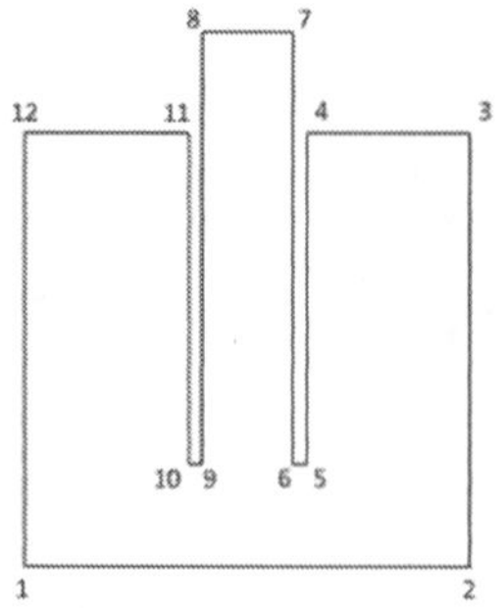

Fig. 2. Sketch map of geometric graph

Fig. 3 is a drawing by ICEM and it is the real geometric graph. The pipe wall is very thin.

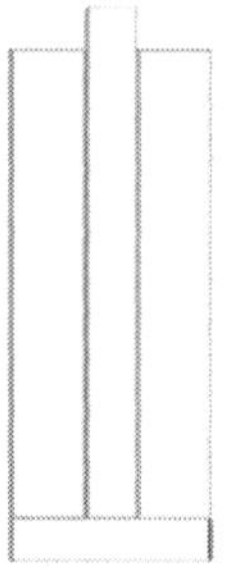

Fig. 3. Real Geometric graph

Step 2 Surface created
Divide geometric graph into some rectangle regions and then create surface. Divided results are shown as Figure 4.

Fig. 4. Surface created

Step 3 Created Part
As shown as Fig. 2, define 7-8 as the fluid enters face. Define 3-4 as the fluid outflow face 1 and 11-12 as fluid outflow face 2.

Step 4 Created Block
Select the structured grid; the blocks need to be created. Using 2D Planer, by means of split block, create some regular square blocks. Delete the surplus blocks.

Step 5 Define mesh parameter
Define global grid 0.01. The node numbers of each edge list are in Table 2.

Table 2. List of node numbers

Edge	7-8	8-16	7-15	11-12	3-4	12-13
Node number	20	30	30	30	30	220
Edge	11-10	16-9	15-6	4-5	3-14	13-1
Node number	220	220	220	220	220	30
edge	2-14	9-10	6-5	/	/	/
Node number	30	6	6	/	/	/

Step 6 Generate the grid
Generate the grid and save the grid file.

2.1.3 Simulation model abstracted

When pouring, there are two kinds of fluid. One is concrete, and the other is slurry. Concrete is poured into pipe and falls on the slurry. The slurry is pushed out the pile hole by the concrete. Actually, concrete and slurry are independent. There is an interface between concrete and slurry. Volume of Fluid model [5] is suitable for this kind of flow.

2.1.4 Simulation and parameters set

Simulation steps and the parameters set are made as follows.

Step 1 Dimension selected
On the start board, select the 2D button, and keep the other button as the default.

Step 2 Read and check grid
Read the grid and check it. Make sure the minimum volume and minimum area are not negative.

Step 3 Set solver parameters

(1) Select gravity. Input -9.8 under the Y coordinate box. Set the time as transient.
(2) In the model plate, select K-epsilon (2eqn) in the drop-down list of Viscous-Laminar. Keep the other selections as the default.
(3) In the material plate, double-click air, and modify the parameters as slurry. Slurry density is 1300 kg/m^3. The dynamic viscosity of slurry is 2.5 kg/m-s. Repeat double-click air, and modify the parameters as concrete. Concrete density is 2400 kg/m^3. The dynamic viscosity of concrete is 15 kg/m-s.
(4) In the model plate, select the VOF (Volume of Fluid) model. Tick Implicit Body Face under Body force formulation. Set the number of Eulerian phases as 2.
(5) Double click primary phase, and set slurry as the main phase. Double click secondary phase, and set the second phase as concrete.
(6) In boundary conditions, set the entrance as velocity. Fluid select concrete. Input flow velocity as 0.1 m/s. In the edit window, set the volume fraction in multiphase as 1. This means concrete pushes out slurry rather than being injected into it. Select the entrance phase as mixture. Set it as velocity, and keep the other set as default. Select entrance phase as slurry. Set it as velocity, and keep the other set as default. Select mixture for exit and set it as outflow.

Select mixture for wall, and set it as wall. Set geometry wall and its shadow as interior.
(7) In the solution control plate, input 0.3 in the momentum box. Input 0.5 in the boxes for turbulent kinetic energy and turbulent dissipation rate.
(8) In the solution initialization plate, input 0.001 in the boxes for turbulent kinetic energy and turbulent dissipation Rate. Click initialization.
(9) In the calculation activities plate, input 30 in auto save every box.
(10) In run calculation, input 1 in the time step size box and input 600 in number of time steps box. This means the flow will continue for ten minutes.
(11) Save the calculation file.
(12) In the run calculation plate, click calculation.

2.2 Laboratory test

2.2.1 Simulation and parameters set

To simulate pouring progress and obtain differential pressure, two pressure transducers were tied on the bottom of a steel stick as shown in the Fig. 5. The transducer type is Series 26 Y. the transducer trademark is Keller. 26 Y transducers are analog signal ones. They can output the electricity value. The measuring range is from 0 mA to 20 mA. The distance between two transducers is approximately 30 centimeters.

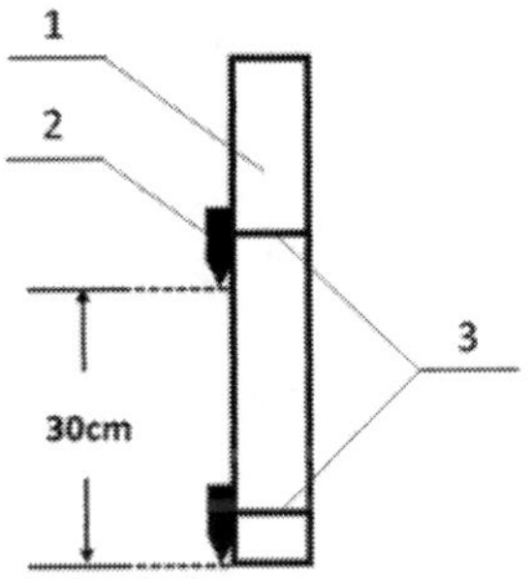

1. Stick 2. Transducer 3. Bundling device

Fig. 5. Transducers arrangement form

2.2.2 Cast-in place pile pouring simulation

A PVC pipe was used to simulate a pile hole. The diameter of PVC pipe is 20 centimeters. The height of PVC pipe is 120 centimeters. The PVC pipe and sticker with transducers compose the pouring simulation system.

Put the stick with transducers into the PVC pipe and then pour concrete and slurry into the PVC pipe in turn. Pull the stick out 10 centimeters at a time. Pressure along the PVC pipe will be obtained. Positions of the stick for different heights of PVC pipe are shown in Fig. 6.

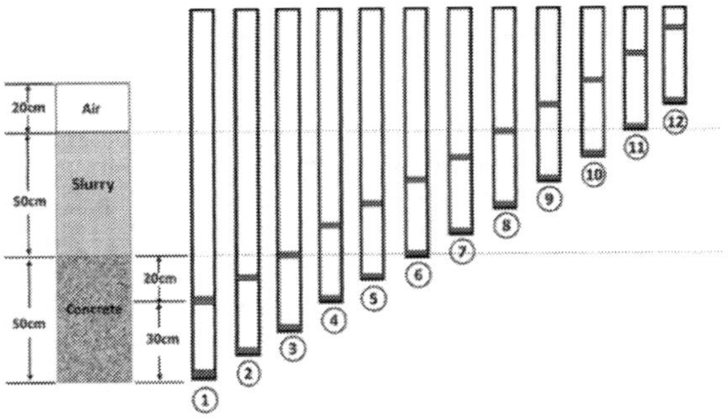

Fig. 6. Positions of stick in pipe

3 Results

3.1 Interface finding by numerical simulation

To find the interface and observe differential pressure changing rules, you can select any pouring time to analyse. For example, select 120s as an example. Pressure along the pipe can be read as shown in Table 3. Below the date minus the above one, differential pressure can be obtained.

Table 3. Pressure in pipe

Depth from pipe bottom /cm	Absolutely pressure /Pa	Differential pressure/Pa
0	98200	/
30	91000	7200
60	83800	7200
90	76600	7200
120	69400	7200
150	62200	7200
180	55000	7200
210	47800	7200
220	42900	3900
250	39000	3900
280	35100	3900
310	31200	3900

Differential pressure is plotted in Fig. 7. 220 centimeters is a demarcation point. Under 220 centimeters, there is concrete around the pipe. Every other 30 centimeters, differential pressure remains constant. Over 220 centimeters, there is slurry around the pipe. Every other 30 centimeters, differential pressure remains constant at another level. At the 220 centimeter point, differential pressure changes suddenly. Therefore, 220 centimeters above the pipe bottom is the interface of concrete and slurry.

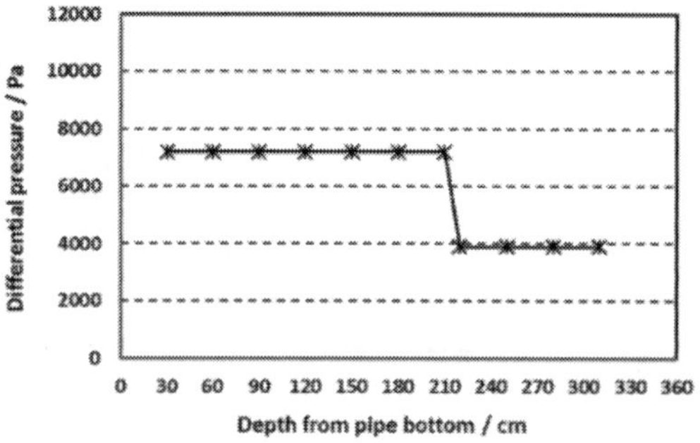

Fig. 7. Differential pressure along pipe by simulation

To support the result that 220 centimeters is the demarcation point of concrete and slurry, the phase of concrete at 120 s is exported as shown in Fig. 8.

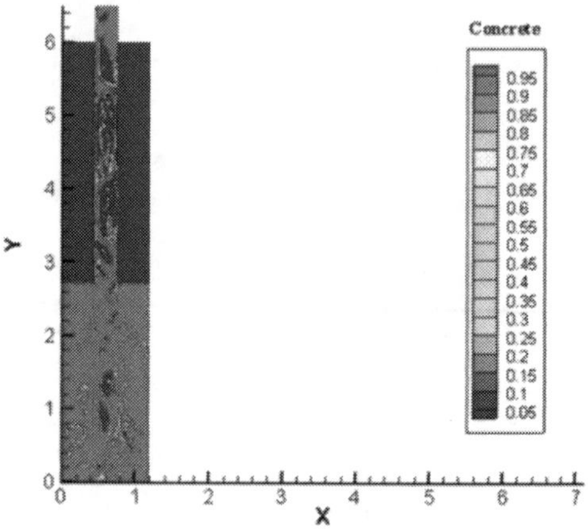

Fig. 8. Phase of concrete in 120

From Fig. 8, at a time of 120s, the interface of concrete and slurry is at a 220 centimeters depth from the pipe bottom. This phase picture supports the Fig. 7 result.

We make the assumption that if we tie some pressure transducers every 30 centimeters along the pipe, pressure can be obtained and the differential pressure figure can be plotted as in Fig. 8. From the differential pressure figure, we can read the interface location of the concrete and slurry.

3.2 Interface finding by Laboratory test

Electricity is detected by pressure transducers on the pipe. Pressure Value and differential pressure are calculated and shown in Table 4.

Table 4. Electricity value recorded by transducers and pressure calculated

Position ID	Electricity		Pressure		Differential pressure
	Upper transducer	Below transducer	Upper transducer	Below transducer	
1	11.73	15.91	24.48	37.71	13.24
2	10.51	14.7	20.61	33.88	13.27
3	9.38	13.58	17.04	30.33	13.30
4	8.27	12.12	13.52	25.71	12.19
5	7.17	10.67	10.04	21.12	11.08
6	6.11	9.37	6.68	17.00	10.32
7	4.99	8.25	3.14	13.46	10.32
8	4.02	7.25	0.06	10.29	10.23
9	4.02	6.18	0.06	6.90	6.84
10	4.02	5.12	0.06	3.55	3.48
11	4.02	4.02	0.06	0.06	0
12	4.02	4.02	0.06	0.06	0

Differential pressure is plotted as shown in Fig. 9.

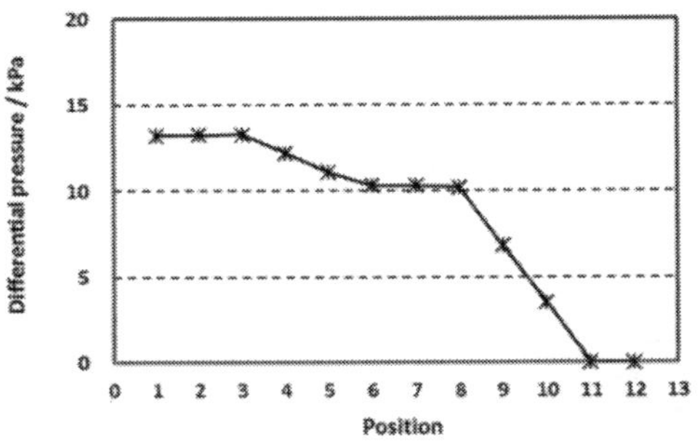

Fig. 9. Differential pressure along pipe by lab test

As shown in Fig. 9, from position 1 to position 3, two transducers are all in concrete and differential pressure remains constant. From position 4 to position 6, the upper transducer is pulled out of the concrete and into the slurry; differential pressure is changed. From position 7 to position 8, two transducers are all in the slurry; differential pressure remains constant at a new level. From position 9 to position 11, the upper transducer is pulled out of the slurry and into the air, differential pressure is changed. After position 11, two transducers are all in the air, differential pressure is almost zero.

Also from Fig. 9, the interface of concrete and slurry can be found, and it is position 3. The interface of slurry and air can be found, and it is position 8.

4 Discussion

By means of detecting pressure along the pipe, differential pressure can be calculated and plotted. From the differential pressure figure, the interface of concrete and slurry can be found. This interface is very important. If the bottom of the pipe is pulled out of this interface, the pile will be separated into two parts. A two parts pile is below construction standard and will be discarded. This causes financial loss. By means of pressure detecting, keeping the pipe in the concrete is good for quality control in cast-in place pile construction.

5 Conclusions

Numerical simulation and laboratory tests are performed to find the interface of concrete and slurry in the cast-in place pile.

Numerical simulation results show that pressure is changing by depth of pipe. In the same medium, differential pressure remains constant for fixed distances. When the interface of two different mediums is stepped over, differential pressure will change immediately.

Laboratory test results show that pressure is changing by depth of stick. Pressure will change when the transducer is pulled out from the concrete and into the slurry. For fixed distance, differential pressure remains constant in the concrete or in the slurry. However, the values in the concrete or in slurry are different. The constant in slurry is lower than that in concrete. When the upper transducer is pulled out of the concrete, the differential pressure changes, and this can be observed in the figure. The interface of concrete and slurry can be found in the differential pressure plot.

Laboratory test results support numerical simulation results. Therefore, by means of setting the transducer on the pipe at fixed distance, pressure can be detected, and a differential pressure plot can be obtained. From the differential pressure plot, the interface of concrete and slurry can easily be found. It can be used to prevent the occurrence of pipe pull out from concrete and ensure the quality of cast-in place pile construction.

Acknowledgements

Thank you Lei Chen for helping with my lab test. Thank you LIU Zhao-lei for proofreading this paper.

This work was supported by the Basic Research Found for National Public Welfare Scientific Research Institutions [grant numbers 2015-9034] and the Ministry of Communications and Enterprise Innovation Item [grant numbers 2015 315 Q13 040].

References

1. Z.Miao, Zhang. *Pile Foundation Project.* (China Building Industry Press, 2007) (In Chinese)
2. G.Qing Zhang, L,Yu, Z.Lei Liu. Journal of Highway Communication technology,**2**,3(2017) (In Chinese)
3. S.Cai Ding. Journal of Highway.**3**,4(2015) (In Chinese)
4. H.Cheng Ding. Guangxi Normal University Journal.31,3(2010)(In Chinese)
5. J.Peng Kang. Ansys Fluent 16.0 Super Study Notebook. (Posts and Telecom Press, 2016) (In Chinese)

Biomimetic preparation of ternary heterogeneous catalyst derived from biopolymer in biological sludge and synergistic effects in Fenton process

Nguyen Thi Oanh[1], Jingyi Zhang[1], Michael Engda Asgodom[1], Mei Yan[1], Huifang Xie[1,a] and Hongxia Qu[2]

[1]*Jiangsu Key Laboratory of Chemical Pollution Control and Resources Reuse, School of Environmental and Biological Engineering, Nanjing University of Science and Technology, Nanjing 210094, Jiangsu Province, China*
[2]*School of Chemical Engineering, Nanjing University of Science and Technology, Nanjing 210094, Jiangsu Province, China*

Abstract. Biopolymer (BioP) was extracted from waste activated sludge and the ternary catalyst BioP@PANI@Fe_3O_4 was prepared by biomimetic method. Control experiments proved the synergistic effects of the components in BioP@PANI@Fe_3O_4. The degradation of high concentration of model pollutant proved its good performance as the heterogeneous Fenton catalyst.The high utilization efficiency of H_2O_2 and wide oprational pH range showed the promising application of BioP@PANI@Fe_3O_4 in the oxidative degradation of organic pollutants. The possible reaction mechanism on the surface of BioP@PANI@Fe_3O_4 lies in that PANI provide the acidic condition and the initially generated ≡Fe(II) species react with H_2O_2 to generate •OH, which destroy the organic dyes adsorbed onto catalyst surface. The redox process Fe(III)→Fe(II) take place easily due to convenient electro transfer in PANI and reduction power of functional groups in BioP.

1 Introduction

Fenton processes were invested to destruct organic pollutants widely since the Fenton was discovered[1] and they performed high degradation efficiency. However, the traditonal Fenton systems have some drawbacks, which limit their application. For Fenton reactions, the optimum pH usually fall in 2.5~3.5[2-4]. However, most wastewater containing dyes is highly alkaline[5] and some environmental media may possess a high buffer capacity[6]. Therefore, the necessity to acidify the reaction medium limits the applicability in the dyeing wastewater treatment. Another inhibition lies in the iron sludge generated during the reaction, which is dangerous solids waste and may cause secondary pollution[3, 7]. Recently, some new types of heterogenerous catalyst were prepared in order to break through the bottlenecks. Georgi et al.[6] found that humic acid (HA) could extend the optimum pH of the Fenton process toward neutral conditions because of chelation. HA coated Fe_3O_4 exhibited high Fenton-like catalytic ability during the mineralization of sulfathiazole[2]. These works indicated that the introduction of biopolymer into nanocomposites can improve their performance. Our previous study showed that composite of polyaniline (PANI) and nano Fe_3O_4 have some advantages such as wide operational pH range, high catalytic activity, high utilization efficiency of H_2O_2 and low iron concentration in water as the heterogenerous Fenton catalyst[8].

Waste activated sludge (WAS) is the major byproduct of wastewater treatment process and the subsequent sludge processing, stabilization and disposal incur the high costs. This WAS problem become more and more pressing because of the current legal constrains, the rising costs and public sensitivity over the past decades[9, 10]. At the same time, the inherent properties of biopolymer elicited interest on developing composite materials that could render these molecular useful[11]. WAS contain over 60~75% organic components, mainly the mixture of microbial cells and secretion. The biological components, such as polysaccharide, proteins, lipid, nucleic acid, uronic acid and humic substances, are rich in carboxyl, hydroxyl, amino and other functional groups[12-14]. So, "we should expect more" out of sludge[10] and new resource recovery-based approach is possible to be developed.

Here, biopolymer (BioP) was extracted from WAS and successfully used for preparing the ternary BioP@PANI@Fe_3O_4 catalyst for the first time. This application of BioP provided a new resource recovery and effective reduction way for large quality of WAS. The catalytic activity and synergistic effects were evaluated using the model dyes and the possible synergistic mechanisms of the components were analyzed in this study.

2 Materials and methods

2.1 WAS and BioP extraction

[a] Corresponding author:huifangxie@njust.edu.cn

WAS from the secondary setting tank in a municipal sewage treatment plant was collected and culticated in our lab for extraction of BioP. Thermal-alkaline extraction was carried out at pH 12.5, temperature 90°C for 60 min. The extracted supernatant was freeze-dried and finely ground to get the BioP power.

2.2 Biomimetic Preparation of ternary heterogeneous Catalyst

For biomimetic preparation of BioP@PANI@Fe_3O_4, 0.5 g BioP power was dissolved in 100 mL citrate-citric acid buffer solution (pH = 4.00) and 0.25 g Fe_3O_4 nanoparticles was added into the solution. After homogenizing by shaking for 10 min, 0.05 M of aniline, 0.05 g/L of hemin was added. Polymerization of aniline was initiated by injecting 0.2 mL 2.0 M H_2O_2 solution at 10 min interval. After total 1.0 mL H_2O_2 solution was used, the polymerization proceeded for 0.5 h at room temperature. Then 0.09 M HCl was added and the mixture was kept stirring for 0.5 h, and allowed to proceed with 12 h. The formed solid was collected via centrifugation at 3500 rpm, washed sequentially with ethanol and water. The final catalyst BioP@PANI@Fe_3O_4 was obtained by drying at 50°C.

2.3 Degradation experiments of dyes

Because Methyl orange (MO) and Rhodamine B (RhB) are typical anionic and cationic azo dyes and often present in wastewater, they were chosen as the target model pollutant in this study. MO and RhB degradation experiments using the ternary catalyst of BioP@PANI@Fe_3O_4 during heterogeneous Fenton-like reactions were preformed, and the optimal parameters were determined upon the basis of degradation efficiency. The effects of reaction condition including catalyst dosage, concentration of dyes, concentration of H_2O_2 and pH value was investigated. The produced catalyst was dosed in 100 mL of dye aqueous solution at certain initial pH values and H_2O_2 was added to initiate the degradation reaction. All experiments were conducted under constant shaking at 150 rpm, 25°C for 30 min. As control, different systems were tested to verify the synergistic effects of the components of catalyst. Sampling was carried out periodically. The collected suspension was filtered through 0.2 μm filters, and the filtrate were then immediately by measuring the absorbance of the solution at 464 nm for MO and 552 nm for RhB, corresponding to the maximum absorption wavelength of dyes.

2.4 Analytical method

The chemical phase structure of the biomimetic products was characterized by X-ray diffraction (XRD). The morphology was characterized by scanning electron microscopy (SEM). The specific surface area and pore diameter distribution was determined using N_2 adsorption-desorption isotherms.

3 Results and discussion

3.1 Synergistic effects of ternary catalyst

Control experiments were conducted to compare the removal efficiency of MO (800 mg/L) and RhB (600 mg/L) without pH adjusting in different systems, and the results were shown in Figure 1.

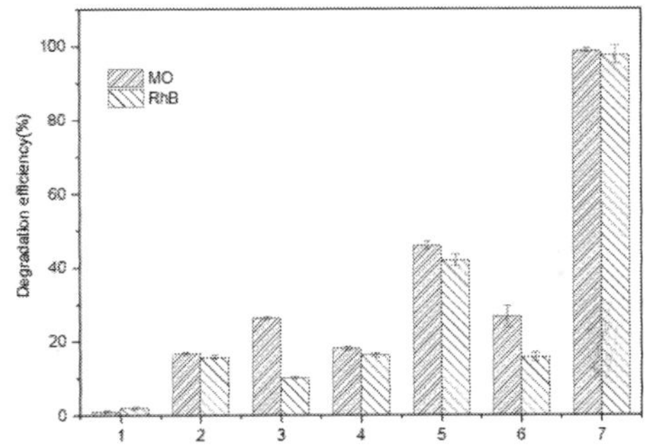

Figure 1 Degradation efficiency of MO (800 mg/L) and RhB (600 mg/L) in different systems. (1) 0.02 M H_2O_2; (2) 0.6 g/L Fe_3O_4+0.02 M H_2O_2; (3) 0.6 g/L PANI+0.02 M H_2O_2; (4) 0.6 g/L BioP@PANI +0.02M H_2O_2; (5) 0.6 g/L PANI@Fe_3O_4+0.02 M H_2O_2; (6) 0.6 g/L BioP@PANI@Fe_3O_4; (7) 0.6 g/L BioP@PANI@Fe_3O_4 +0.02 M H_2O_2.

As shown in Figure 1, 0.02 M H_2O_2 only led to a very slight removal of dyes within 30 min. The small amount of dyes diminished in Fe_3O_4/H_2O_2, PANI/H_2O_2 and BioP@PANI/H_2O_2 system is ascribed to the catalytic activity of nano-Fe_3O_4 and PANI for H_2O_2[2, 15, 16]. About 26.7% MO and 15.8% RhB removal were observed mainly due to adsorption of 0.6 g/L BioP@PANI@Fe_3O_4. In the presence of H_2O_2, the composite of PANI@Fe_3O_4 shows higher catalytic activity than pure PANI and Fe_3O_4. While the removal of dyes using BioP@PANI@Fe_3O_4 was notably higher than that for PANI@Fe_3O_4, implying that the activity was enhanced by the introduction of BioP. These data indicated BioP@PANI@Fe_3O_4 performance well beyond those of its components and there are obvious synergistic effects in the ternary composite. The high activity makes it an efficient heterogeneous Fenton-like catalyst. The similar enhancement of catalytic activity was also found in the catalyst Fe_3O_4/HA due to the electron transfer between the complexed Fe(II)-HA or Fe(III)-HA[2].

3.2 Affecting factors of dyes degradation in heterogeneous Fenton process catalyzed by BioP@PANI@Fe_3O_4

In heterogeneous Fenton-like processes, the reaction between H_2O_2 with active sites of ≡Fe(II) take place on the surface of catalyst and active radicals generate[17, 18],

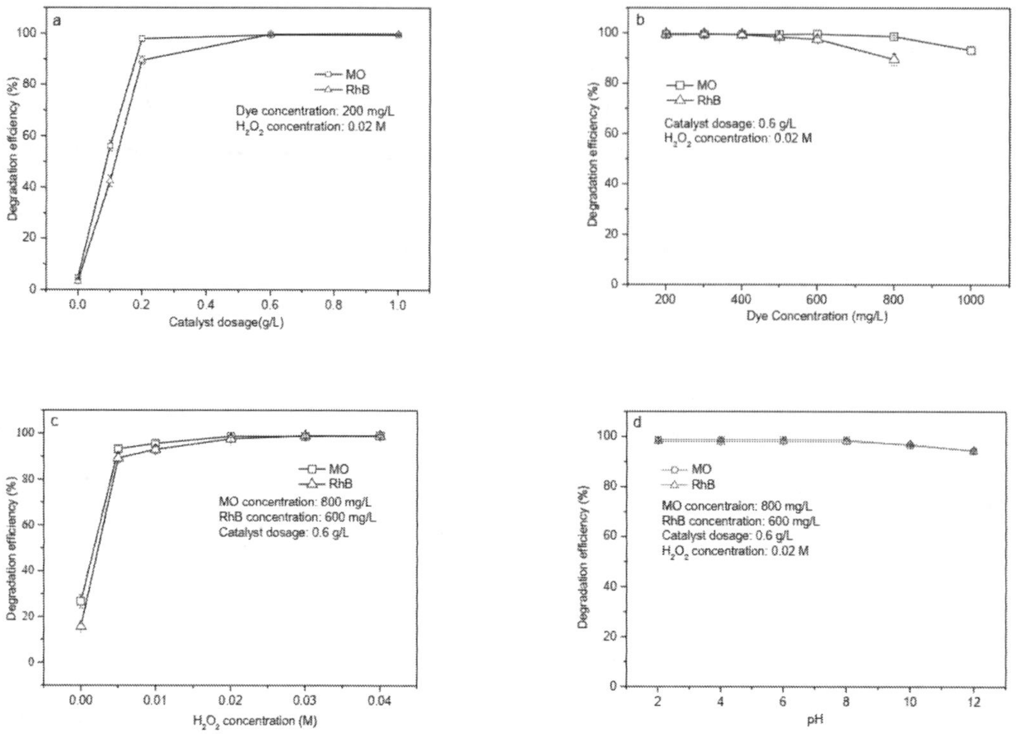

Figure 2 Degradation of MO and RhB by catalyst under different a) catalyst dosage ; b) dye original concentration ; c) H_2O_2 concentration ; d) pH value

so the suitable dosage of catalyst is essential for the degradation of dyes. As shown in the Figure 2a, when the catalyst dosage increased from 0.1 g/L to 0.6 g/L in 200 mg/L dyes solution, the degradation efficiency of MO and RhB increased from 56.09% and 42.41% to both above 99% after 30 min reaction. When catalyst dosage and H_2O_2 were fixed at 0.6 g/L and 0.02 M respectively, the maximum initial concentration of MO and RhB should be lower than 800 mg/L and 600 mg/L respectively if above 99% degradation efficiencies were obtained (Figure 2b). This can be attributed to the inductive effects caused by high dyes concentration, as available active sites were predominantly covered by the dye molecules[3]. Figure 2c depicts the relationship between dyes degradation and H_2O_2 concentrations. The presence of H_2O_2 increased dramatically the degradation of dyes. When the concentration increased to 0.02 M, the degradation efficiency got to about 99% for both MO and RhB. These results suggest that H_2O_2 plays the very important role in the degradation of dyes, and more active radicals are formed when H_2O_2 concentration increased. However, the optimized H_2O_2 concentration is relatively low, which showed that the H_2O_2 can be used efficiently in the ternary catalyzed system. It should be pointed out that the catalyst showed very wide operational pH range as shown in Figure 2d. For the original pH 2~8 in MO and RhB solution, the degradation efficiency all kept over 98%. Even at pH 10 and 12, the efficiency also maintained about 97% and 95%. The results proved this ternary BioP@PANI@Fe_3O_4 can overcome the drawback of narrow pH range of traditional Fenton catalyst, which makes it a promising candidate as the high efficient catalyst for treatment of various dyes-containing wastewater.

3.3 Catalyst characterization

Figure 3 shows the XRD patters of BioP@PANI@Fe_3O_4 and the three components, Fe_3O_4, BioP and PANI. The main characteristic peaks of PANI centered at $2\theta = 20°$ and 25°[19, 20], the peaks of Fe_3O_4 at $2\theta = 35.7°$[21] and the peaks of BioP at $2\theta = 31.7°$ and 45.3° are remained in the produced catalyst with weaker summit and broader range. The changes indicate that the main structure of PANI, Fe_3O_4 and BioP retains in the composite, but their crystalline is distorted during the prepare.

The SEM image of catalyst showed that the particles about tens of microns aggregate into porous lumps and the fine granules with diameter ranging from 100 nm to a few microns are well dispersed throughout the surface of larger particles. The presence of BioP effectively prevented the aggregation of PANI and Fe_3O_4. The surfaces of the composite are very rough, which provide good possibilities to trap and adsorb the pollutants in water.

The specific surface area and pore diameter distribution were calculated by the Brunauer-Emmett-Teller (BET) and Barrett-Joyner-Halenda (BJH) methods, respectively, using N_2 adsorption-desorption isotherms.

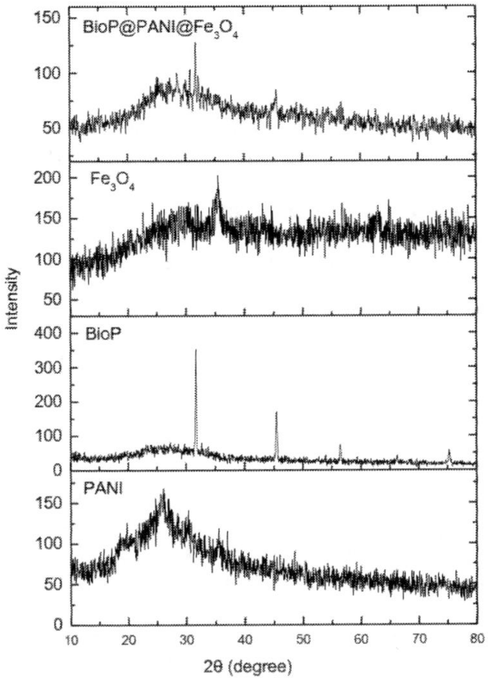

Figure 3 XRD patters of catalyst and three components

Figure 4 SEM image of catalyst

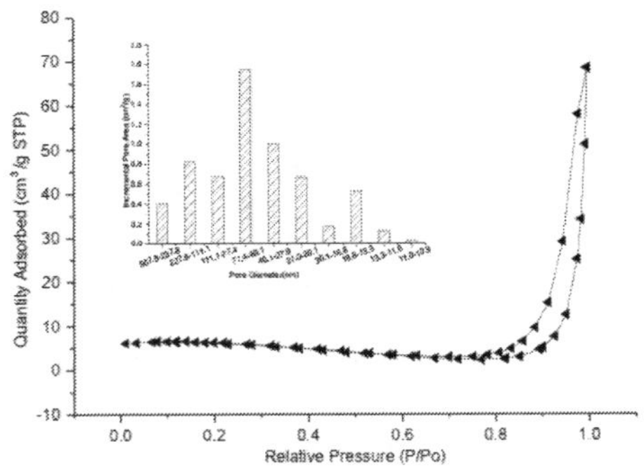

Figure 5 Nitrogen adsorption-desorption isotherms (Insert : Pore area distribution)

As shown in Figure 5, the nitrogen adsorption-desorption isotherm exhibited multi-layer adsorption (Type IV and V isotherms) and hysteresis loops type H1 according to the Brunauer-Deming-Deming-Teller classification, which indicates that BioP@PANI@Fe_3O_4 is agglomerates (assemblages of rigidly joint particles) and relatively high pore size uniformity and facile pore connectivity[22]. The pore size distribution indicates that the catalyst contains macro- and meso-porous and the pore diameter of 77.4-40.1 nm provide the maximal pore area. The BET surface area of this composite are 19.89 m^2/g, which much smaller than that reported for iron-loaded activated carbon used as adsorbent with BET surface area over 300 m^2/g[23, 24]. This result indicated that the absorption might not be the most important removal reason of dyes, which is consistent with the results in Figure 1. It is speculated that degradation of dyes was due to catalysis instead of adsorption.

3.3 Reusability and stability of catalyst

Reusability and stability are vital for a catalyst used in pollutant degradation, so the repeated MO and RhB degradation tests were performed using the ternary catalyst in the Fenton-like reaction. As illustrated in Figure 6, BioP@PANI@Fe_3O_4 was stable and maintained high catalytic efficiency in the six rounds with above 90% degradation efficiency. Additionally, the leaching test of iron indicated that the leached iron in the first and sixth runs were about 7 mg/L and less than 0.4 mg/L. Low iron dissolution means that heterogeneous catalysis was mainly responsible for the fast dye degradation[25, 26]. Meanwhile, the released low iron concentration can greatly reduce the treatment cost of subsequent iron sludge, which is one of the bottlenecks of application of Fenton process for the treatment of wastewater. These results demonstrate that this catalyst has a strong potential for wastewater treatment because of its excellent stability, reusability and safety.

https://doi.org/10.1051/matecconf/201823803001

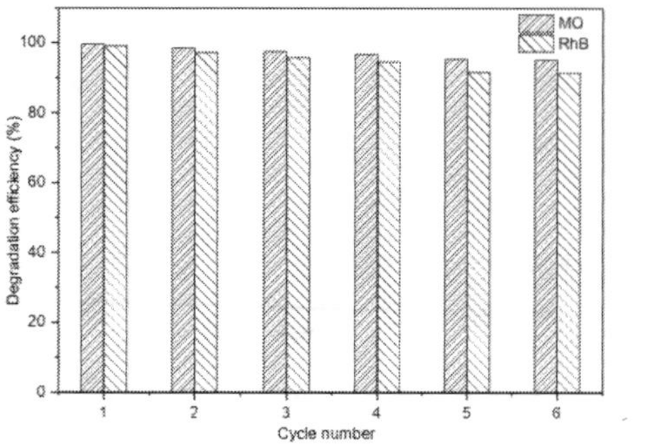

Figure 6 Reusability of catalyst for the degradation of dyes

3.4 Possible mechanism

3.4.1 Verification of ·OH

Radical ·OH are thought the main active oxidant in the Fenton reaction and tertbutanol[27] and isopropanol[28] are known as ·OH scavengers. In our experiments, varying concentrations of tertbutanol and isopropanol were added to the system in order to verification the roles of ·OH. The results summarized in Figure 7 clearly showed that the degradation of dyes was dramatically influenced by ·OH radical scavengers and were almost completely inhibited in the presence of 100 mM tertbutanol or isopropanol. So it is obvious that ·OH radicals are the oxidant that initiates the degradation of dyes.

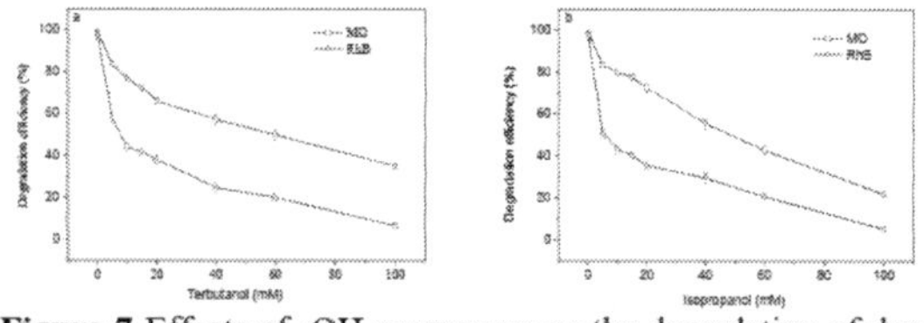

Figure 7 Effects of ·OH scavengers on the degradation of dyes (MO 800mg/L, RhB 600mg/L, Catalyst 0.6 g/L, H_2O_2 0.02 M)

3.4.2 Possible synergistic effects in the degradation reaction

The ternary catalyst BioP@PANI@Fe_3O_4 exhibit high activity for degradation of organic dyes and there are obvious synergistic effects among the three components. The possible interaction is schemed in Figure 8.

Nano-Fe_3O_4 in the BioP@PANI@Fe_3O_4 plays the main role in the production of ·OH. As reported[18, 21], the initially generated ≡Fe(II) species react with H_2O_2 to generate ·OH as reaction (1) and the ·OH destroy dyes:

$$Fe^{2+}+H_2O_2+H^+ \rightarrow \cdot OH+Fe^{3+}+OH^- \quad (1)$$

The introduction of PANI and biopolymer can protect iron from leaching, which was proved by the low leaching iron concentration. Just as shown in Figure 8, the nitrogen-containing functional groups in PANI and BioP, the carboxyl in BioP act as active sites[6, 29] for forming complexes with ≡Fe(II, III). The ligand and the net-like structure influence the leaching of iron and resulting in the low dissolution into bulk solution.

In the traditional Fenton reactions, the reduction of Fe^{3+} is the rate-limiting step:

$$Fe^{3+}+H_2O_2+H^+ \rightarrow HO_2\cdot+Fe^{2+}+H^+ \quad (2)$$

Fe^{3+} easily form insoluble precipitates, which lead to the turbidity and color of the solution. In the BioP@PANI@Fe_3O_4/H_2O_2 system, this redox process Fe(III)↔Fe(II) on the catalyst surface can take place when an excess of H_2O_2 is added into the reaction solution. However, the iron redox cycle can be easily accomplished even without H_2O_2. The redox process Fe(III)↔Fe(II) is facilitated at the solid-water interfacein Fe_3O_4[30, 31] and the electron transfer between the complexed Fe(II) or Fe(III) can lead to more rapid regeneration of Fe(II) species. Moreover, PANI is like an "electron pool"[21] and then improve the electron transfer on the interface of catalyst. At the same time, there are plenty of hydroxyl and carboxyl groups in the polysaccharide and protein in BioP. These organic matters can also exhibit reducing power to produce Fe(II). Based on the analysis, the high catalytic ability of BioP@PANI@Fe_3O_4 may be caused by the electron transfer, leading to rapid regeneration of Fe(II) active sites and the production of ·OH radicals at low H_2O_2 concentration.

Usually reaction (1) is considered as the pH-independent below pH 3[32]. However, the BioP@PANI@Fe_3O_4 shows wide work pH range from 2 to 12 for removal of dyes, which should be attributed to PANI. The PANI in the catalyst is the semi-oxidized state of PANI (Emeraldine form), and the number of protons can be varied depending on the medium[33]. In the reaction system with higher pH, de-doping process of PANI leads to the release of protons and decreases pH values. On the other hand, the Fe(II)-complex with BioP may change the optimum pH for its reaction with H_2O_2 towards neutral conditions[6, 34].

Based on the above discussion, the possible reaction mechanism on the surface of BioP@PANI@Fe_3O_4 is proposed as follows: PANI provide the acidic condition in the reaction system and the initially generated ≡Fe(II) species react with H_2O_2 to generated •OH on the catalyst surface. The organic dyes adsorbed onto the BioP@PANI@Fe_3O_4 surface are destroyed by •OH radicals. The redox process Fe(III)→Fe(II) take place either in the presence of excess H_2O_2 (Reaction 2) or by electro transfer. A small amount of dissolved iron and •OH disperse into bulk solution and the chain reactions same as in homogeneous Fenton system occur[32] and organic dyes are destroyed.

3.5 Application in synthetic dyeing wastewater

It was proved that BioP@PANI@Fe_3O_4 was an effective catalyst in the degradation of the model pollutant MO and RhB. However, verifying its promise as a degradation catalyst for wastewater treatment is more important. Thus, more complex synthetic dyeing wastewater containing 100 mg/L MO, RhB, Methyl Blue (MB) and Coomassie Brilliant Blue (CCB) respectively were treated.

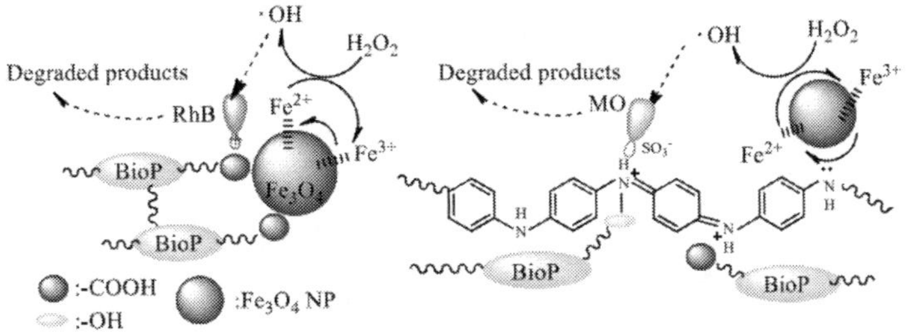

Figure 8 Possible mechanism on the surface of BioP@PANI@Fe_3O_4 composite

As shown in Figure 9, a decrease of 52.7% in COD achieved after treating for 60 min and the COD removal increased to 87.6% after 360 min. The results showed the high oxidation degree of organic dyes can be obtained catalyzed by BioP@PANI@Fe_3O_4.

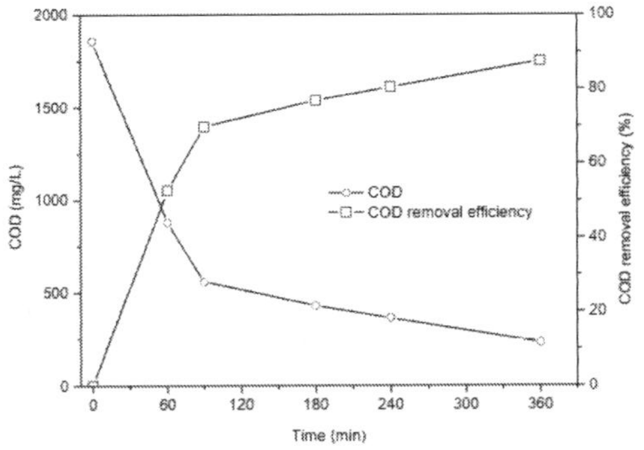

Figure 9 Profile of COD and COD removal of synthetic dyeing effluent (Catalyst 0.6 g/L, H_2O_2 0.02 M)

4 Conclusion

A new kind of ternary heterogeneous catalyst derived from activated sludge was prepared by biomimetic method and showed outstanding catalytic performance on the degradation of high concentration dyes in Fenton reaction. The components in the catalyst have obvious synergistic effects and the possible mechanism was postulated. The excellent results provide a new way of resource recovery and mass reduction for WAS, and BioP@PANI@Fe_3O_4 is a potential synergic heterogeneous catalyst for oxidative degradation of organic pollutants in the presence of low concentration H_2O_2.

Acknowledgments

This work was supported by the National Natural Science Foundation of China (grant numbers 21575066 and 51778296).

References

1. H.J.H. Fenton, J. Chem. Soc., Trans. **65** (1984) 899-910
2. H. Niu, D. Zhang, S. Zhang, X. Zhang, Z. Meng, Y. Cai, J. Hazard. Mater. **190** (2011) 559-565
3. N.A. Zubir, C. Yacou, X. Zhang, J.C. Diniz da Costa, J. Environ. Chem. Eng. **2** (2014) 1881-1888
4. J.J. Pignatello, E. Oliveros, A. MacKay, Crit. Rev. Env. Sci. Technol. **36** (2006) 1-84
5. Y. Yao, Y. Mao, Q. Huang, L. Wang, Z. Huang, W. Lu, W. Chen, J. Hazard. Mater. **264** (2014) 323-331
6. A. Georgi, A. Schierz, U. Trommler, C.P. Horwitz, T.J. Collins, F.D. Kopinke, Appl. Catal., B Environ. **72** (2007) 26-36
7. S. Guo, N. Yuan, G. Zhang, J.C. Yu, Microporous Mesoporous Mater. **238** (2017) 62-68
8. M. Yan, Q. Zhang, H.-F. Xie, J.-M. Kong, H.-X. Qu, China Environ. Sci. **37** (2017) 1394-1400
9. W.W. Li, H.Q. Yu, B.E. Rittmann, Nature, **528** (2015) 29-31
10. J. Peccia, P. Westerhoff, Environ. Sci. Technol. **49** (2015) 8271-8276
11. R. Nagarajan, W. Liu, J. Kumar, S.K. Tripathy, F.F. Bruno, L.A. Samuelson, Macromol. **34** (2001) 3921-3927
12. V. Janaki, B.-T. Oh, K. Vijayaraghavan, J.-W. Kim, S.A. Kim, A.K. Ramasamy, S. Kamala-Kannan, Carbohydr. Polym. **88** (2012) 1002-1008
13. S.K. Das, P. Ghosh, I. Ghosh, A.K. Guha, Colloids Surf. B Biointerfaces **65** (2008) 30-34
14. J. Merrylin, S. Kaliappan, S.A. Kumar, I.T. Yeom, B.J. Rajesh, Int. J. Environ. Sci. Technol. **10**(2013) 85-92
15. J. Zhang, J. Zhuang, L. Gao, Y. Zhang, N. Gu, J. Feng, D. Yang, J. Zhu, X. Yan, Chemosphere **73** (2008) 1524-1528
16. A.H. Gemeay, R.G. El-Sharkawy, I.A. Mansour, A.B. Zaki, Appl. Catal., B: Environ. **80** (2008) 106-115
17. G. Sheng, G. Zhang, J. Wang, J. Colloid Interface Sci. **433C** (2014) 1-8
18. S. Zhang, X. Zhao, H. Niu, Y. Shi, Y. Cai, G. Jiang, J. Hazard. Mater. **167** (2009) 560-566
19. S.S. Umare, B.H. Shambharkar, R.S. Ningthoujam, Synth. Met. **160** (2010) 1815-1821

20. T. Abdiryim, X.G. Zhang, R. Jamal, Mater. Chem. Phys. **90** (2005) 367-372
21. L. Xu, J. Wang, Environ. Sci. Technol. **46** (2012) 10145-10153
22. M.K. And, M. Jaroniec, Chem. Mater. **13** (2001) 3169-3183
23. T.A. Saleh, Naeemullah, M. Tuzen, A. Sarı, Chem. Eng. Res. Des. **117** (2017) 218-227
24. Q. He, J. Dai, L. Zhu, K. Xiao, Y. Yin, J. Alloys Compd. **687** (2016) 326-333
25. H. Zhang, G. Xue, H. Chen, X. Li, Chemosphere, **191** (2018) 64-71
26. G. Zhou, Z. Chen, F. Fang, Y. He, H. Sun, H. Shi, J. Environ. Sci. **35** (2015) 20-26
27. J. Peller, O. Wiest, P.V. Kamat, J. Phys. Chem. A **105** (2001) 3176-3181
28. M. Yan, H. Xie, Q. Zhang, H. Qu, J. Shen, J. Kong, J. Mater. Sci. Chem. Eng. **4** (2016) 26-34
29. Q. Li, L. Sun, Y. Zhang, Y. Qian, J. Zhai, Desalination **266** (2011) 188-194
30. A.G. Williams, M.M. Scherer, Environ. Sci. Technol. **38** (2004) 4782-4790
31. N. Wang, L. Zhu, D. Wang, M. Wang, Z. Lin, H. Tang, Ultrason. Sonochem. **17** (2010) 526-533
32. I. Magario, F.S. García Einschlag, E.H. Rueda, J. Zygadlo, M.L. Ferreira, J. Mol. Catal. A: Chem. **352** (2012) 1-20
33. P. Zhao, J. Chem. Soc. Faraday Trans. **92** (1996) 3063-3067
34. B.M. Voelker, B. Sulzberger, Environ. Sci. Technol. 30 (1996) 1106-1114

The low infrared emissivity of $Ce_{1-x}Y_xO_{2-x/2}$ samples at high-temperature contributed by enhanced conductivity

Honghan Bu, Chuyang Liu*, Guoyue Xu, Yong Jiang, Tengchao Guo and Jianchao Zhang

College of Material Science and Technology, Nanjing University of Aeronautics and Astronautics, Jiang Jun Street 29, Nanjing 211106, China

Abstract: The $Ce_{1-x}Y_xO_{2-x/2}$ samples were synthesized with different doping amount by solid state reaction method. Phase structures of the samples were analyzed and characterized by XRD. The effects of different doping amount on infrared emissivity properties at 3-5μm wavebands were investigated systematically. It is found that the powder with doping molality of x=0.2 exhibits the lowest infrared emissivity value ca. 0.21 when the test temperature is 600 °C. The mechanism for the low infrared emissivity is explored thoroughly. The reduction of infrared emissivity with increasing temperature is ascribed to the enhancement of electrical conductivity. Our results suggest that $Ce_{1-x}Y_xO_{2-x/2}$ powder with x=0.2 is a potential candidate for stealth material of aircraft engines.

1. Introduction

The infrared (IR) radiation of an aircraft is one of the most important factors which affect the detection range of an IR detection system **[1]**. In general, IR radiation from a specific exhaust system can be attributed to two parts: hot plume IR radiation and high temperature engine parts IR radiation, such as that of nozzle, and turbine blades **[2]**. As for Aircraft engine, it usually produces strong infrared radiation because of its high working temperature, which renders the aircraft easily exposed to the enemy **[3]**. Therefore, in order to improve the survivability of military targets on the battlefield, it is important to develop the materials with low emissivity at high temperature.

As well known, CeO_2 with fluorite structure exhibits high refractive index and dielectric constant **[4]** and, therefore, it is a promising material with low infrared emissivity **[5]**. As reported, the value of infrared emissivity reaches 0.3 at 600°C after heat treatment **[6]**. However, as a stealth material, its infrared emissivity of 0.3 is still slightly higher, which is far from sufficient for engineering applications. The main factor influencing infrared emissivity is the conductivity of materials **[7, 8]**, so improving the conductivity of the material is beneficial to reducing the infrared emissivity of the material.

Doping can improve the conductivity of metal oxides **[9, 10]**. Therefore, in order to improve the conductivity of CeO_2 samples, doping is the most effective and convenient method. The ionic radius of Ce^{4+} is 92 pm **[11]**. In order to reduce the effect of lattice distortion on the conductivity of CeO_2 samples, Y^{3+} ions (90 pm) with similar radii to Ce^{4+} ions are the appropriate doping ions **[12]**. Herein, we proposed the Y^{3+} ions doped CeO_2 samples via solid state reaction method in this work, and the influences of doping molality on conductivity and infrared emissivity are discussed in detail.

*Corresponding author: liuchuyang@nuaa.edu.cn

2. Experimental details

2.1 The preparation of $Ce_{1-x}Y_xO_{2-x/2}$ samples

The doped oxides $Ce_{1-x}Y_xO_{2-x/2}$ were prepared by solid state reaction method. CeO_2 samples and Y_2O_3 samples were mixed before calcination. This mixture was gently ground in an agate mortar for homogenization. The obtained samples were calcined for 4 h in furnace at 1500 °C with a step of 100 °C with a constant heating rate of 4 °C/min until reach the annealing temperature. After annealed, the samples were cooled to room temperature gradually in furnace. Finally, the pre-treated $Ce_{1-x}Y_xO_{2-x/2}$ samples with different x values were obtained. The values of x in $Ce_{1-x}Y_xO_{2-x/2}$ were 0.05, 0.10, 0.20 and 0.30.

2.2 The resistance test of $Ce_{1-x}Y_xO_{2-x/2}$ samples

2.2.1 The preparation of test samples

8% polyvinyl alcohol solution was added into the prepared $Ce_{1-x}Y_xO_{2-x/2}$ samples. Then the samples was pressed into Φ = 10 mm cylindrical semi-finished product, as seen in Fig. 1. To reduce the impact of contact resistance on test results, a layer of conductive silver pulp was coated on the surface of test samples. The test samples were put into the muffle furnace at 800 °C for 30 min, making silver layer firmly adhere to the sample surface.

2.2.2The resistance test of samples

The resistance values of doping samples were examined by YX-381TR pointer type digital multimeter. In order to ensure good contact between wire and sample at high temperature, a set of resistance testing molds were designed by me. The structure diagram is shown in Fig.1

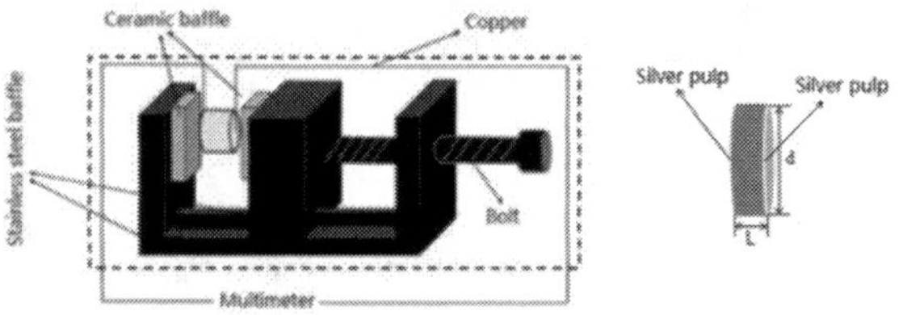

Fig.1 Resistance test mold and the sample of test

2.2.3 The calculation of conductivity ***[13]***

$$\sigma = \frac{1}{R} * \frac{L}{S} = \frac{4L}{\pi R d^2} \qquad (1)$$

Where R is the obtained resistance, L is the sample thickness, S represents electrode area. According to a typical Arrhenius type behavior which is a thermally activated process, the electrical conductivity is used to perform a plot of the Arrhenius equation.

$$\sigma = \sigma_0 exp\left(-\frac{E_a}{kT}\right) \qquad (2)$$

Where σ_0 is the pre-exponential factor, Ea is activation energy, k is the Boltzmann constant.

2.3 The infrared emissivity testing of samples

The infrared emissivity in the range of 3-5μm in different temperature was detected by an infrared emissometer (IR-2 double-band emissometer, Shanghai Institute of Technical Physics, CAS, China). The testing temperature ranges from 25 °C to 600 °C with a heating rate of 20 $K\cdot min^{-1}$.

3. Results and discussion

3.1. X-ray diffraction study (XRD)

Fig. 2a shows the XRD diffraction spectra of samples with different doping amounts. Clearly, the diffraction of peaks corresponded to (111), (200), (220) and (311) planes are observed in all samples, in good agreement with the face-centered cubic fluorite structure of CeO_2 with space group Fm–3m (JCPDS card No. 34-0394) **[14, 15]**. After the introduction of Y^{3+} to CeO_2, no characteristic diffraction peaks assigned to Y_2O_3 phase is observed, indicating that Y^{3+} could enter into the CeO_2 lattice to form $Ce_{1-x}Y_xO_{2-x/2}$ solid solution oxides **[16]**.

Fig. 2b is an enlarged figure of the XRD diffraction spectra. It can be seen that the main diffraction peak at (111) have a little shift to higher angles, and with the increasing doping amount, the shift gradually increases.

The lattice constant (a) of crystalline powder can be obtained by removing the systematic errors of both un-doped and doped CeO_2 through the construction of Nelson-Riley (NR) plots **[17]**. The lattice constant of both un-doped and Y-doped CeO_2 samples measured in the present experiments is in good agreement with the theoretical values reported by Vanpoucke et al. **[18]**, shown in Table 1. The substitution of Ce^{4+} ions with Y^{3+} ions diminishes the lattice constant from 5.48 Å to 5.36 Å, and this is due to Y^{3+} ion have smaller ionic radius than that of Ce^{4+} ion **[19]**.

For Fluorite structure of CeO_2, it has the relation for the cubic system **[20]**

$$\frac{n\lambda}{2sin\theta} = \frac{a}{\sqrt{h^2+k^2+l^2}} \quad (3)$$

Where, n = positive integer, λ = wavelength of the incident wave, (h k l) = Miller indices, a = lattice constant. The lattice constant (a) is inversely proportional to diffraction angle θ. As a result, the decreasing lattice constant a results in an increasing diffraction angles θ.

In a word, the result strongly confirms that the Y^{3+} does replace the Ce^{4+} in the CeO_2 lattice.

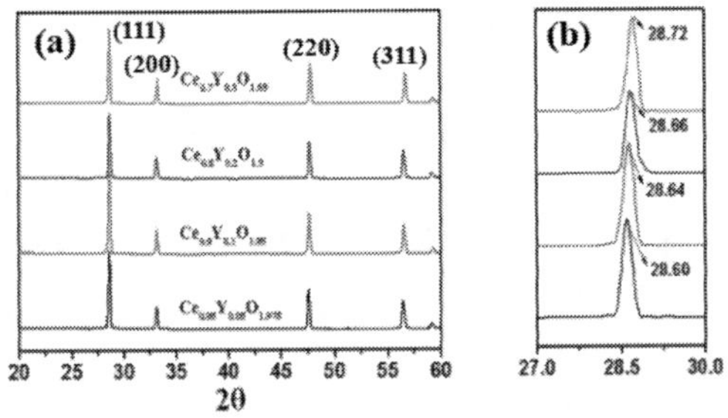

Fig.2 The XRD diffraction spectra of $Ce_{1-x}Y_xO_{2-x/2}$ samples (a) 20-60° of 2θ (b) 27-30° of 2θ

Table 1 Lattice constant of samples with different doping amounts

X values	Lattice Constant a (Å)
0	5.48
0.05	5.46
0.1	5.43
0.2	5.41
0.3	5.36

3.2 The infrared emissivity and conductivity of $Ce_{1-x}Y_xO_{2-x/2}$.

Fig. 3 summarizes the infrared emissivity values of the samples in 3-5μm waveband in the different temperature ranges. It can be clearly found that the infrared emissivity of all samples decreases significantly with the temperature rising from 30°C to 600°C. Further, the infrared emissivity values reduce gradually from x=0 to x=0.2, while increases gradually from x=0.2 to x=0.3.

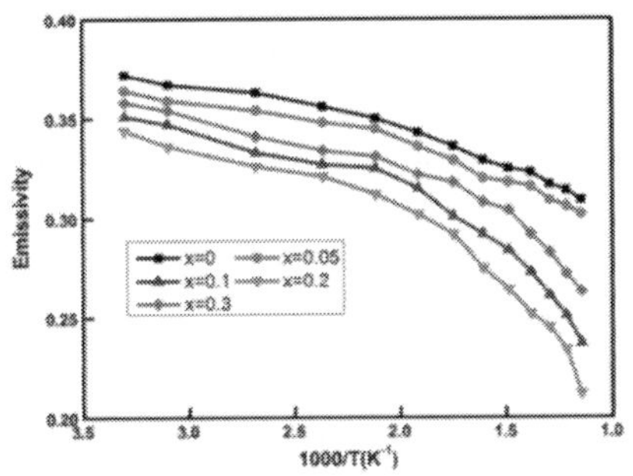

Fig. 3The infrared emissivity values of $Ce_{1-x}Y_xO_{2-x/2}$ samples

Simultaneously, the electrical conductivity (Fig. 4) of all samples was tested at above temperature range. The results show that change tendency of electrical conductivity values is contrary to the change tendency of infrared emissivity. With the increasing temperature, electrical conductivity of all samples increases significantly. Additionally, the electrical conductivity values increase gradually from x=0 to x=0.2, while decrease from x=0.2 to x=0.3.

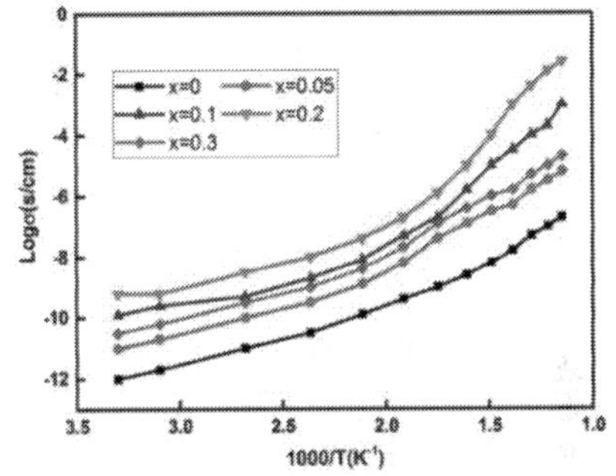

Fig. 4 The electrical conductivity of $Ce_{1-x}Y_xO_{2-x/2}$ samples

3.3 The influences of doping amount on the conductivity of $Ce_{1-x}Y_xO_{2-x/2}$ samples

When the Y^{3+} replaced the Ce^{4+} in the CeO_2 lattice, oxygen vacancy (V_o) defects occurred at once. The defect reaction follows the principle of charge conservation. The effective negative charge of Y'Ce can be compensated in three different ways, which are closely related to the intrinsic defect reaction:

$$M_2O_3 \xrightarrow{2CeO_2} 2M'_{Ce} + 3O_o + V_o^{\bullet\bullet}$$

$$2M_2O_3 \xrightarrow{CeO_2} 3M'_{Ce} + M_i^{\bullet\bullet\bullet} + 6O^{\times}_o$$

$$Ce^{\times}_{Ce} + O^{\times}_o + 2M_2O_3 \xrightarrow{CeO_2} 4M'_{Ce} + Ce_i^{\bullet\bullet\bullet\bullet} + 8O^{\times}_o$$

As reported, the V_o compensation form is the optimal compensation form as for rare earth element cation with large radius **[21]**. With the increasing doping amount, more Y^{3+} replaced the Ce^{4+}, therefore, more V_o were produced, which greatly increases the ionic conductivity of the materials **[22]**. However, when the doping molarity reaches a certain level, the conductivity begins to deteriorate, which is ascribed to the changes in transition types of V_o, as shown in Fig. 5 **[23]**. The corresponding transition energy was calculated according to the literature **[24]** and the results are summarized in Table 2. It is pointed that lower transition energy of V_o can lead to easier transition of V_o.

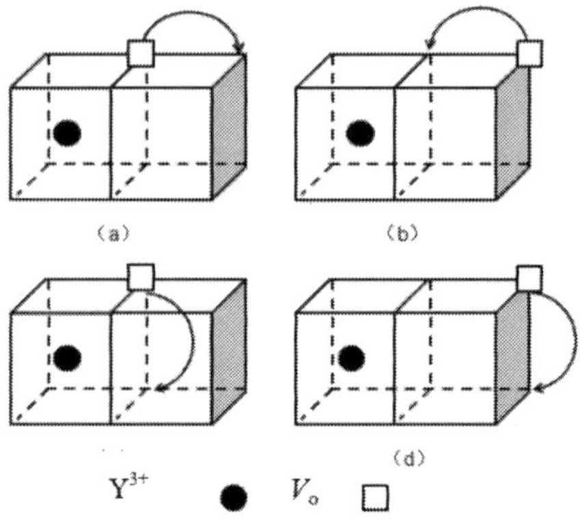

Fig. 5 Transition type of oxygen vacancy

Table 2 The transition types and transition energy of oxygen vacancy

	transition types of V_o	transition energy of V_o
(a)	$1^{st} \rightarrow 2^{nd}$	0.31
(b)	$2^{nd} \rightarrow 1^{st}$	0.22
(c)	$1^{st} \rightarrow 1^{st}$	0.57
(d)	$2^{nd} \rightarrow 2^{nd}$	0.20

When CeO_2 has lower Y^{3+} doping amount, V_o is mainly located in the 2nd and it results in more easier transition of V_o. In this case, more V_o were produced with the increasing doping amounts, which contributes to the improvement of conductivity of doped samples. When Y_2O_3 doping amount reaches a certain value, the Vo located in 2nd up to the maximum value, which results in the maximum conductivity. Afterwards, Y_2O_3 doping amount continue to increase, more Y^{3+} occupied Ce^{4+} ion in lattice position, causing the original V_o located in 2nd gradually converting into the V_o located in 1st of new Y^{3+}. Obviously, it is difficult for V_o to transition because of higher transition energy, which leads to the decreasing of conductivity.

3.4 Mechanism for enhanced conductivity of $Ce_{1-x}Y_xO_{2-x/2}$ samples under elevated temperature

As shown in Fig. 4, the curves of lgσ -1000/T of all doped samples show an increasing linear relationship, which indicates that the conductivity of doped samples improves gradually with the increasing temperature. As reported, the electrical conductivity can be expressed as the following equation **[25]**

$$\sigma = nq\mu \qquad (4)$$

Where n is the concentration of the intrinsic charge carriers, μ is the mobility of charge carriers. With the temperature rising, μ has no obvious change, while the intrinsic excitation of CeO_2 increases rapidly and the concentration of the thermally activated intrinsic charge carriers increases rapidly, which results in the sharp increasing of electrical conductivity **[26]**.

4. Conclusions

The $Ce_{1-x}Y_xO_{2-x/2}$ samples are prepared successfully by solid state reaction method. The X-ray diffraction study confirms that the synthesized $Ce_{1-x}Y_xO_{2-x/2}$ samples have cube fluorite structure. The suitable doping of Y^{3+} ions is conducive to improving the conductivity of metal oxide CeO_2, which leads to the reduction of the infrared emissivity in CeO_2. When the doping amount of Y^{3+} increases from 0.05 to 0.2, the electrical conductivity of $Ce_{1-x}Y_xO_{2-x/2}$ samples rises gradually. But the conductivity of $Ce_{1-x}Y_xO_{2-x/2}$ samples diminished afterwards with the doping amount increasing from 0.2 to 0.3. The conductivity value is the lowest at the doping amount of x=0.2, which results in the lowest infrared emissivity reaching 0.21. Obviously, the $Ce_{1-x}Y_xO_{2-x/2}$ powder with x=0.2 has great potential applications for adaptive camouflage in infrared waveband.

Acknowledgements

This research is financially supported by the National Natural Science Foundation of China (Grant Nos: 51173079) and a project funded by the Priority Academic Program Development of Jiangsu Higher Education Institutions (PAPD).

References

[1] E. Coiro, J. Aircr., **50**, 103 (2013)

[2] T. Zhou, Q. Wang, T. Li, Chin. J. Aeronaut., **30**, 651 (2017)

[3] O. Michailovich, D. Adam, IEEE T. Med. Imaging, **22**, 368 (2003)

[4] J. P. Huang, Y. B. Li, X. D. He and et al, Appl. Surf. Sci., **259**, 301(2012)

[5] R. Siegle, J. R. Howell. Thermalradiation heat transfer. McGraw-Hill Book Company. 1 (1972)

[6] Y. Jiang, G. Y. Xu, T. C. Guo and et al, Infrared Technol., **33**, 699 (2011) in Chinese.

[7]T. G. Guo, G. Y. Xu, S. J. Tan and et al, Mater. Res. Bull. **95**, 354 (2017).

[8] X. X. Yan, Y. T. Cai, R. Lu and et al, Int. J. Polym. Sci., **2017**, 5017356 (2017).

[9] U. Godavarti, V. D. Mote, M. Dasari, J. Asian Ceram. Soc., **5**, 391 (2017)

[10] M. Gupta, S. Shirbhate, P. Ojha and et al, Solid State Ionics, **320**, 199 (2018)

[11] H. Muta, Hi. Kado, Y. Ohishi and et al, J. Nucl. Mater., **483**, 192 (2017)

[12]P. Gupta, R. Nagarajan, Materialstoday Chem., **7**, 15 (2018)

[13]W. Lerdprom, C. Li, D. D. Jayaseelan and et al, J. Eur. Ceram. Soc., **37**, 343 (2017)

[14] Y. N. Liao, L. F. He, C. G. Man and et al, Chem. Eng. J., **256**, 439 (2014)

[15]T. Désaunay, G. Bonura, V. Chiodo and et al, J. Catal., **297**, 193 (2013)

[16] X. Y. Feng, J. X. Guo, X. R. Wen and et al, Appl. Surf. Sci., **445**, 145 (2018)

[17] J. B. Nelson, D. P. Riley, Proc. Phys. Soc., **57**, 160 (1945)

[18] D. E. P. Vanpoucke, P. Bultinck, S. Cottenier and et al, J. Mater. Chem. A, **2**, 13723 (2014)

[19] N. Y. Mostafa, A. Badawi, S. I. Ahmed, Results Phys., **10**, 126 (2018)

[20] M. J. K. Kumar, J. T. Kalathi, J. Alloys Compd., **748**, 348 (2018)

[21] J. A. Kilner, B. C. H. Steele. Mass Transport in Anion-Deficient Fluorite Oxides. in: O. T. Sorensen. Ed. Nonstoichiometric Oxides. New York: Academic Press, 237 (1987)

[22] M. G. Shahraki, S. Ghorbanali, Microelectron. Reliab., **82**, 153 (2018)

[23] Z. F. Ma, Hebei University of Technology, Heibei (2005) in Chinese.

[24] R. E. Williford, W. J. Weber, R. D. Devanathan and et al, J. Electroceram., **3**, 409 (1999)

[25] E. K. Liu, B. S. Zhu, J. S. Luo. Semiconductor physics, Publishing House of Electronics Industry, Beijing, **7**, 126 (2017)

[26] Š. Csáki, J. Ondruška, V. Trnovcová and et al, Appl. Clay Sci., **157**, 19 (2018)

MATEC Web of Conferences **238**, 03003 (2018) https://doi.org/10.1051/matecconf/201823803003
ICNFM2018

Study on the effect of electrocatalytic oxidation treatment of 2,4,6-Trinitrophenol wastewater

Yaling Li, Wenqiang Jiang* and Ruyu Li

College of environmental science and engineering, Qilu University of Technology (Shandong Academy of Sciences), Jinan 250353, People's Republic of China

Abstract. 2,4,6-Trinitrophenol is a toxic aromatic nitro-compounds that widely used in pharmaceutical, chemical and pesticide production. Due to its stable structure and poor biodegradability, advanced electrocatalytic oxidation technology was selected to treat simulated wastewater. The goal of the present work is to optimize the electrolysis conditions such as current density, electrolysis pH, and electrolyte concentration. A Pt modified TiO2 electrode was chosen as the anode accompanied with a titanium electrode of the same size as the cathode The results showed that the removal efficiency of 2,4,6-Trinitrophenol was the highest when the current density was 20mA/cm2, electrolyte pH=5, electrolyte concentration was 2 g/L. Under the optimal condition, the removal rate of 2,4,6-Trinitrophenol reached 99.76% after 120 minutes electrolysis. The decay of TNP could also be described by the pseudo-first-order kinetics formula with respect to TNP concentration. Therefore, electrocatalytic oxidation technology might provide an effective method for the degradation of nitroaromatic organic compounds.

1 Introduction

2,4,6-Trinitrophenol (TNP, also known as picric acid) is a kind of toxic aromatic nitro-compounds that widely used in the manufactures of explosives, pesticides and dyes [1]. This pollutant is characterized by large chroma, complex composition, and poor biodegradability, which make it difficult to handle. Baljinder Singh et al. isolated B. cereus strain PU from soil contaminated with explosive to degraded TNP wastewater, the optimal degradation rate of 68% was obtained after 12 hours of incubation [2]. S.Chai et al. pointed out the degradation mechanism of TNP may be considered as three steps, methyl group oxidation, decarboxylation and mineralization of nitro-groups [3].

In recent years, many researchers have been focusing on advanced oxidation processes such as Fenton, ozone oxidation, photocatalytic oxidation. It would appear that the most promising technology for the treatment of waste water containing TNP is advanced electrocatalytic oxidation processes (AEOP) [4]. The principal advantages of AEOP include environmental compatibility, energy efficiency, safety, selectivity and cost effectiveness. For AEOP the formation of hydroxyl radical (·OH) is especially desired, since it is the strongest oxidizer and can react non-selectively with various types of organics [5]. With the development of AEOP, new electrode materials with higher Faraday potentials and coated electrodes are in demand. The coated electrode is an electrode that immobilizes a molecule, an ion, or a polymer which having excellent characteristics on the surface of the electrode to improve the electrocatalytic oxidation performance of the electrode [6].

In this study, the high performance liquid chromatographic method (HPLC) was applied to determine the concentration of TNP. The goal of the present work is to optimize the electrolysis conditions such as current density, electrolysis pH and electrolyte concentration to find the optimal electrolytic conditions for the degradation of TNP in water. The results might provide a theoretical basis for the potential application of electrochemical degradation of aromatic nitro-compounds.

2 Experimental materials and methods

2.1 Experimental materials

In this work 2,4,6-Trinitrophenol (98.5% purity) was selected as a model organic compound and it was obtained from Sigma-Aldrich (in America). Its structural formula and general characteristics are shown in Table 1.

Table 1. General characteristics of 2,4,6-Trinitrophenol

Name	2,4,6-Trinitrophenol
Chemical structure	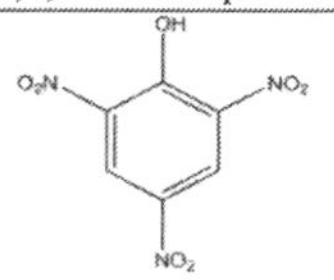
Molecular Formula	$C_6H_3N_3O_7$

Wenqiang Jiang:13001709819@163.com

CAS Number	88-89-1
Molecular mass	229.11 g/mol

Anhydrous sodium sulfate (99.9%) as electrolyte (supply by Panre Applichem). Other chemicals were all of analytic grade and used as received without further purification. All experiments were performed at room temperature.

2.2 Experimental Methods

2.2.1 Electrolysis

Pt modified TiO_2 electrode of 50.0 cm^2 area was used as the anode accompanied with a titanium electrode of the same size as the cathode at a distance of 1 cm between electrodes. In order to investigate the factors affecting electrocatalysis, the experiments were carried out in a plexiglass reactor using 500 ml of TNP solution.

The apparatus of the electrocatalytic oxidation experiments were mainly consisted of a DC power supply (0–30 V, 0–10 A), plexiglass reactor equipped a magnetic stirrer. The anode and cathode were positioned vertically and parallel to each other. During the experiments, samples were drawn from the reactor at certain intervals and the concentration of 2,4,6-Trinitrophenol was quantitatively analyzed by a high performance liquid chromatography (HPLC) equipped with a UV spectrophotometer. An Ultimate XB-C18 column (4.6mm×150mm×5μm), filled with ultra-pure fully porous spherical silica gel was used. The mobile phase was a mixture of 80% methanol, 20% water and 0.5% phosphoric acid, the flow rate was1 mL/min under 8.5 MPa pressure and column temperature was 303 K. The detection UV wavelength was 353 nm.

2.2.2 Linear regression equation of TNP

0.1000g 2,4,6-Trinitrophenol was accurately weighed and diluted to 1000 mL with distilled water to obtain a standard solution of 2,4,6-Trinitrophenol at a concentration of 100 mg/L. The standard solution was diluted to different concentrations, such as 1 mg/L, 5 mg/L, 10 mg/L, 20 mg/L, 50 mg/L. A series of gradient concentrations of solution were filtered with 0.25μm nylon Whatman filters before injected into the HPLC to determine the concentration of TNP. The linear relationship between 2,4,6-Trinitrophenol concentration and peak area was analyzed and shown in Figure 1. This graph represents a consistent linear relationship

The peak area and the concentration of TNP solution in the range of 0-100 mg/L, and the R^2 value was 0.9991.

In the electrolytic experiments, concentrations of 2,4,6-Trinitrophenol could be calculated by equation (1).

$$C = \frac{A - 44077}{73734} \qquad (1)$$

C and A are TNP concentration and the peak area respectively

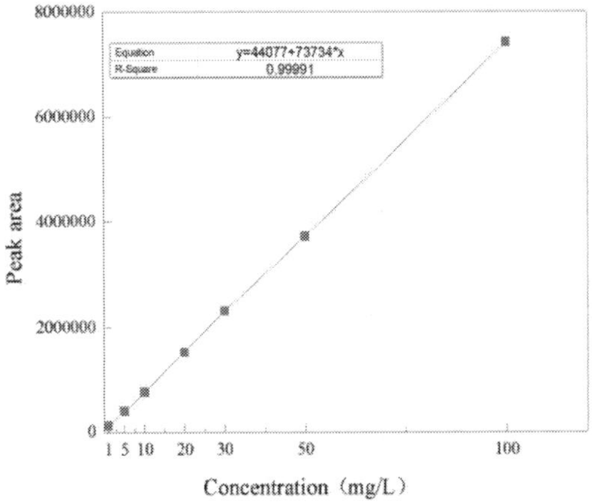

Figure 1. Linear relationship between peak area (A) and the concentration of 2,4,6-Trinitrophenol solution (C)

3 Results and discussion

The rate of TNP degradation (R) was calculated by Eq.(2) to estimate electrolytic efficiency:

$$R = \frac{C_0 - C_i}{C_0} \times 100\% \quad (2)$$

where C_0 and C_i are the TNP concentration of the original wastewater and the treatment liquid, respectively.

3.1Current density effects

Current density is the current through which the unit electrode area passes, and is generally expressed in mA/cm^2. It is a key factor that affects the electrocatalytic oxidation process because it directly affects ·OH production on the electrode, which determines the degradation efficiency of organic matter in an aqueous solution [7].

Electrolysis experiments were performed under various current densities, and results are shown in Figure 2. In the electrolysis experiments, the degradation rate was roughly identical at current densities of both 30 and 40 mA/cm^2, where the degradation rate reached 95.45% and 96.40%, respectively, after electrolyzing for 120 minutes. High current densities accelerate the rate of organic matter oxidation on the electrode's surface. On the contrary, when the applied current density is too high, the formation of side products may hamper the generation and availability of ·OH radicals [8]. Therefore, as current density increases from 30 to 40 mA/cm^2, the rate of TNP removal does not significantly improve. At a current density of 15 mA/cm^2, the rate of TNP degradation was only 88.85%. At such a weak current density, electrode activity may not have been fully activated. The economic consideration in operations as observed under experimental conditions is that a current density of 20 mA/cm^2 is optimal for 2,4,6-Trinitrophenol electrolysis.

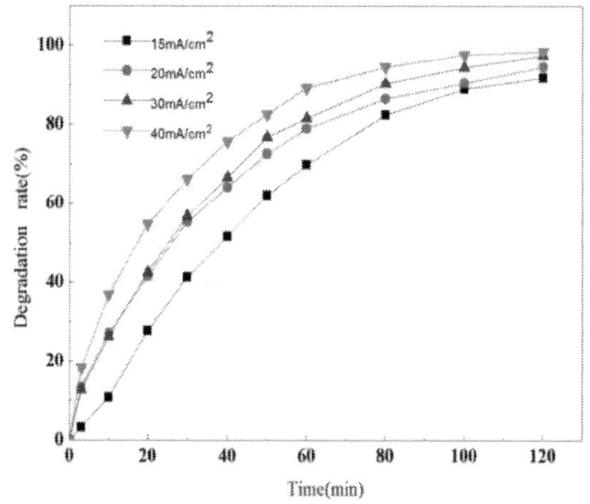

Figure 2. Current density effect on TNP electrolysis.

3.2 The effects of electrolyte concentration

Anhydrous sodium sulfate was added to the TNP electrolyte as an electrolyte, which increases anion and cation electrolyte concentrations, thus enhancing conductivity. An increased number of ions that move freely in the electrolyte, increases conductivity and reduces energy consumption during reactions. [9].

Electrolysis experiments were carried out at various electrolyte concentrations to study concentration's effect on TNP degradation. As shown in Figure 3, the effect of degradation varies due to the different electrolyte concentrations added. As the electrolyte concentration increases from 1 to 2 g/L, reactions exhibit improved degradation speed and degradation rate. When electrolyte concentration is 2 g/L, the TNP solution has the highest degradation rate.

However, by continuously increasing the electrolyte concentration, the rate of TNP degradation decreases due to excess electrolytes that allow ions to move freely in solution, which causes ion collision and subsequently reduces the degradation rate [10]. Therefore, in subsequent experiments, we selected an electrolyte concentration of 2 g/L due to its higher degradation rate and current efficiency.

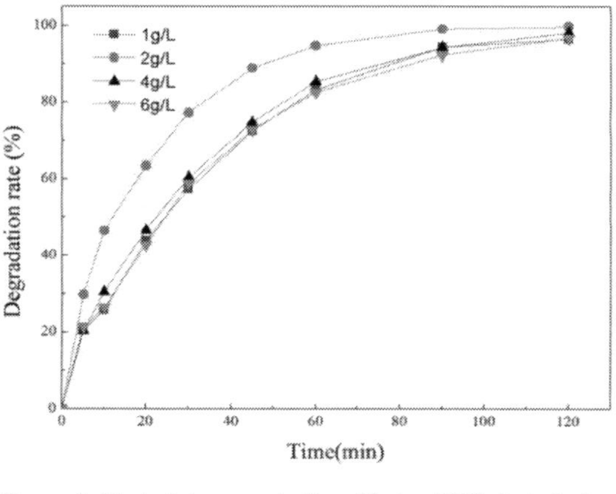

Figure 3. Electrolyte concentration effect on TNP electrolysis.

3.3 The influence of electrolyte pH

A solution's pH is also an important parameter that affects electrochemical performance. The relationship between the degradation rate and pH is shown in Figure 4. Analyzing the concentration curves and reaction times indicate that TNP decay may also be described by the pseudo-first-order kinetics formula with respect to TNP concentration. The following equations describe the reaction kinetics when pH values are at 2.0, 5.0, 7.0, and 9.0:

$$ln(c_0/c_t) = 0.0357t - 0.0758 \quad (3)$$

$$ln(c_0/c_t) = 0.0307t - 0.0253 \quad (4)$$

$$ln(c_0/c_t) = 0.0271t - 0.0259 \quad (5)$$

$$ln(c_0/c_t) = 0.0273t - 0.0171 \quad (6)$$

At a pH of 2.0 and 5.0, the reaction rate constant was higher than that at a pH of 7.0 or 9.0. In Figure 4, we observe that the rate of TNP degradation and current efficiency were highest at a pH of 2.0. When solution pH was higher than 5.0, the rate of TNP degradation and the reaction rate constant decreased gradually with increasing pH values. This can be attributed to the fact that the Pt/TiO_2 electrode redox potential increases under acidic conditions, which increases the production of ·OH [11]. However, electrolyte pHs that are too low to shorten the electrode's service life [12]. Hence, for these electrolysis experiments, the optimum reaction pH was 5.0.

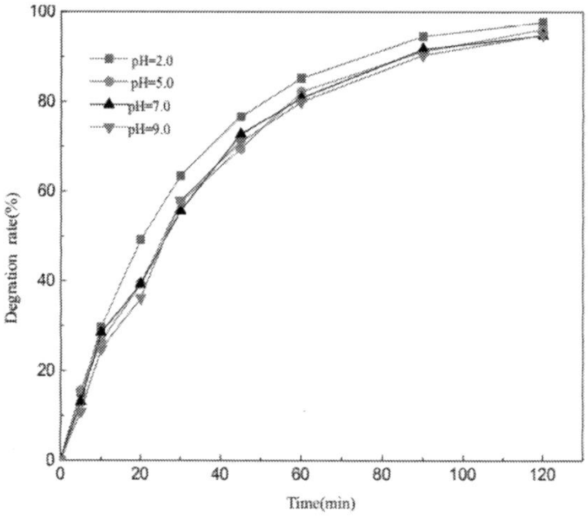

Figure 4. The electrolyte pH effect on TNP electrolysis.

4 Conclusions

The proposed study was focused on the degradation of TNP under AEOP conditions. We established methods to determine TNP in a solution using HPLC. Increasing the current density and electrolyte concentration within a certain range can increase the rate of TNP degradation. By optimizing current density, electrolysis pH and electrolyte concentration, a Pt modified TiO_2 electrode can be successfully applied to the treatment of TNP wastewater. The rate of TNP degradation may exceed 99.96% under optimal conditions, i.e., a current density of 20 mA/cm^2, an electrolyte pH of 5.0, and an electrolyte concentration of 2 g/L. Future research would be focused on new electrode materials development and TNP wastewater degradation mechanism.

Acknowledgment

This work was supported by the Natural Science Foundation of Shandong Province, China.

References

1. S.Yi, W.Q.Zhuang, B.Wu et al. Environ. Sci. Technol. **40** (2006)
2. B.Singh, J. Kaur, K. Singh. Biotechnol lett. **33** (2011)
3. S.Chai, G.Zhao, R. Sedlak, et al. Chem, Environ Sci Technol. **46** (2012)
4. Y.Y.Shao, J.Wang, H.Wu, J. Liu. Electroanalysis. **22** (2010)
5 G. Zhang, W. Choi, Z, Yu, et al. Hazard. Mater. **188** (2011).6
6. H.An, Q.Lin, DJ.Tao.Appl Sur Sci. **258** (2011)
7. D.Bejan, DJ. Malcolm, L.Morrison, et al Electrochim Acta. **54** (2009)
8. CB.Liu, K.Wang, LY.Chen, et al. Rev Anal Chem. **31** (2012)
9. V.Mani, A.P.Periasamy, SM.Chen. Soc. **133** （2012）
10. M.A.Quiroz, J.Lobato, R.Paz, et al Water.Res.**39** (2005)
11. M.C.Tomei, M.C.Annesini, Environ. Sci.Technoi. **34** (2014)
12. B.Chen, C.Yang, N.k.Goh. Phsy Chen. **111** (2007)

Catalytic Dehydrofluorination of Hydrofluoroalkanes to Fluorinated Olifein Over Ni/AlF$_3$ Catalysts

Wenzhi Jia [1], Xia Cai [1], Yong Zhang [1], Xiaohua Zuo [1], Juanjuan Yuan [2], Xinhua Liu[3], Zhirong Zhu [2] and Xiangyi Deng[1, a]

[1]*School of Chemistry and Chemical Engineering, Hubei Polytechnic University, 435003, Hubei, China*
[2]*School of Chemical Science and Engineering, Tongji University, Shanghai 200092, China*
[3]*Dyson school of design engineering, Imperial college London, London W2 1PG, UK*

Abstract. The hydrofluoric acid-resisting aluminum compounds (AlF$_3$, AlPO$_4$, AlN) supported with Ni catalyst were prepared by the wetness impregnation and tested for dehydrofluorination of hydrofluoroalkane to synthesize fluoroolefins. It is found that Ni/AlF$_3$ catalyst has the best catalytic performance, CF$_3$CFH$_2$ conversion of 29.3% after the reaction at 430 °C for 30 h, CF$_2$HCH$_3$ conversion of 31.8% after the reaction at 250 °C for 30 h, respectively. Comparatively, dehydrofluorination temperature of CF$_3$CFH$_2$ is higher than CF$_2$HCH$_3$ over the aluminum compounds catalyst, and the activity of catalysts is related with Lewis acidity. For the aluminum compounds catalyst, addition of Ni had promoted the activity and stability of Lewis acidic catalysts, it is attributed to synergistic catalysis of Lewis acid sites and Ni.

1. Introduction

Fluorine-containing olefins are mainly used as monomer raw materials for fluoropolymer, such as vinyl fluoride for the PVF, vinylidene fluoride for the PVDF, and also as intermediates for the fluorine-containing fine chemicals such as trifluorobromoethylene, etc. Fluoroolefins are a kind of important fine chemicals, the synthesis route has been extensively studied in recent years. In many production methods of fluorine-containing olefins, dehydrofluorination (simplified to De-HF) of hydrofluorocarbons (HFCs) is simple and efficient. At present, there are three De-HF ways: pyrolysis [1-3], liquid alkali [4] and heterogeneous catalysis [5]. Pyrolysis usually needs a high temperature, rigorous equipment, many by-products. Liquid alkali way requires the use of strong alkali and produces large amount of by-product salts, those will bring about serious environmental problems. The catalytic De-HF with low temperature and high selectivity is focus on the synthesis of fluorine-containing olefins.

Most of the transition metal (such as Pd, Cu, Ni, Fe, Ir, Ru, etc.) has the role of activated C-H bond, so the transition metal and its supported catalyst in the catalytic dehydrogenation have a very important application. Pd/AlF$_3$ catalyst has the high reactivity for dehydrofluorination of 1, 1-difluorocyclohexane and 1, 1, 1, 3, 3-Pentafluoropropane [6,7], respectively. However, the noble metal catalyst due to its high price and gradually replaced by a large number of cheap non-noble metal, which also meet the objectives of modern chemical sustainable development [8]. Relatively inexpensive Ni is considered to be one of a ideal metal for replacing noble metal. Alonso et al. [9] pointed out that nano-Ni particles can promote the H transfer reaction (such as alkane dehydrogenation, alcohol oxidation dehydrogenation, amine dehydrogenation). Shimizu et al. [8] studied the reaction mechanism of propanol dehydrogenation on Ni/Al$_2$O$_3$ catalyst, and believed that Ni can activate C-H bond and promote the dehydrogenation reaction with the acid-base site on Al$_2$O$_3$. Osaki et al. [10] reduced NiO-Al$_2$O$_3$ aerogels to obtain Ni/Al$_2$O$_3$ catalyst, TEM found that Ni particles with less than 6 nm were highly dispersed on the surface of Al$_2$O$_3$, which had a very good catalytic performance for CH$_4$ reforming. Aluminum compounds supported with Ni catalysts were investigated in the De-HF of HFCs in this paper.

The study related to De-HF mechanism of HFCs is few up to now. Kamiguchi et al. [11] gives a schematic representation of the typical dehydrohalogenation mechanism of HFCs, as shown in Scheme 1. It is generally believed that De-HF is carried out by the E1 mechanism. The reaction initial step is cleavage of C-F bond by activation. Usually, the fluorine of HFCs with strong electronegativity has partially negativity charged, it can be chemically adsorbed by Lewis acid sites and accelerate the cleavage of the C-F bond. Teinz et al. [12] argued that AlF$_3$ belonged to Lewis acid was an effective De-HF catalyst, and the interaction between Lewis acid sites and fluorine atom initiates the De-HF. Li et al. [13] suggested that the weak Lewis acid sites on the P$_2$O$_7^{4-}$

[a] Corresponding author: deng29606@sina.com

salt is the reaction center, it proposed that the carbonium-ion mechanism over the $Mg_2P_2O_7$ catalyst for De-HF of 1, 1, 1-trifluoroethane(CF_3CH_3) to vinylidene fluoride. Dehydrofluorination of HFCs should involve the cleavage of C-F and C-H bond. For the above studies, they focus the C-F bond activation, however, the synergistic activation of C-F and C-H bond is not proposed up to now. At present, there is no clear conclusion about De-HF mechanism for HFCs on the catalyst surface. In this paper, Aluminum compound supported Ni catalysts were used to catalytic De-HF of CF_3CFH_2 and CF_2HCH_3, respectively. BET, XRD, NH_3-TPD and FT-IR characterization were conducted to catalysts, and the relationship between activity and structure of catalyst was revealed. The synergistic catalysis mechanism of dehydrofluorination of HFCs was discussed.

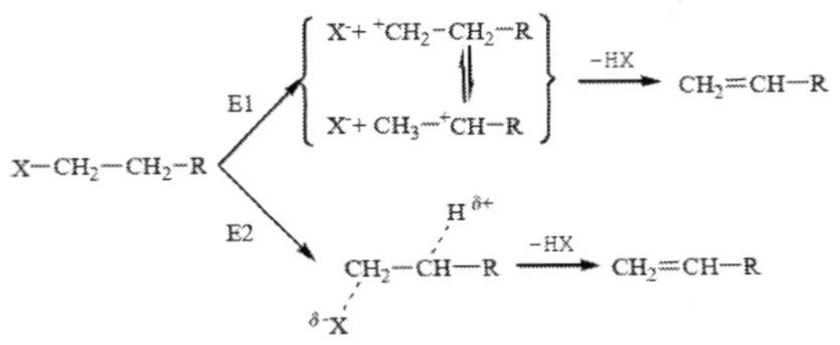

Scheme 1 The De-HF mechanism of HFCs.

2. Materials and Methods

2.1 Catalyst preparation

The Ni/AlF_3, Ni/AlN and $Ni/AlPO_4$ catalyst were prepared by the wetness impregnation in aqueous solution of $Ni(NO_3)_2{\cdot}6H_2O$. The raw materials included $Ni(NO_3)_2{\cdot}6H_2O$, AlF_3, AlN and $AlPO_4$ were from the Aladdin. The aluminum compounds impregnated with nickel nitrate was dried at 100 °C overnight. Finally, it was calcined at 500 °C for 3 h at air. Then prepared catalyst was subject to reduction in H_2 flow (40 mL/min) at 500 °C for 2 h. Prior to use, prefluorination was carried out in a stainless steel tubular reactor with a diameter of 3 cm and a length of 40 cm.

2.2 Dehydrofluorination of HFCs

Dehydrofluorination of HFCs (CF_3CFH_2, CF_2HCH_3) was carried out in a continuous-flow fixed-bed metal reactor (15 mm i.d.) operated in the downflow mode at atmospheric pressure. The reaction temperature was determined with a thermocouple in contact with the 2.0 g catalyst. The flow rates of the gas HFCs and N_2 were carefully regulated by mass flow controllers. The molar ratio of N_2/HFCs was 10 in a total flow rate of 22.5 ml min^{-1} and the gas hourly space velocity (GHSV) was 675 h^{-1}. To remove the product HF, the reaction effluent passed an aqueous KOH solution, then, it was analyzed by a gas chromatograph (HP-5890) using a HP-PLOT/Al_2O_3 column (I. D. 0.53 mm × 50 m) with a FID detector.

2.3 Characterization

Surface areas of the catalysts were determined by the modified BET method from the N_2 sorption isotherms at 77 K on an automatic NOVA 2200 eapparatus.

X-ray diffraction (XRD) patterns were collected on a Foucs D8 powder diffractometer operating at 40 kV and 30 mA using Cu-*Ka* radiation, in the 2θ range from 10 to 70° with a scan rate of 0.2° min^{-1}.

Pyridine (Py) absorbed infrared spectrum (Py-IR) was also conducted at 200 and 400 °C, under a high vacuum. The IR spectrum a self-supported wafer of the sample was also recorded in the spectral range 1800–1400 cm^{-1} with 64 scans and at a resolution of 4 cm^{-1}.

Ammonia temperature-programmed desorption (NH_3-TPD) was conducted on an automatic Altamira-100 Characterization System. 100 mg of catalysts was loaded into a quartz tublar reactor and was heated from room temperature to 500 °C and kept for 30 min in a flow of N_2 (30 ml min^{-1}). Then it was cooled down to 50 °C. A flow of NH_3 (20 ml min^{-1}) was then introduced for 10 min. The gaseous or physical sorption NH_3 was removed by purging N_2 flow (30 ml min^{-1}) at 80 °C for 1 h. Then the sample was heated to 600 °C with a ramp of 10 °C min^{-1}. The desorbed NH_3 was monitored continuously via a TCD detector. The total amount of NH_3 desorbed was exactly determined by the adsorption with an excess of dilute HCl and following the titration with NaOH solution. For the titration, the mixture indicator was adopted, containing 0.1% brom-cresol green ethanol solution and 0.2% methyl red ethanol solution.

High resolution transmission electron microscopy (HRTEM) of the sample was obtained on a JEOL JEM 2100F equipment with a field emissive gun, operating at 200 kV and with a point resolution of 0.24 nm.

3. Results and discussion

3.1. Reactivity

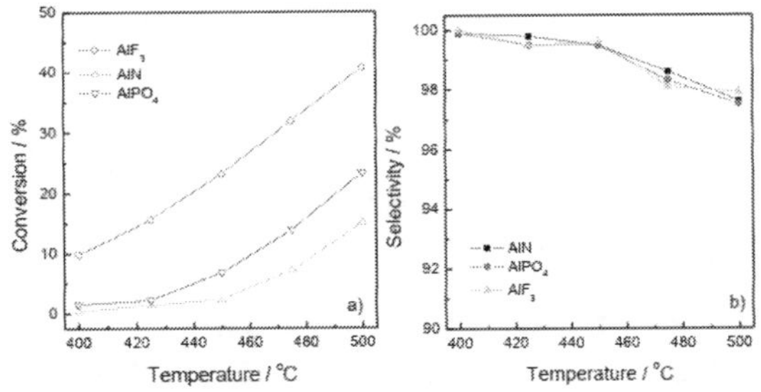

Figure 1 a) Conversion and b) selectivity of AlF_3, AlN and $AlPO_4$ catalysts for De-HF of CF_3CFH_2.

Since the HF with strong corrosion produced during dehydrofluorination (De-HF) of hydrofluorocarbons (HFCs), choice of catalyst is harsh. Generally, fluorides, phosphates and nitrides have the hydrofluoric acid-resistance. It is selected the AlF_3, AlN and $AlPO_4$ catalysts for De-HF of HFCs and investigated their catalytic performance. 1, 1-difluoroethane (CF_2HCH_3) and 1, 1, 1, 2-tetrafluoroethane (CF_3CFH_2) are considered as typical HFCs molecular for De-HF reaction. Figure 1a) is the activity results of the AlF_3, AlN and $AlPO_4$ catalyst in the reaction temperature range of 400~500 °C. It is found that the activity of AlF_3 is higher than that of the AlN and $AlPO_4$. The AlN and $AlPO_4$ have little catalytic activity when the reaction temperature is less than 425 °C. Fig.1b) is the relationship between selectivity to trifluoroethylene and reaction temperature. At 480 °C, selectivity of three catalysts is not much different, all above 98%. From Figure 1b), the selectivity of three catalysts to trifluoroethylene is gradually decreased as the reaction temperature increases, due to presence of by-product CFH=CFH, CF_2HCF_3 and CF_2HCFH_2 derived from the cracking or isomerization of CF_3CFH_2 under the high temperature [14,15]. The temperature is also an important factor affecting the performance of the catalyst. Among the three catalysts, AlF_3 catalyst has a good catalytic performance for De-HF of CF_3CFH_2.

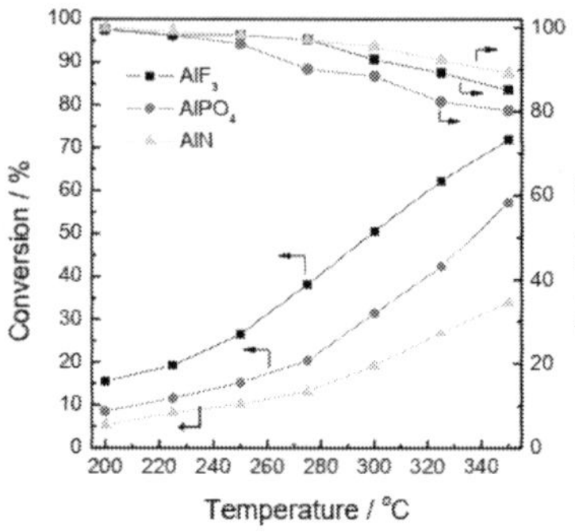

Figure 2 Reactivity of AlF_3, AlN and $AlPO_4$ catalyst for dehydrofluorination of CF_2FCH_3.

Then, the AlF_3, AlN and $AlPO_4$ catalysts used to dehydrofluorination of CF_2HCH_3 to vinyl fluoride. It can be seen from the Figure 2 that AlF_3 has the highest activity for De-HF of CF_2HCH_3, compared with the $AlPO_4$ and AlN catalysts. When the temperature is low (200~250 °C), the selectivity of the three catalysts to vinyl fluoride is not very different (close to 100%). When the reaction temperature is more than 280 °C, selectivity (S) of three catalysts is: $S_{AlN} > S_{AlF3} > S_{AlPO4}$. Similarly, the temperature has a significant effect on De-HF of CF_2HCH_3, and the AlF_3 catalyst has a high activity.

Table 1. Dehydrofluorination of CF_3CFH_2, CF_2HCH_3.

Catalyst	Conversion of CF_3CFH_2 / % (430 °C)				Conversion of CF_2HCH_3 / % (250 °C)			
	Time on stream / h				Time on stream / h			
	5	10	20	30	5	10	20	30
AlF_3	25.6	25.2	24.3	22.5	27.8	27.7	25.2	24.3
AlN	3.8	3.7	3.6	3.4	10.2	10.1	9.5	9.2
$AlPO_4$	8.6	8.5	8.4	8.2	15.2	15.0	14.5	14.1
Ni/AlN	5.2	5.1	5.0	4.8	13.4	13.2	12.5	12.1
Ni/AlF_3	30.2	29.8	29.2	29.3	32.5	32.1	31.5	31.8
Ni/$AlPO_4$	11.8	11.3	10.5	11.2	17.5	17.4	16.2	16.7

Supported Ni has a high activity for alkyl dehydrogenation, ammonia selective oxidation, alkylated steam reforming [16], etc., which has attracted much attention in the field of heterogeneous catalysis. The Ni/AlF_3, Ni/AlN and Ni/$AlPO_4$ catalysts used to De-HF of CF_3CFH_2 and CF_2HCH_3, respectively, and the reactivity results are shown in Table 1. As seen from Table 1 that the activity of the AlF_3, AlN and $AlPO_4$ catalysts has decreased with the reaction time for De-HF of HFCs (CF_3CFH_2 and CF_2HCH_3). It is found that activity of three catalysts is improved after 2%Ni supported on the AlF_3, AlN and $AlPO_4$ catalysts. Although the activity of catalyst supported Ni also slows down, the decrease rate is obviously lower. So, the addition of Ni can improve activity for De-HF of HFCs. Dehydrofluorination temperature of CF_3CFH_2 is significantly higher than that of CF_2HCH_3, which is due to CF_3CFH_2 molecules with more strong electron-withdrawing F and higher dissociation of the C-F bond.

3.2 Characterization

The surface area, pore diameter, pore volume of catalysts listed in Table 2. Surface area of the AlF_3, $AlPO_4$ and AlN is 52 $m^2 g^{-1}$, 29 $m^2 g^{-1}$, 85 $m^2 g^{-1}$, respectively. After supported Ni, their surface areas have little decrease, with no difference of pore diameter and vlolume. AlN catalytst has the low activity, it may due to the small surface area (29 $m^2 g^{-1}$).

Table 2 Physical properties of Al-based and Ni/AlF_3 catalyst.

Catalyst	S_{BET} / $m^2 \cdot g^{-1}$	Pore diameter / nm	Pore volume / $cm^3 \cdot g^{-1}$	Acid sites / $mmol \cdot g^{-1}$
AlF_3	52.0	9.0	0.646	0.326
$AlPO_4$	85.0	17.9	0.483	0.154
AlN	29.0	25.8	0.563	0.048
Ni/AlN	20.0	26.5	0.545	0.041
Ni/AlF_3	50.0	14.4	0.226	0.322
Ni/$AlPO_4$	64.0	13.2	0.046	0.147

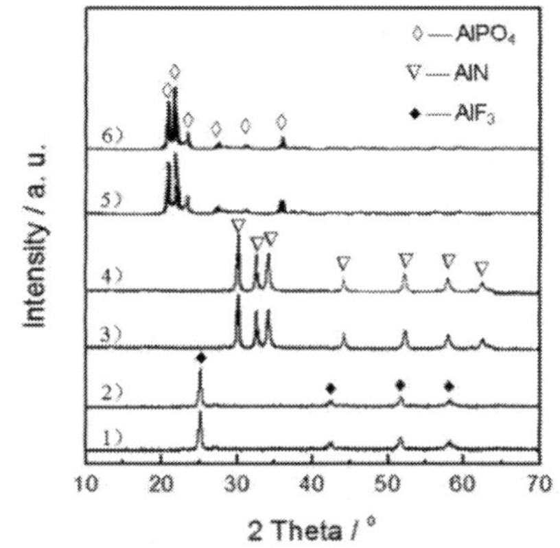

Figure 3 XRD patterns. 1) AlF_3, 2) Ni/AlF_3, 3) Ni/AlN, 4) AlN, 5) $AlPO_4$, 6) Ni/$AlPO_4$

Figure 3 shows the XRD results of catalysts, AlF_3 (PDF: 44-0231), AlN (PDF: 25-1133) and $AlPO_4$ (PDF: 45-0177). As seen from Fig. 3 that the addition of Ni in the AlF_3, AlN and $AlPO_4$ catalyst has no obvious change in the XRD diffraction peak. No peaks of Ni shown in the

Figure 3, this indicates that Ni present in the catalyst in the form of highly dispersion.

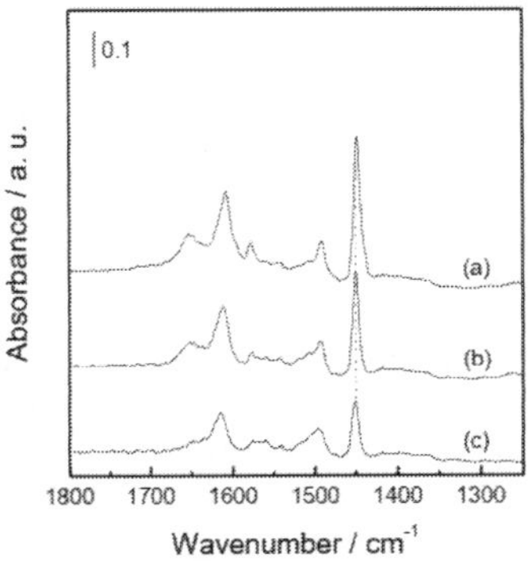

Figure 4 Py-IR spectra of (a) AlF_3, (b) $AlPO_4$ and (c) AlN

In this paper, Py-IR technique was used to detect the surface acid types of the AlF_3, AlN and $AlPO_4$ catalysts. Figure 4 shows the Py-IR results of the AlF_3, AlN and $AlPO_4$ catalysts. The vibrational peaks are attributed to Lewis acids at 1450 cm^{-1}, 1608 cm^{-1} and 1652 cm^{-1} (ν_{19b} and ν_{8a}, respectively), and 1540 cm^{-1} (ν_{19b}) attributed to Brönsted acid [17,18]. It can be seen from Figure 4 that the vibration peaks (ν_{19b}) of the AlF_3, AlN and $AlPO_4$ catalysts appear at 1450 cm^{-1}, indicating that the surface of these catalysts is predominantly Lewis acid. After Ni was loaded, the surface acid types of the Ni/AlF_3, Ni/AlN and Ni/$AlPO_4$ catalysts did not change significantly.

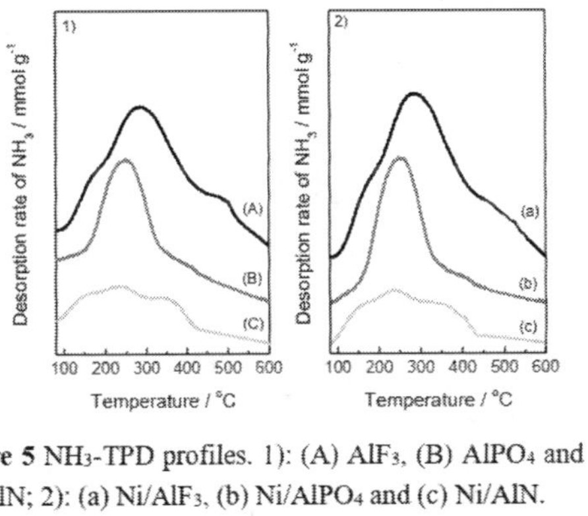

Figure 5 NH_3-TPD profiles. 1): (A) AlF_3, (B) $AlPO_4$ and (C) AlN; 2): (a) Ni/AlF_3, (b) Ni/$AlPO_4$ and (c) Ni/AlN.

In Figure 5.1) is the NH_3-TPD curve for the AlF_3, AlN and $AlPO_4$ catalysts. Compared with three catalysts, the AlN_3 has the largest NH_3 desorption peak area, followed by the $AlPO_4$, that of AlN is the smallest, so the acidity (A) order of the three catalyst is: $A_{AlF3} > A_{AlPO4} > A_{AlN}$, the acid amounts are listed in Table 2. Figure 5.2) is the NH_3-TPD results for the Ni/AlF_3, Ni/$AlPO_4$ and Ni/AlN catalysts. By comparing the catalyst doped and undoped Ni from Figure 5, it was found that the distribution of acid sites (NH_3 desorption peak position) and the peak area were not obviously different. Their acidity results are shown in Table 2. The acid amount of the catalyst before and after the loading of Ni did not change significantly. Thus, the loading Ni did not significantly alter the Lewis acid properties of catalysts.

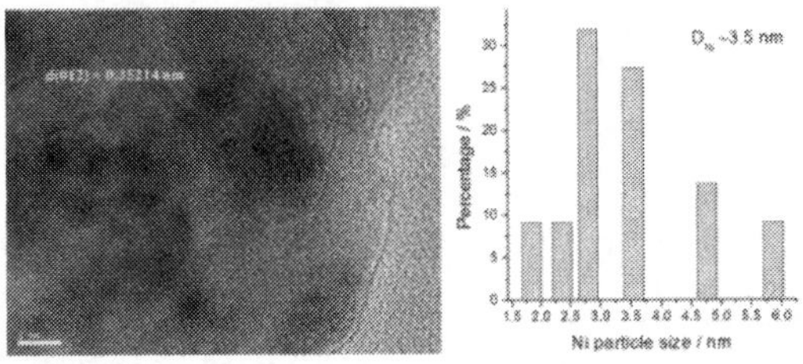

Figure 6 TEM images of Ni/AlF_3 catalyst

As can be seen from Table 1, the addition of Ni improved the activity of catalysts. As the Ni/AlF_3 catalyst has good activity for De-HF of HFCs, Ni supported on the catalyst surface are characterized by TEM, as shown in Figure 6. The lattice spacing of AlF_3 was calculated to be 0.35214 nm, which was attributed to the 012 crystal plane of AlF_3 according to the standard card of XRD. Distribution of Ni Particle Size As shown in the histogram in Figure 6, the size of Ni particles on the surface is mainly distributed at about 3 nm, which perform the excellent activity for De-HF of HFCs in this region of Ni particles.

3.3 Effect of Lewis acid sites and Ni

The Lewis acid sites play an important role to dehydrofluorination (De-HF) of hydrofluorocarbons (HFCs). De-HF of HFCs requires more stringent conditions of reaction, such as high temperatures. The dissociation energy of C-F is 522.0 ± 8.4 kJ mol^{-1}, which is the higher than 414 kJ mol^{-1} of C-H, so the cleavage of C-F bond is considered to be a key step of catalytic De-HF of HFCs. It is well known that Lewis acid are the reactivity centers of the activated C-X (X=F, Cl, Br, I) bonds [12]. AlF_3 catalyst has a high activity for De-HF of HFCs due to the fact that AlF_3 has large number of Lewis acid sites (0.326 mmol g^{-1}). For the AlF_3, AlN and $AlPO_4$ catalyst, it was also found that Lewis acid sites were proportional to activity of De-HF, the same results reported in our previous works [19].

According to the result of Table 1, for the AlF_3 catalyst, conversion of CF_2HCH_3 is 22.5% at 430 °C, conversion of CF_3CFH_2 is 24.3% at 250 °C. For the Ni/AlF_3 catalyst, conversion of CF_2HCH_3 is 29.3% at 430 °C, conversion of CF_3CFH_2 is 31.8% at 250 °C. The addition of Ni in the AlF_3, AlN and $AlPO_4$ catalyst improved the De-HF activity of HFCs (CF_2HCH_3 or CF_2HCH_3). Besides, the stability of AlF_3 catalyst obviously improved by addition of Ni as seen from the Table 1. The activity of AlF_3 catalyst decline with time on stream, the long-chain fluorocarbons and/or coke coving the Lewis acid centres might result in deactivation of catalysts. The long-chain fluorocarbons and coke derived from the coupling and cracking of carbon-positive ion intermediates [13], carbon-positive ion intermediate generated by cleavage of C-F bond over strong Lewis acid sites which considering as the first step

[12]. The Ni/AlF_3 catalysts possessed good activity and stability, it may be due to the fact that Ni restrain the formation of carbon-positive ion intermediate through the synergistic catalysis.

Scheme 3 shows the De-HF process of HFCs to fluoroolefins over the bi-functional Ni/AlF_3 catalyst. Carmichael Group [20,21] proposed two De-HF mechanism of CF_3CH_3 and CF_2HCH_3, which is the homolytic and heterolytic cleavage in the C-F and C-H bond fracture. For the AlF_3 catalyst, the catalytic mechanism due to the carbon-positive ion (heterolytic cleavage), the activation of C-F bond over Lewis acid sites is the initial step. However, for the Ni/AlF_3 catalyst, the synergistic catalysis took place on the surface of catalyst (homolytic cleavage), the synergetic activation of C-F and C-H bond by Lewis acid sites and Ni showed in schematic diagram (Scheme 3). So, in the E2 elimination mechanism of HFCs, Lewis acid sites and Ni activated C-F and C-H bond, respectively. This mechanism could avoid the carbon-positive ion intermediates as much as possible. That may be main reason that Ni/AlF_3 catalyst has a good catalytic performance.

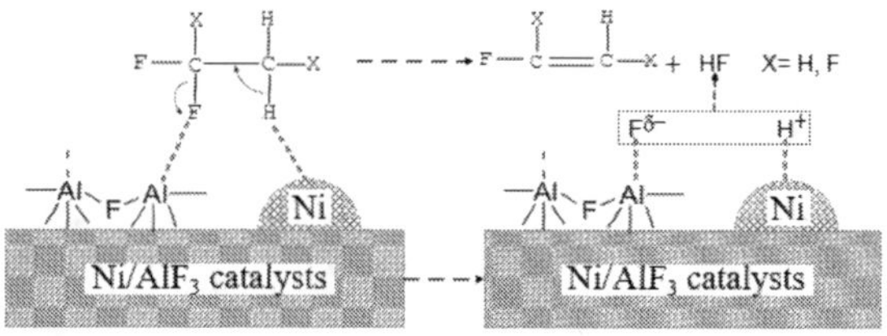

Scheme 3 Reaction mechanism of De-HF of HFCs over the Ni/AlF_3 catalyst.

4. Conclusion

Catalytic performance of AlF_3, AlN and $AlPO_4$ catalyst were investigated for dehydrofluorination of CF_3CFH_2 and CF_2HCH_3, respectively. By the IR and NH_3-TPD characterization, the activity of AlF_3, AlN and $AlPO_4$ catalyst were related to their Lewis acid sites. After additive Ni in the three Al catalysts, catalytic performance of catalysts were improve obviously, and of that Ni/AlF_3 catalyst is the best, the Ni particle is about 3 nm. The dehydrofluorination of HFCs to fluoroolefin over the Ni/AlF_3 catalyst were subjected to a synergistic catalysis of Lewis acid sites and Ni.

Acknowledgments

This work was financially supported by National Natural Science Foundation of China (No. 51372077 and 21603069), Education Foundation of Hubei Province (No. Q20184502), the Science Foundation from Hubei Polytechnic University (No. 16xjz05R), National Undergraduate Training Program for Innovation and Entrepreneurship (201610920008).

References

1. B.A. Mahler, M.J. Nappa, J.P. Knapp, US Patent, 20090278075. 2009
2. V.N.M. Rao, M. J. Nappa, A.C. Sievert, US Patent, 20090264689. 2009
3. J. Harmon, US Patent, 2599631. 1952
4. A.C. Sievert, N.A. Carlson, US Patent, 20100174123. 2010
5. K. Takahshi, T. Chaki, Y. Shiotani, WO Patent, 2010021406. 2010
6. M. Tojo, S. Fukuoka, H. Tsukube, Bull. Chem. Soc. Japan. **84**, 333 (2011)
7. F. Wang, W.X. Zhang, Y. Liang, Y.J. Wang, J.Q. Lu, M.F. Luo, Chem. Res. Chin. Univ. **31**, 1003 (2015)
8. K. Shimizu, K. Kon, K. Shimura, S.M.A. Hakim Sidikki, J. Catal. **300**, 242 (2013)
9. F. Alonso, P. Riente, M. Yus, Acc. Chem. Res. **44**, 379 (2011)
10. T. Osaki, T. Mori, J. Non-Crystal. Solids **355**, 1590 (2009)
11. S. Kamiguchi, M. Watanabe, K. Kondo, M. Kodomari, T. Chihara, J. Mol. Catal. A: Chem. **203**, 153 (2003)
12. K. Teinz, S. Wuttke, F. Börno, J. Eicher, E. Kemnitz, J. Catal. **282**, 175 (2011)
13. G.L. Li, H. Nishiguchi, T. Ishihara, Y. Moro-oka, Y. Takita, Appl. Catal. B **16**, 309 (1998)
14. K.U. Niedersen, E. Schreier, E. Kemnitz, J. Catal. **167**, 210 (1997)
15. E. Kemnitz, K.U. Niedersen, J. Catal. 155, 283 (1995)
16. G. Garbarino, I. Valsamakis, P. Riani, G. Busca, Catal. Comm. **51**, 37 (2014)
17. G. Busca, Physical Chemistry Chemical Physics, **1**, 723 (1999)
18. G. Ramis, L. Yi, G. Busca, Catal. Today, **28**, 373 (1996)
19. W.Z. Jia, M. Liu, X.W. Lang, C. Hu, J.H. Li, Z.R. Zhu, Catal. Sci. Technol. **5**, 3103 (2015)
20. B.D. Neely, H. Carmichael, J. Phys. Chem. **77**, 307 (1973)
21. B. Noble, H. Carmichael, C.L. Bumgardner, J. Phys. Chem. **76**, 1680 (1972)

Research on modified silicone compound photocatalyst by doping N

DaWei Pan[1], GuangXu Shao[1] and QingWen Zhang[2]
[1]*Department of Environmental Engineering, School of Chemical Engineering, University of Science and Technology Liaoning, Anshan, Liaoning, China*
[2] *Anshan Iron and Steel Group Engineering Technology Co., Ltd., Anshan, Liaoning, China*

Abstract. In order to improve the visible light response and reaction efficiency of nitrogen-doped TiO_2 set urea as nitrogen source, and silica gel as the load reagent adopt sol-gel method to prepare powder of nitrogen-doped TiO_2 supported on SiO_2 , use equal volume impregnation method to load MnO_2 on catalyst, and then obtain the composite photocatalyst after drying, and roasting .The prepared composite catalysts were characterized by XPS, TEM, SEM, XRD and other methods. Moreover, the photo-catalytic activity of the composite catalysts under visible-light region was tested, and the influences of nitrogen content and calcinations temperature on the photo-catalytic activity were investigated. The results show that:(1) the range of light response of modified composite catalyst was expanded from ultraviolet region to visible-light region, which results in high visible-light catalytic activity in the degradation of methyl orange.(2) Through the mechanism of photo-catalytic reaction and the treatment effect analysis, MnO_2 as catalytic resulting O_2 can be served as a good electronic capture agent and improve the reaction efficiency.(3)With the reduction of nitrogen content,and the increase of calcination temperature, the visible-light catalytic activity was weakened.(4) The activity of catalyst was reused six times, without significant reduction and had excellent efficiency and stability.

Introduction

With the increasingly serious energy shortage and environmental pollution, fanding the suitable photocatalyst to realize the effective utilization of solar energy, has become a hot issue in the field. Therefore, TiO_2 with excellent catalytic performance attracts more attention in the environmental, [1] which has the advantages of low cost, good chemical stability, non-toxic,and no secondary pollution. But there are also shortcomings, e.g.,the forbidden band of TiO_2 is wide (Eg=3.2eV)and can only absorb UV-light; In addition, the hole and electron stimulated by light are easily compounded so as to reduce the photo-quantum efficiency [2-3]. Many domestic and international experts and scholars have tried to use the nonmetal ion doping method to extend the visible light absorption of TiO_2, where nitrogen-doping can effectively hence the absorption of the catalyst in the visible-light region, as well as the efficiency of photo-catalytic oxidation [4]. SiO_2 as the load reagent can effectively control the growth of TiO_2 particles so as to get smaller size, and larger specific surface area, which can improve the separation performance, and make them inactivation to prolong the service life [5-6]. Researches show that quantum yield is the most important factor influencing treatment effect of photocatalyst, so improving the light quantum yield of photo-catalyst, promoting the separation of photon-generated carriers and reducing the recombination rate are key factors, in this paper, by preparing the nitrogen-doped TiO_2/MnO_2 supported on SiO_2 composite photocatalyst the new ecological O_2 resolved from hydrogen peroxide which are produced in the process of catalytic hydrolysis can be used as electron capture agent to effectively promote the separation of photon-generated carriers and to reduce its compound probability [7]. This kind of photo-catalyst has advantages of visible light response, and smaller and equally distributed particle size, and can efficiently restrain the compound of electronic - hole pairs, which greatly improve the treatment effect of the pollutants.

[a]Corresponding author :Tel: 86 13998091973; E-mail:wxq_as@163.com

Experiment

1.1 preparation of Nitrogen-Doped TiO_2/MnO_2 supported on SiO_2 composite photo-catalyst

The preparation was made as follow : After adjusting the PH value of Solution B (10mL of deionized water and 20mLof anhydrous ethanol) to about 2, dropwise add Solution A(10mL of actic acid,10mL of dibutyl phthalate and 30mL anhydrous erhanol) into it with strong stirring, and then continue to strongly stir for 30 minutes after dropping, add different volumes of urea whose concentration is 1mol/L to the mixed Solution with N/Ti molar ratio of 10%, 15%, 20% and 25% respectively, and continue to stir for 30 minutes, add a certain amount of SiO_2 as load reagent and strongly stir for 60 minutes, stand and air dry the mixture at room temperature until it become xerogel and then grind it in agate mortar, calcine it in muffle furnace for 3hours at different temperature of 350℃,400℃,450℃ and 500℃ respectively, grind it again and screen it by standard sieve with 300 meshes, adopt the equal volume impregnation method to load a certain concentration of manganese nitrate on the calcined sample of catalyst in ultrasonic instrument, after 1 hour's drying, and calcination a series of nitrogen-doped TiO_2/MnO_2 supported on SiO_2 composite photo-catalyst with different calcination temperature is obtained.

1.2 performance of Nitrogen-Doped TiO_2/MnO_2 supported on SiO_2 composite photo-catalyst

The adsorption ability and photo-catalytic performance of nitrogen-doped TiO_2/MnO_2 supported on SiO_2 composite photo-catalyst was evaluated by the degradation of methyl orange, and the reaction was carried out in the self-made additional stirring temperature cycle optical instrument with 150w iodine tungsten lamp as light source. The reaction device keeps the temperature of the reaction system at 25 ℃, and the magnetic stir keeps uniform concentration and temperature of solution in the system ,Take a certain quality of nitrogen-doped TiO_2/MnO_2 supported on SiO_2 composite photo-catalyst in reaction device, add 100mL of methyl orange solution with certain concentration, and make it adsorption-desorption balance through 30 minutes' magnetic stirring under the condition of no light, and then turn on the iodine tungsten lamp for illumination. In the process of aphotic adsorption - desorption balance and illumination, take a sample every 20 minutes, use the principle of centrifugal separation to remove the photo-catalyst, adopt ultraviolet-visible spectrophotometer to measure the absorbance of supernatant fluid at the longest absorption wave, and then calculate the concentration according to the standard working curve based on the concentration(C_t) of the supernatant fluid at the degradation time(t) and the initial concentration(C_0) of degradation, calculat the degradation rate (D) of methyl orange to evaluate the photocatalytic activity of nitrogen-doped TiO_2/MnO_2 supported on SiO_2 composite photo-catalyst:

$$D=(c_t - c_0)/c_0 \times 100\%$$

Results and Discussion

2.1 Characterization of Catalysts

2.1.1 X-ray diffraction (XRD) analysis of the catalyst

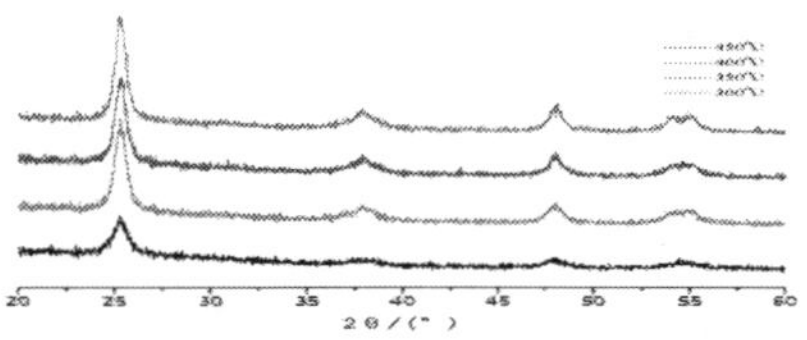

Figure1. The XRD spectrum of 20%N at different calcination temperatures

The X-ray diffraction (XRD) spectrum in Figure 1 shows that: all nitrogen doped TiO_2/MnO_2 supported on SiO_2 composite photocatalysts with N/Ti molar ratio of 20% prepared at 300℃, 350℃, 400℃ and 450℃ respectively, are anatase phases, and the Ti-N diffraction peak does not appear, because high temperature made N volatilize, resulting in low N content, or larger Ti - N clusters fail to form in the complex catalyst; Meanwhile the movement of characteristic peak of anatase type TiO_2 does not appear, because of nitrogen-doping, it shows that nitrogen doping does not change its crystal tructure. Using **Scherrer** formula to calculate the average particle size, the results are 13.8nm, 14.6nm, 12.3nm, and 17.2nm. With the increase of calcination temperature, the characteristic peak of anatase crystal phase becomes more keen-edged, which means the crystal shape is more and more regular and the size becomes larger and larger, because the increase of calcination temperature causes agglomeration of particles.

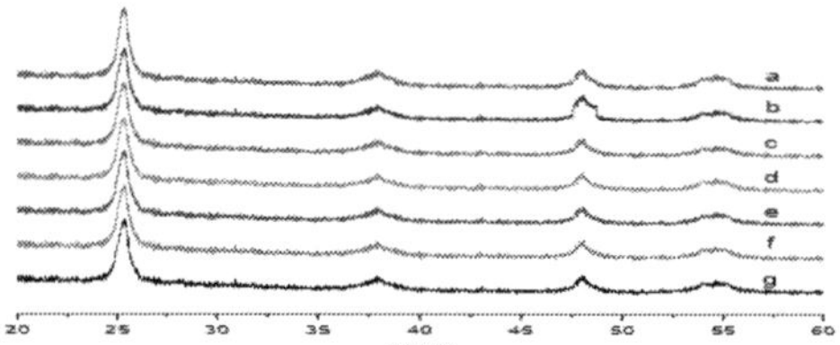

(a. 400℃TiO_2 b.400℃TiO_2/SiO_2 c. 400℃N/TiO_2 d. 400℃ 10%$N/TiO_2/SiO_2/Mn^{2+}$ e. 400015%$N/TiO_2/SiO_2/Mn^{2+}$ f. 400℃ 20%$N/TiO_2/SiO_2/Mn^{2+}$ g.400℃25%$N/TiO_2/SiO_2/Mn^{2+}$)

Figure 2. XRD spectrum of nitrogen doped samples with different molar ratios at calcination temperature of 400℃

Figure 2 shows that: all nitrogen-doped TiO_2/MnO_2 supported on SiO_2 composite photo-catalysts with different N/Ti molar ratio prepared at temperature of 400℃ are anatase phases. Using the **Scherrer** formula to calculate the average grain size, the results are 9.1nm, 8.4nm, 14.7nm, 12.4nm, 13. 2 nm, 12.3nm ,and 14.2nm. Nitrogen-doping enlarges the catalyst particles, and the size has a minimum value with the change of N/Ti molar ratio, which is corresponding to the catalyst with molar ratio of 20%. It is consistent with the verification results according to theoretical basis that more keen-edged diffraction peak and smaller full width at half maximum means smaller particle and better crystal.

2.1.2 Surface morphology analysis of Catalyst

Scanning electron microscope (SEM)

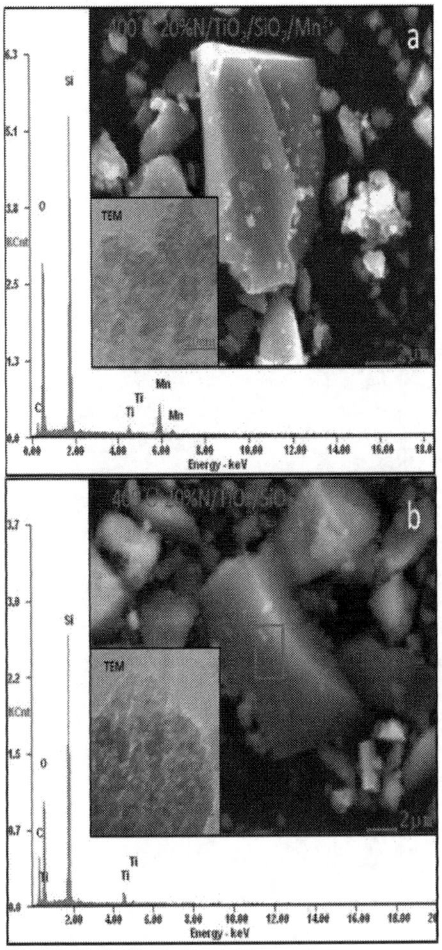

(a)20%N/TiO_2/SiO_2/Mn^{2+} (b)20%N/TiO_2/SiO_2

Figure 3. The SEM images of TiO_2 photo-catalysts

Figure 3 is the SEM images and energy spectrum diagrams of nitrogen-doped TiO_2 supported on SiO_2 composite photocatalyst, to which Mn^{2+} is added. It can be seen from the figure: TiO_2 particle size ranges from few nanometers to dozens of nanometers which is successfully loaded on the surface of SiO_2 particle which relatively uniform distribution; the comparison of energy spectrum diagram shows that Mn^{2+} has been successfully incorporated into and loaded on the surface of SiO_2 ,which slightly increases TiO_2 particle size. Thus it shows SiO_2 as loaded reagent can effectively improve the distribution of TiO_2 particles, and enhance the dispersing performance; and the Mn^{2+} has been successfully incorporated and loaded on the surface of SiO_2.

Transmission electron microscope (TEM)

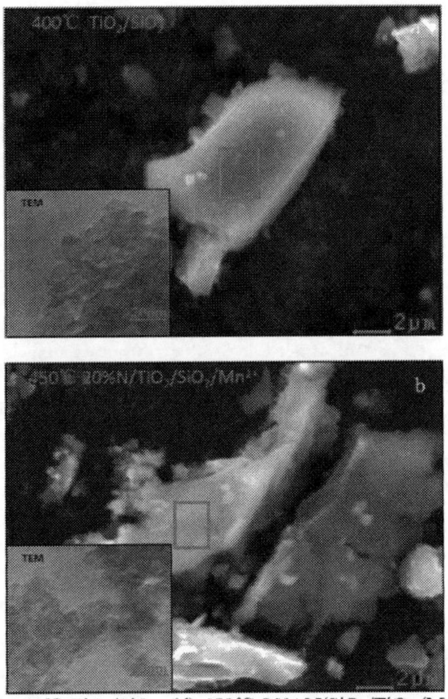

(a) 400℃ TiO_2/SiO_2 (d) 450℃ 20% N/SiO_2/TiO_2 /Mn^{2+}

Figure 4. TEM microgarghs of complex photo-catalyst

Figure 4 is the transmission electron microscope (TEM) images of different composition photo-catalysts. We can see from Figure 4 (a) that the TiO_2 particles are uniformly distributed on the surface of SiO_2, indicating that SiO_2 loading can effectively improve the dispersibility of TiO_2 particles; through the comparison of Figure 3 (a) and Figure 4 (b) it can be seen that the particle size increases with the increase of calcination temperature, which is due to the agglomeration of the grain caused by the increase of calcination temperature, verifying the analysis conclusion in XRD.

2.1.3 UV-vis diffuse reflectance.

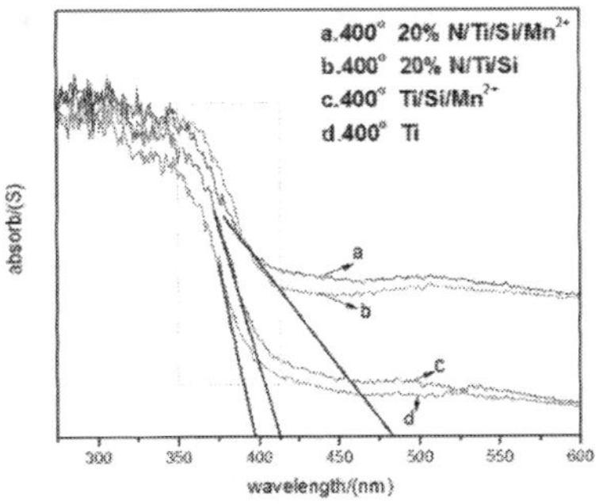

Figure 5 . UV-Vis Diffuse reflectance spectra of complex photo-catalyst

https://doi.org/10.1051/matecconf/201823803005

The UV-Vis diffuse reflectance spectra of Fig. 5 show that the absorption range of the modified catalyst is extended to visible light region from 390 nm to 480 nm, and the absorption wavelength range is increased by 90 nm, which indicates that nitrogen doping can effectively improve the visible light response activity of the catalyst. According to the tangent comparison of the curves (c) and (d), the addition of Mn only enlarges the absorption range of the catalyst by about 20 nm (from 390 nm to 410 nm). This shows that Mn is helpful to increase the number of photons involved in the photocatalytic reaction and enhance the absorption intensity of the composite photocatalyst, but does not produce obvious visible light response. According to the tangent comparison of curves a and (c), it is shown that doping N can effectively improve the visible light response activity of the catalyst, but Mn can not produce this effect. From the slope of the light absorption threshold region of the curves (a) and (d), the visible light response range of the composite photocatalyst is obviously enlarged, which indicates that the N_{2p} orbital replaces part of the O_{2p} orbital and forms a new hybrid orbital. The formation of Ti-O-N bonds on the surface of titanium dioxide forms a new energy level structure, which reduces the band gap of titanium dioxide and thus obtains visible light activity.

2.1.3 X-ray photoelectron spectroscopy of Catalyst

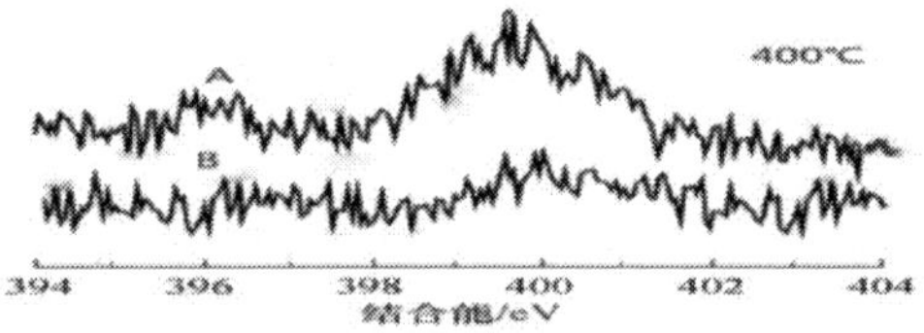

Figure 6. XPS energy spectrum diagram of N_{1s} of TiO_2 photocatalyst at calcination temperature of 400℃

Figure 6 is the XPS energy spectrum diagram of composite catalysts at calcination temperature of 400 ℃,where A and B respectively represent the catalysts with and without nitrogen-doping. It can be seen from the figure that, for the nitrogen-doped catalyst samples, there is a new N_{1s} peak at 396.1eV, and a N_{1s} peak of the catalyst surface adsor bed N_2 at 401.7eV, thus the N atoms and TiO_2 molecules are combined by chemical bonds. Asahi [8] et al, attributed the peak appearing in the 396eV to N atoms mixed with the crystal lattice, and attributed peaks appearing in the 400eV and 402eV to the molecular adsorption of N_2 on the catalyst surface. They pointed out that the catalyst with visible light activity is associated with the N atoms, and this is because the hybridization of N_{2p} orbital and O_{2p} orbital happened, which can decrease the forbidden band width of TiO_2, so as to be motivated by the visible light.

2.2 Photo-catalytic activity of Nitrogen-Doped TiO_2/MnO_2 supported on SiO_2 composite photo-catalyst

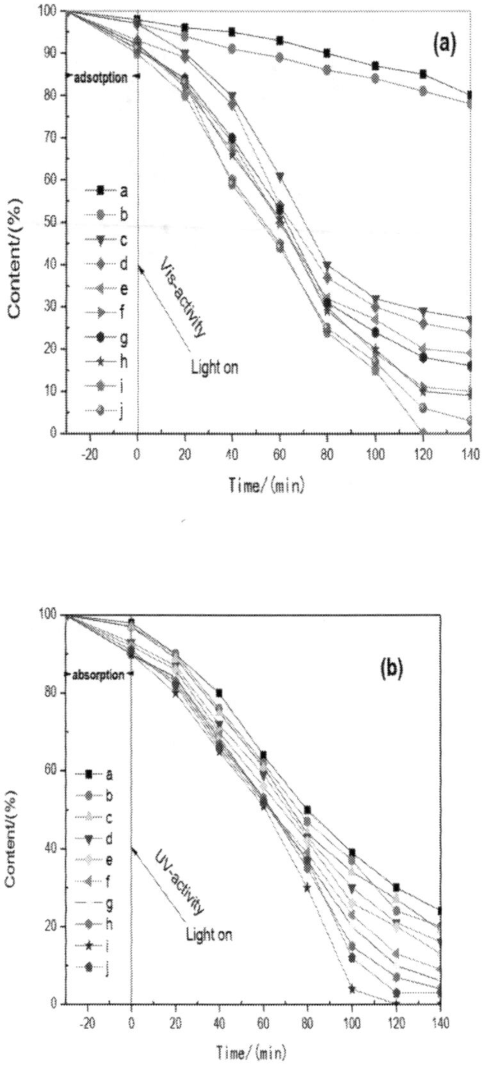

(1)Visible light (2)Uv light)(a. 400℃TiO_2 b.400℃TiO_2/SiO_2 c. 300℃ 20%N/$TiO_2/SiO_2/Mn^{2+}$ d. 400℃20%N/TiO_2/SiO_2 e. 350℃ 20%N/$TiO_2/SiO_2/Mn^{2+}$ f. 40010%N/$TiO_2/SiO_2/Mn^{2+}$ g.400℃15%N/$TiO_2/SiO_2/Mn^{2+}$ h.450℃ 20%N/$TiO_2/SiO_2/Mn^{2+}$ i.400 20%N/$TiO_2/SiO_2/Mn^{2+}$ j. 40025%N/$TiO_2/SiO_2/Mn^{2+}$)

Figure 7. The photo-catalytic activity of TiO_2 under different conditions

Figure 7 (1) shows test results of visible-light activity of catalyst. The comparison between *d* and *a* /*b* shows that the catalytic activity of visible light of nitrogen-doped TiO_2/SiO_2 has been significantly improved. This is because the doped N atoms partly replaced O atoms enter into crystal lattice of TiO_2, and formed an independent N_{2p}narrowband above the valence band, causing the response of composite catalysts to the visible light[8]. It can be seen from f, g, i and j in the figure, the catalytic activity of the composite photo-catalysts gradually

 https://doi.org/10.1051/matecconf/201823803005

increases with the increment of nitrogen-doping content, It will be optimal when the percentage of nitrogen doping content is 20%, at this moment the composite catalyst can completely get rid of methyl orange within 120 minutes, and that with different doping amount will not have the same effect under the same conditions. The reason may be that the newly created energy level after nitrogen-doping and the bulk phase of TiO_2 form a compound semiconductor structure, hence the electric potential difference of both valence bands can make the hole of TiO_2's valence band migrate to the newly formed energy levels, so as to effectively reduce the recombination probability of photo-generated electron hole, and improve the activity of catalyst [6]. can be seen from I and h in the figure that the removal rate decreases with the increase of temperature, and the reason may be that under high temperature, N element was oxidized and volatilizated in the heating process, leading to the decrease of N content in TiO_2 lattice, and the enlargement of size, and decrease of specific surface area, can also lower the activity of visible-light response. Figure 6(1) shows the catalyst activity test results of UV-light. It can be seen from the figure that the composite catalyst in ultraviolet region also has a good degradation effect. simultaneously, Figure 7 shows, Mn^{2+} has obviously improved the processing efficiency of catalyst. This is because the Mn^{2+} is oxidized into MnO_2 through calcination and loads on the surface of catalyst. It is used as catalyst to catalyze H_2O_2 to_produce O_2 which is served as electron capture agent to effectively promote the separation of photo-generated carrier and reduces its compound probability.

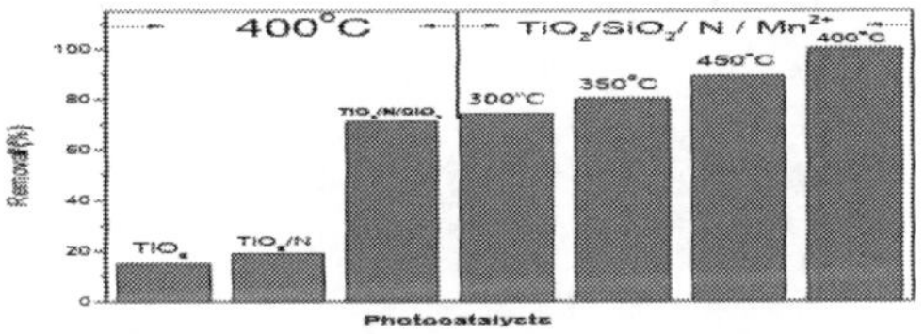

Figure 8. Photo-catalytic activity of TiO_2 with different condition of Mn^{2+} under Visible light

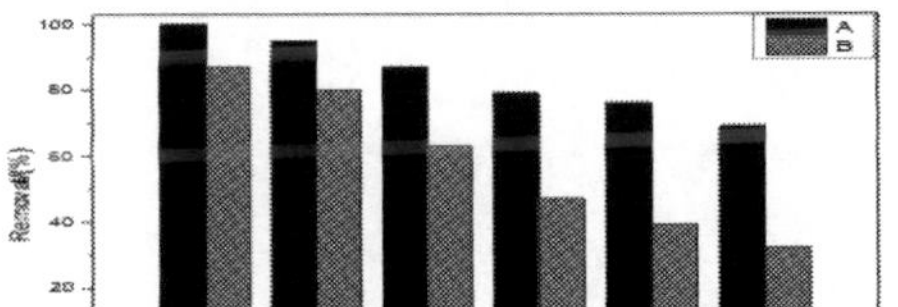

(A.400℃ 20%N/TiO_2/SiO_2/Mn^{2+} B. 400℃ 20%N/TiO_2/ Mn^{2+})

Figure 9 Effect of run times on the photo-catalytic activity of TiO_2

Figure 9 shows the experimental results of repeated use of catalyst. Under the same cycle times, the catalytic activity of the composite catalyst adding SiO_2 as lode reagent is obviously higher. After reused six times, the degradation rate of composite catalyst is 75%, while that without SiO_2 is only 27% under the same condition. In conclusion, SiO_2 as load reagent can effectively inhibit the deactivation of catalyst, so as to improve its service life.

In conclusion, for nitrogen-doped TiO_2/MnO_2 supported on SiO_2 composite photo-catalyst, nitrogen-doping can expand the range of light response of TiO_2 and improve the activity in visible and ultraviolet region. SiO_2 as load reagent has smaller particle size, and larger specific surface area, which can improve the separation performance of catalyst, so as to prolong the service life of catalyst. The mechanism shows that holes migrating to the surface of the semiconductor TiO_2 react with H_2O or OH^- absorbing on the surface and the generated ·OH, is free radical with strong oxidability, which can oxidize a variety of organic pollutants, and plays a main role in photo-catalytic system as oxidizing agent. The Mn^{2+} is oxidized into MnO_2 by calcination, which loads on catalyst surface, and catalyzes ·OH. The generated O_2 can react with electrons which migrate to the surface of the semiconductor TiO_2 to generate ·O_2, and other active-oxygen free radicals. The reactive group will also be able to participate in oxidation-reduction reactions, and the generated O_2 is served as electron capture agent to promote the separation of photo-generated carrier and reduce the recombination probability, which significantly improve the photo-catalytic activity of the catalyst. At the same time, according to the reaction dynamics, this process can promote the generation of· OH which is served as the capture agent of hole.

2.3Photocatalytic reaction mechanism diagram

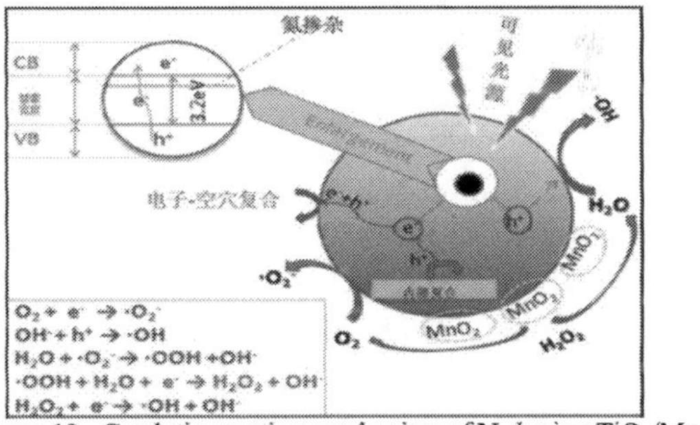

Figure 10. Catalytic reaction mechanism of N-doping TiO_2/Mn O_2 load on SiO_2 composite photo-catalyst

According to the simulated catalytic reaction mechanism of Fig.10, the N_{2p} orbital and O_{2p} orbital in the bulk phase of TiO_2 were hybridized by nitrogen doping, thus the width of the forbidden band of TiO_2 was reduced and therefore could be excited by visible light. Electrons and holes migrate to the surface of catalyst particles under the action of electric field or through diffusion to react with the adsorbed substances on the surface. Holes are usually trapped by OH^- and surface-adsorbed H_2O molecules to form OH^-, and hydroxyl radicals are the main active substances in photocatalysis. Electrons are usually trapped by O_2 molecules adsorbed on the surface of particles, resulting in free radicals such as O_2, ·OOH, which can participate in many redox processes. Electron-hole recombination is the most important reason for the serious decrease of reaction efficiency. Electron-hole recombination is very fast, but the rate of trapping carriers is relatively slow. Electrons are more active than holes. Therefore, O_2 produced by high temperature oxidation of Mn^{2+} to MnO_2 is used as electron trapping agent in advance. Adsorption on the catalyst surface is very important, it can improve the surface electron migration rate to inhibit carrier recombination, thereby increasing the quantum efficiency of photocatalytic reaction, but also promote the formation of hydrogen peroxide molecules adsorbed on the catalyst surface. Therefore, the prepared composite catalyst can achieve visible light response and obtain high treatment effect.

Conclusions

Make full use of the good performance of visible-light response of nitrogen doped TiO_2,great adsorption of SiO_2 and excellent catalytic activity of MnO_2, through the effective combination of sol-gel method [9] and incipient-wetness impregnation method to prepare nitrogen-doped TiO_2/MnO_2 supported on SiO_2 composite photo-catalyst. The results show that the urea in the form of N atom replaces O atoms in the TiO_2 lattice, form Ti - O - N bong on the surface of TiO_2, and constitute the new energy level structure, which made the absorption of catalyst cause red-shift and induce the visible light response, while the composite catalysts in both the ultraviolet and visible region show higher catalytic activity. SiO_2 as load reagent can not only control the growth of TiO_2 particles, to get smaller size and larger specific surface area, but also improve the separation performance of catalyst, so as to expend its service life. The Mn^{2+} is oxidized to MnO_2 and loads on catalyst surface to make O_2 which migrated to the surface of the semiconductor TiO_2 as electron capture agent to promote the separation of photo-generated carrier and reduce the recombination probability. Then the generated active oxygen free radicals participate in oxidation-reduction reaction of dye pollutants to make them degrade, which significantly improves the photo-catalytic activity of the catalyst.

References

[1]Vinodgopal, Kam. EnvironmentalScience.Technology, **29** 3 :841-845(1995)
[2]J.W. J i, W.H. Ma, Y.P. Huang, etc. Chinese Science Bulletin,**48** 21 : 2199-2204(2003)
[3]W.H. Yuan, H.Q. Bi, C.H. Wei. Journal of south China university of technology, **32** 3 :29-32(2004)
[4]S. Yi , H. Yamaki, Q. W. Zhang ,et. Solid State Ionics, **172** 1-4 :205~209(2004)
[5]C.H. Ao, S.C. Lee. Appl.Catal.B:Environ, **44** 3 :191-205(2003)
[6]X.Y. Chen, S.L. Liu, X. Chen, et . Acta,Phys.Chim.Sin. **22** 5 :517-522(2006)
[7]W.Y. Choi , et. .J.Phys.Chem. , **98**:1369-1379(1994)
[8]Asahi R, Morikawa T, Ohwaki T, et al.Science, **293** **5528** :269-271(2001)
[9]Burda. C, Y. Lou, X. Chen, et. Nano Letters **3** 8 :1049-1051(2003)
[10]D.M.Shi,Z.X.Li,J.H.Duo. Environmental Science and Technology. **9**:46-4(2002)

Photocatalytic oxidation degradation mechanism study of methylene blue dye waste water with GR/iTO$_2$

Xi-quan WANG[1], Song-fu HAN[1],Qing-wen Zhang[2], Nan Zhang[1], and Dan-dan ZHAO[1]
[1]*Dept. of Environmental Engineering, School of Chemical Engineering, Liaoning University of Science and Technology, Anshan 114051, China*
[2]*Dept. of Angang Group Engineering Technology, Anshan 114051, China*

Abstract. Graphene/titania multiplex photo-catalyst was prepared through hydrothermal method. Methylene blue with initial concentration of 20 mg/L in aqueous solution was treated by photocatalytic oxidation and the degradation process was investigated by UV/Vis spectrum, FTIR and GC-MS Spectra analysis. The results indicated that the conjugate structure of N-S heterocyclic compound was broken and aromatic ring was oxidized to open the ring. The methylene blue molecules were finally mineralized to H_2O and CO_2 in the photocatalytic degradation process.

1 Introduction

Dye wastewater is hard to degrade due to its high content of organic compounds, complexity of the components and high chroma[1]. In view of the difficulty in dealing with dye wastewater by the traditional methods, a lot of research work and trials have been done to seek appropriate technology. Photocatalytic oxidation is regarded as a future advanced oxidation technology for the removal of toxic organic dye from water. Among various semiconducting materials, most attention has been focused on TiO_2 because of its characters of environmental-protecting, sound stability, non-secondary-pollution and effective degradation[2]. The hydroxyl radicals caused by the excitation of TiO_2 under UV irradiation could oxidize and mineralize organic pollutants into H_2O and CO_2. However, the photocatalytic activity of TiO_2 is unsatisfactory because of the recombination of light-generated electron-hole pairs and the agglomerate phenomena[3]. It has been proved that the TiO_2-C compound materials have more outstanding catalyzed effect[4-5]. Graphene is regarded as a good carrier for photocatalyst owing to its large specific surface, fast-paced electronic conduct and large mechani- -cal strength. The unique physical properties could be favorate to separate electron-hole pairs and avoid agglomeration that in turn leading to the enhanced photocatalytic activity[6-7]. Methle- -ne blue is a widely used organic toxic dye in a range of different fields, which can cause environmental contaminations to the harm environment and the human health. The adsorption characteristics of MB on various adsorbents have previously been extensively investigated, nevertheless, the degradation mechanism of methylene blue in photocatalytic oxidation process is still poorly known owing to its complex chemical structure[8-9]. In this paper, Graphene/titania (GR/TiO_2) multiplex photocata lyst was prepared through hydrothermal method. Following previous studies, the intermediate products during the photocatalytic reaction were analysised with UV/Vis, IR and GC-MS spectra to propose the detailed degradation pathways between photocatalytic oxidation and methylene blue solution.

*Corresponding author :Tel: 86 18340797937; E-mail:2920816646 @qq.com[a]

2. Materials and methods

2.1. Dye materials

The cationic dye, methylene blue (analytically pure) was purchased from Sinopharm chemical reagent Co. LTD. Methylene blue, 3,7-bis(dimethylamino)-phenothiazin-5-ium chloride, is a heterocyclic aromatic chemical compound with the molecular formula $C_{16}H_{18}N_3SCl$[10].The stock solution was prepared by dissolving accurately weighted dye in deionized water to the concentration of 1000 mg/L. The working solutions were obtained by diluting the dye stock solution to the concentration of 20 mg/L. The pH variation during the course of reaction was not significant, so the solutions were conducted at natural conditions.

 https://doi.org/10.1051/matecconf/201823803006

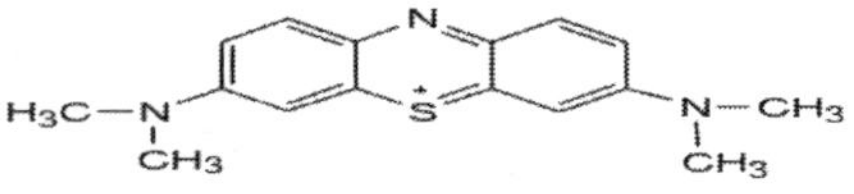

Structural formula of methylene blue

2.2. Chemical reagent

Butyl titanate, nitric acid, glacial acetic acid, Anhydrous ethanol, phosphoric anhydride , sulphuric acid, graphite powder, methylene blue, potassium persulfate, potassium permanganate, The chemical reagents listed are analytically pure.

2.3. Photocatalytic oxidation experim-

-ents

Graphite oxide was prepared by modified Hummers method[11]. A predetermined amount of graphite oxide was slowly added into TiO_2 sol which was prepared by sol-gel method. Stirring the mixture was continued for about 2h, and then the solution was aged at room temperature. Then the gel was hydrothermally treated at 200℃ for 12h . After this hydrothermal process, the product was washed by deionized water to neutrality, then dried at 70℃. Finally, the GR-TiO_2 photocatalyst was obtained. The preparation method of TiO_2 was similar to that of Gr-TiO_2, except there was no introduction of graphene oxide into the preparation system.

The Photocatalytic oxidation experiments were carried out in photocatalytaic reactor at room temperature. The reactor consists of ultraviolet lamps (main wavelength is 254 nm) and magnetic stirrers. In our study, a volume of 100 ml of the aqueous solution of methylene blue with the concentration of 20 mg/L was introduced in reactor with 100 mg photocatalyst, and then the mixture was stirred for 30 min in the darkness, after which the reaction suspension was irradiated under UV light. The reaction mixture was stirred vigorously using magnetic stirrer for during the entire reaction time. Samples were collected at appreciate intervals and were centrifuged to remove catalyst particles with high-speed centrifugal.

2.4 The analysis method

The X-ray diffractometer (XRD) was carried out to identify the crystal phase. The surface morphology of samples was determined by scanning electron microscopy (SEM) and transmission electron microscopy (TEM). A UV/VIS spectrometer recording the spectra over the 200nm–800 nm range was used to measure methylene blue solution samples at 10min interval during photocatalytic oxidation. On the other hand, the samples were dried completely and then pressed as KBr pellets to acquire the FTIR spectra by an FTIR spectrometer from 400 to 4000cm^{-1}. Intermediate compounds of methylene blue solution were also identified by a GC–MS system. Before analyzing with the GC-MS system, the products were extracted using trichloromethane in a separating funnel for three times.

3. Results and discussion

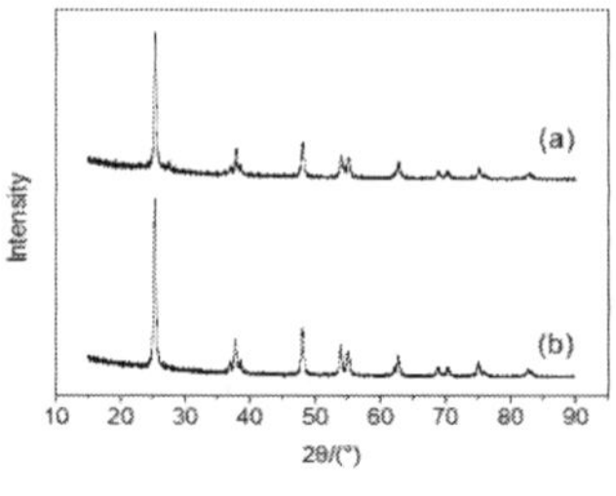

Fig.1 XRD spectras of GR/TiO_2 (a) and TiO_2 (b)

Figure 1 shows the XRD patterns of GR/TiO_2 (a) and TiO_2 (b). As shown in Fig.1, 25.3°, 37.8°, 48.0°, 53.9°, 55.0°, 62.8°, 68.8°, 70.3°, 75.1°, 76.1° and 82.8° were all characteristic peaks of anatase, indicating that the TiO_2 in GR/ TiO_2 were anatase crystal which had better photocatalytic active relative to brookite and rutile. Combared Fig.1(a) with Fig.(b), the XRD spectras of GR/TiO_2(a) and TiO_2(b) were basically the same, illustrating that the peaks of GR were covered and crystal phase of TiO_2 remained unchanged.

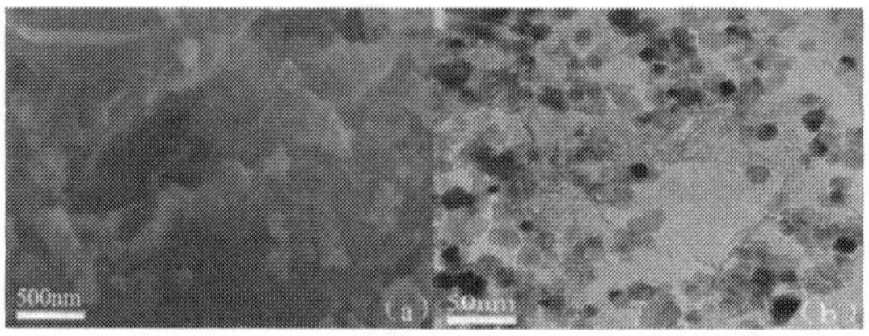

Fig.2 SEM (a) and TEM (b) photo of microstructure of GR/TiO_2

Figure 2 shows the SEM (a) and TEM (b) images of GR-TiO_2. As shown in Fig.2(a), the TiO_2 powers had uniform loaded into GR that had high specific surface area. The images indicated that the TiO_2 in composite catalyst was approximately 25nm in diameter and scattered on the film-shaped GR. With loose agglomeration and favorable dispersibility, that is, dye molecules and catalysts contact area was increased, and resulting in the enhancement of photocatalytic activity.

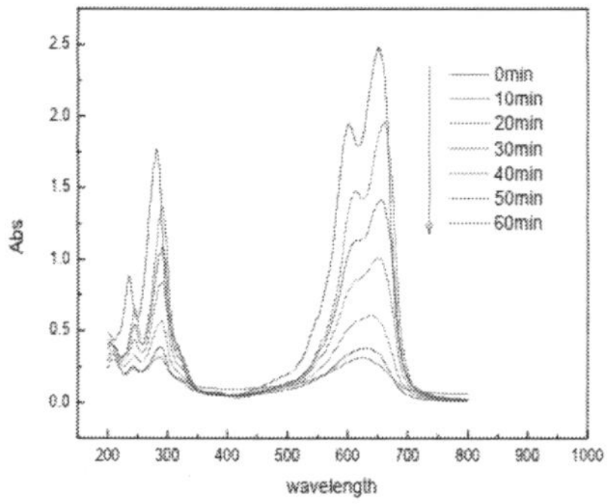

Fig.3 The UV - Vis absorption spectra of methylene blue

(t=0~1h) in photo-catalytic oxidation process

Fig.3 shows the UV-Vis absorption spectra of methylene blue solution at different photocatalytic oxidation reaction time. For the UV-Vis absorption spectra, 612nm and 665nm in the visible region is the characteristic peak of large conjugated system of N-S heterocycle group. The 291nm characterizes phenothiazine structure and the peak of 245nm in UV region belong to aromatic and polycyclic aromatic. Methylene blue is a chromogenic agent whose N-S heterocycle group attached to benzene includes lone pair electron, in which sulfhydryl group is the main chromophoric group. According to the UV-Vis absorption spectra during reaction time, the strong absorption peaks in the visible region decreased obviously, in view of the absorption peaks become sharper with the extending of the conjugated unsaturated system, indicating that conjugated structure of heterocycle was destroyed. The results in Fig. 3 stated that the destruction of the conjugate structure was almost synchronization during the processing of decolorization. At the same time, the characteristic absorption peaks between 200 and 400nm were gradually reduced, which stated that phenothiazine structure and benzene were attacked by the hydroxyl radicals and oxidized to open the ring. Thus, it is concluded that strong oxidizing substance hydroxyl radicals which generated by ultraviolet dissociating TiO_2 directly broke directly unsaturated bonds of conjugated system and change the structure of dye molecule, then the intermediate would be decomposed into small molecules during photocatalytic oxidation.

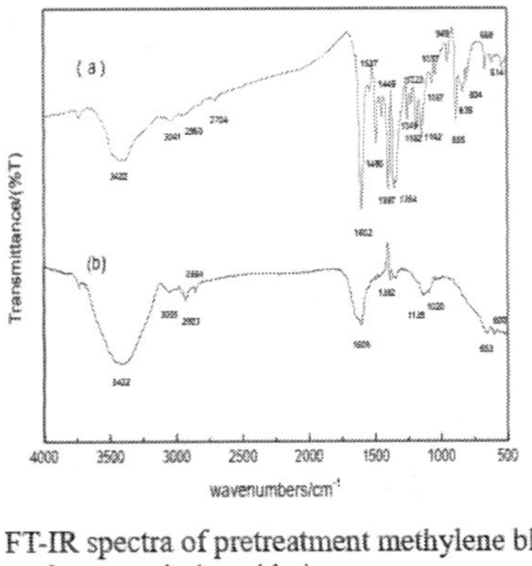

Fig.4 The FT-IR spectra of pretreatment methylene blue in photo-catalytic oxidation process

The transformation of the methylene blue molecular structure was further evaluated by measurement of the FTIR spectra at 0 and 30min. The results shown in Fig. 4(a) are the FTIR spectra of methylene blue solution without Photocatalytic oxidation and Fig. 4(b) displays FTIR spectra of methylene blue after Photocatalytic oxidation for 30min. According to the UV-Vis absorption spectra of methylene blue, it was illustrated that methylene blue had N-S heterocycle group conjugate structure and phenothiazine structure and aromatics. The results in Fig. 3(a) showed that the broad peak of 3422cm^{-1} was determined as the O-H stretching vibration of water molecules. It could be seen the peak corresponding to C-H absorption of benzene ring occured at 3041 cm^{-1}. The peak of 2960 cm^{-1} belongs to the stretching vibration peak of methyl group and the other typical flexural vibration of methyl group occurs at 1397 cm^{-1}. The C=C framework vibration of benzene ring vibration and the C=N stretching vibration were found at 1602 cm^{-1}[12], the absorption peaks at and 1490 cm^{-1} and 1445 cm^{-1} were also related to C=C stretchings. The characteristic absorption peak of C-N bond linked aromatic amine occurs at 1354 cm^{-1}. The peak of 1250cm^{-1} is the C=S stretching vibration absorption peak. The absorption peaks at 1222 cm^{-1}、1181 cm^{-1}、1143 cm^{-1}, 1066 cm^{-1} and 1037 cm^{-1} are assigned to the in-plane bending vibration peaks of aromatic C-H, while the absorption peaks at 885 cm^{-1}, 836 cm^{-1} and 667 cm^{-1} are attributed to out-plane bending vibration peaks of aromatic C-H. The absorption peak of 614 cm^{-1} is caused by C- S-C framework vibration[13].

As shown in Fig.3b, the number of absorption peaks had decreased significantly and the characteristic peak of C=N and C=S had decreased significantly after photocatalytic oxidation, thus it could be concluded that the N-S heterocyclic compound was broken during the degradation process. The characteristic peak of benzene and C-H bending vibration of aromatic C-H had fallen substantially indicating that phenyl structure was destroyed. Owing to the ring opening reaction of benzene and the rupture of dye molecules, the absorption peaks at 2923 cm^{-1} which belonged to the stretching vibration peak of methyl group became stronger. Thus, it is summarized that the conjugate structure of N-S heterocyclic compound was broken and aromatic ring was oxidized to open the ring, the dye molecules were degraded and formed micro-molecular organics during photo catalysis reaction.

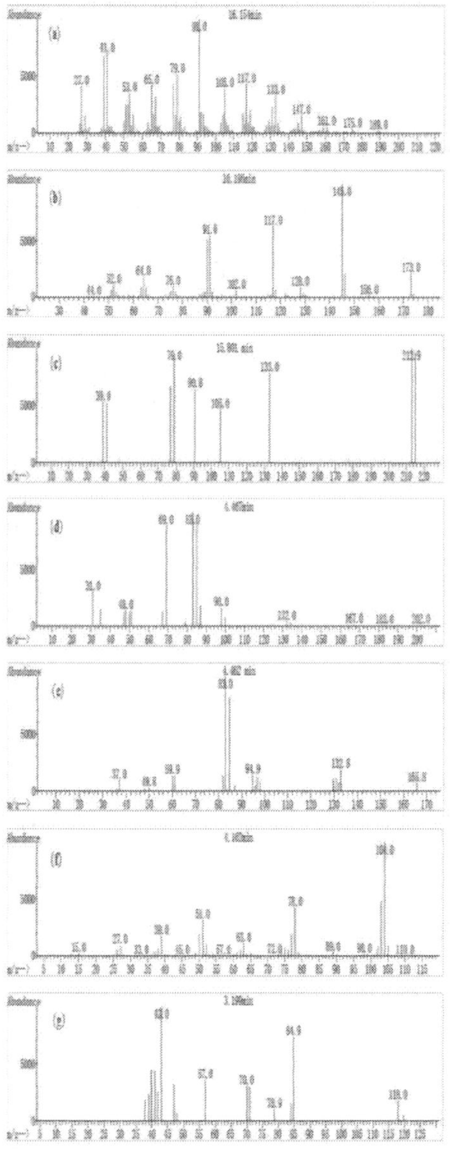

Fig.5 The GC-MS spectra of identified compounds in methylene blue solution of Photocatalytic oxidation:(a) Product at 16.186min, (b) Product at 16.154 min, (c) Product at 15.801min, (d) Product at 4.467 min, (e) Product at 4.462min, (f) Product at 4.162 min, (g)Product at 3.199min

In order to verify the above deductions, a further attempt was made to identify the mineralization during the degradation of methylene blue through GC-MS analyses. The qualitative results of the GC-MS analysis in Fig.5 showed that some volatile intermediate products of methylene blue solution were identified during Photocatalytic oxidation. The m/z at 215.0 could be easily identified to 2-Amino-5-dimethylamino-benzenesulfonic acid anion (Fig. 5c) with the molecular formula being $C_8H_{11}N_2SO_3$ speculated from the ion abundance ratios which suggested that N-S heterocyclic compound of methylene blue were broken. The m/z value of 173.0 and 189.0 corresponding to 4-Amino-benzenesulfonic acid and 2-Amino-5-hydroxy-benzenesulfonic acid were likely to be the degradation intermediate of 2-Amino-5-dimethylamino-benzenesulfonic acid attacked by hydroxyl radical, whose molecular formula were $C_6H_7NSO_3$ (Fig. 5a) and $C_6H_7NSO_4$ (Fig. 5c) respectively. At the same time, the molecular ion peak of m/z 202.0 corresponding to 4-Nitro-benzenesulfonic acid anion was the oxidized resultant of 4-Amino-benzenesulfonic with the molecular formula being $C_8H_{10}N_2O_2$ (Fig. 5d). The molecular ion peak of m/z 165.8 related to Dimethyl-(4-nitro-phenyl)-amine with the molecular formula being $C_6H_{10}N_2O_2$ (Fig. 5e) was the products after hydroxyl radical attacked the N-S heterocyclic of dye molecules, while the m/z value observed at 110.0 belonging to p-Dihydroxybenzene with the molecular formula being $C_6H_6O_2$ (Fig. 5f) was the oxidized resultant of benzene ring. The m/z value of 118.0 belonged to the small molecular products of degradation with the molecular formula being $C_4H_6O_4$ (Fig.5g), corresponding to succinic acid generated by open-loop benzene. It further showed that the broken position of methylene blue treated by photocatalytic oxidation was the conjugate structure of the N-S heterocycle group which is in line with the analysis results by the UV and IR spectrogram.

Based on the intermediate and above analytical results, the degradation pathways of methylene blue is described, as shown in Fig. 5. In the process of photocatalytic oxidation treatment of methylene blue, hydroxyl radicals firstly attacked the N-S heterocycle group conjugate structure because the electron density of sulfhydryl was higher, the methylene blue had been degraded to 2-Amino-5-dimethylamino-benzenesulfonic acid anion and Dimethyl-(4-nitro-phenyl)-amine. And then the Dimethyl-(4-nitro-phenyl)-amine was captured by hydroxyl radicals and formed p-Dihydroxybenzene. Furthermore, 2-Amino-5-dimethylamino-benzenesulfonic acid anion continued to be broken into 4-Amino-benzenesulfonic acid and 2-Amino-5-dimethylamino-benzenesulfonic acid. 4-Amino-benzenesulfonic acid was oxidized to 4-Nitro-benzenesulfonic acid anion. Finally, 4-Nitro-benzenesulfonic acid anion was directly attacked and generated then continued to be degraded CO_2 and H_2O through a series of reactions. This is similar with the degradation mechanism of methylene blue by atmospheric pressure dielectric barrier discharge plasma[14].

Fig.6 The degradation pathways of methylene blue reacting with Photocatalytic oxidation.

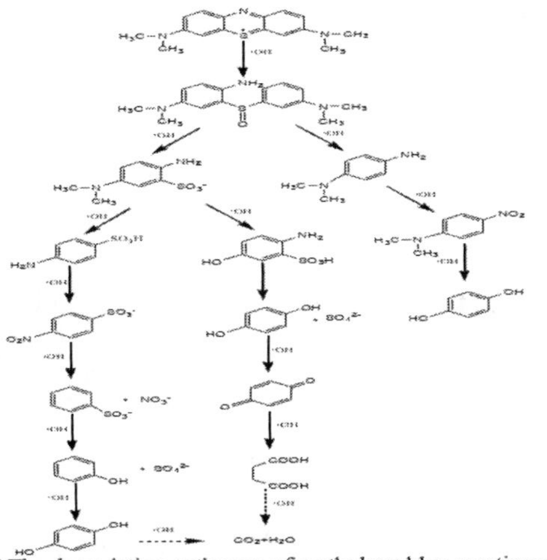

Fig.6 The degradation pathways of methylene blue reacting with Photocatalytic oxidation.

4. Conclusions

(1)The crystalline of TiO_2 in GR/ TiO_2 prepared by hydrothermal method is anatase and the introduction of graphene oxide into the preparation system could enhance the photocatalytic activity because of loose agglomeration and favorable dispersibility which increasing the contact area of dye molecules and catalysts.

(2)The results obtained from the variety of spectrums and chromatograms shows that the degradation of MB solution is that hydroxyl radicals firstly attacked the N-S heterocycle chromophore group, then the resultants were oxidized and mineralized into small-molecule products which continued to be degraded CO_2 and H_2O through a series of reactions.

5. Reference

[1]Doğan M, Özdemir Y, Alkan M. Adsorption kinetics and mechanism of cationic methyl violet and methylene blue dyes onto sepiolite[J]. Dyes and Pigments, 2007, 75(3): 701-713.

[2]Hashimoto K, Irie H, Fujishima A. TiO_2 photocatalysis: a historical overview and future prospects[J]. Japanese journal of applied physics, 2005, 44(12R): 8269.

[3] Kumar S G, Devi L G. Review on modified TiO_2 photocatalysis under UV/visible light: selected results and related mechanisms on interfacial charge carrier transfer dynamics [J]. The Journal of Physical Chemistry A, 2011, 115(46): 13211-13241.

[4]Irie H, Watanabe Y, Hashimoto K. Carbon-doped anatase TiO_2 powders as a visible-light sensitive photocatalyst[J]. Chemistry Letters, 2003, 32(8): 772-773.

[5]Zaleska A. Doped-TiO_2: a review[J]. Recent Patents on Engineering, 2008, 2(3): 157-164.

[6] Geim A K, Novoselov K S. The rise of graphene[J]. Nature materials, 2007, 6(3): 183-191.

[7] Neto A H C, Guinea F, Peres N M R, et al. The electronic properties of graphene[J]. Reviews of modern physics, 2009, 81(1): 109.

[8]Doğan M, Abak H, Alkan M. Adsorption of methylene blue onto hazelnut shell: Kinetics, mechanism and activation parameters[J]. Journal of Hazardous Materials, 2009, 164(1): 172-181.

[9]Özer D, Dursun G, Özer A. Methylene blue adsorption from aqueous solution by dehydrated peanut hull[J]. Journal of Hazardous Materials, 2007, 144(1): 1

[10]Tayade R J, Natarajan T S, Bajaj H C. Photocatalytic degradation of methylene blue dye using ultraviolet light emitting diodes[J]. Industrial & Engineering Chemistry Research, 2009, 48(23): 10262-10267.

[11] Hummers Jr W S, Offeman R E. Preparation of graphitic oxide[J]. Journal of the American Chemical Society, 1958, 80(6): 1339-1339.

[12]Imamura K, Ikeda E, Nagayasu T, et al. Adsorption behavior of methylene blue and its congeners on a stainless steel surface[J]. Journal of colloid and interface science, 2002, 245(1): 50-57.

[13]Xiao G N, Man S Q. Surface-enhanced Raman scattering of methylene blue adsorbed on cap-shaped silver nanoparticles[J]. Chemical Physics Letters, 2007, 447(4): 305-309.

[14]Huang F, Chen L, Wang H, et al. Analysis of the degradation mechanism of methylene blue by atmospheric pressure dielectric barrier discharge plasma[J]. Chemical Engineering Journal, 2010, 162(1): 250-256.

Visible light $Bi_2S_3/BiFeO_3$ photocatalyst for effective removal of Rhodamine B

Xiquan Wang[1], Nan Zhang[1,a] and Gao Wang[1]

[1] *School of Chemical Engineering, University of Science and Technology Liaoning, High-Tech District,185 Qian shan Middle Road, Anshan 114051, China.*

Abstract: Bi_2S_3-sensitized $BiFO_3$ (BFO) photocatalyst (Bi_2S_3/BFO) was successfully synthesized through a facile and environmental ion exchange method between BFO and Thiourea (H_2NCSNH_2, TU). The photocatalysts were characterized by X-ray diffraction (XRD), scanning electron microscopy (SEM), energy-dispersive X-ray spectroscopy (EDS) and UV-vis diffuse reflection spectroscopy (DRS). The obtained Bi_2S_3/BFO composites showed excellent photocatalytic performance for decomposing Rhodamine B (RhB) compared with pure BFO under visible light irradiation (λ>400nm). 5% Bi_2S_3/BFO exhibited the highest photocatalytic activity and excessive amount of Bi_2S_3 would result in the decrease of photocatalytic activity of BFO. The mechanism of enhanced photocatalytic activity was proposed on the basis of the calculated energy band positions.

1 Introduction

Photocatalysis has received much attention owing to current environment concerns and energy demand[1]. Visible-light-responding photocatalysts have attracted increasing interest in the past years due to their potential applications in splitting of water and degradation of organic pollutants under visible light irradiation. Recently, a growing number of investigations on exploitation of Bi-containing compounds, such as $BiVO_4$[2, 3], Bi_2MoO_6 [4], Bi_2WO_6[5, 6], have been undertaken. A series of researches have revealed that whether Bi^{3+}-containing or Bi^{5+}-containing compounds have exhibited high photocatalytic ability in visible light range, which is ascribed to the hybridized valence band (VB) of O 2p and Bi 6s that narrows the band gap[7]. As a member of Bi-based multi-component oxides, typical Rhombohedral structure BFO has been widely studied because of its special properties, such as narrow band gap (2.0-2.7eV), ferroelectric and ferromagnetic properties[8]. BFO had been considered as one of the third-generation visible-light responsive photocatalyst. Gao et al.[9] synthesized perovskite-type BFO photocatalysts via a facile hydrothermal method and used it to degradat Methyl Orange under visible-light irradiation. Gao et al.[10] also synthesized perovskite phase material BFO by simple sol-gel route and evaluated the photocatalytic performance on visible-light degradation of Methyl Orange. However, research concerning the improvement of visible light photocatalytic properties of BFO is still scarce.

Up to now, previous studies had showed that semiconductor combine with other semiconductors could extend the spectral responsive range, separate the charge carriers effectively and further enhance photocatalytic activity of single-component materials dramatically[11, 12]. Especially the construction of Bi_2S_3 heterostructured composites has considered as an attractive and promising method. Bismuth sulfide (Bi_2S_3) is a typical layered band-gap semiconductor with a narrow band gap (1.3-1.7eV) and large absorption coefficient[13], which has been used for electrochemical hydrogen storage[14], hydrogen sensor[15], biomolecule detection[16], X-ray computed tomography imaging[17], and photo-responsive materials[18]. Because of its narrow band-gap, it was used as a potential visible light photocatalyst through combine with Bi_2O_3[19, 20], $BiVO_4$[21], $Bi_2Sn_2O_7$[22], $Bi_4Ti_3O_{12}$[23] and BiOCl[24, 25]. However, Bi_2S_3/BFO composite photocatalysts have never been reported.

Herein, we mention a new combination route preparation of sol-gel and solvothermal process for Bi_2S_3/BFO composites. Rhodamine B (RhB) as a model pollutant under visible light (lambda >400nm) evaluated the photocatalytic performance, results showed that the photocatalytic activity of Bi_2S_3/BFO composite is better than pure BFO. In addition, also studied the mechanism of Bi_2S_3/BFO composites.

[a] Corresponding author: Tel: 86 18341277889;Email:2387923823@qq.com

2 Experimental

2.1 Chemical and materials

All the chemical materials without further processing used during experiment were analytically pure . Iron nitrate ($Fe(NO_3)_3 \cdot 9H_2O$), Bismuth Nitrate ($Bi(NO_3)_3 \cdot 5H_2O$), Thiourea (CN_2H_4S, TU), Ethylene Glycol ($C_2H_6O_2$), Hydrochloric Acid(HCl) and absolute Ethyl Alcohol (C_2H_6O) were obtained from Quan Rui Chemical Reagent Co., Ltd (Liao Ning, China). Rhodamine B ($C_{28}H_{31}ClN_2O_3$, RhB) was provided by Dong Xing chemical (Liao Ning, China). The wastewater was simulated by dissolving 10 mg/L RhB to the deionized water.

2.2. Preparation of BFO nanoparticles

The BFO nanoparticles sample was prepared through a sol-gel method according to previous studies[10]. In a typical synthesis, BFO powders were synthesized by a simple sol-gel route. $Bi(NO_3)_3 \cdot 5H_2O$ and $Fe(NO_3)_3 \cdot 9H_2O$ of 0.02 mol (1:1 molar ratio) were dissolved in 50 mL Ethylene Glycol as precursor solution, and the final concentration was 0.4 mol/L. The mixture was stirred for about 1.5 h at 80°C to obtain the sol. The gel was kept at 120°C for 4 days to form a xerogel powder. Then the powder was heated at 300°C for 3 h as a pretreatment. The final powder was calcined at the temperatures of 500°C for 2 h under air condition.

2.3. Synthesis of Bi_2S_3/BFO composites

In order to synthesize Bi_2S_3/BFO composites, 0.1564 g of as-prepared BFO was dispersed in 20mL absolute Ethyl Alcohol under ultrasonic.A certain amount of TU dissolved in absolute Ethyl Alcohol was added dropwise to the BFO solution.Finally, transferred the appropriate absolute Ethyl Alcohol to the mixture to make the volume reach 24mL.After sonication for 30 min, put 30mL of solution into a Teflon-lined stainless steel autoclave and heated at 180°C for 6 h. Finally, the precipitates of Bi_2S_3/BFO with different theoretical molar percentage of Bi_2S_3 and initial BFO (from 2.5 to 50%) were collected, washed and dried at 80°C.

For comparison, the bulk Bi_2S_3 sample was prepared through a hydrothermal process according to Ref[26]. $Bi(NO_3)_3 \cdot 5H_2O$ and TU were mixed in deionized water (1:2.5 molar ratio) at 140°C for 12 h.

2.4 Characterization of catalyst

The phases of the samples were examined by X-ray diffraction (XRD) on a Philips X'pert PRO diffractometer with Cu K_α radiation. The size and morphology of the prepared samples were observed by scanning electronmicroscope (SEM, Carl Zeiss EVO LS-15). Chemical analysis of the photocatalyst was performed by X-ray analysis (EDS) in a CarlZeiss EVO LS-15 scanning electronmicroscopy (SEM). The UV–Vis diffuse reflectance spectrum (UV-DRS) of the powders were measured by a UV–Vis spectrophotometer (Gucun ZF-I) equipped.

2.5. Degradation experiment

The photocatalytic performance of the photocatalysts were evaluated by the degradation of RhB under visible light irradiation. The visible light source was a 150 W halogen tungsten lamp (λ_{max}= 588 nm) with the combination of a cut-off filter (λ> 400 nm) to eliminate UV radiation during visible light experiments. The system was cooled by wind and water and maintained at room temperature. In every run, 100mg Bi_2S_3/BFO was added to 100mL RhB solution (10mg/L) in a Pyrex vessel. Before the experiment, the suspension was magnetically stirred in the dark for 1 h to ensure the establishment of an adsorption–desorption equilibrium of the dye on the catalysts' surfaces. Under visible light irradiation, samples were regularly extracted and the catalyst powders were centrifugally removed. The residual RhB concentration was monitored by measuring the absorbance of the solution at the characteristic absorption wavelength of 554nm for RhB with a UV-Vis spectrophotometer (UV-5100, Metash) at room temperature. The photocatalytic efficiency was calculated by the decolouration percentage: $\eta = C/C_0 \times 100\%$, where the C is the final dye concentration, C0 is the initial concentration of dye.

3 Result and discussion

3.1. Crystalline property

In this study, BFO nano-materials was prepared by a sol-gel process with $Bi(NO_3)_3 \cdot 5H_2O$ and $Fe(NO_3)_3 \cdot 9H_2O$ in ethylene glycol solvent followed by heating, calcining and collecting in the first place. Then, the process of product synthesis is as follows.

$$4Fe(NO_3)_3 \rightarrow 2Fe_2O_3 + 12NO_2 + 3O_2 \quad (1)$$

$$4Bi(NO_3)_3 \rightarrow 2Bi_2O_3 + 12NO_2 + 3O_2 \quad (2)$$

$$Fe_2O_3 + Bi_2O_3 \rightarrow 2BiFeO_3 \quad (3)$$

The composition and phase structures of the as-prepared samples were examined by XRD. Fig. 1 presents the XRD patterns of pure BFO, Bi_2S_3 and Bi_2S_3/BFO with different Bi_2S_3 contents. It can be observed that all the samples are well crystallized. The diffraction peaks of pure BFO could be indexed to the standard PDF card of rhombohedral BFO (JCPDS No.86-1518) , as well as the feature peaks of the pure Bi_2S_3 are clearly corresponding to the orthorhombic structure of Bi_2S_3 (JCPDS No.17-0320). No impurity peaks were detected, implying that the final products of Bi_2S_3 and BFO are pure phases.

Furthermore, when the amount of Bi_2S_3 is below 10%, there are no diffraction peaks of Bi_2S_3 that can be observed, which may be due to the low content of Bi_2S_3 below the XRD detection limit. However, when the amount of Bi_2S_3 reaches 25%, the peaks of (130) and (211) planes of orthorhombic Bi_2S_3 appear. Moreover, when the amount of Bi_2S_3 reaches 50%, the peak of (240) planes of orthorhombic Bi_2S_3 is appearance. In addition, with the molar percentage of Bi_2S_3 increasing, the intensities of diffraction peaks of Bi_2S_3 increase whereas those of BFO decrease. Therefore, the above results suggest that Bi_2S_3/BFO composites can be successfully synthesized by a solvothermal method.

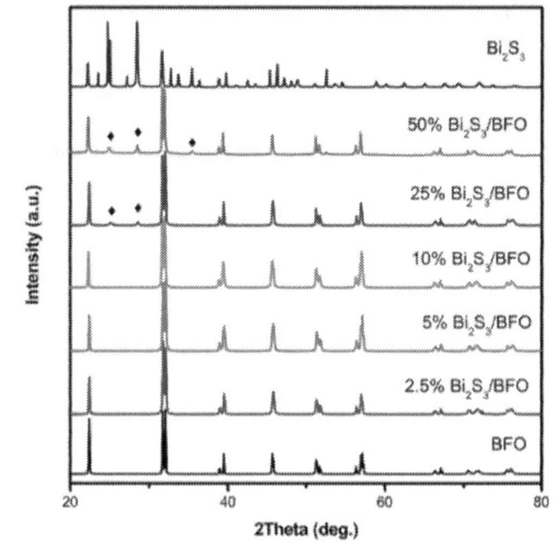

Figure 1. XRD patterns of BFO, Bi_2S_3 and Bi_2S_3/BFO composites with different Bi_2S_3 contents.

3.2. Morphology analysis

Fig. 2 shows SEM images of BFO and B_2S_3/BFO composites with different Bi_2S_3 contents. For comparison, the morphology of BFO nanoparticles which is annealed at 500°C with the concentration of 0.4 mol/L precursor solution was shown in fig. (a), revealing that the size of BFO materials pose irregular morphology with mean sizes below 100 nm. The Bi_2S_3/BFO samples exhibit two structures: a nanorod shape and a irregular shape particles, as shown in Fig. 2(b), (c), and (e-h). According to previous studies and these images, we can deduce that nanorod shape particles and polygonal shape particles are Bi_2S_3 and BFO respectively. At the same time, EDS analysis of the nanorods (Fig. 2d) suggests that the nanorods are composed of Bi and S elements, further confirming the result of XRD. All these above confirm the formation of Bi_2S_3 on BFO.

3.3 UV-vis DRS analysis

Fig. 3 displays the UV-vis diffuse reflection spectrum (DRS) of the as-prepared samples. It can be observed that all the samples displayed absorption in the visible light region and pure Bi_2S_3 appeared strong absorption over the whole visible light region. For Bi_2S_3/BFO composites, the absorption band is similar to the pure BFO, and the edge of absorption band moves towards long wavelength gradually. The absorption intensity increases with the increase of Bi_2S_3 content in 500-800 nm region, which is due to the high visible light response of Bi_2S_3.

For a semiconductor, theoretically, the optical absorption near the band edge follows the formula: $\alpha h\nu=A(h\nu-E_g)^{n/2}$[27], where α, ν, E_g, and A are the absorption coefficient, light frequency, band-gap energy, and a constant, respectively. The value of n depends on

Element	Weight%	Atom%
O K	5.79	36.31
S K	3.88	12.14
Fe L	6.21	11.16
Bi M	84.12	40.38
Total	100.00	100.00

Figure 2. SEM images of (a) BFO and also B_2S_3/BFO composites with different Bi_2S_3contents: (b) 2.5%, (c) 5%, (e) 10%, (f) 25%, and (g & h) 50%. Also EDS spectrum of the Bi_2S_3/BFO composites.

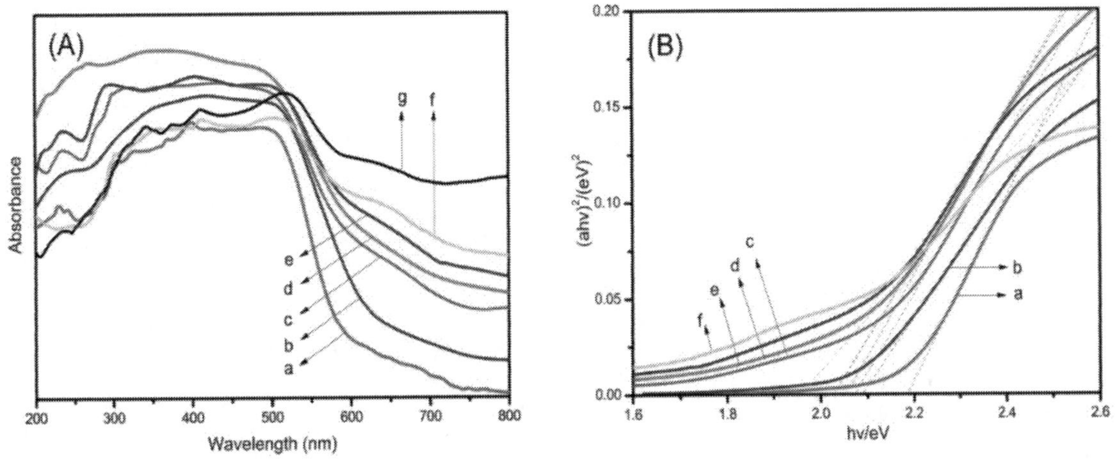

Figure 3. (A) UV–vis diffuse reflectances spectra and (B) Plot of $(\alpha h\nu)^2$ vs. energy ($h\nu$) for the band-gap energy of (a) BFO, (g) Bi_2S_3 and Bi_2S_3/BFO samples with different Bi_2S_3 contents: (b) 2.5 %, (c) 5 %, (d) 10 %, (e) 25 %, (f) 50 %.

whether the transition is direct (n = 1) or indirect (n = 4) in a semiconductor. For BFO and Bi_2S_3, they pertain to direct transition. Thus, n is equal to 1. Fig. 6b shows the curve of $(\alpha h\nu)^2$ versus $h\nu$. On the basis of the extrapolation, the band gap energy (E_g) of BFO and 2.5, 5, 10, 25, 50% Bi_2S_3/ BFO are respectively estimated to be about 2.18, 2.09, 2.07, 2.06, 2.03 and 1.96 eV, which is attributed to the quantum confinement of Bi_2S_3 nanocrystals. Otherwise, the well-crystallized Bi_2S_3 samples have a band gap close to 1.3 eV, and exhibit considerable absorption strength in almost all visible light ranges.

3.4 Photocatalytic activity test

The photocatalytic activities of the prepared products were evaluated by photodegradation of RhB under visible light irradiation (λ>400nm). Fig. 4a shows the change of absorption spectra of RhB aqueous when the Bi_2S_3/BFO content is 5% after 3 h. It can be seen from the spectra that the major absorption peaks of RhB at 554 nm gradually decreased as the irradiation time increases and blue shifted step by step. The results show that the ethyl groups are removed one by one which is consistent with the literature description. And compared with pure BFO and Bi_2S_3, 5% Bi_2S_3/BFO shows the excellent photocatalytic activity for the RhB degradation. Furthermore, the stepwise blue shift of the main peak can be attributed to the formation of the de-methylated of RhB[28].

Fig. 4b shows the photocatalytic activities of different catalysts. For comparison, carried out the same degradation experiment in the absence of photocatalyst which shows no

appreciable degradation of RhB after irradiating for 3 h. Also, in the presence of pure BFO and Bi_2S_3 only about 62.6% and 11.1% of RhB concentration were decomposed after irradiation for 3 h, which could attributed to the high stability and negligible self-photolysis of RhB. The dark adsorption abilities for RhB were 14.4%, 12.1%, 32.5% and 19.6% for 2.5%, 5%, 10%, 25% Bi_2S_3/BFO samples, respectively. The activities of the four types of Bi_2S_3/BFO photocatalysts were better than that of BFO, arising from their visible light responsive behaviors. The highest

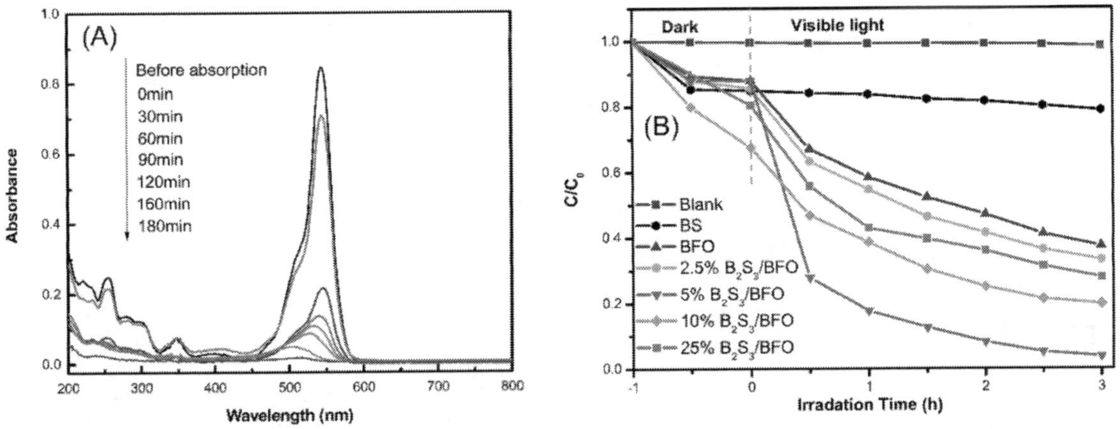

Figure 4. (A) UV-Vis spectral changes during the photodegradation of RhB mediated by 5% Bi_2S_3/BFO sample and (B) Photocatalytic activities of different catalysts.

activity was obtained over the 5% Bi_2S_3/BFO sample, on which more than 96.3% of RhB was degraded within 3 h. Clearly, the content of Bi_2S_3 dramatically affect the photocatalytic activities of Bi_2S_3/BFO catalysists even though the content was very low. RhB was decomposed into 66.8%, 80.3% and 72.1% after 3 h irradiation with 2.5% Bi_2S_3/BFO, 10% Bi_2S_3/BFO and 25% Bi_2S_3/BFO respectively. Therefore, it can be observed that the photocatalytic activities of the Bi_2S_3/BFO composites increase and then decline with the increase of irradiate time. According to the Langmuir-Hinshelwood reaction kinetics model, the photocatalytic oxidation of Bi_2S_3/BFO is a first-order reaction.

3.5 The energy band structure and possible photocatalytic mechanism

Whether Bi_2S_3/BFO heterostructure is conducive to the separation of photo-generated carriers or not is closely related to their band-edge positions. The valence band (VB) edge and the conduction band (CB) edge positions of Bi_2S_3/BFO were calculated empirically according to formula[27]:

$$E_{VB}=\chi-E^{e}+0.5E_g \quad (4)$$

$$E_{CB}=E_{VB}-E_g \quad (5)$$

where EVB is the valence band-edge potential, χ is the electronegativity of the semiconductor, which is defined as the geometric mean of the absolute electronegativity of the constituent atoms, Ee is the energy of free electrons on the hydrogen scale (ca. 4.5 eV), Eg is the band gap energy of the semiconductor. The χ values of BFO and Bi_2S_3 were 5.74 eV and 5.28 eV[22, 29], respectively. The band-gap energies of BFO and Bi_2S_3 are 2.18 eV and 1.3 eV. Given the equation above, the top of the valence band (VB) and the bottom of the conduction band (CB) of BFO and Bi_2S_3 are calculated to be 2.33, 0.14 and 1.48, 0.09 eV. Both the CB and the VB of Bi_2S_3 are higher than those of pure BFO. Obviously, the difference between energy bands of BFO and Bi_2S_3 demonstrated the heterostructure of Bi_2S_3/BFO is propitious to the separation and transportation of charge carriers.

Based on the above results, the energy band levels of Bi_2S_3/BFO and the possible charge-separation process are

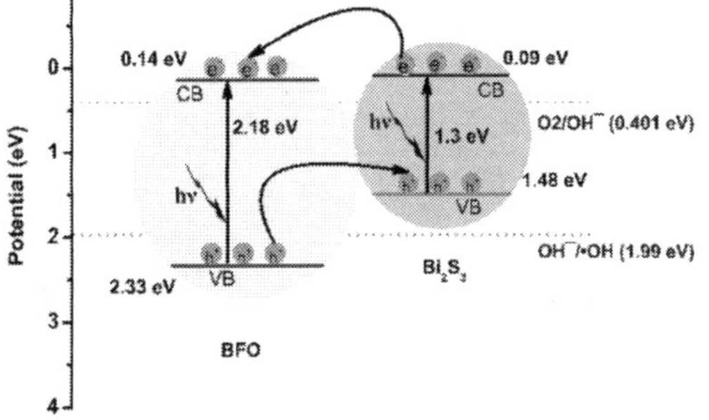

Figure 5. Diagram for energy band levels of Bi_2S_3/BFO composites and the possible charge separation process.

shown in Fig. 5 Under visible-light irradiation, Bi_2S_3 and BFO are both easily excited and photo-induced electrons and holes are generated correspondingly. Due to the higher CB position of Bi_2S_3 than that of BFO photo-generated electrons in Bi_2S_3 could transfer to the CB of BFO and the photo-generated holes in the VB of BFO migrate to Bi_2S_3 because of the less positive VB of Bi_2S_3 than that of BFO. Thus, the photo-induced electrons and holes can be efficiently separated and RhB will be oxidized to the final products. The redox potential of O_2/OH^- is 0.401 eV, the electrons located on the CB of BFO (E_{CB}=0.14 eV) can reduce O_2 to OH^-. However, the photo-generated holes at the VB of Bi_2S_3 with potential of 1.48 eV, which is more negative than the standard reduction potential of $OH^-/\cdot OH$ (1.99 eV)[22], can't oxidize OH^- to ·OH. Therefore, this result indicated that the active species are photogenerated holes that play more important role in the photodegradation process of RhB rather than ·OH.

4 Conclusions

In summary, Bi_2S_3/BFO composites have been successfully synthesized through a facile and economical ion exchange method. The prepared Bi_2S_3/BFO photocatalyst with high photocatalytic activity of RhB under visible light irradiation, is about two times higher than the pure BFO photocatalyst. The content of Bi_2S_3 played an important role to the photocatalytic activities and the optimal content of Bi_2S_3 was 5% with maximal photocatalytic degradation efficiency of 96.3%. This result could be attributed to the good visible light absorption of Bi_2S_3 and the effective separation of electron-holes pairs due to the formation of heterojunction between the two semiconductors. The composites displayed much higher activity than the individual components towards the degradation of RhB under visible-light irradiation, suggesting that they are new types of visible-light-driven photocatalysts for water purification application and environmental remediation. Therefore, this study provides a new and simple way to prepare composite nanomaterials.

References

1. H. Tong, S. Ouyang, Y. Bi, N. Umezawa, M. Oshikiri, J. Ye, Nano-photocatalytic materials: possibilities and challenges, Advanced materials (Deerfield Beach, Fla.), **24** (2012) 229-251.
2. T.W. Kim, K.S. Choi, Nanoporous $BiVO_4$ Photoanodes with Dual-Layer Oxygen Evolution Catalysts for Solar Water Splitting, Science, **343** (2014) 990-994.
3. Y.H. Ng, A. Iwase, A. Kudo, R. Amal, Reducing Graphene Oxide on a Visible-Light $BiVO_4$ Photocatalyst for an Enhanced Photoelectrochemical Water Splitting, Journal of Physical Chemistry Letters, **1** (2010) 2607-2612.
4. G. Tian, Y. Chen, W. Zhou, K. Pan, Y. Dong, C. Tian, H. Fu, Facile solvothermal synthesis of hierarchical flower-like Bi_2MoO_6 hollow spheres as high performance visible-light driven photocatalysts, Journal of Materials Chemistry, **21** (2011) 887-892.
5. L. Ge, C.C. Han, J. Liu, Novel visible light-induced g-C_3N_4/Bi_2WO_6 composite photocatalysts for efficient

degradation of methyl orange, Applied Catalysis B-Environmental, **108** (2011) 100-107.
6. H.L. Wang, L.S. Zhang, Z.G. Chen, J.Q. Hu, S.J. Li, Z.H. Wang, J.S. Liu, X.C. Wang, Semiconductor heterojunction photocatalysts: design, construction, and photocatalytic performances, Chemical Society reviews, **43** (2014) 5234-5244.
7. R.a. He, S. Cao, P. Zhou, J. Yu, Recent advances in visible light Bi-based photocatalysts, Chinese Journal of Catalysis, **35** (2014) 989-1007.
8. Y. Li, X. Fang, M. Cao, Thermal frequency shift and tunable microwave absorption in $BiFeO_3$ family, Scientific reports, **6** (2016) 24837.
9. T. Gao, Z. Chen, F. Niu, D. Zhou, Q. Huang, Y. Zhu, L. Qin, X. Sun, Y. Huang, Shape-controlled preparation of bismuth ferrite by hydrothermal method and their visible-light degradation properties, Journal of Alloys and Compounds, **648** (2015) 564-570.
10. T. Gao, Z. Chen, Y. Zhu, F. Niu, Q. Huang, L. Qin, X. Sun, Y. Huang, Synthesis of $BiFeO_3$ nanoparticles for the visible-light induced photocatalytic property, Materials Research Bulletin, **59** (2014) 6-12.
11. C. Chen, W. Ma, J. Zhao, Semiconductor-mediated photodegradation of pollutants under visible-light irradiation, Chemical Society reviews, **39** (2010) 4206-4219.
12. X. Chen, S. Shen, L. Guo, S.S. Mao, Semiconductor-based Photocatalytic Hydrogen Generation, Chemical Reviews, **110** (2010) 6503-6570.
13. M.T.S. Nair, P.K. Nair, Photoconductive bismuth sulphide thin films by chemical deposition, Semiconductor Science and Technology, **5** (1990) 1225.
14. L. Li, N. Sun, Y. Huang, Y. Qin, N. Zhao, J. Gao, M. Li, H. Zhou, L. Qi, Topotactic Transformation of Single-Crystalline Precursor Discs into Disc-Like Bi_2S_3 Nanorod Networks, Advanced Functional Materials, **18** (2008) 1194-1201.
15. K. Yao, W.W. Gong, Y.F. Hu, X.L. Liang, Q. Chen, L.M. Peng, Individual Bi_2S_3 Nanowire-Based Room-Temperature H_2 Sensor, The Journal of Physical Chemistry C, **112** (2008) 8721-8724.
16. O. Rabin, J. Manuel Perez, J. Grimm, G. Wojtkiewicz, R. Weissleder, An X-ray computed tomography imaging agent based on long-circulating bismuth sulphide nanoparticles, Nature materials, **5** (2006) 118-122.
17. K.L. Ai, Y.L. Liu, J.H. Liu, Q.H. Yuan, Y.Y. He, L.H. Lu, Large-Scale Synthesis of Bi_2S_3 Nanodots as a Contrast Agent for In Vivo X-ray Computed Tomography Imaging, Advanced Materials, **23** (2011) 4886-4891.
18. T. Wu, X.G. Zhou, H. Zhang, X.H. Zhong, Bi_2S_3 Nanostructures: A New Photocatalyst, Nano Research, **3** (2010) 379-386.
19. Y. Yan, Z. Zhou, X. Zhao, J. Zhou, A controlled anion exchange strategy to synthesize core-shell beta-bismuth oxide/bismuth sulfide hollow heterostructures with enhanced visible-light photocatalytic activity, Journal of colloid and interface science, **435** (2014) 91-98.
20. L. Chen, J. He, Q. Yuan, Y. Liu, C.-T. Au, S.-F. Yin, Environmentally benign synthesis of branched Bi_2O_3–Bi_2S_3 photocatalysts by an etching and re-growth method, J. Mater. Chem. A, **3** (2015) 1096-1102.
21. X. Gao, H.B. Wu, L. Zheng, Y. Zhong, Y. Hu, X.W. Lou, Formation of Mesoporous Heterostructured $BiVO_4/Bi_2S_3$ Hollow Discoids with Enhanced Photoactivity, Angewandte Chemie, **126** (2014) 6027-6031.
22. W. Xu, J. Fang, Y. Chen, S. Lu, G. Zhou, X. Zhu, Z. Fang, Novel heterostructured $Bi_2S_3/Bi_2Sn_2O_7$ with highly visible light photocatalytic activity for the removal of Rhodamine B, Materials Chemistry and Physics, **154** (2015) 30-37.
23. Y. Liu, M. Zhang, L. Li, X. Zhang, In situ ion exchange synthesis of the $Bi_4Ti_3O_{12}/Bi_2S_3$ heterostructure with enhanced photocatalytic activity, Catalysis Communications, **60** (2015) 23-26.
24. S. Jiang, K. Zhou, Y. Shi, S. Lo, H. Xu, Y. Hu, Z. Gui, In situ synthesis of hierarchical flower-like Bi_2S_3/BiOCl composite with enhanced visible light photocatalytic activity, Applied Surface Science, **290** (2014) 313-319.
25. H. Cheng, B. Huang, X. Qin, X. Zhang, Y. Dai, A controlled anion exchange strategy to synthesize Bi_2S_3 nanocrystals/BiOCl hybrid architectures with efficient visible light photoactivity, Chem Commun (Camb), **48** (2012) 97-99.
26. C. Tang, C. Wang, F. Su, C. Zang, Y. Yang, Z. Zong, Y. Zhang, Controlled synthesis of urchin-like Bi_2S_3 via hydrothermal method, Solid State Sciences, **12** (2010) 1352-1356.
27. H.P. Jiao, X. Yu, Z.Q. Liu, P.Y. Kuang, Y.M. Zhang, One-pot synthesis of heterostructured Bi_2S_3/BiOBr microspheres with highly efficient visible light photocatalytic performance, Rsc Advances, **5** (2015) 16239-16249.
28. W. Wang, H. Cheng, B. Huang, X. Lin, X. Qin, X. Zhang, Y. Dai, Synthesis of $Bi_2O_2CO_3/Bi_2S_3$ hierarchical microspheres with heterojunctions and their enhanced visible light-driven photocatalytic degradation of dye pollutants, Journal of colloid and interface science, **402** (2013) 34-39.
29. S.J.A. Moniz, R. Quesada-Cabrera, C.S. Blackman, J. Tang, P. Southern, P.M. Weaver, C.J. Carmalt, A simple, low-cost CVD route to thin films of $BiFeO_3$ for efficient water photo-oxidation, Journal of Materials Chemistry A, **2** (2014) 2922-2927.

Fabrication of Co^{3+} and B co-doping $BiVO_4$ with improved photocatalytic performance for organic degradation

Zhilin Li, Chongyue Jin and Min Wang[a]

School of Mechanical Engineering and Automation, Northeastern University, Shenyang 110189, China

Abstract. Co^{3+} and B co-doping $BiVO_4$ photocatalysts were successfully prepared for organic pollutants degradation through sol-gel and impregnation two-step process. Their structure, surface chemical composition, and optical absorption were characterized. All as-prepared samples consisted of the monoclinic phase. It was found that the Co dopants was existed in the form of Co_3O_4 in B-$BiVO_4$, thus a p–n heterojunction formed in $BiVO_4$. The optical absorption results reveal that Co_3O_4 could greatly enhance the absorption and decrease in the band-gap energy of $BiVO_4$. The 0.5% Co^{3+} doping B-$BiVO_4$ sample exhited the highest photocatalytic activity for methyl orange photocatalytic degradation, being approximately 92% of MO degradation rate under 50 min visible light irradiation.

1 Introduction

As one of the non-titania visible-light-driven photocatalysts, $BiVO_4$ has recently attracted considerable attention, as it is commonly used as a photocatalyst in water splitting and oxidative decomposition of organic pollutants under visible- light irradiation [1-5]. It has been found that the photocatalytic activity of $BiVO_4$ is determined by its crystal phase. $BiVO_4$ with a monoclinic scheelite structure can show photocatalytic properties and is commonly used as a photocatalyst in water splitting and oxidative decomposition of organic contaminants under visible light irradiation [6-8]. The photocatalytic activity of pure $BiVO_4$ still needs improvement due to its high rate of photogenerated electron-hole pair recombination. The introduction of dopant elements causes the formation of charge traps for electrons and/or holes and thus prolongs the recombination time [9].

In the current study, the strategy of Co^{3+} and B co-doping was employed to improve the photocatalytic performance of $BiVO_4$ photocatalyst. The Co^{3+} doping B-$BiVO_4$ photocatalysts were successfully prepared for organic pollutants degradation through sol-gel and impregnation two-step process. The effects of Co^{3+} doping on the structure, light absorption ability and photocatalytic performance of B-$BiVO_4$ were studied in detail.

2 EXPERIMENT

2.1 Photocatalyst preparation

The $BiVO_4$ and B-$BiVO_4$ precursors were synthesized through a sol-gel method as reported in reference [10].The Co^{3+} and B co-doping $BiVO_4$ samples were prepared by the impregnation method. The B-$BiVO_4$ precursor powder (3.0 g) was added to 2 mL of distilled water containing an appropriate amount of $Co(NO_3)_2$ in a ceramic dish. Water was evaporated at 353 K. The suspension was stirred using a glass rod during evaporation. The resulting powder was collected and calcined in air at 500 °C for 5 h, and then cooled to room temperature to obtain Co^{3+} doping B-$BiVO_4$ samples, labeled as xCo-B-$BiVO_4$ nanoparticles. The Co doping concentration (x wt %) was chosen as 0.3, 0.4, 0.5, 0.6, and 0.7, which is the weight percentage of $Co(NO_3)_2$ to $BiVO_4$. 0.5Co doping $BiVO_4$ was prepared with the same impregnation method except that its precursor is pure $BiVO_4$. For comparison, pure $BiVO_4$ and B-$BiVO_4$ were prepared according to reference [10].

2.2 Characterization of photocatalyst

The X-ray diffraction (XRD) spectra of the catalysts were obtained by X-ray diffraction (XRD) with Cu Kα radiation (model D/max RA, Rigaku, Japan). The accelerating voltage and the current were 40 kV and 150 mA, respectively. The morphologies of the samples were observed with a field emission scanning electron microscope (SEM, S-3000N, Hitachi, Japan), and the chemical composition of the photocatalysts was determined using an energy dispersive X-ray spectrometer (EDS) attached to the SEM. The binding energies of Bi, V, Co, B, and O were measured at room temperature using an X-ray photoelectron spectroscope (XPS, VGESCALAB MARK II) using Mg Kα radiation. The diffuse reflectance spectra (DRS) were obtained for

[a] Min Wang:wangmin@me.neu.edu.cn

the dry-pressed disk samples using a UV-vis spectrophotometer (TU-1901, Puxi, China) equipped with an integrating sphere assembly using $BaSO_4$ as the reflectance standard. The spectra were recorded in the wavelength range from 230 to 800 nm at 25 ± 1 °C.

2.3 Photocatalytic performance test

The photocatalytic activities of the samples were determined by measuring the degradation of methyl orange (MO) in an aqueous solution under visible light irradiation. In the activity test, a 250 W tungsten halogen lamp with a < 420 nm cutoff filter was used as visible light source. The photo-degradation experiment was performed according to the following process: The photocatalyst (0.01 g) was suspended in 50 mL aqueous solution of methyl orange with the initial concentration of 15 mg/L. The suspension was vigorously stirred with a magnetic stirrer in the dark for 60 minutes prior to illumination to achieve adsorption/desorption equilibrium, as well as in the photoreactor throughout the experiment. At given time intervals, the collected samples were filtered through a 0.45 μm Millipore filter to remove catalyst particles. The filtrate concentration was monitored by recording the absorbance at 464 nm, the maximum absorption wavelength for methyl orange, using a UV-1800 UV-Vis spectrophotometer (Puxi, China).

3 Results and discussion

3.1 XRD analysis

The XRD patterns of as-prepared samples are shown in Fig. 1. All the as-prepared samples exhibited typical peaks of monoclinic $BiVO_4$ (JCPDS cards No. 75-1866)[2,5]. Although some studies in the literature have reported that $Co(NO_3)_2$ can be transformed into Co_3O_4 during the calcination process [11], no Co_3O_4 peaks were found in XRD analysis, which might be attributed to the low Co content. Fig. 1 (B) shows the shift of the peaks at approximately 28.8 and 30.5^0 corresponding to the (121) and (040) planes, respectively. The peaks shift toward the left in the case of B single doping, but our results show a gradual shift toward the right as the Co doping content increased. The right shift suggests that the Co cation was successfully doped into the $BiVO_4$ and the increased amount of Co dopant in the samples lead to a change in the local crystal structures by a few degrees.

3.2 SEM analysis

The morphology of the synthesized samples was analyzed by SEM. Fig. 2 shows the three typical SEM images of the $BiVO_4$, B-$BiVO_4$, and B-0.5Co-$BiVO_4$ photocatalysts. All three samples displayed a sphere-like morphology, but there is a minor difference in the morphologies and particle shapes of the samples between pure $BiVO_4$ and B-$BiVO_4$. After Co doping, the sizes of some particles decreased, which could cause an increase in the surface area of the sample. The composition of the B-0.5Co-$BiVO_4$ sample is determined by energy dispersive X-ray spectroscopy (EDX) experiments. As shown in Fig. 2d, we can see the clear signals for Bi, V, O, and Co.

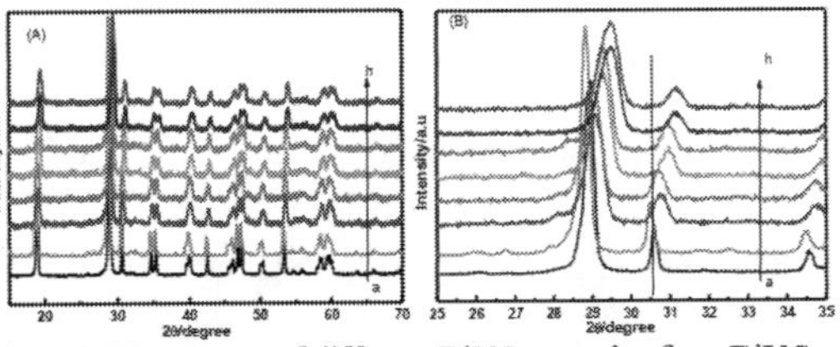

Fig. 1 XRD patterns of different $BiVO_4$ samples for a.$BiVO_4$, b.B-$BiVO_4$, c. 0.5Co-$BiVO_4$, d. 0.3Co-B-$BiVO_4$ e.0.4Co-B-$BiVO_4$, f. 0.5Co-B-$BiVO_4$, g. 0.6Co-B-$BiVO_4$, h. 0.7Co-B-$BiVO_4$

Fig. 2 SEM images of a. pure $BiVO_4$, b.B-$BiVO_4$, c. 0.5Co-B-$BiVO_4$, and d. EDX of 0.5Co-B-$BiVO_4$

3.3 XPS analysis

The chemical states and chemical compositions of the samples are investigated through XPS. Fig. 3 shows the Co 2p (a), Bi 4f (b), B 1s(c,d), V2p(e), and O1s(f) XPS spectra of the as-prepared samples. The Co 2p high-resolution XPS scan spectrum is shown in Fig. 3 (a). Two peaks at 780.7 and 797.3 eV, which can be assigned to the Co 2p3/2 and Co 2p1/2, respectively, were observed. These peaks are very close to those of Co_3O_4 [12], for which corresponding peaks were reported at 780.8 and 797.1 eV.

Fig. 3(b) shows the Bi 4f high-resolution XPS scan spectrum of pure $BiVO_4$, B-$BiVO_4$ and 0.3 B-0.5Co-$BiVO_4$; the binding energies of Bi 4f appear at 159.1~159.5 eV and 164.4 ~165.2 eV for Bi $4f_{7/2}$ and Bi $4f_{5/2}$, respectively, which are characteristic of the Bi^{3+} species [2-7]. The B 1s XPS spectra of the B-doped and B-Co codoped samples are shown in Fig. 3(c) and (d). Each XPS spectrum of B 1s in Fig. 3(c) and (d) exhibits an asymmetric broad peak from 184 to 191 eV, which is a typical characteristic peak for B 1s species. The asymmetric peak was deconvolved into two components at E_b = 184.2 and 190.8 eV for B-$BiVO_4$, and

E_b = 184.3 and 188.8 eV for B-0.5Co-$BiVO_4$. The peak at 184.2 and 185.5 eV is assigned to B_4C [13], which showed no photocatalytic activity.

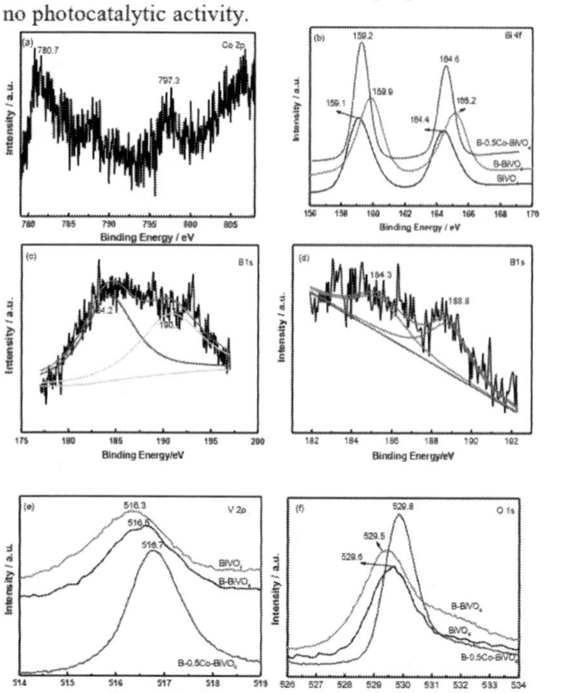

Fig. 3 (a) Co 2p XPS spectra for B-0.5Co-$BiVO_4$; (b) Bi 4f XPS spectra for $BiVO_4$, B-$BiVO_4$, and B-0.5Co-$BiVO_4$ sample; (c) B 1s XPS spectra for B-$BiVO_4$, and (d) B 1s XPS spectra for B-0.5Co-$BiVO_4$ samples; (d) V 2p and (e) O 1s XPS spectra of $BiVO_4$, B-$BiVO_4$, and B-0.5Co-$BiVO_4$ sample

The standard binding energies for B1s in B_2O_3, H_3BO_3 (B-O bond), and VB_2 (V-B bond) are 193.6 eV [13], 193.0 eV[14], 188.3 eV[15], respectively. It may be concluded that a portion of the B atoms might be doped in the $BiVO_4$ lattice, replacing the O to form a B-Bi-O bond for this B-$BiVO_4$ sample. After further doping with Co, the binding energy of B1s changed to 188.8 eV. This is possibly due to the doping Co ions replacing some Bi ions to form B-Co-O-bonds considering that the standard binding energies for B1s in CoB is 188.5 eV [12]. However, we can deduce that there is not some CoB phase appearing in the $BiVO_4$ because the binding energy of Co in 0.5Co-B-$BiVO_4$ sample is 780.7 eV, as compared with CoB which is approximately 778 eV (as shown in Fig. 3(a)). The XPS results show that both B and Co are successfully doped into the $BiVO_4$ lattice. Fig. 3 (e) shows the V 2p XPS spectra of pure $BiVO_4$, B-$BiVO_4$, and 0.5Co-B-$BiVO_4$. The V $2p_{3/2}$ signals have peaks at E_b = 516.3 eV for $BiVO_4$, E_b = 516.5 eV for B-$BiVO_4$, and E_b = 516.7 eV for 0.5Co-B-$BiVO_4$, which correspond the V^{5+}[7].The peaks at binding energies of 529.6 eV for pure $BiVO_4$, 529.6 eV for the B-$BiVO_4$, and 529.8 eV 0.5Co-B-$BiVO_4$ correspond to the O1s, as shown in Fig. 3(f). It is worth noting that the binding energy of O 1s for B-0.5Co-$BiVO_4$ shifts to higher position (529.8 eV). It is known that the smaller electronegativity of Co (1.88) compared with the Bi (2.02), it led to stronger chemical binding of Co-O bond and increased O1s binding energy [12].

3.4 UV-Vis DRS analysis

The UV-Vis DRS of the pure $BiVO_4$, B-$BiVO_4$, Co-$BiVO_4$, and B-xCo-$BiVO_4$ catalysts are shown in Fig. 4. All samples exhibit different absorption spectra in the visible light region. The absorption edge of pure $BiVO_4$ is determined to be 529 nm. Band gap energy (*Eg*) can be used to evaluate the optical absorption performance of $BiVO_4$. The *Eg* was determined from the equation of $(\alpha h\nu)2 = A(h\nu - Eg)n$, where the *A*, α, and $h\nu$ represent a constant, absorption coefficient, and incident photon energy, respectively. In the case of direction transition for $BiVO_4$, the value of n is 1[6]. The Eg value of $BiVO_4$ samples can be estimated from a plot of $(\alpha h\nu)^2$ versus photon energy ($h\nu$), as shown in Fig. 4. The intercept of the tangent to the X-axis is approximately equal to the Eg value. Thus, the Eg can be estimated to be approximately 2.40. For B-$BiVO_4$, due to the introduction of small amounts of boron, red shift of the absorption spectrum occurred. Some authors proposed that the visible light response was due to the narrowing of the band gap by mixing B1s and O 2p states [13]. The corresponding band-gap energy for the B-$BiVO_4$ catalyst is 2.36 eV. The presence of Co_3O_4 results in an increase in the capacity for visible light absorption, which should be attributed to the small band gap of Co_3O_4. Co_3O_4 is a p-type semiconductor with direct transition at 1.45 and 2.07 eV [12], corresponding, respectively, to edges of $O^{2-} \rightarrow Co^{3+}$ excitation and $O^{2-} \rightarrow Co^{2+}$ charge transfer. The latter is the basic optical band gap energy for interband transitions. Co_3O_4 has absorption in nearly the entire visible range of the spectrum and induces an extension of the light absorption spectrum of the composite semiconductor even with low cobalt content. Thus, with the increase of Co doping concentration, the absorption is strengthened in the visible region, and the band gap energies decrease. The band gaps of Co-doped and xCo-B-$BiVO_4$(x is 0.3, 0.4, 0.5, 0.6, 0.7) are 2.33, 2.31, 2.26, 2.23, 2.21 and 2.11, respectively.

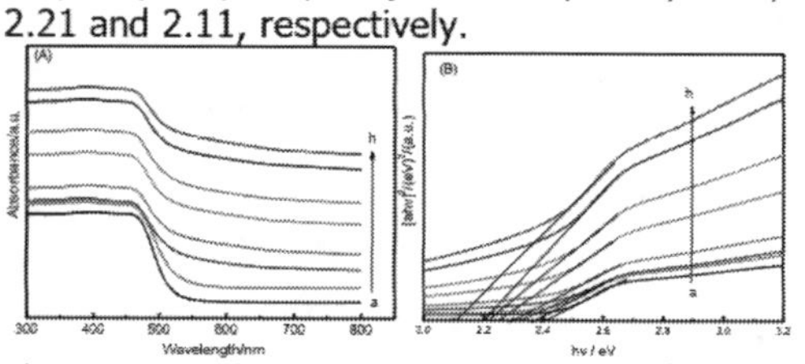

Fig. 4 UV-Vis absorption spectra (A) and $(ah\nu)^2 - h\nu$ curves (B) of some samples- a.$BiVO_4$, b.B-$BiVO_4$, c. 0.5Co-$BiVO_4$, d. B-0.3Co-$BiVO_4$ e.B-0.4Co-$BiVO_4$, f. B-0.5Co-$BiVO_4$, g. B-0.6Co-$BiVO_4$, h. B-0.7Co-$BiVO_4$

3.5 Photocatalytic perfomance

The photocatalytic degradation of MO dye on the as

 https://doi.org/10.1051/matecconf/201823803008

prepared $BiVO_4$ catalysts was carried out under visible-light irradiation. The results are shown in Fig. 5. It shows that only a small amount of MO could be degraded through either direct photolysis in the absence of catalyst or adsorption in the dark. The evidence indicates that the pure $BiVO_4$ a catalysts exhibited low visible light photocatalytic efficiency for the degradation of MO within 50 min irradiation. Compared with pure $BiVO_4$, B- and Co- single doped $BiVO_4$ can enhance the photocatalytic activity to some degree. However, it can be obviously observed that B-Co codoping can increase $BiVO_4$ photocatalytic activity effectively. The MO photocatalytic degradation rate for B-Co codoped $BiVO_4$ increases with Co doping content when the doping Co is lower than 0.5%, but decreases when the amount of Co is higher than 0.5%. The degradation rate of MO can be up to 92 % in 50 min. So the optimal mole percentage of Co doping in B-Co codoping $BiVO_4$, is 0.5%. These results confirmed that Co and B elements exhibited a synergetic effect on the improvement of photocatalytic activity efficiency. Moreover, as displayed in Fig. 5(B), the photocatalytic degradation of MO solution fitted well with the pre-first-order kinetics according to the Langmuir-Hinshelwood (L-H) model within 50 min.

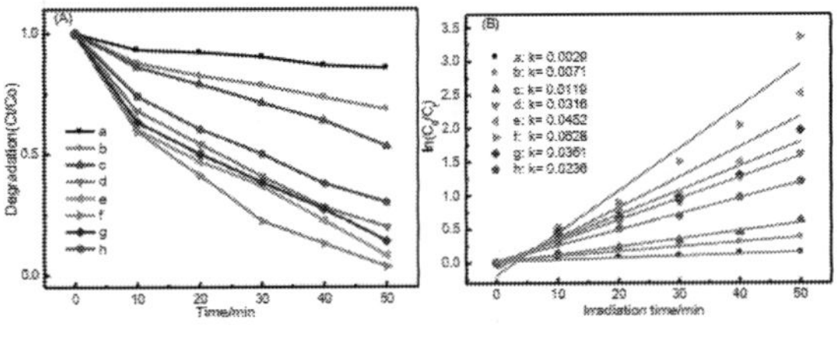

Fig. 5 (A) MO degradation under visible light illumination for 50 min in the presence of B-$BiVO_4$ with various nickel doping, pure $BiVO_4$ and without photocatalyst; (B) *ln* (C_0/C_t) versus time for B-$BiVO_4$ doping concentrations of a.$BiVO_4$, b.B-$BiVO_4$, c. 0.5Co-$BiVO_4$, d. B-0.3Co-$BiVO_4$ e.B-0.4Co-$BiVO_4$, f. B-0.5Co-$BiVO_4$, g. B-0.6Co-$BiVO_4$, h. B-0.7Co-$BiVO_4$

Based on the above analysis, the scheme that the as prepared Co^{3+} and B codoped $BiVO_4$ catalysts exhibited higher photocatalytic efficiency can be explained as following scheme (Fig. 6): in the case of Co^{3+} and B codoped $BiVO_4$, on the one hand, the impurity energy levels are introduced above the valence band due to B dopants, thus causing the absorption edge to shift into the visible region. The impurity levels create another possible transition route for excited electrons during visible light irradiation, which electrons are promoted from the B1s level to the conduction band of $BiVO_4$. On the other hand, because the Co_3O_4 [23] is a p-type semiconductor and $BiVO_4$ is an n-type material [1], a p-n heterojunction formed at the interface of the two materials in the composite photocatalysts. The photoinduced electrons on the $BiVO_4$ particle surface can easily migrate to Co_3O_4 via interfaces; similarly, photoinduced holes on the Co_3O_4 surface can transfer to $BiVO_4$ owing to the different valence band edge potentials. As a result, a larger amount of electrons on the Co_3O_4 surface and holes on the $BiVO_4$ surface can participate in photocatalytic reactions to directly or indirectly decompose MO, thus enhancing the photocatalysis efficiency.

However, more Co_3O_4 can also act as recombination centers for the electrons and holes. Therefore, an appropriate amount of the doped Co in B-$BiVO_4$ catalysts is necessary so that the recombination of photogenerated electrons and holes can be suppressed effectively.It is interesting to note that the photocatalytic activities of B and Co_3O_4 codoped $BiVO_4$ catalysts enhance with the increase in Co doping content up to 0.5 % and then decrease. A similar result was also reported by Zhang et al. [30]. It is hypothesized that high Co_3O_4 content is harmful to the photocatalytic activity because of the excess coverage of active sites on the $BiVO_4$ surface by Co_3O_4 particles and can also act as recombination centers for the electrons and holes [16].

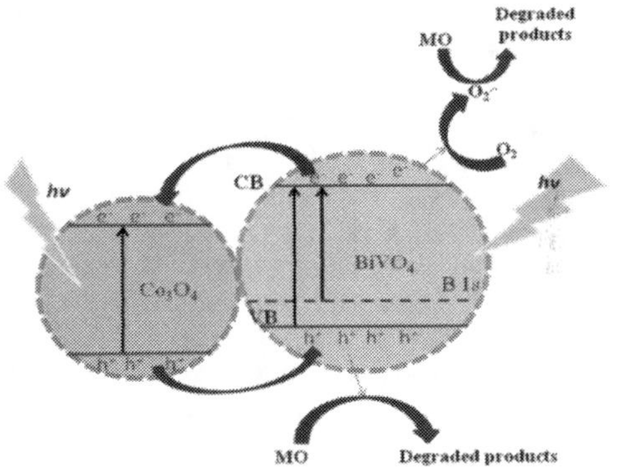

Fig.6 Proposed degradation scheme of organic pollutants by B-Co-$BiVO_4$ under visible light irradiation.

4 Conclusions

Co^{3+} and B codoped $BiVO_4$ catalysts were successfully prepared through a sol-gel and impregnation two-step process. Compared with the pure $BiVO_4$, B-, Co- single-doped $BiVO_4$, theCo^{3+} and B codoped $BiVO_4$ exhibits higher photoactivity under visible light. These samples show higher photocatalytic activity for methyl orange photocatalytic degradation, and the highest degradation rate for B-0.5Co-$BiVO_4$ can reach to be approximately 92% under visible light irradiation in 50 min. The B dopants led to impurity energy levels above the valence band and a p-n heterojunction formed due to Co_3O_4 being a p-type semiconductor and $BiVO_4$ an n-type. Codoping can narrow the band gap. The synergetic effects of B and Co_3O_4 may efficiently promote the separation of photogenerated holes and electrons and are the main reasons for the high photodegradation rate of dye pollutants under visible light irradiation.

References

1. F. Chen, Q.Yang, X.M. Li, G.M. Zeng, D.B. Wang, C.G. Niu, Appl. Catal. B: Environ. 200 (2017) 330-342.

2. R.Z. Sun, Q.M. Shi, M. Zhang, L.H. Xie, J.S. Chen, X.M. Yang, M.X. Chen, W.R. Zhao, J. Alloys Compd. 714 (2017) 619-626.
3. M. Wang, Y.S. Che, C. Niu, M.Y. Dang, D. Dong, J. Hazard. Mater. 262 (2013) 447-455.
4. H.J. Kong, H.W. Da, J. Kim, S.I. Woo, Chem. Mater. 28 (2016) 1318-1324.
5. Z.B. Xiang, Y Wang, D. Zhang, P. Ju, J. Ind. Eng. Chem. 40 (2016) 83-92.
6. X. Chang, T. Wang, P. Zhang, J.J Zhang, A. Li, J. L. Gong. J. Am. Chem. Soc. 137 (2015) 8356-8359.
7. S.S. Xue, H.B. He, Z. Wu, C.L. Yu, Q.Z. Fan, G.M. Peng, K. Yang, J. Alloys Compd. 694 (2017) 989-997.
8. M. Wang, P. Y. Guo, T. Y. Chai, Y. H. Xie, J. Han, M. Y. You, Y. Z. Wang , T. Zhu, J. Alloys Compd. 691(2017)8-14.
9. L. Yan ,W. Zhao, Z. Liu, Dalton Trans, 45 (2016) 11346-11352.
10. M. Wang, H. Y. Zheng, Q. Liu, C. Niu, Y. S. Che, M. Y. Dang, Spectrochim. Acta, Part A, 114(2013) 74-79
11. B. Zhou, X. Zhao, H.J. Liu, J.H. Qu, C.P. Huang, Sep. Purif. Technol. 77 (2011) 275-282.
12. C. L. Yu, K. Yang, J. C. Yu, F. F. Cao, X. Li , X. C. Zhou. J. Alloys Compd. 509 (2011) 4547-4552.
13. H. J. Long, Q. J. Meng, J. Yuan, W. S. Yang, Y.A. Cao, Acta. Chim. Sinica., 66 (2008) 657-661.
14. C. D. Wagner, W. M. Riggs, L.E. Davis, J.F. Moulder, in:G.E. Muilenberg(Ed.). Handbook of X-Ray Photoelectron Spectroscopy, Perkin-Elmer Co., Minnesota, 1979.
15. F. Y. Wei, L.S. Ni, .Chin. J Catal., 28(2007) 905-909.
16. C. L. Yu, K. Yang, J. C. Yu, F. F. Cao, X. Li , X. C. Zhou. J. Alloys Compd. 509 (2011) 4547-4552.

Effect of calcination temperature on the catalytic performance of $CoFe_2O_4$/Nitrogen doped sludge based activated carbon in activation of peroxymonosulfate for degradation of coking wastewater

Yongkang Tao[1], Lihua Li[1,a], Lixiong Ren[1], Yu Liang[1] and Xin Wang[1]

1. School of Chemical Engineering, University of Science and Technology Liaoning, Anshan Liaoning 114051, China

Abstract. A novel supported heterogeneous magnetic catalyst $CoFe_2O_4$/N-doped sludge based activated carbon ($CoFe_2O_4$/N-SAC) was prepared by polymer network gel method for the first time. The physicochemical properties of the materials were characterized by means of XRD, SEM, TEM, VSM and XPS techniques. The prepared catalyst is applied to the heterogeneous activation of peroxymonosulfate for degradation of coking wastewater, and the effect of calcination temperature on the catalytic activity was investigated. The result reveals that the catalyst shows the highest catalytic activities under the calcination temperature is 800 °C with the TOC removal rate of coking wastewater is 84.31%.

1. Introduction

Over the past period of time, the advanced oxidation processes (AOPs) based on peroxymonosulfate (PMS) have been widely concerned for degradation of coking wastewater because the sulfate radical (SO_4^{-}•) is an oxidizing agent with a high ability to oxidize organic pollutants due to its higher oxidation-reduction potential and longer half-life period (the oxidation-reduction potential and half-life period of SO_4^{-}• are 2.6-3.1eV and 4s, respectively), and its use avoids some shortcomings of traditional Fenton and ozone, which both use hydroxyl radicals (•OH) to degrade organic pollutants with the half-life of •OH is short and the redox potential is low (the half-life and redox potential of •OH were 1μs and 1.8~2.7 eV, respectively) [1-2]. Transition metal ions combined with persulfate to form SO_4^{-}• is a highly efficient method for the degradation of organic pollutants and Co^{2+} is the best catalyst for activating persulfate [3]. However, the detrimental impact of Co^{2+} lost with the effluent on environment and human health is a serious concern. Thus it is essential to develop heterogeneous Co catalysts [1].

Effluent treatment applications of multifarious heterogeneous Co-base catalysts such as Co_3O_4, $CoFe_2O_4$ coupled with PMS have been investigated [4-5] and the spinel-structured nano-ferrite $CoFe_2O_4$ has been widely used in wastewater treatment because of its excellent catalytic performance [6]. Activated carbon has been used as a carrier for heterogeneous catalysts because of its huge specific surface area [7-12]. In addition, the research has been shown that doping heteroatoms such as N into the carbon network can create more active sites and bring new properties such as higher selectivity towards reduction reaction [13-14]. Sludge contains a large amount of organic matter, which can be used as raw material for the preparation of sludge based activated carbon (SAC) [15]. Supporting $CoFe_2O_4$ on SAC can not only solves the problems of easy agglomeration and poor stability of unsupported $CoFe_2O_4$, but also solves the conundrum of SAC separation, which is beneficial to the practical application of the catalyst.

In this study, the $CoFe_2O_4$/N-SAC was prepared and used to activate PMS to generate SO_4^{-}• for degradation of coking wastewater. The effect of calcination temperature was investigated.

[a] Corresponding author: lilh2011@163.com

2. Methods

2.1 Materials

Sewage sludge used in this study is collected from coking plant of Anshan iron and Steel Group (Anshan, China). Coking wastewater is taken from the aerobic tank effluent of Anshan Iron and Steel coking plant (Anshan, China). $HKSO_5 \cdot 0.5K_2SO_4 \cdot 0.5HKSO_4$ was purchased from Sigma-Aldrich, $Co(NO_3)_2 \cdot 6H_2O$, $Fe(NO_3)_3 \cdot 9H_2O$, Citric acid, acrylamide and ammonium persulfate were analytical grade and purchased from rchased from Sinopharm Chemical Reagent Co., Ltd. (Shanghai, China), and DI water was used throughout.

2.2 Preparation of catalysts

Sewage sludge of biological wastewater treatment plant and agricultural waste were used as the raw material to product SAC via chemical activated process with $ZnCl_2$ as activation agent within a muffle furnace under 600 °C for 1.5 h with the protection of N_2 before cooled to room temperature.

$CoFe_2O_4$/N-SAC was prepared by Polymer network gel method using $Co(NO_3)_2 \cdot 6H_2O$ and $Fe(NO_3)_3 \cdot 9H_2O$ as the precursors and Citric acid as the complexant. The appropriate amounts of $Co(NO_3)_2 \cdot 6H_2O$, $Fe(NO_3)_3 \cdot 9H_2O$ and citric acid (Co: Fe: citric acid=1:2:3,by mole) were dissolved in 100 mL distilled water under vigorous stirring in a 80°C water bath with the concentrations of precursor solution is 15%, the ammonia was used to regulate pH at 7, 10 g SAC was immersed in the solution and then 7.5 g acrylamide and 0.075 g ammonium persulfate was added to the system. After a few minutes, the uniform wet gel formed by polymerization was dried at 80°C for 12h. Finally, it was transferred to muffle furnace and calcined at 600-900 °C for 2 h with the absence of oxygen condition to obtain the catalysts.

2.3 Degradation experiment of coking wastewater

The oxidation of coking wastewater was carried out in conical bottles. The catalyst and oxidant were added into the coking wastewater. The reaction was conducted in the 25 °C constant temperature water bath oscillator. 50 mL suspension samples were taken out at specific time intervals and quenched with 1 mL of ethanol to prevent further reaction. The TOC was determined with a total organic carbon analyzer (Sievers InnovOx, GE, America).

2.4 Characterization of catalyst

The crystal structure of catalysts was analyzed by a powder X-ray diffractometer (XRD, RigakuD/max-2000) with monochromatic Cu Ka radiation (45 kV, 50 mA). The morphology of the catalysts was characterized by a Jeol 2100 TEM with field emission gun at 200 kV. The surface morphology of the catalysts was investigated using a scanning electron microscopy (SEM, Evo 18, Gemarny). The room temperature magnetic properties of the samples were investigated using vibrating sample magnetomertry (VSM, 730T, America) at magnetic fields up to 15KOe. The elemental compositions of $CoFe_2O_4$/N-SAC were detected by XPS on a Thermo Escalab 250.

3. Results and discussions

3.1 Characterization of SAC and $CoFe_2O_4$/N-SAC

3.1.1 XRD

Figure 1 shows the XRD patterns of $CoFe_2O_4$/N-SAC with various calcination temperatures, all the catalysts exhibited the typical patterns of cubic phase of $CoFe_2O_4$ with spinel structure and the space group Fm3m (JCPDS No.00-003-0864).

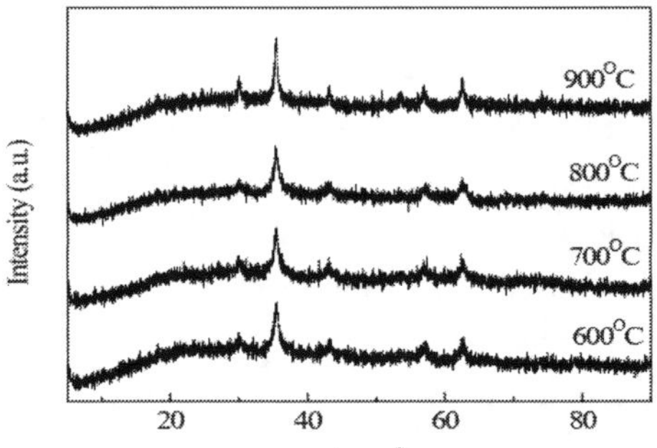

Figure 1. XRD patterns of $CoFe_2O_4$/N-SAC calcined at different temperatures

It can be seen from the diagram that with the increase of temperature, the diffraction peaks of $CoFe_2O_4$/N-SAC became more narrow and intense, indicating that the particle size of $CoFe_2O_4$ becomes bigger. When the temperature is further increased from 800 to 900 °C, the diffraction peak intensity of each crystal plane increased remarkably, which caused by the agglomeration of $CoFe_2O_4$ on the surface of catalyst [16].

3.1.2 XPS

In order to investigate chemical states and the compositions in the content on the surface of the SAC and $CoFe_2O_4$/N-SAC which calcined at different temperatures, the XPS measurements were carried out. As shown Figure 2, the XPS spectra of Co 2p, Fe 2p, C 1s, N1s and O 1s were obtained with the characteristic absorption peaks appear at 284.62 eV, 399.59 eV, 532.53 eV, 711.27 eV and 780.87 eV, respectively, and the corresponding element contents in various samples are shown in Table 2. Compared with SAC, $CoFe_2O_4$/N-SAC prepared at different calcination temperatures showed characteristic absorption peaks of N1s, indicating that nitrogen was successfully doped into sludge-based activated carbon. Moreover, the absorption peak of O1s decreased obviously, which indicated that some oxygen-containing groups reacted with the amino group of polyacrylamide in the process of nitrogen doping.

In order to further study the existing forms and changes of N in $CoFe_2O_4$/N-SAC at different calcination temperatures, the peaks of N1s in these materials was fitted. The specific results are as follows:

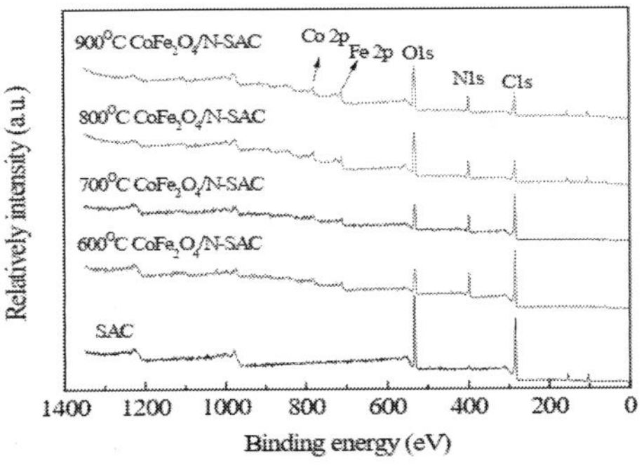

Figure 2. XPS spectra analysis of the different samples

Figure 3 shows the peak fitting spectra of N1s at different calcination temperatures from 400 °C to 800 °C. The peaks locate at 398.3 eV, 399.3 eV, 400.5 eV, 401.2 eV and 406.4 eV correspond to pyridine nitrogen, amino or imino ($-NH_2$/-NH-), pyrrole nitrogen, graphitized nitrogen and nitrogen oxides [17], respectively. In addition, the content of pyridine nitrogen increases with the increase of temperature when the calcination temperature is lower than 800 °C, but declines when the calcination temperature further increases to 900 °C, while the content of pyrrolic nitrogen and graphitized nitrogen increases with the increase of calcination temperature. This may be attributed to the transformation from pyridine nitrogen to pyrrolic nitrogen at higher temperature [18].

Table 1 Contents of elements in different samples

Sample	C(wt%)	O(wt%)	N(wt%)	Fe(wt%)	Co(wt%)	Co/Fe (mole ratio)
SAC	59.80	28.48	-	-	-	-
600°C $CoFe_2O_4$/N-SAC	58.15	16.78	15.86	12.46	4.07	1:3.23
700°C $CoFe_2O_4$/N-SAC	53.08	18.62	14.48	12.52	5.94	1:2.22
800°C $CoFe_2O_4$/N-SAC	31.70	24.30	13.49	13.49	7.17	1:1.98
900°C $CoFe_2O_4$/N-SAC	26.56	27.38	12.33	14.30	7.66	1:1.97

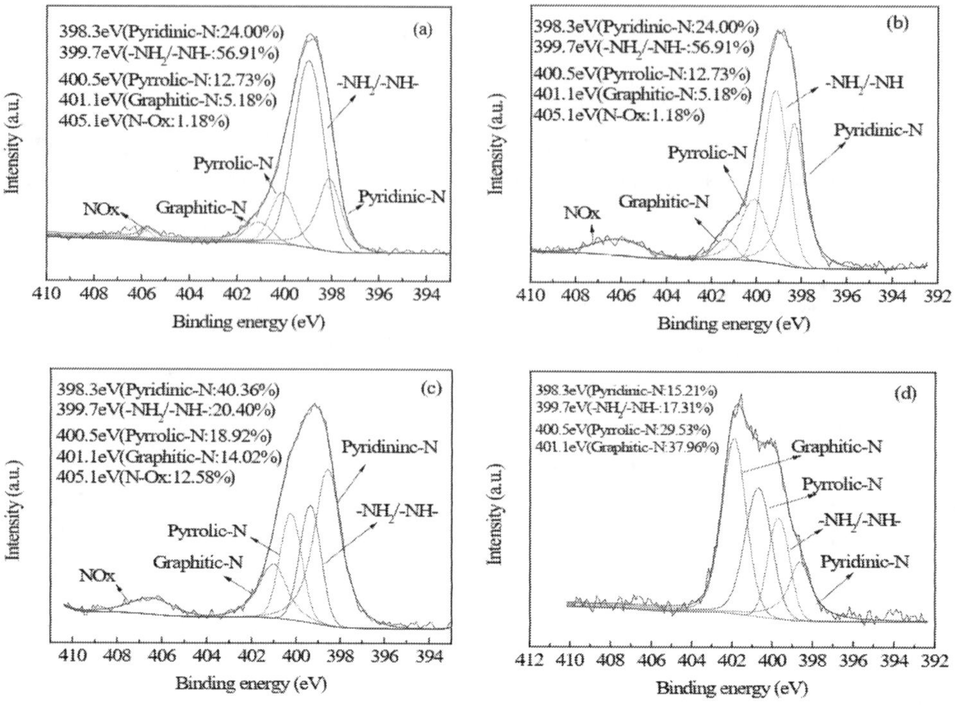

Figure 3. XPS spectrum of N1s at (a) 600°C, (b) 700°C, (c) 800°C and (d) 900°C

3.1.3 VSM

The magnetization curve of $CoFe_2O_4$/N-SAC is shown in Fig.4. That catalyst possessed a typical ferromagnetic hysteresis, with the Saturation magnetization (Ms) of 32.564 emu/g, residual magnetization (Mr) of 14.217 emu/g and coercive force of 134.11 G. Nonzero coercive force and residual magnetization proves the non superparamagnetic properties of the synthesized composite [19]. The minimized coercivity ensures that the $CoFe_2O_4$/N-SAC does not become permanently magnetized after exposure to an external magnetic field, which in turn permits the composite particles to be re-dispersed without aggregation when the magnetic field is removed [20]. The inset in Fig.4 affirms that the composite could be easily separated and reused from solution by applying an external magnetic field, which is beneficial to its practical application in wastewater treatment.

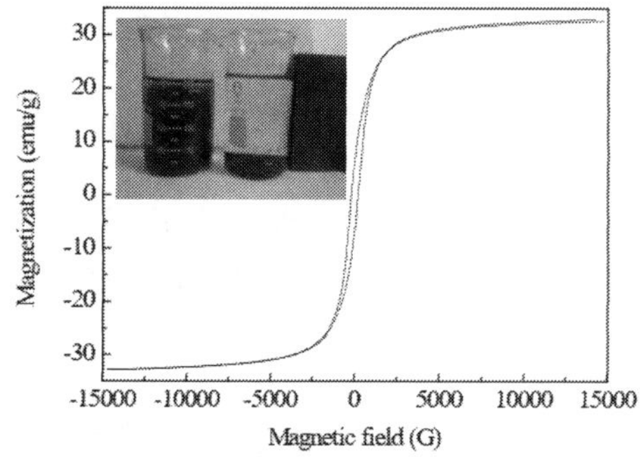

Figure 4. Magnetization curve of $CoFe_2O_4$/N-SAC

3.2 Influence of Calcination temperature on catalytic effect of catalyst

As shown in Fig.5, the effect of calcination temperature on the catalytic activity of $CoFe_2O_4$/N-SAC was studied. The results show that the effect of the calcination temperature on TOC removal is significant. The TOC removal rate of the coking wastewater increases from 50.34% to 80.53% with the increase of temperature from

500 to 800°C, but the removal rate of TOC decreases when the temperature is further increased to 900 °C.

Cobalt ferrite has an ideal inverse spinel structure that contains different types of crystal faces in which the (111) plane exposes exclusively tetrahedral Fe^{3+} sites, the (100) face expose the octahedral Co^{2+}/Fe^{3+} sites. The chemical components exposed on the surface depended on the exposed crystal face of mixed oxides which dictated by the differences in surface energy between the two component oxides. It has been demonstrated that the surface energy of iron oxide to be lower than that of cobalt oxide, suggesting the Fe^{3+} terminated (111) surfaces would be advantaged and it therefore seems reasonable to indicate that the small $CoFe_2O_4$ particles expose more Fe^{3+} (111) surfaces, while the larger particles formed with increased calcination temperature expose more (100) facets thus increasing the Co/Fe ratio [21]. The results of XPS in this study show that the ratio of Co/Fe (as shown in Tab.1) in the catalyst increases with the increase of calcination temperature, which is consistent with the results reported in the literature. Increasing surface Co^{2+} content is beneficial to catalysis. The removal rate of TOC increased with the increase of calcination temperature from 500 to 800 °C mainly ascribe to the larger $CoFe_2O_4$ nanoparticles at higher calcinated temperature have more Co rich surfaces and are more active towards organic pollutant decomposition than their smaller counterparts at lower calcinated temperature. When the calcination temperature further increased to 900 °C, the ratio of Co/Fe does not increase remarkably but the removal rate of TOC decreased due to the agglomeration of $CoFe_2O_4$.

Moreover, the influence of calcination temperature on the carrier is mainly manifested in the change of nitrogen structure. It can be seen from the XPS spectra of N1s (As shown in Fig.9) at different calcination temperatures that nitrogen with different chemical structures appears at different calcination temperatures. Previous studies have shown that nitrogen-doped carbon materials exhibit higher catalytic activity in catalytic reactions. But there is still a great deal of controversy about which type of nitrogen plays a major role in catalysis. Luo et al [22] demonstrated that pyridine nitrogen, containing a pair of lone electron, could effectively convert the valence band structure of carbon materials and enhance the π state. Long et al [23] found that graphitized nitrogen was the active site of catalysis. Sun et al [13] proved that graphitized N play a dominant role for PMS activation by theoretical calculation. Guo et al [14] proved that pyridine nitrogen was the main active site in the oxygen reduction reaction of nitrogen-doped carbon materials by template catalyst.

In this experiment, the result shows that pyridine nitrogen was the highest at the calcination temperature of 800 °C, and the catalytic activity is best at that calcination temperature, which could be concluded that the pyridine-N is the main active site in this catalytic reaction and this conclusion is consistent with Guo's report.

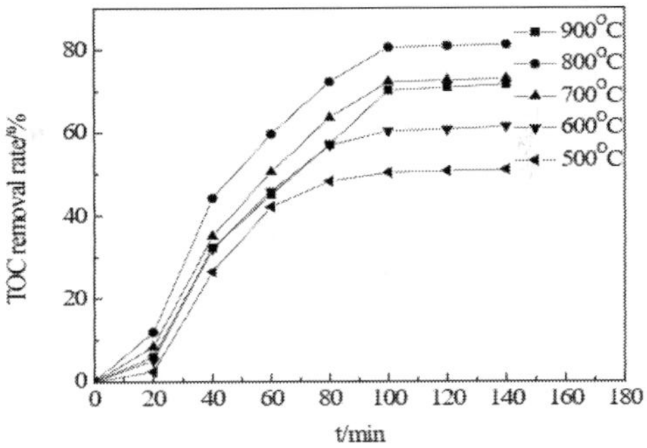

Fig.5 Effect of calcination temperature on TOC removal rate

4. Conclusions

A heterogeneous catalyst $CoFe_2O_4$/N-SAC was synthesized and characterized. That catalyst is ferromagnetic with the Saturation magnetization (Ms) of 32.564 emu/g, residual magnetization (Mr) of 14.217 emu/g and coercive force of 134.11G, which could be easily separated and reused from solution by applying an external magnetic field. The effect of the calcination temperature on the catalytic performance of $CoFe_2O_4$/N-SAC was investigated. The results indicated that the chemical components exposed on the surface and the chemical structures of N were influenced by the calcination temperature.

Acknowledgments

The authors gratefully acknowledge the financial support from the National Nature Science Foundation of Liaoning

Province of China (Grant No. 20170540452), and Major Science and Technology Platform Program of Engineering Research Center of Advanced Coal & Coking Technology and Efficient Utilization of Coal Resources, the Education Department of Liaoning Province (Grant No. USTLKFZD 2016340)

Reference

1. G.P. Anipsitakis, D.D. Dionysiou, Environ. Sci. Technol. **37**, 4790-4798 (2003).
2. H. Longxing, Y. Fan, Z. Lianpei, Chin. Catal. **36**, 1785–1797 (2015).
3. G.P. Anipsitakis, D.D. Dionysiou, Environ. Sci. Technol. **38**, 3705-3713 (2004).
4. G.P. Anipsitakis, E. Stathatos, D.D. Dionysiou, Phys. Chem. B. **109**, 13052-13055 (2005).
5. W. Bian, Z. Yang, P. Strasser, Power Sour. **250**, 196–203 (2014).
6. X. Yujiao, B. Weiyong, W. Jiao, Electrochim. Acta. **151**, 276–283 (2015).
7. H. Zhuang, H. Han, B. Hou, Bioresour. Technol. **166**, 178–186 (2014).
8. A. Abdedayem, M. Guiza, A. Ouederni, C. R. Chim. **18**, 100–109 (2015).
9. L. Hao, D. Huiping, S. Jun, Mol. Catal. **364**, 101–107 (2012).
10. P. Rai, R.K. Gautam, S. Banerjee, Environ. Chem. Eng. **3**, 2281–2291 (2015).
11. M. Wang, P. Zhang, J. Li, Chin. Catal. **35**, 335–341 (2014).
12. J. Wu, H. Gao, S. Yao, Sep. Purif. Technol. **147**, 179–185 (2015).
13. D. Xiaoguang, A. Zhimin, S. Hongqi, ACS Appl. Mate. Interfaces. **7**, 4169-4178 (2015).
14. G. Donghui, R. Shibuya, C. Akiba, Science. **351**, 361-365 (2016).
15. E. Kacan, Environ. Manag. **166**, 116–123 (2016).
16. F. Huang, C. Chen, F. Wang, Catal. Surv. Asia. **21**, 143–149 (2017).
17. Y. Xia, R. Mokaya, Adv. Mater. **16**, 1553-1558 (2004).
18. M. Zhigang, C. Xiaoye, D. Yukou, Appl. Surf. Sci. **2**, 1704–1710 (2011).
19. X. Lejin, W. Jianlong, Environ. Sci. Technol. **46**, 10145−10153 (2012).
20. J. Dong, Z. Xu, S.M. Kuznicki, Adv. Funct. Mater. **19**, 1268-1275 (2009).
21. Z. Luo, S. Lim, Z. Tian, Mater. Chem. **21**, 8038-8044 (2011).
22. J. Long, X. Xie, J. Xu, ACS. Catal. **2**, 622-631 (2012).
23. Y. Qiujing, C. Hyeok, R. Souhail, Appl. Catal. B-Environ. **88**, 462-469(2009).

MATEC Web of Conferences **238**, 04001 (2018) https://doi.org/10.1051/matecconf/201823804001
ICNFM2018

Optimization of Fermentation Process Parameters for Ginsenoside Re Bioconversion by Plackett-Burman and Box–Benhnken Design

Xiuhong Wu*, Lina Zhang, Ailin Sui, Boqu Qu, Shaoyan Wang

Department of Chemical and Engineering, University of Science and Technology of Liaoning, Anshan, Liaoning, 114051, China

Abstract. In this work, Plackett-Burman (PB) and Box-Behnken (BB) experiment design were used to investigate the medium components of fermentation forth of ginsenoside Re bioconversion to ginsenoside Rh1. Firstly, nine variables including cultivation time, temperature, pH, reactant concentration, inoculum's size, peptone, $MgSO_4 \cdot 7H_2O$, K_2HPO_4 and $NaNO_3$ were screened. The most important effective factors were estimated according to the statistical regression. Subsequently, the steepest ascent experiment was used to approach the optimum response range for each effective factor. Then Box-Benhnken Design (BBD) was finally used to analyze the response surface of the screened factors for further optimization. The optimized conditions for the influential factors were: ginsenoside Re 1.0 g/L, pH 4.5 and 7 days. The bioconversion of ginsenoside Re to Rhl could reach up to 51.76 %.

1 Introduction

The history of people utilized natural product has been gone through nearly for thousands of years, especially in Asia. For a few decades, people concerned increasingly the definite component of the natural product as it has been reported that these main components have the properties of neuroprotection, cardiovascular system protection, anti-oxidant, anti-diabetic, anti-cancer et al. [1-4]. Nowadays, minor components such as taxol, ginsenoside Rg3, Rhl are gaining increasingly attention as they have more physiological activities [5-8]. Bioconversion catalyzed by either microorganism or isolated enzyme nowadays has become an attracted approach to produce pharmaceuticals, especially such kind of minor component [9-13]. Though biotransformation has the characteristic of the mild reaction condition, relative higher selectivity, the target transformation ratio was still in a relative lower level due to the very complex condition of bio-reaction system. The common methods "one factor at a time" would not account for the interactions between the factors. In recent years, the statistical methods served as an effective way to screen for the important factors among many variables was the use of the Plackett-Burman design. The key is forming various combinations of the factors with various combinations of the factors with various amounts and could limit the factors for optimization. Subsequently, the actual values of these factor could achieve by Box-Behnke design which could understanding the interactions among various process parameters [14-18].

In our previous work, PB and BBD methods once succeeded in optimizing the biotransformation of glycyrrhizin [19]

In this study, PB and BBD experimental designs were used to optimize the parameters of fermentation forth of ginsenoside Re bioconversion to Rh1. The variables were screened, and the actual value of the important factors were estimated.

2 Material and methods

2.1 Material

2.1.1 Microorganism

Aspergillus tubingensis was isolated at The Department of Microorganism of the University of Science and Technology Liaoning and identified according to its morphology and rDNA gene sequence (KX664401.1) by Sangon Biotech (Shanghai) Co. Ltd.

2.1.2 Media and cultivation

Slope plate medium (g/L): Potato juice, 200; glucose, 20.00; peptone, 10.00; agar, 20.00; pH 5.50; Seed culture (g/L): Potato juice, 20.00; raw extraction of GL, 1.00; glucose, 10.00; peptone, 10.00; pH 5.50; in a shaker for cultivation; Basal fermentation cultivation medium （g/L）: 98% GL, 1.20; NaNO3, 1.50; K2HPO4, 1.00;

[a] Xiuhong Wu: wxh93@163.com

All the media and cultivation were sterilized under the temperature of 121 ℃ and keep for 20 minutes

2.2 Methods

2.2.1 Preparation of Aspergillus tubingensis seed culture

Under the case of sterilization, an *A. tubingensis* spore suspension was prepared by washing spores from slope plates by sterile water. The suspension was inoculated for 72 h in a rotary shaker under the temperature of 28℃ and the rate of 160 r/min. Then seed cultivation medium was filtered with gauze and was washed for several times with sterile water until the prepared mycelia suspensions contained no ginsenoside Re. The seed culture was then used for further experiment.

2.2.2 Preparation of fermentation culture

10 ml seed medium was inoculated in the 100 ml fermentation culture medium in a 250 ml Erlenmeyer flask and then cultured in a rotary shaker at 28℃and 160 r/min for 120 h. Finally, the fermentation culture was centrifuged at 4000 r/min for 10 min. Then the transformation yield was calculated with the following formula:

$$A\% = \frac{C_1 V_1}{C_0 V_0} \times 100\% \quad (1)$$

Where A is the transformation ratio, C_0 is the initial concentration of ginsenoside Re at the initial time of the fermentation, mg/L; V_0 is the initial volume of transformation forth, mL; C_1 is the concentration of ginsenoside transformed, V_1 is the final volume of fermentation forth. Ginsenoside Re, Rh1 were measured using Agilent 1100 HPLC (Agilent technology, USA) with Agilent Extend C_{18} (4.6 mm×250 mm, 5 μm). Chromatographic conditions for analysis as: mobile phase, acetonitrile: H_2O=70:30 (v/v); column temperature, 25.0 ℃; mobile phase rate, 1.00 ml/min; wavelength, 203 nm; UV detector.

2.3 Experiment design

2.3.1 Steepest ascent design

Plackett-Burman experiment design was first used to estimate the influence factors. And then steepest ascent design was used to approach the optimal range of each factor. Steepest ascent design takes the variation direction of the gradient as the ascent direction to make the process of optimization rapidly and economically

2.3.2 Box-Benhnken design (BBD)

BBD is a statistical method to solve multivariable problem. Compared with other general methods such as single-factor experimental and orthogonal design, BBD was characterized by its less experiment time, and a relative higher accurate regression equation [20-23].

3 Result and discussion

3.1 Estimation of the main influence factors

The software Design-Expert 8.0.6 experimental design was used in this work for designing PB experiment. According to the numbers of factors of the biotransformation system, experimental runs was set to N=12. nine influential factors include: cultivation time (A), C_{Re} (B), $MgSO_4{\bullet}7H_2O$ (D), $NaNO_3$ (E), K_2HPO_4 (F), peptone (G), inoculation amount (H), cultivation temperature(I), and pH (J) were investigated. Besides, (C) and (K) were two virtual variables to meet the running of the software. Each factor was considered by low level and high level, respectively. Bioconversion ratio was served as the response value, and the experiment was carried out duplicate. Detailed experiment design and the results of running were listed in Table 1.

Table 1. Plackett-Burman design and the corresponding response measurements variable (coded value) (N=12)

	A	B	C	D	E	F	G	H	I	J	K	A%
1	1	1	-1	-1	-1	1	-1	1	1	-1	1	14.4
2	1	-1	1	1	-1	1	1	1	-1	-1	-1	27.5
3	1	-1	-1	-1	1	-1	1	1	-1	1	1	19.5
4	-1	1	-1	1	1	-1	1	1	1	-1	-1	16.0
5	-1	1	1	-1	1	1	1	-1	-1	-1	1	16.8
6	1	1	1	-1	1	-1	1	-1	1	1	-1	26.3
7	-1	-1	-1	1	-1	1	1	-1	1	1	1	37.2
8	1	1	-1	1	-1	1	-1	-1	-1	1	-1	6.6
9	1	-1	1	1	1	-1	-1	-1	1	-1	1	22.4
10	-1	1	1	1	1	-1	-1	1	-1	1	1	8.0
11	-1	-1	1	-1	-1	1	-1	1	1	1	-1	32.9
12	-1	-1	-1	-1	1	-1	-1	-1	-1	-1	-1	13.7

The experimental results of PB was input to the software SPASS for statistical analysis. The effects of various factors on response variable are cultivate time (A), Re concentration (B) and pH(J). In Table 2, item B indicated the positive or negative of effect of the factor. Therefore, from Table 2, cultivation time (A) and pH (J) need increase, while Re concentration (B) may need decreased in the following experiment.

3.2 Estimation optimal factor levels of by steepest ascent method Estimation optimal factor levels of by steepest ascent method

Based on the results showed in Table 2. A steepest ascent experiment was designed, and the range of the influential factors was shown in Table 3. The other factors were maintained at their optimal levels. The results obtained were also shown in Table 3.

Table 2. Regression coefficients and their significance level

Factor	Effect (B value)	Standard error	*t* ratio	*p*-value
Total		28.36	14.09	0.005*
A	25.29	12.64	6.28	0.050*
B	-42.02	-21.01	-10.44	0.016*
D	6.54	3.27	1.63	0.131*
D	-25.36	-12.68	-6.30	0.268*
E	-2.42	-1.21	-0.60	0.402*
F	-10.77	-5.39	-2.68	0.076*
H	29.09	14.55	7.23	0.056*
I	13.41	6.71	3.33	0.508*
J	36.63	18.32	9.10	0.002*

*Significant at $p<0.05$

Table 3. Steepest ascent design and the corresponding values

Trial number	C_{Re} (mg/L)	pH	Cultivate time(days)	*A* %
1	200	3.5	4	19.65
2	500	4.0	5	17.92
3	1000	5.0	6	22.32
4	1500	7.0	7	11.89

From Table 3, we can see that the highest biotransformation ratio was obtained in trial 3. Hence, the factor levels of this trial were used as the central point of response surface optimization in the subsequent BBD experiment.

3.3 Estimation of optimal levels of the main factors by Box-Benhnken Design

The results of the steepest ascent design were used for further optimization in BBD to estimate factors and levels. Three influence factors i.e. cultivation time, pH and concentration of Re with three level's response surface analysis experiment were carried out. Two types of experimental data points were selected: the factorial point and the zero point which was the central point of the investigation range. The results of seventeen trials are shown in Table 4.

3.4 Analysis experiment data of BBD

The biotransformation ratio was served as the response variable. Multi-variable regression analysis was carried out using the software Design-Expert 8.0.6. The analysis results are shown in Table 5. After regression and fitting, a quadratic regression polynomial was obtained. The response value verse variables A, B and C in term of code is as followed:

$$A\% = 48.24 + 4.51A - 7.77B - 4.58C - 8.85AB + 5.02AC - 2.40BC - 8.82A^2 - 13.73B^2 - 3.12C^2 \quad (2)$$

Where A was the dependent variable, the arithmetic mean response of the 17 runs was 48,24, and the estimated coefficients for factors A, B, and C were 4.51, -7.77 and -8.85, respectively. The main effects (A and B) represented the average result of changing one factor at a time from low to high value. The interactive terms (AB. BC, and AC) showed how the response changes when three factors were simultaneously changed. The polynomial terms (A^2, B^2, and C^2) were included to investigate nonlinearity. The regression model needed to be evaluated for model significance and the lack-of-fit. Analysis of variance (ANOVA) was used for these evaluations.

Table 4. Box-Benhnken experimental design conditions and the corresponding values

Trial number	A: Cultivation time	B: pH	C: Ginsenoside Re	*A* %
1	0	0	0	39.6
2	1	0	-1	38.6
3	-1	0	1	24.0
4	-1	1	0	21.6
5	0	0	0	53.6
6	-1	-1	0	17.7
7	0	-1	1	37.6
8	0	0	0	49.5
9	1	0	1	39.9
10	0	-1	-1	42.4
11	1	-1	0	47.5
12	0	0	0	48.7
13	1	1	0	16.0
14	0	0	0	49.8
15	0	1	1	15.5
16	-1	0	-1	42.7
17	0	1	-1	30.0

Table 5. ANOVA for BB Design quadratic model

Source	Sum of Square	DF	Mean Square	F Value	P Value	Significance
Model	2509.81	9		14.60	0.0009	significant
A	162.72	1		8.52	0.0224	
B	482.52	1		25.26	0.0015	
C	167.72	1		8.78	0.0210	
AB	313.29	1		16.40	0.0049	
AC	100.80	1		5.28	0.0552	
BC	22.99	1		1.20	0.3089	
A^2	327.90	1		17.17	0.0043	
B^2	794.00	1		41.57	0.0004	
C^2	40.91	1		2.14	0.1867	
Residual	133.69	7	19.10			
Lack of Fit	25.97	3	8.66	0.32	0.8109	not significant
Pure Error	107.72	4	26.93			
Cor Total	2463.51	16				
R-Squared 0.9494			AdjR-Squared 0.8844			

As shown in Table 5, both linear and quadratic terms of the equation were significant, indicating a nonlinear relationship between the influence variables and biotransformation ratio. The coefficient of the regression equation is R^2=0.9494 which means that the model may use for fitting the experimental data. The adjusted R^2 indicates that the three influence factors identified in the PB experiment explained 88.44 % of the variance in the response variable. In addition, the lack of fit P>F was 0.8109; therefore, the model has a practical significance [20-21].

According to the regression equation and the analysis summarized in Table 5, fitted surfaces of each pairs of factors in the reaction analysis can be derived; the three-dimensional plots of interactions are shown in Figure 1(a)-(c).

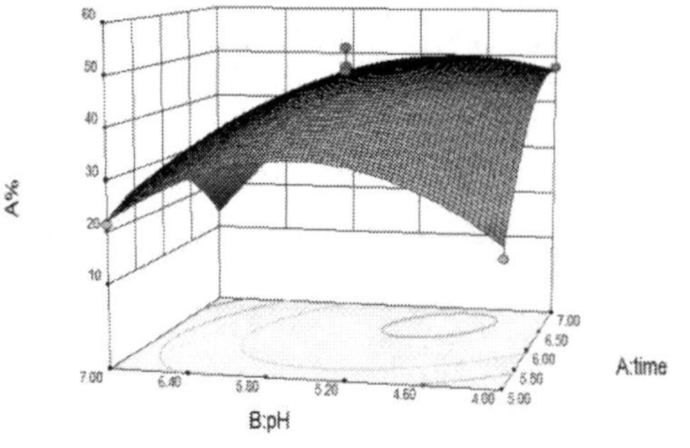

(a)

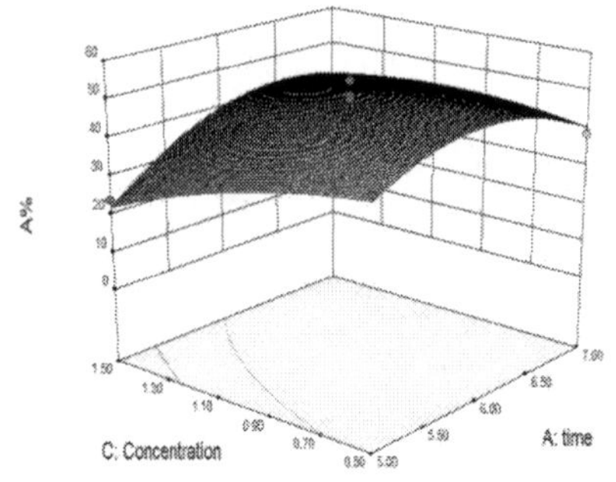

(b)

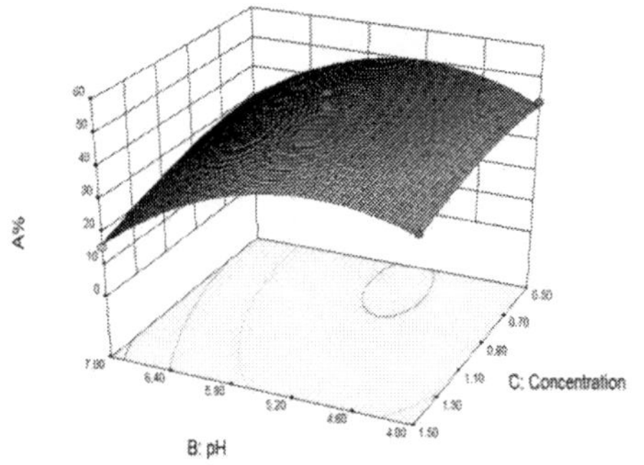

(c)

Figure 1. Fitted surface for the transformation ratio as a function of cultivation time (A) and pH (B) in (a), cultivate time (A) and ginsenoside Re (C) in (b) and pH (B) and ginsenoside (C) in (c)

Figure 1. (a)-(c) showed that the influence between two factors presented a relationship of parabolic curve. Figure 1(a) showed that the percent biotransformation more significantly than time. The optimum pH was located between 2.0 and 4.0. As can be seen from Figure (b) and (c), pH and time affected the percent biotransformation more significantly than Re concentration. There is a maximum point in these three figures. The optimization results of BBD derived from the regression equation is Re 1.0 g/L, pH 4.5 and cultivate time 7 days.

3.5 Verification test

To verify the accusation of the model established, the biotransformation of Re was carried out duplicate under the optimized condition i.e. Re 1.0 g/L, cultivation time 7 days and pH 4.50. The average percent transformation got reach to 51.76 %, a satisfied result for acquiring ginseenoside Rh1.

4 Conclusion

In this study, based on basal fermentation medium, PB design method was applied to estimate the most important influence factors among nine which may affect biotransformation ratio of Re to Rh1. The results showed that Re, pH and cultivate time were the most obviously influence factors. To search the central point of the response surface experiment in a reasonable and simply way, steepest ascent experiment was used. The optimum fermentation conditions achieved by BBD are: Re 1.0 g/L, $NaNO_3$ 3.0 g/L, $MgSO_4 \cdot 7H_2O$ 1.5 g/L, K_2HPO_4 3.0 g/L, peptone 2.0 g/L, pH 4.50, inoculums size 30.0 mL, temperature 28.0 ℃ and cultivate time 7 days. 51.76 % transformation ratio obtained based on above conditions.

References

1. L. W. Qi, C. Z. Wang, C. S. Yuan. Phytochemistry, **72** (2011), 689.
2. H. W. Park, G. In, S. T. Han, M.W. Lee, S.Y. Kim, K.T. Kim, B.G. Cho, G.H. Han, and I.M. Chang. J. Ginseng. Res., **37**(4) (2013), 457.
3. A. Dey, R. Bhattacharya, A. Mukherjee, D. K. Pandey. Biotechnol. Adv. ,**35** (2017), 178.
4. Z. P. Li, G. E. Ji. Food. Sci. Biotechnol., **26**(4) (2017), 987.
5. K. H. Chang, M. N. Jo, K.T. Kim, H.D. Paik. J. Ginseng. Res., **38** (2014), 47.
6. G.T. Chen, M. Yang., Z. Q. Lu, J.Q. Zhang, H. L. Huang, Y. Liang, S.H. Guan, Y. Song, L.J. Wu, and D.A. Guo. J. Nat. Prod., **70**(2007), 1203.
7. S.R. Ko, K.J. Choi, K. Uchida, Y. Suzuki.. Plata. Med., **69**(2003), 285.
8. L.H. Quan, C. Wang, Y. Jin, T. R. Wang, Y. J. Kim, D. C. Yang. Antonie van Leeuwenhoek, **104** (2013),129.
9. J. Upadhyaya, M.J. Kim, Y. H. Kim, S.R. Ko, H.W. Park, M.K. Kim. J. Ginseng. Res., **40** (2016), 105.
10. L.P. Wu, Y. Jin, C.R. Yin, L.L. Bai. J. Ind. Microbiol. Biotechnol., **39** (2012), 521.
11. Gao, W.J. Xu, Q. Fang, F. Liang, R.T. Jin, D. Wu, G.H. Tai and Y.F. Zhou.. Ann. Microbiol., **63** (2013),139.
12. G.W. Lee, K. R. Kim, D. K. Oh. Biotechnol. Lett., **34** (2012),1679.
13. S.A. Palaniyandi, B. M. Son, K. Damodharan, J.W Suh, and S. H. Yang. Biotechnol. Biopro. E., **21**(2016), 587.
14. P. F. Yue, Q. Zheng, B. Wu, P.Y. Hu, Z. F. Wu and M. Yang. J. Disper. Sci. Technol., **33**(2) (2012), 213.
15. B. Gullóan, G. Garrote, J. L. Alonso, and J.C. Parajó. J. Agric. Food. Chem., **55**(2007), 5580.

16. H. S. Oberoi, P. V. Vadlani, R. L. Madl, L. Saida, and J.P. Abeykoon. J. Agric. Food. Chem., **58**(2010), 3422
17. S. Chatterjee, A. Kumar, S. Basu, S. Dutta. Chem. Eng. J., **181-182** (2012), 289.
18. S.S. Lin, W.F. Dou, H.Y. Xu, H. Z. Li, Z.H. Xu and Y.H. Ma. Appl Microbiol Biotechnol., **75** (2007),1015
19. Y. L. Quan, L. Wang, Y. S. Liu, J.X. Cong, S.Q Xie, and X.H. Wu. Korean Chem. Eng. Res., **53**(3) (2015), 321.
20. Namita Gupta, Gauri Mehra, and Rani Gupta. Can. J. Microbiol., **50**(5) (2004), 361.
21. S. Ali and I. Haq. BMC Biotechnol., **10** (2010), 86
22. J.N. Hu, J.H. L. X.M. Zhu, J.A. Shin, P. Adhikari, J.K. Kim, and K.T. Lee. J. Agric. Food. Chem., **56**(2008), 10988.
23. T. Palvannan and P. Sathishkumar, J. Basic. Microb., **50**(2010), 325.

A new rhodamine fluorescent OFF-ON-OFF probe for selective sensing of Fe^{3+} and AcO^- in aqueous media and living cells

Qinglei Liu[a], Huimin Liu [a], Yanan Lei [a], Yan Gao[b*] and Bing Zhao [c*]

[a] *School of Perfume and Aroma Technology, Shanghai Institute of Technology, Shanghai 200235, P.R. China*
[b] *Department of Chemical Engineering, Liaoning University of Science and Technology, Anshan, 114051, P.R. China*
c Chemistry and Chemical Engineering Institute, Qiqihar University, Qiqihar 161006, China

Abstract. A new fluorescent chemosensor for Fe3+ was developed based on a rhodamine platform. L displayed highly selective and sensitive "OFF-ON" fluorescence response and naked-eye color change to Fe3+ in aqueous solution. The resulting L-Fe3+ complex was found to act as a selective" ON-OFF" fluorescence probe for AcO- against common anions and cations with a Fe3+ displacement approach. The detection limits of L to Fe3+ and L-Fe3+ complex to AcO- were estimated to be 6.04×10-8 mol/L and 7.51×10-8 mol/L, respectively. The good biocompatibility of L enabled the investigation of fluorescent response for Fe3+ and AcO- in living Ana-1 cells by confocal microscope.

1.Introduction

As an essential element for life, iron plays an important role in a wide range of biochemical processes [1], several biological functions depend directly or indirectly on the proper concentration and oxidation states of iron to maintain the homoeostatic mechanism of biosystem[2,3]. Its deficiency and excess can induce a variety of diseases [4, 5]. Therefore, it is highly desirable to develop fluorescent probes for the real-time monitoring of iron ions in environmental and living systems [6, 7]. Due to great sensitivity, fluorescent chemosensors combining ionophores and fluorophores are commonly considered superior [8-10]. In recent years, rhodamine has been attracted a great deal of attention due to its excellent properties, including good photostability, long emission wavelength, high fluorescence quantum and so on[11-13]. However, for the majority of rhodamine fluorescent probes, their anti-interference ability is poor, which are usually chelate with two or more metal ions simultaneously[14]. Due to the interference of Cr^{3+}, Al^{3+}, Cu^{2+} in the identification process of Fe^{3+}[15-18], rhodamine fluorescent probe exhibited low selectivity for Fe^{3+}[19,20].

The recognition and sensing of anions by proper design of anion receptors is currently of major interest because anions play a fundamental role in chemical and biological processes. Various kind of anions such as F^-, Cl^-, I^-, PO_4^{3-}, CH_3COO^-, etc. play a major role both in environmental and biological systems [21-23]. Among the above anions, AcO^- received the most attention from chemists because of its unique properties, the detection of AcO^- with the help of easily synthesized receptor and minimal instrumental assistance is desirable towards practical applications [24, 25]. Recently, rhodamine-cation complex solution has been documented for the recognition of various anions [26-28]. This intrigued us to design a novel Fe^{3+}-fluorophore complex to develop a AcO^--specific fluorescence chemosensor.

In this work, the allyl group was introduced into p-Hydroxybenzaldehyde to improve the charge distribution of probe, herein we designed and synthesized a new turn-on fluorescence sensor based on rhodamine for the quantification of Fe^{3+}. The in situ formed **L**-Fe^{3+} complex solutions exhibited a sensitive response to AcO^- via a turn-off response by the displacement strategy. The reversible "off-on-off" fluorescence sensor in solution showed the good application in live Ana-1 cell imaging.

2. Results and discussion

The synthetic route for compound **L** was shown in Scheme **1**. The structure of **L** was confirmed by NMR, mass spectra (Fig.S1-S5). The detailed experimental procedures and the characterization of the compound **L** were described in the experimental section.

The recognition profiles of compound **L** were investigated in CH_3CN/H_2O solutions by UV-vis spectroscope toward various heavy metal and transition

Corresponding author: gys20080901@163.com

metal cations, including Ca^{2+}, Cd^{2+}, Cu^{2+}, Hg^{2+}, Mn^{2+}, Ni^{2+}, Pb^{2+}, Zn^{2+}, Fe^{3+}, Na^{+}, Al^{3+}, Mg^{2+}, Ag^{+}, Cr^{3+}. As presented in Fig. 1, free probe **L** showed an absorption band at 558nm, when 10 equiv. of Fe^{3+} was added to the solutions of probe **L**, the absorption intensity was significantly enhanced, the change in spectroscope was come along obvious color change from colorless to pink, Differently, nearly no significant adsorption changes could be observed in the presence of other interested metal ions. These results clearly suggested that **L** can serve as a "naked-eye" Fe^{3+} indicator in aqueous media.

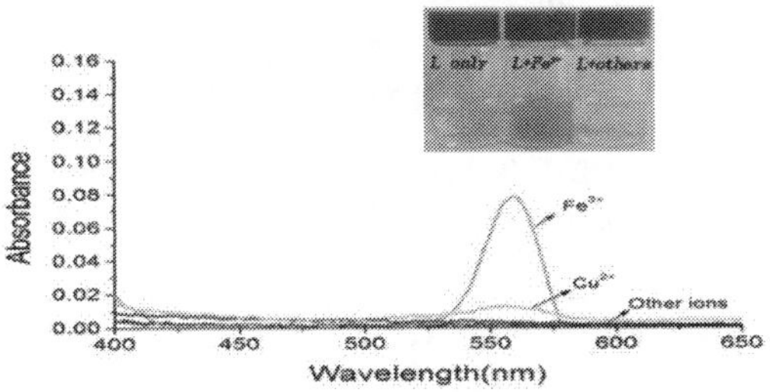

Fig. 1 Absorbance spectra of **L** (10μM) in the absence and presence of 10 equiv. different metal ions in CH_3CN/H_2O(v/v=3:7, 10mM Tris-HCl buffer, pH =7.2) solution.

In addition, we have tested the fluorescence properties of **L** toward a series of metal ions such as Cu^{2+}, Pb^{2+}, Fe^{3+}, Fe^{2+}, Hg^{2+}, Al^{3+}, Cr^{3+}, Ni^{2+}, Mn^{2+}, Co^{2+}, Zn^{2+}, Ag^{+}, Ba^{2+}, Cd^{2+} and Na^{+} (100μM) in CH_3CN/H_2O buffer (3:7, v/v, 20mM, pH=7.2). As shown in Fig. 2, fluorescence probe **L** alone and other metal ions all displayed weak fluorescence emission bands except for Fe^{3+}, the addition of Fe^{3+} induced a remarkable fluorescence enhancement (59-fold), whereas the addition of other metal ions to the solution **L** exhibited no obvious change in the fluorescence spectra. Meanwhile, the emission color of solutions can be clearly observed to change from colorless to red-orange under illumination with 365 nm light in presence of Fe^{3+}, indicating that there were some interaction between **L** and Fe^{3+}. The specific response of **L** towards Fe^{3+} should be attributed to be based on the opening function of the spirolactam ring [29-31]. Compound **L** can be considered as a new off-on fluorescent chemosensor for Fe^{3+}.

The interference of the potentially competing metal ions toward Fe^{3+} detection was also investigated. As shown in Fig. 3, chemosensor **L** was treated with 10 equiv. of Fe^{3+} and 10 equiv. of other coexistent metal ions such as Cu^{2+}, Pb^{2+}, Fe^{2+}, Hg^{2+}, Al^{3+}, Cr^{3+}, Ni^{2+}, Mn^{2+}, Co^{2+}, Zn^{2+}, Ag^{+}, Ba^{2+} , Cd^{2+} and Na^{+}. As a result, no interference was observed for the detection of Fe^{3+} by **L**, these results manifest the recognition of Fe^{3+} was not interfered by these competing metal ions.

Fluorescence titration of **L** (10μM) with Fe^{3+} was performed in CH_3CN/H_2O (v/v=3:7) solution. As shown in Fig.5, the fluorescence intensity increased gradually in the presence of increasing amounts of Fe^{3+} (0-10equiv.). Job plot was utilized to determine the binding stoichiometry of the **L**-Fe^{3+} complex, it revealed that **L** and Fe^{3+} formed a complex in 1:1 stoichiometry (Fig.4) [32]. Hildebrand-Bebesiplots gave binding constants of 2.9×10^4 M^{-1} in HEPES buffer [33] (Fig. 5). The detection limit for Fe^{3+} was estimated to be 6.04×10^{-8}M based on a 3σ/slope analysis (Fig. S6) [34].Which was sufficiently low to detect submicromolar concentration of the Fe^{3+}, implying that **L** had great potential for use in the development of Fe^{3+} sensor materials.

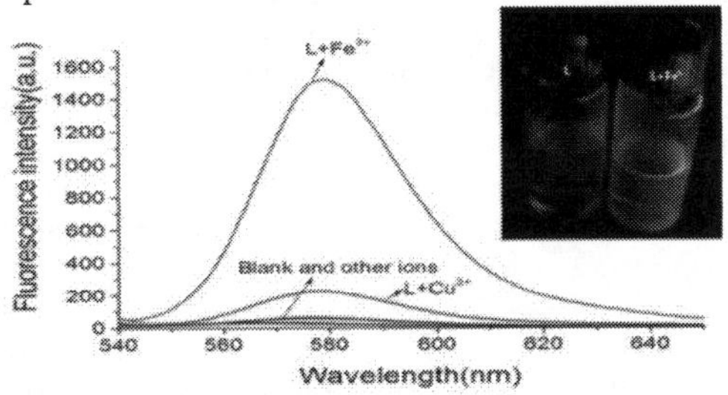

Fig. 2 Fluorescence spectra of **L** (10μM) upon addition of 10 equiv. of various metal ions in CH_3CN/H_2O(v/v=3:7, 10mM Tris-HCl buffer, pH=7.2) solution, λ_{ex}= 520nm.

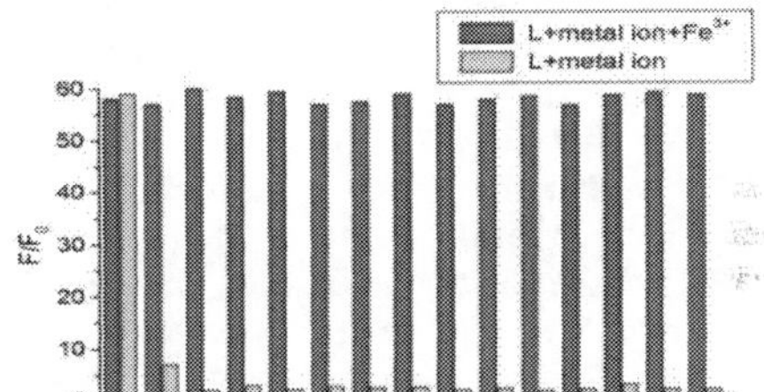

Fig. 3 Fluorescence responses of **L** (10μM) to various metal ions (10 equiv.) in CH_3CN/H_2O (v/v=3:7, 10mM Tris-HCl buffer, pH=7.2) solution. Bars represent the final integrated fluorescence response (F) over the initial integrated emission (F_0). Green bars represent the addition of an excess of the appropriate metal ion (10 equiv.) to a 10μM solution of **L**. Red bars represent the addition of Fe^{3+}(10 equiv.) to the solution (λex=520nm). (1)Fe^{3+}; (2)Cu^{2+}; (3)Pb^{2+}; (4)Fe^{2+} ; (5) Hg^{2+}; (6)Al^{3+}; (7) Cr^{3+}; (9)Ni^{2+}; (9) Mn^{2+}; (10) Co^{2+}; (11) Zn^{2+}; (12) Ag^{+}; (13)Ba^{2+} ; (14)Cd^{2+}; (15)Na^{+}.

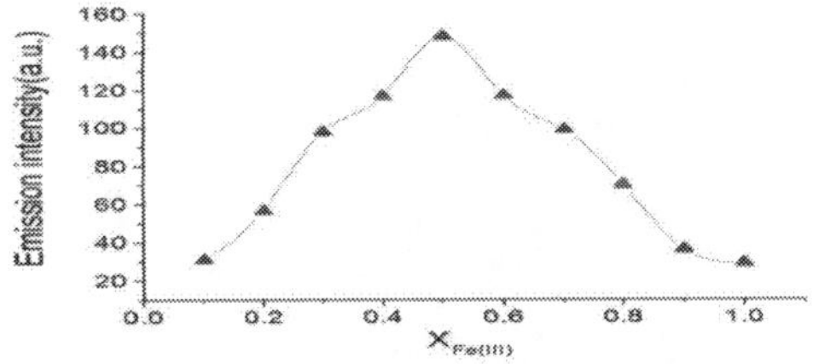

Fig. 4 Job's plot evaluated from the fluorescence spectra of **L** and Fe^{3+} in buffered aqueous (λ_{ex}=520nm)

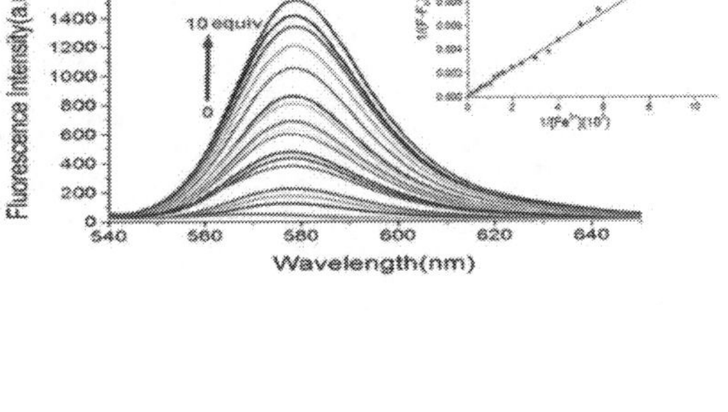

Fig. 5 Fluorescence emission of a solution of **L** (10μM) in CH_3CN/H_2O (v/v=3:7) upon addition of increasing concentrations of Fe^{3+} (0-10equiv) (λ_{ex}=520nm)

Furthermore, the probe **L** was found to have a short respone time towards Fe^{3+}. As shown in Fig. S7, the fluorescence intensity of **L** rapidly enhanced after Fe^{3+} (100μM) was added, reached its maximum value in 1 min and stabilized thereafter, indicating that the **L** could be used for the real-time detection of Fe^{3+}.

The recognition of **L**-Fe^{3+} complex towards anions and nucleotides were then evaluated. As shown in Fig. 6, the fluorescence intensity of the **L**-Fe^{3+} at 580nm was gradually decreased with the addition of increasing concentration of AcO^- (0-2equiv.), and the color of receptor **L**-Fe^{3+} solution showed a dramatic color change from pink to colorless, indicating that the decomplexation of Fe^{3+} by AcO^-. However, other anions such as S^{2-}, F^-, Cl^-, Br^-, NO_3^-, HSO_4^-, $H_2PO_4^-$, PO_4^{3-}, SCN^-, HPO_4^{2-} and Cys were found to hardly induce any variation either in visible color change or in the absorption spectra(Fig. 7). In addition, fluorometric titration assays detected the reversible detection of AcO^-. As shown in Fig.8, "OFF-ON-OFF" fluorescence changes were observed by the alternative additions of Fe^{3+} and AcO^- to the solution of **L** and **L**-Fe^{3+} respectively, indicating that **L** can be developed as a reversible fluorescence chemosensor for Fe^{3+} and AcO^-. The detection limit of **L**-Fe^{3+} for AcO^- was determined to be 7.51×10^{-8} mol/L (Fig. S8).

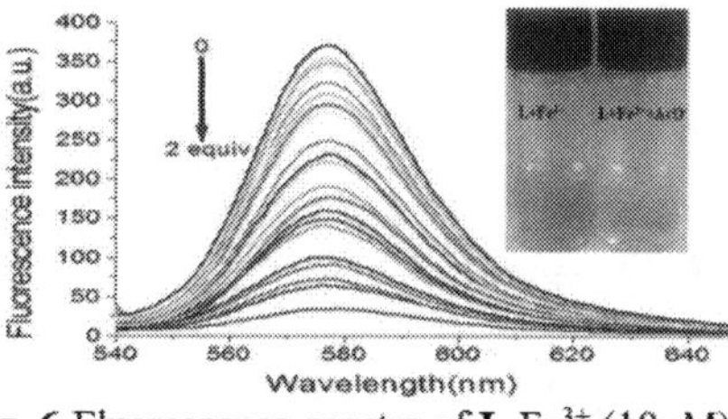

Fig. 6 Fluorescence spectra of **L**-Fe^{3+} (10μM) in Tris-HCl aqueous buffer (CH_3CN/H_2O =3:7, 10mM, pH=7.2) in the presence of different amounts of AcO^- (0-20μM). Inset: normalized fluorescence intensities of **L**-Fe^{3+} (10μM) at 580 nm as a function of AcO^- (0-20μM).

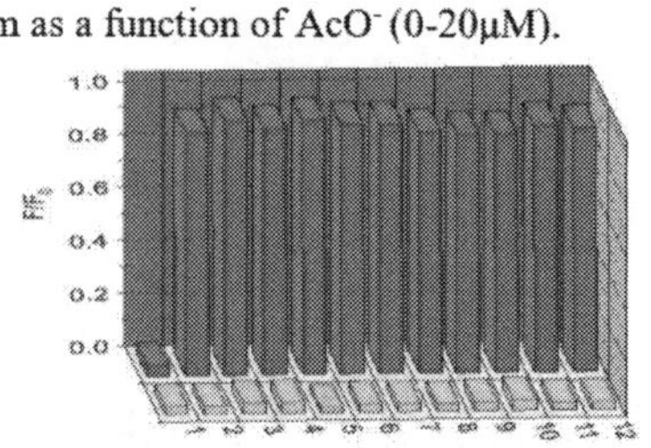

Fig. 7 Normalized fluorescence responses of **L**-Fe^{3+} (10μM) to various anions in Tris-HCl aqueous buffer (CH_3CN/H_2O =3:7, 10mM, pH=7.2). The red bars represent the emission changes of L-Fe^{3+} in the presence of all kinds of analytes (20μM): 1. AcO^- , 2. S^{2-}, 3. F^-, 4. Cl^-, 5. Br^-, 6. NO_3^-, 7. HSO_4^-, 8. $H_2PO_4^-$, 9. PO_4^{3-}, 10. SCN^-, 11. HPO_4^{2-}, 12. Cys. The green bars represent the changes of the emission that occurs upon the subsequent addition of 20μM of AcO^- to the above solution.(λ_{ex}=520nm)

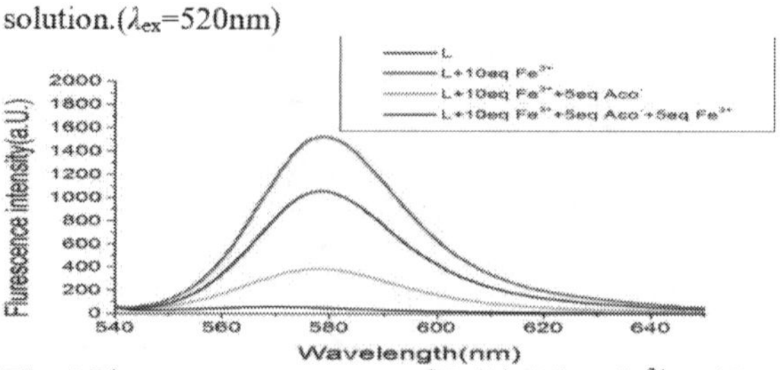

Fig. 8 Fluorescence responses of **L** (10μM) to Fe^{3+} and **L**-Fe^{3+} complex to AcO^- in CH_3CN/H_2O(v/v=3:7, 10mM Tris-HCl buffer, pH=7.2) solution (λ_{ex}=520nm)

The effect of pH on the fluorescence of probe **L** to Fe^{3+} and **L**-Fe^{3+} complex to AcO^- was evaluated. As shown in Fig. 9, **L** and **L**-Fe^{3+} were all weakly fluorescent over a wide pH range (pH=4.0-10.0). Meanwhile, probe **L** displayed an efficient fluorescence response to Fe^{3+} in the pH range of 5.0-8.0. It can be seen that the fluorescence intensity of **L**-Fe^{3+} is evidently enhanced (pH≥5) and reaches its maximum at pH 7. The results indicated that the **L** can be used as a chemosensor for Fe^{3+} detection and the **L**-Fe^{3+} ensemble could work for the detection of AcO^- in physiological conditions.

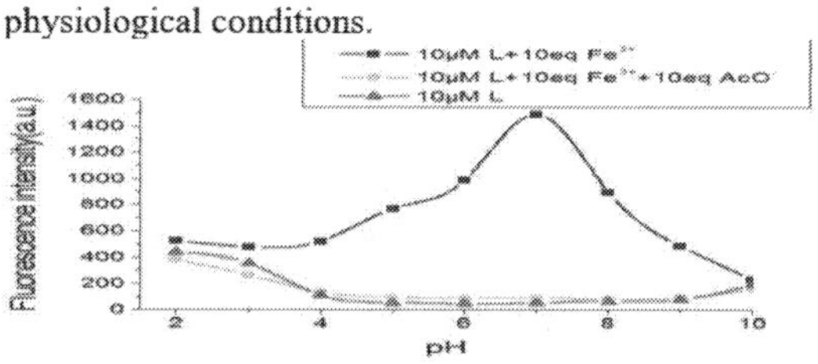

Fig. 9 Effect of pH to the reaction of probe **L** (10μM) to Fe^{3+} (10equiv.) and **L**-Fe^{3+} complex to AcO^- (10equiv.) in CH_3CN/H_2O(v/v=3:7) solution.

The proposed displacement strategy for the AcO^- detection was further supported by the results of mass spectra studies, FT-MS spectra of **L** showed a molecular-ion peak $[L+H^+]^+$ at m/z 601.31426(Fig. S3). When the Fe^{3+} ion was added into the solution of **L**, the peak at m/z 718.4273 is assignable to $[L\text{-}2H^++Fe^{3+}+NO_3^-+H^+]^+$ species (Fig. S4). This result confirmed that 1:1 stoichiometry complex between **L** and Fe^{3+}. In the presence of AcO^-, FT-MS spectra displayed a molecular-ion peak at m/z 601.31695 again (Fig. S5), indicating that the decomplexation of Fe^{3+} ion in the presence of AcO^- led to the liberation of fluorophore, **L**.

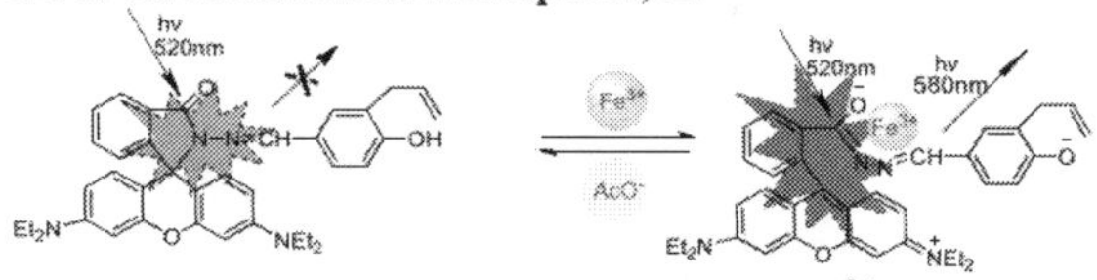

Scheme 2. The proposed mechanism for Fe^{3+} and AcO^- detection by **L**

The cytotoxicity of **L** toward the Ana-1 cells was investigated by the reduction activity of the methyl thiazolyl tetrazolium (MTT [3-(4,5-dimethylthiazo-l-2-yl)-2,5-diphenyltetrazolium bromide]) assay. As shown in Fig. S9, the cellular viability was estimated to be approximately 89.5% after incubation with **L** (20μM) for 24 h, which demonstrates that **L** has low cytotoxicity and biocompatibility to living cells. Due to the good spectroscopic properties and biocompatibility of the probe, **L** was applied for fluorescence imaging in living Ana-1 cells. As shown in Fig. 10, Ana-1 cells incubated with 5μM **L** for 30 min at 37°C gave no intracellular background fluorescence[35](Fig. 10a). Moreover, no intracellular fluorescence was detected for Ana-1 cells with 50μM $FeCl_3$ in the growth medium for 3 h at 37°C. However, as shown in Fig. 10b, bright intracellular fluorescence was observed for Ana-1 cells with 50μM $FeCl_3$ in the growth for 3 h at 37°C and then staining with L under the same loading conditions. After treating the cells with 25μM NaOAc for another 3 h at 37°C, the intense red fluorescence of the cells disappeared (Fig. 10c). It can be seen from the bright field transmission measurement after Fe^{3+} and **L** incubation (Fig. 10d) that the Ana-1 cells were viable throughout the imaging experiments. These facts implied that **L** was membrane-permeable and could be used for detecting Fe^{3+} in biological samples.

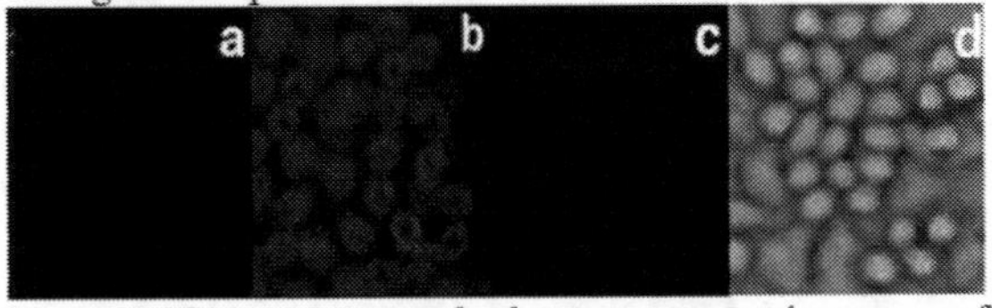

Fig. 10 Fluorescence and phase contrast images of living Ana-1 cells: (a) fluorescence image of Ana-1 cell incubated with L (5μM) for 30 min at 37℃ in the red channel; (b) fluorescence image of Ana-1 cell incubated with L (5μM) for 30 min, and then incubation with Fe^{3+} (50μM) for 3 h at 37℃ in the red channel; (c) On the basis of panel b, 25μM of NaOAc for another 3 h at 37℃. (d) bright field image of living Ana-1 cells shown in panel b to confirm viability.

3. Experimental

3.1. Materials and instrumentation

All reactants and chemicals were obtained from Aladdin and used without further purification. Thin layer chromatography (TLC) was carried out on alumina backed plates coated with Merck gel GF254. Mass spectral data were obtained on a Solaril X70 FT-MS. ^1H NMR and ^{13}C NMR were recorded at room temperature on a Bruker Avance-500 NMR spectrometer. Chloroform ($CDCl_3$) was used as solvent and tetramethylsilane (TMS) as internal standard. Fluorescence spectra measurements were recorded with a Hitachi F-4500 spectrofluorometer. A Techcomp UV-8500 spectrophotometer was used for Absorption measurements.

3.2. Synthesis and Characterization

The substituted benzaldehyde (**3**) was prepared according to the literature method [36]. The unique ligand **L** was prepared *via* the reaction of rhodamine hydrazine and **3** in methanol. (Scheme 1)

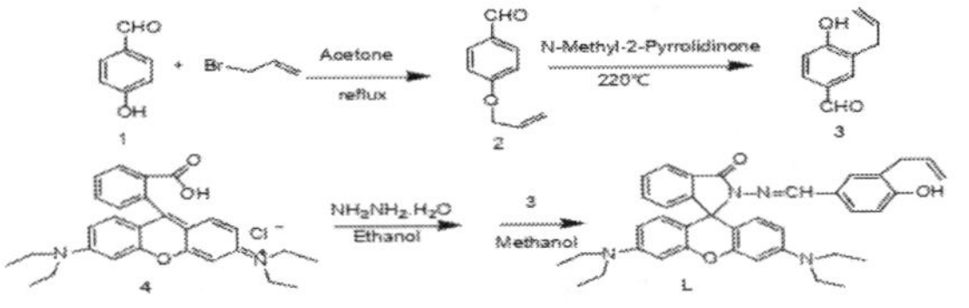

Scheme 1. Synthetic route of **L**

A solution of Rhodamine B (2.00 g, 4.18 mmol) in methanol (80 mL) was added Hydrazine hydrate (1.00 mL, 16.72 mmol), and the resulting mixture was stirred and refluxed for 4h, The solution changed from dark purple to light orange. Then the mixture was cooled and solvent was removed under reduced pressure to give Rhodamine hydrazide. The crude Rhodamine hydrazide was dissolved in 80 mL absolute ethanol. Then intermediate **3** (12mmol) was added the mixture was refluxed for 6 h. The solvent was then removed and the crude product was purified by column chromatography on silica gel using dichloromethane/methanol=25:1 (v/v) as eluant to get compound L (1.12 g, 44.5%) as pink solid. ^1H NMR ($CDCl_3$, 500MHz, ppm): 8.42(s, 1H, CH=N), 7.98 (d, 1H, Ar-H), 7.51 (m, 2H, Ar-H), 7.37 (s, 1H, Ar-H), 7.30(d, 1H, Ar-H), 7.19 (d, 1H, Ar-H), 6.79 (d, 2H, Ar-H, J=8.5 Hz), 6.56 (d, 1H, Ar-H, J=3.5), 6.45(d, 2H, Ar-H, J=2.2 Hz), 6.27(d, 2H, Ar-H, J=2.2Hz), 6.0(m, 1H, =CH), 5.32 (s, 1H, OH), 5.14(d, 2H, $=CH_2$, J=6.5), 3.35(m,10H, NCH_2CH_3), 1.18 (t, 12H, NCH_2CH_3, J=7.1Hz); ^{13}C NMR ($CDCl_3$, ppm): 12.61, 29.32, 35.05, 44.32, 65.92, 97.98, 106.15, 108.12, 115.78, 116.57, 123.30, 123.70, 125.28, 127.75, 128.22, 129.25, 129.70, 133.15, 136.12, 147.29, 148.95, 151.99, 153.05, 155.97, 164.89. FT-MS(ESI) m/z: calcd for $C_{38}H_{40}N_4O_3$ $[M+H]^+$: 601.3127, found 601.3142;

4. Conclusion

In summary, a new fluorescent probe **L** was designed and synthesized. **L** exhibited extreme selectivity for Fe^{3+} over other metal ions in aqueous environment. Furthermore, the *in situ* generated nonfluorescent **L**-Fe^{3+} complex was employed for the specific recognition of AcO^- by fluorescence turn-off signaling through a displacement process. Confocal microscopy imaging implied that **L** can be potentially used as a promising candidate for Fe^{3+} and AcO^- imaging in Ana-1 cell.

Acknowledgments

The project was supported by National Natural Science Foundation of China (21506106) and the Fund for Shanghai Institute of Technology(10259).

References and notes

1. P. Aisen, E. A. Leibold, Curr. Opin. Chem. Biol. 1999, 3, 200.
2. H. Kozlowski, J. Brasun, E. Gaggelli, D.Valensin, G.Valinsin, Coord. Chem. Rev. 2009, 253, 2665.
3. N. Zhang, J. S. Suleiman, M. He, B. Hu, Talanta. 2008, 75, 536.

4. R. Meneghini, Free Radic. Biol. Med.1997, 23, 783.
5. C. Brugnara, Clin. Chem. 2003, 49, 1573.
6. S. Z. Topal, A.G. Gurek, K. Ertekin, D. Atilla, B.Yenigul, V. Ahsen, Sensor Lett. 2010, 8, 336.
7. Z. Zhang, B. Xu, J. Su, L. Shen, Y. Xie, H. Tian, Angew. Chem. Int. Ed. 2011, 50, 11654.
8. B. Zhao, Y. C. Zhou, M. J. Fan, Z. Y. Li, L.Y. Wang, Q. G. Deng, Chin. Chem. Lett. 2013, 24, 257.
9. A. K. Singh, R. Singh, P. Saxena, Sensors. 2004, 4, 187.
10. S. S. M. Hassan, M. S. EL-Shahawi, A. M. Othman, M. A. Mosaad, Anal. Sci. 2005, 21, 673.
11. C. C. Wang, D. Zhang, X. Y. Huang, Talanta. 2014, 128, 69.
12. M. Zhang, Y. H. Gao, Tetrahedron Letter.2007, 48, 3709.
13. C. C. Wang, D. Zhang, Sens. Actuators B. 2014, 198, 33.
14. Y. Xiang, Org. Lett. 2006, 8, 1549.
15. R. K. Pathak, V. K. Hinge, P. Mondala, C. P. Rao, Dalton Trans. 2012, 41, 10652.
16. V. K. Gupta, A. K. Singh, M. R. Ganjali, P. Norouzi, F. Faridbod, N. Mergu. Sens. Actuators B. 2013, 182, 642.
17. N. Mergu, V. K. Gupta, Sens. Actuators B. 2015, 210, 408.
18. L. K. Kumawat, N. Mergu, A. K. Singh, V. K. Gupta, Sens. Actuators B. 2015, 212, 389.
19. D. P. Murale, S. T. Manjare, Y. S. Lee, D. G. Churchill, Chem. Commun. 2014, 50, 359.
20. Y. Yang, K. Yu, L. Yang, J. Liu, K. Li, S. Luo, Sensors. 2015, 15, 49.
21. V. Amendola, D. Esteban-Gomez, L. Fabbrizzi,M.Licchelli, Acc. Chem. Res. 2006, 39, 343.
22. P.D. Beer, P.A. Gale, Angew. Chem. Int. Ed. 2001, 40, 486.
23. P.A. Gale, S.E. Garcia-Garrido, J. Garric, Chem. Soc. Rev. 2008, 37, 151.
24. W. Gong, B. Gao, S. Bao, J. Ye, G. Ning, J. Incl. Phenom. Macrocycl. Chem. 2012, 72, 481.
25. J. Shao, H. Lin, M. Yu, Z. Cai, H. Lin, Talanta. 2008, 75, 551.
26. J. F. Zhang, Y. Zhou, J. Yoon, Y. Kim, S.J. Kim, J. S. Kim, Org. Lett. 2010,12,3852.
27. C. Kaewtong, J. Noiseephum, Y. Uppa, N. Morakot, N. Morakot, B. Wanno, T. Tuntulani, B. Pulpoka, New J. Chem. 2010, 34, 1104.
28. X.Y. Hu, J. Wang, X. Zhu, D. P. Dong, X.L. Zhang, S. Wu, C.Y. Duan, Chem. Commun. 2011, 47, 11507.
29. D. A.Hinckley, P. G. Seybold, Spectrochimica Acta Part A: Molecular Spectroscopy. 1998, 44, 1053.
30. J. S. Wu, I. C. Hwang, K. S. Kim, J. S. Kim, Org. Lett. 2007, 9 , 907.
31. S. L. Shen, X. P. Chen, X. F. Zhang, J.Y. Miao, B. X. Zhao, J. Mater, Chem. B. 2015, 3, 919.
32. W. C. Vosburgh, Am. Chem. Soc. 1941, 63, 437.
33. (a) H. A. Benesi, Am. Chem. Soc. 1949, 71, 2703; (b) M. Barra, C. Bohne, J. Am. Chem. Soc. 1949, 112, 8075.
34. R. Koteeswari, P. Ashokkumar, E. J. P. Malar, V. T. Ramakrishnan , P. Ramamurthy, Chem. Commun. 2011, 47, 7695.
35. M. Zhang, Y. H. Gao, Tetrahedron Letters. 2007, 48, 3709.
36. W.X. Gu, R. B. Silverman, Am. Chem. Soc.2003, 5, 415.

Influence of the combination of last sense codon and stop codon on expression efficiency of green fluorescent protein gene in *Escherichia coli* when the expression vector pKK223-3

Xing Zhang, Aiping Fei, Yong Wang, Zhigang Fang, Yingxue Teng, Xiaoliang Hao*

University of Science and Technology Liaoning, Anshan China, 114000

Abstract. In this paper, the influence of the combination of last sense codon and stop codon on expression efficiency was studied. By our study, for the last sense codon CCG, the RFI of GFP(CCG) was 2.1 fold when the stop codon was UAA, but in comparison, the RFI was 1.1 fold when the stop codon was changed from UAA to UAG. For last sense codon TAG, the RFI of GFP(TAG) with the stop codon UAG was stronger than that with the stop codons UAA and UGA.

1 INTRODUCTION

Green fluorescent protein (GFP) was discovered as a companion protein to aequorin, the famous chemiluminescent protein from Aequorea jellyfish[1]. GFP is composed of 238 amino acids and hence a rather small protein with a molecular weight of roughly 27 kDa[2]. Excitation at 396 nm results in an emission maximum at 508 nm[3]. The discovery of GFP led to a new revolution in molecular biology, whose different mutants had been engineered over the last few years[4-5]. Due to its favourable features, GFP rapidly became a popular tool in various applications in biology research. During the last decade, it has been introduced into a wide range of organisms, including bacteria, yeasts[6-7].

Meantime, *Escherichia coli* (*E.coli*) is a convenient host for protein expression and one of the organisms of choice for the production of recombinant proteins in high quantities and low production costs, it has become the most popular expression platform. *E.coli* derived gene has become one of the most widely used. Its popularity is due to its lack of interference with plant metabolism, the ease and sensitivity of the assay and its stability both in vivo and in vitro[8-17].

As we all know, gene contains a specific sequence of nucleotides which gives the instructions for the specific sequence of amino acids, if the gene is mutated, may change the formation of amino acids, and then influence the structure and function of the protein[18]. To master the relationship between the gene sequence and protein expression is helpful for understanding principles of gene expression and effectively controlling the production of protein. GFP and its some mutants have many useful applications, such as real-time detection, no disruption or toxicity to the host cells, no requirement for cofactors, and feasibility of fusion with the target proteins, have become one of the most used fluorescent probes in cell biology and molecular biology[19-21].

Fluorescent proteins have proven to be excellent tools for live cell imaging. A mutation which results in an exchange of the amino acid serine at position 65 to threonine dramatically increased the intensity of green fluorescence[22].

Some texts about the influecne of the codon on the protein expression were greatly researched. Codon usage bias was considered to influence the elongation rate because each codon was decoded at a different rate[23-24].

The C-terminal region contains all the information required for efficient translocation and can therefore be used as a signal sequence for recombinant protein targeting[25]. In addition, the sequence context motif surrounding the ATG initiation codon (ATG-context) is an important factor that increases protein production[26-32]. Introduction of the rare codons GCC, CGG, and ACC for alanine, arginine, and threonine reduced GFP production 2.1, 3.3, and 1.7 fold in comparison to the favored codons GCU, CGU, and ACA, respectively[33].

During our former researches, my collaborators and I inserted 64 kinds of last sense codons at the 5' end of the stop codon of the GFP gene and studied on influence of last sense codon on expression in *E.coli*[34]. In the process, in order to study the influence of the combination of last sense codon and stop codon on expression efficiency of GFP gene in *E.coli* when the expression vector pKK223-3, I changed the stop codon from UAA to UAG and UGA, selected 18 kinds of last sense codons for each stop codon, and studied the influence of the combination of last sense codon and stop codon on expression efficiency of GFP gene in *E.coli* when the expression vector pKK223-3.

[a] Corresponding author: Haoxiaoliang1980@163.com

2 MATERIALS AND METHODS

2.1 Constructions of Plasmid and Cloning Vectors

In order to research the combination of last sense codon and stop codon on expression efficiency of GFP gene, 18 kinds of GFP genes with special features were selected, such as had the higher or lower fluorescence intensities, medium fluorescence intensities, the same name with the stop codon, etc. NNN was the insertion of 18 kinds of last sense codons. The stop codon was changed from UAA to UGA and UAG (Figure 1).

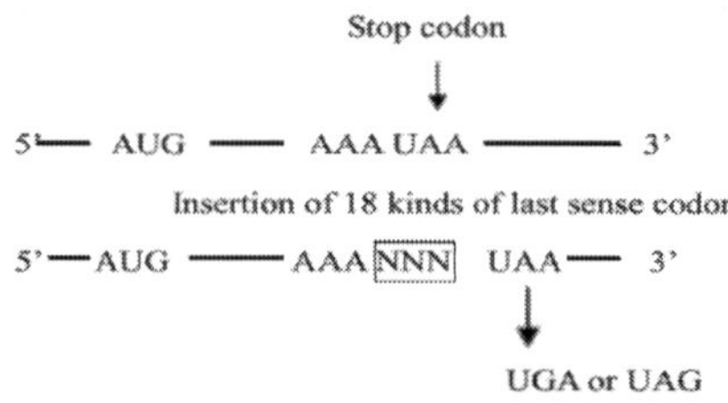

Figure 1. The insertion of last sense codons and change of stop codon in GFP gene.

pKK223-3 was used for the expression vector in this study. In order to research the relationship between the last sense codon and the protein expression efficiency, polymerase chain reaction (PCR) was used for removing these last sense codons of the modified GFP genes. PCR was performed at 94 °C for 10 min, followed by 25 cycles of 94 °C for 45 s, 48 °C for 1 min and 68 °C for 45 s. The principle of PCR was seen in Figure 2. After PCR, the plasmids were constructed by introduction of the modified GFP genes into expression vector. *E.coli* JM109 was transformed with the ligation mixture, the resulting clones were analyzed by sequencing. Primers (PCR) were purchased from Sigma-Aldrich (Tokyo, Japan). Restriction enzymes (EcoRI, HindIII), a PCR amplification kit and T4 DNA ligase had been purchased from TaKaRa (Otsu, Japan). We inserted the last sense codons by PCR, DNA fragments were acquired by enzyme digestion, and recombined by transformation and ligation. The flow diagram of experiment was seen in Figure 3.

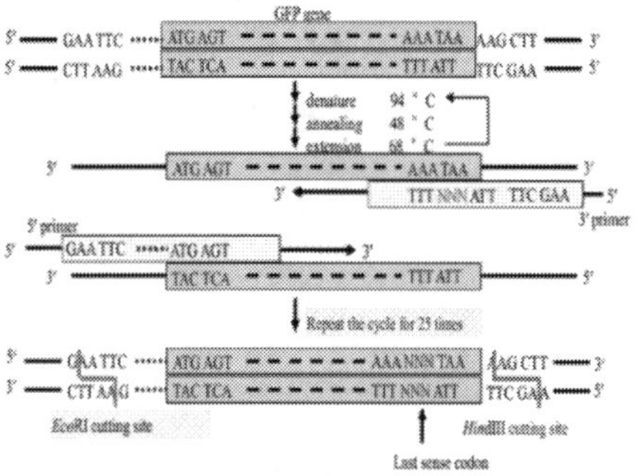

Figure 2. The principle of PCR.

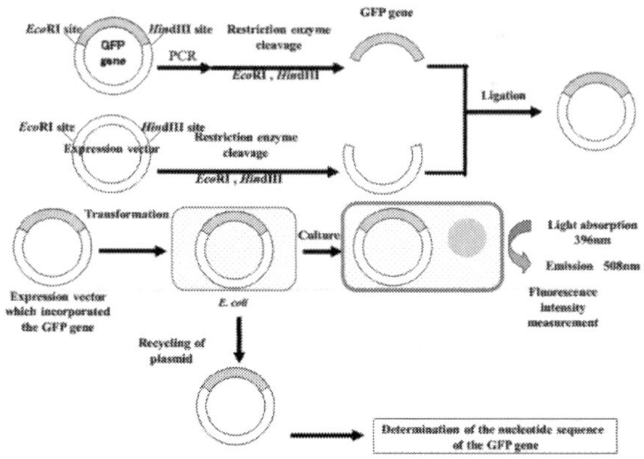

Figure 3. Flow diagram of experiment.

When the expression vector was pKK223-3, 18 kinds of last sense codons were chosen and were inserted into GFP genes by PCR (Figure 2). The used primers of 5' site was the same :

GFP EcoRI Primer (5'Primer)

5'-CCCGAATTCTTTAACTTTAGGAAACACAATT CATGAGTAAAGGAGAAGAACTT-3'

The used primers of 3' site was shown in Table 1. The NNN part was the stop codon (UAA, UGA, UAG). The shaded parts indicated restriction enzyme digested sites corresponding to the primer description.

2.2 Measurement of GFP Relative Fluorescence Intensity in *E.coli*

GFP fusion fluorescence intensity was an excellent indicator of over-expression potential[35]. Because fluorescence was one of the most convenient ways to follow a protein expression and purification procedure[36], so the fluorescence intensity was used to analyze the expression efficiency of proteins. The cells were cultivated in LB medium supplemented with 0.1 mg/mL ampicillin and 40 μM isopropyl-β-D-thiogalactopyranoside (IPTG) at 37 °C for 18 h[37]. The culture of *E.coli* was measured its absorbance at 600 nm, and the fluorescence intensity at 508 nm was excited at 396 nm of the same culture and measured 3 times by the Gemini fluorescence microplate reader (Nihon Molecular Devise). The expression efficiency of GFP gene was compared as the value of the fluorescence intensity / the absorbance at 600 nm. The calculation formula of RFI was shown by formula (1).

$$RFI = \frac{\text{Value of fluorescence intensity of modified GFP gene}}{\text{Value of fluorescence intensity of wild type}} \quad (1)$$

2.3 SDS-PAGE Analysis of Extracts of *E.coli* Having last sense codon of GFP Gene

After the fluorescence intensities of acquired GFP variants were known, the GFP genes were expressed in the proteins, the quantities of proteins were compared, and the influences of the last sense codon and stop codon on protein expression were analyzed. In order to be easily compared, sodium dodecyl sulfate polyacrylamide gel electropheresis (SDS-PAGE) experiment was done. *E.coli* JM109 was transformed by the expression vector including GFP gene, and cultured in the LB medium of 20 mL under the presence of 40 μM IPTG. DNase was added into the solution and the solution was incubated at 37 °C for 1 h to remove the remained DNA of the solution. The insoluble parts were separated from the solution by centrifugation (4°C, 13,000 rpm, 10 min) and soluble proteins were analyzed by SDS-PAGE.

3 RESULTS AND DISSCUSSION

3.1 Comparison of three kinds of stop codons without insertion of last sense codon by the expression vector pKK223-3

The stop codon had three kinds (UAA, UAG, and UGA), the wild types of three kinds of stop codons under the same conditions were determined the fluorescence intensities, it was found that the sequence of fluorescence intensity for three kinds of GFP genes of stop codons was wild(UGA) > wild(UAA) > wild(UAG). The result of comparison was seen in Figure 4.

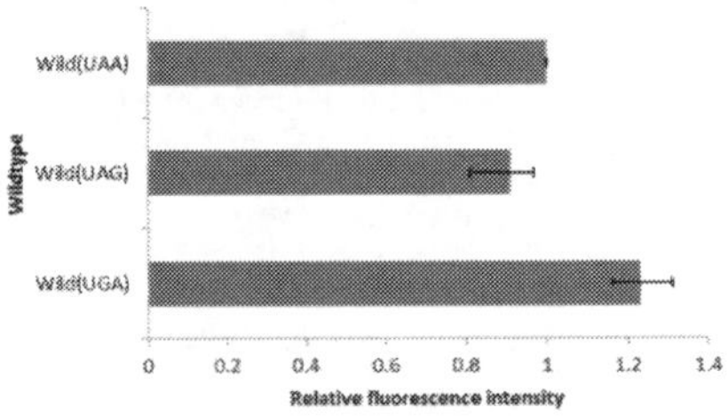

Figure 4. Comparison of the RFIs by three kinds of stop codons without insertion of last sense codon by the expression vector pKK223-3.

3.2 Comparison of the RFI by the stop codon was changed by the expression vector pKK223-3

From Figure 5, it was seen that the last sense codon TCG had been inserted into GFP gene, the stop codon UAA was marked in square.

As the results (Figure 4), for most of the last sense codons, with one kind of last sense codon, when the stop codon was replaced, the change of fluorescence intensity was small. But some specific phenomena were seen, for example, for the last sense codon CCG, the RFI of GFP(CCG) was 2.1 fold when the stop codon was UAA, but in comparison, the RFI was 1.1 fold when the stop codon was changed from UAA to UAG. For last sense codon TAG, the RFI of GFP(TAG) with the stop codon UAG was stronger than that with the stop codons UAA and UGA.

Comparison to wild type GFP, some GFP genes had the stronger or lower RFIs when the three kinds of stop codons. For example, when the expression vector was pKK223-3, GFP(CCG), GFP(GGA) always had the stronger fluorescence intensities and GFP(GTT), GFP(TTG), GFP(CTA) always had the lower fluorescence intensities when the three stop codons (UAA, UAG, and UGA) comparison to wild type GFP.

For the last sense codons which belonged to the same amino acid, the GFP genes with these last sense codons had the nearly same RFIs, for example, the last sense codons GTT and GTG belonged to Valine, the last sense codon ACC and ACA belonged to Threonine, the last sense codon AGT and AGC belonged to Serine. Their fluorescence intensities of GFP genes with these last sense codons were separately the same.

There were also special cases, for example, the last sense codon CCC, CCA and CCG belonged to Proline, GFP(CCC) and GFP(CCA) had the near fluorescence intensities, but for the last sense codon CCG, when the stop codon was UAA, the RFI of GFP(CCG) was particularly strong (2.1 fold).

The recognition of stop codons by release factors 1 (RF-1) and 2 (RF-2) might lead to peptide chain termination during translation. Stop codons (UAG and UAA) are recognized by RF-1 while stop codons (UAA and UGA) are recognized by RF-2. The difference of the influence on expression efficiency among stop codons might be caused by the change of the release factors, this needed further verification.

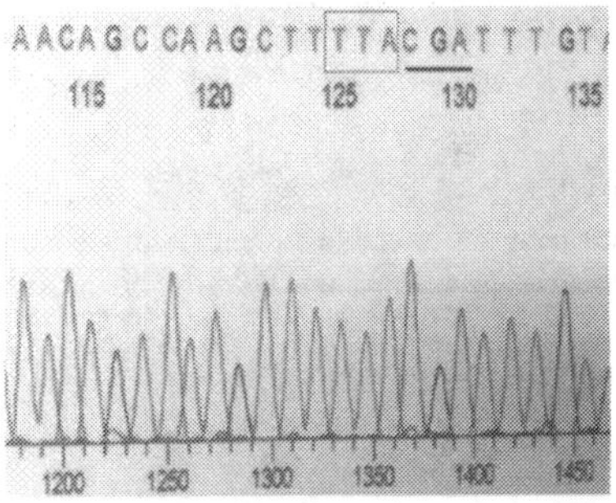

Figure 5. The insertion ot last sense codon TCG.

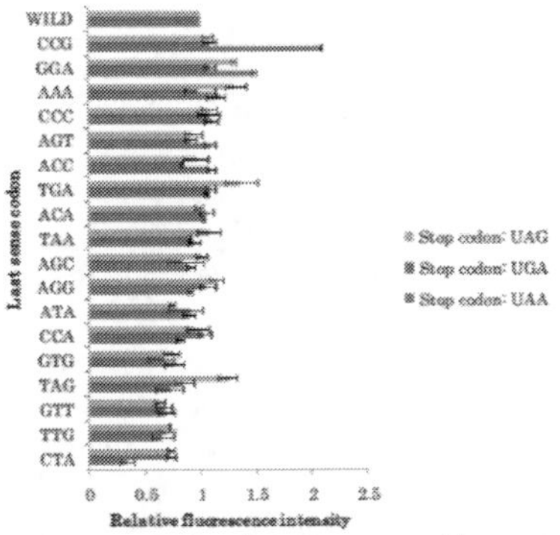

Figure 6. RFIs of GFP expressed by genes with insertion of last sense codons by the expression vector pKK223-3.

3.3 SDS-PAGE of soluble proteins extracted from *E.coli*

To confirm that the RFI was related to the expression efficiency of GFP gene, SDS-PAGE of soluble proteins extracted from *E.coli* was done. The result of SDS-PAGE was seen in Figure 7 by the expression vector pKK223-3. Three kinds of last sense codons CCG, AGT, CTA were selected, and the RFIs by three kinds of stop codons were compared. As shown in Figure 6, the results of RFIs were: CCG(UAA) 2.1 fold, CCG(UAG) 1.1 fold, CCG(UGA) 1.1 fold, AGT(UAA) 1.1 fold, AGT(UAG) 0.92 fold, AGT(UGA) 0.91 fold, CTA(UAA) 0.33 fold, CTA(UAG) 0.73 fold, and CTA(UGA) 0.71 fold. Molecular weight of GFP was about 27,000, from the band of GFP, it was shown that the band of CCG(UAA) was thickest, the bands of CTA(UAA), CTA(UAG), CTA(UGA) were thinner, the thicknesses of other bands were nearly the same. The result of RFIs had the same tendency with that of SDS-PAGE (Figure 7).

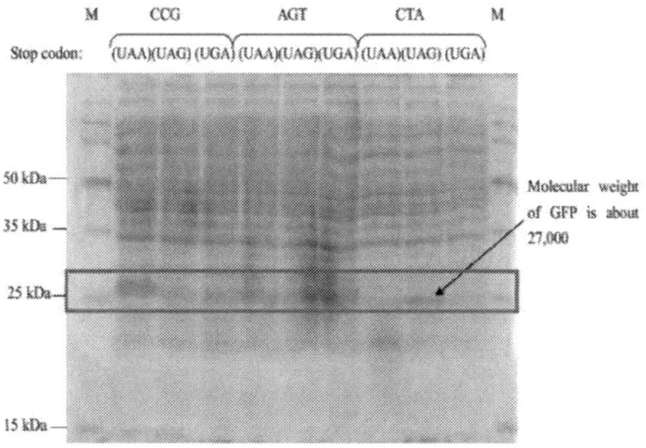

Figure7. SDS-PAGE of soluble proteins extracted from *E.coli* in GFP gene M: size marker. MW of GFP: ~ 27kDa

Conclusions

In our study, the influence of the combination of last sense codon and stop codon on expression efficiency of GFP gene in E.coli was researched. Besides the last sense codon, the stop codon had the effects on the expression of genes. When the stop codon was changed, for most of GFP genes which were inserted last sense codons, the fluorescence intensities were not changed. Some GFP genes always had the stronger expression efficiencies, such as GFP(CCG), GFP(GGA), when these last sense codons were inserted, the quantities of proteins would be increased. It was confirmed that the combination of stop codon and the last sense codon had influenced the expression efficiency of GFP gene. By insertion of last sense codon, we changed some fluorescence intensities of GFP variants, that meaned that this way could control the increase and decrease of the protein expression efficiency. In the future research, it is expected to insert the codon from different site of GFP gene, compare the different influence of GFP variants by insertion the codon, especially some GFP variants which could greatly improved the protein quantities, finally achieve that the protein could be expressed by artificial control.

Acknowledge

This work was financially supported by Liaoning Natural Science Foundation Project (project number: 20170520093, 20170540466), Liaoning Provincial Education Office Service Local Project (project number: 2017FWDF05).

References

1. O. Shimomura, F. H. Johnson, and Y. Saiga, "Extraction, purification, and properties of aequorin, a bioluminescent protein from the luminous hydromedusan, Aequorea," *J. Cell. Comp. Physiol*, vol. 59, pp. 223-239 (1962)
2. D. C. Prasher, V. K. Eckenrode, W. W. Ward, F. G. Prendergast, and M. J. Cormier, "Primary structure of the Aequorea victoria green-fluorescent protein," *Gene*, vol. 111, pp. 229-233 (1992)
3. J. C. March, G. Rao, and W. E. Bentley, "Biotechnological applications of green fluorescent protein," *Appl Microbiol Biotechnol*, vol. 62, pp. 303-315(2003)
4. A. Miyawaki and R Y. Tsien, "Monitoring protein conformations and interactions by fluorescence resonance energy transfer between mutants of green fluorescent protein," *Methods Enzymol*, vol. 327, pp. 472-500 (2000)
5. Y. Wu, Y. Zhou, J. Song, X. Hu, Y. Ding, and Z. Zhang, "Using Green and Red Fluorescent Proteins to Teach Protein Expression, Purification, and Crystallization," *Biochemistry and molecular biology education*, vol. 36, pp. 43-54 (2008)
6. J. Morschhauser, S. Michel, J. Hacker, "Expression of a chromosomally integrated, single copy GFP gene in Candida albicans, and its use as a reporter of gene regulation," *Mol Gen Genet*, vol. 257, pp. 412-420, February (1998)
7. M. Chalfie, M, Y. Tu, G. Euskirchen, W. W. Ward, D.C. Prasher, "Green fluorescent protein as a marker for gene expression," *Science*, vol. 263, pp. 802-805 (1994)
8. N. H. Tolia and L. Joshua-Tor, "Strategies for protein coxpression in *Escherichia coli*," *Nature methods*, vol. 3, pp. 55-64 (2006)
9. G. L. Rosano and E. A. Ceccarelli, "Recombinant protein expression in *Escherichia coli*, advances and challenges," *Frontiers in microbiology*, vol. 5, pp. 1-17 (2014)
10. G. Hannig and S. C. Makrides, "Strategies for optimizing heterologous protein expression in *Escherichia coli*," *Trends in Biotech*, vol. 16, pp. 54-60 (1998)
11. F. Baneyx, "Recombinant protein expression in *Escherichia coli*," *Curr Opin Biotechnol*, vol. 10, pp. 411-421 (1999)
12. A. K. Panda, "Bioprocessing of therapeutic proteins from the inclusion bodies of *Escherichia coli*," *Adv Biochem Eng Biotechnol*, vol. 85, pp. 43-93(2003)
13. L. A. Palomares, S. Estrada-Mondaca, and O. T. Ramirez, "Production of recombinant proteins, challenges and solutions," *Methods Mol Biol* ch.2, pp. 15-52 (2004)
14. F. J. Mergulhao, D. K. Summers, and G. A. Monteiro, "Recombinant protein secretion in *Escherichia coli*," *Biotechnol Adv*, vol. 23, no. 3, pp. 177-202 (2005)
15. J. Yin, G. Li, X. Ren, and G. Herrler, "Select what you need: A comparative evaluation of the advantages and limitations of frequently used expression systems for foreign genes," *J Bacteriol*, vol. 127, pp. 335-347 (2007)
16. RA. Jefferson , TA. Kavanagh, MW. Bevan, "GUS fusions: glucuronidase as a sensitive and versatile gene fusion marker in higher plants," *EMBO J*, vol. 6, pp. 3901-3907 (1987)
17. RA. Jefferson, "Assaying chimeric genes in plants: the GUS gene fusion system," *Plant Mol Biol Rep*, vol. 5, pp. 387-405 (1987)
18. G. Kudla, A. W. Murray, D. Tollervey, and J. B. Plotkin, "Coding-Sequence Determinants of Gene Expression in *Escherichia coli*," *Science*, vol. 324, pp. 255-257 (2009)

19. B. P. Cormack, R. Valdivia, and S. Falkow, "FACS-optimized mutants of the green fluorescent protein (GFP)," *Gene*, vol. 173, pp. 371-374 (1996)
20. C. Nolte, M. Matyash, T. Pivneva, C. G. Schipke, C. Ohlemeyer, U. K. Hanisch, F. Kirchhoff, and H. Kettenmann, "GFAP promoter controlled EGFP-expressing transgenic mice: a tool to visualize astrocytes and astrogliosis in living brain tissue," *Glia*, vol. 33, pp. 72-86 (2001)
21. H. J. Cha, M. Q. Pham, G. Rao, and W. E. Bentley, "Expression of Green Fluorescent Protein in Insect Larvae and Its Application for Heterologous Protein Production," *Biotechnology and bioengineering*, vol. 56, pp. 239-247 (1997)
22. R. Heim, A. B. Cubitt, and R. Y. Tsien, "Improved green fluorescence," *Nature*, vol. 373, pp. 663-664 (1995)
23. A. Fluitt, E. Pienaar, H. Viljoen, "Ribosome kinetics and aa-tRNA competition determine rate and fidelity of peptide synthesis," *Comput Biol Chem*, vol. 31, pp. 335-346 (2007)
24. JF. Curran , M. Yarus, "Rates of aminoacyl-tRNA selection at 29 sense codons *in vivo*," *J Mol Biol*, vol. 209, pp. 65-77 (1989)
25. FJM. Mergulhão, DK. Summers, GA. Monteiro, "Recombinant protein secretion in *Escherichia coli*," *Biotechnol Adv*, vol. 23, pp. 177-202 (2005)
26. M. Kozak, "Point mutations close to the aug initiator codon affect the efficiency of translation of rat preproinsulin *in vivo*," *Nature*, vol. 308(5956), pp. 241-246 (1984)
27. DR. Cavener, "Comparison of the consensus sequence flanking translational start sites in Drosophila and vertebrates," *Nucleic Acids Res*, vol. 15(4), pp. 1353-1361 (1987)
28. DR. Cavener , SC. Ray, "Eukaryotic start and stop translation sites," *Nucleic Acids Res*, vol. 19(12), pp. 3185-3192 (1991)
29. A. Ranjan, SE. Hasnain, "Influence of codon usage and translation initiation codon context in theAcNPV-based expression system: computer analysis using homologous and heterologous genes," *Virus Genes*, vol.9(2), pp. 149-153 (1991)
30. T. Sugio, H. Matsuura, T. Matsui, M. Matsunaga M, "Effect of the sequence context of the AUG initiation codon on the rate of translation in dicotyledonous and monocotyledonous plant cells," *J Biosci Bioeng*,vol.109(2), pp. 70-173 (2010)
31. S. Agarwal, S. Jha, I. Sanyal, DV. Amla, "Effect of point mutations in translation initiation context on the expression of recombinant human a1-proteinase inhibitor in transgenic tomato plants," *Plant Cell Rep*, vol.28(12), pp. 1791-1798 (2009)
32. L. Cherbas, P. Cherbas, "The arthropod initiator: the capsite consensus plays an important role in transcription," *Insect Biochem Mol Biol*, vol.23, pp. 81-90 (1993)
33. C. Finger, M. Gamer, S. Klunkelfuß, B. Bunk, R. Biedendieck, "Impact of rare codons and the functional coproduction of rate-limiting tRNAs on recombinant protein production in Bacillus megaterium," *Appl Microbiol Biotechnol*, vol. 99, pp. 8999-9010 (2015)
34. X. L. Hao, S. Inoue, and M. Ishikawa, "Influence of insertion of the last sense codon on expression efficiency of green fluorescent protein gene in *Escherichia coli*," Journal of Materials Science and Chemical Engineering, vol.3, pp.13-18 (2015)
35. J. Hammon, D. V. Palanivelu, J. Chen, C. Patel, and D. L. Minor, "A green fluorescent protein screen for identification of well-expressed membrane proteins from a cohort of extremophilic organisms," Protein Sci, vol. 18, no. 1, pp. 121-133 (2009)
36. G S. Waldo, B. M. Standish, J. Berendzen, and T. C. Terwilliger, "Rapid protein-folding assay using green fluorescent protein," Nat Biotechnol, vol. 17, pp. 691-695 (1999)
37. K. R. Siemering, R. Golbik, R. Sever, and J. Haseloff, "Mutations that suppress the thermosensitivity of green fluorescent protein," Current Biology, vol. 6, pp. 1653-1663 (1996)

Studied on the extraction, response surface analysis and antioxidant activity of flavonoids in the Momordia cochinchinensis seeds

Junyu Jiang, Aiping Fei, Yong Wang, Zhigang Fang, Yingxue Teng, Xiaoliang Hao*

University of Science and Technology Liaoning, Anshan China, 114000

Abstract. This text was mainly studied the extraction flavonoids technology from Momordia cochinchinensis seeds. The optimum extraction condition was as follows: when the ratio of liquid to solid was 20:1, the ethanol concentration was 90% and the extraction time was 2h. Under this condition, the actual average yield of flavonoids could reach 0.363%.

1 INTRODUCTION

More and more attentions have been paid to the natural products, especially the products derived from medicinal plants. Some plant natural products have many advantages, which could be used in many fields, such as drug discovery, anticancer, immunity, etc. Many diseases of human beings are closely related to the free radicals in the human body[1]. There are some certain relationships between free radicals and antioxidation. Potential sources of antioxidant compounds have been searched in several types of plant materials such as vegetables, fruits, leaves, oilseeds, and many other sources[2]. Compared with synthetic antioxidants, natural plant antioxidants have many advantages, especially safety[3].

Antioxidants are compounds that inhibit or delay the oxidation of other molecules by initiation or propagation of oxidizing chain reactions[4]. Natural products contain many functional compounds, and the exertion of these natural products depends to a large extent on these functional compounds[5]. Flavonoids are parts of these functional compounds and play a preventive role in the development of cancer and heart disease[6-7].

Momordia cochinchinensis is a kind of medicine plant whose seeds, roots and leaves have been used as food. Momordia cochinchinensis contains many functional compounds, such as the flavonoids compounds[8].

Ultrasound-assisted extraction (USAE) is an important method to obtain high valuable compounds and could contribute to the increase in the value of some food[9]. In our study, USAE method was used to extract the Momordia cochinchinensis seed. The procedure condition was optimized by using response surface. The infrared spectrum and high performance liquid chromatography (HPLC) were also used to confirm the existence of flavonoids.

2 MATERIALS AND METHODS

2.1 Materials

Momordia cochinchinensis seeds were purchased from the local supermarket. Anhydrous ethanol, sodium nitrite, aluminum sulfate, sodium hydroxide were obtained from Beijing North Chemical Fine Chemicals Co., Ltd. Rutin, quercetin, EDTA (disodium ethylenediamine tetraacetic acid), saffron red, o-phenol, disodium hydrogen phosphate, sodium dihydrogen phosphate, and sodium dihydrogen phosphate were purchased from National drug group chemical reagents Co., Ltd. All the reagents were analytically pure.

2.2 Instruments

Dfy-400 high-speed universal mill (Wenling Linda machinery manufacturing co., LTD); 00000792-1 electronic balance (Zhuji chaoze equalizer equipment co., LTD); Kq-250b, ultrasonic cleaner (Kunshan ultrasonic instrument co., LTD); ReactlR 15 infrared analyzer (German mettler company); 1100 chromatographic analyzer (Agilent company).

2.3 Experimental

2.3.1 Extraction procedure of flavonoids

Momordia cochinchinensis seeds were ground to powder and weighed to 1.00 g, the powder and a certain volume ethanol solution were added into the triangular flask. The flask was placed into the ultrasonic cleaner, temperature was set as 50 °C, power was set as 400 W. After extraction, the solution was filtered by filter paper.

a Corresponding author: Haoxiaoliang1980@163.com

2.3.2 Standard curve of rutin

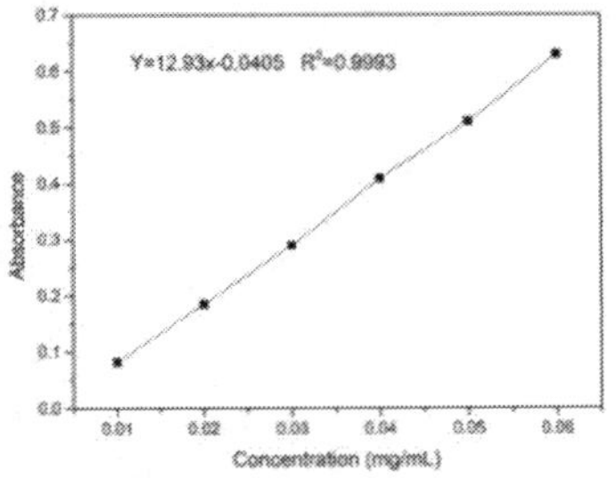

Figure 1. Standard curve of rutin.

Momordia cochinchinensis seeds were dried at 105 ℃. After the powder was smashed and weighed, 95% ethanol was added, the final volume of the solution was 10 ml.

2.3.3 Calculation of flavonoids extraction rate

The flavonoids extraction rate C% was calculated according to the formula (1).

$$C(\%)=\frac{V\times X}{100\times A\times W}\times 100\% \qquad (1)$$

X: flavonoids extraction rate (mg/ml)
V: extraction volume（ml）
A: sample volume (ml)
W: sample weight (g)

2.3.4 Antioxidant analysis

(1) Scavenging of superoxide anion free radical

In this experiment, the phthalic three phenol self oxidation method was used. The absorbance value was determined at 325 nm. The self oxidation rate of o-phenylene three phenol was calculated by using the Tris-HCl buffer solution (pH=8.2) as blank. Inhibition rate was calculated according to the formula (2).

$$\text{Inhibition rate}(\%)=(A_0-A_1)/A_0\times 100 \qquad (2)$$

A_0: absorbance of blank sample
A_1: absorbance of flavonoids solution

（2）Scavenging effect of hydroxyl radical

The first test tube was added by 0.2 mol/l phosphoric acid buffer (pH=7.4), and 0.2 ml 520 mg/l reddish red solution was added; the other 8 test tubes were added by 1.0 ml 0.2 mol/l phosphate buffer (pH=7.4), 0.2 ml 520 mg/l reddish red solution and 1.0 ml $EDTANa^{2}$-Fe^{2+}. Then the distilled water (7, 6, 5, 4, 3, 3, 3, 0 ml) was added, respectively. Finally 0.8 ml 6% H_2O_2 solution was added, with heated 30 min in water bath pot, the absorbance value at 520 nm was measured. Scavenging rate (S) was calculated according to the formula (3).

$$S(\%)=(A_{Sample}-A_{Blank})/(A_{Contrast}-A_{Blank})\times 100\% \qquad (3)$$

3 RESULTS AND DISSCUSSION

3.1 Single factor experiment

3.1.1 Effect of ratio of liquid to solid on flavonoids extraction

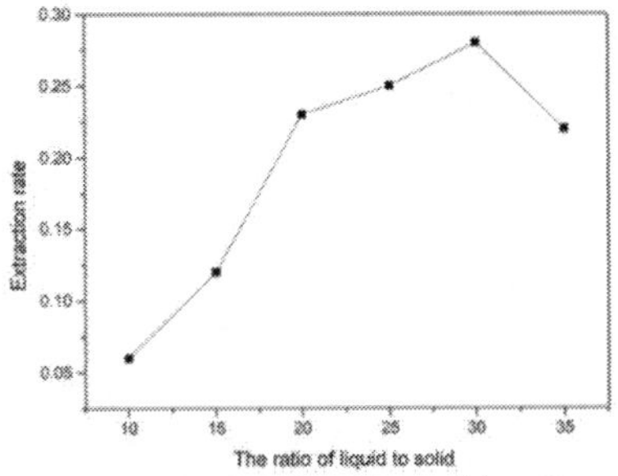

Figure 2. Effect of ratio of liquid to solid on flavonoids extraction.

As shown in Figure 2, the increase in the amount of solvent was conducive to the precipitation of flavonoids. However, under a certain amount, the flavonoids extraction rate was increased slowly, when the ratio reached 30:1, the extraction rate began to decline.

3.1.2 Effect of ethanol concentration on flavonoids extraction

As shown in Figure 3, at first, the flavonoids extraction rate was increased and then decreased with the increase of ethanol concentration. The reason might be that with the increase of concentration, the solvent polarity was decreased and the dissolution rate of flavonoids was decreased.

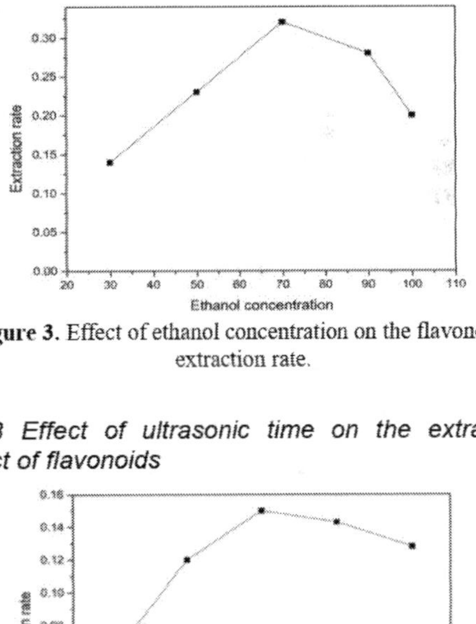

Figure 3. Effect of ethanol concentration on the flavonoids extraction rate.

3.1.3 Effect of ultrasonic time on the extraction effect of flavonoids

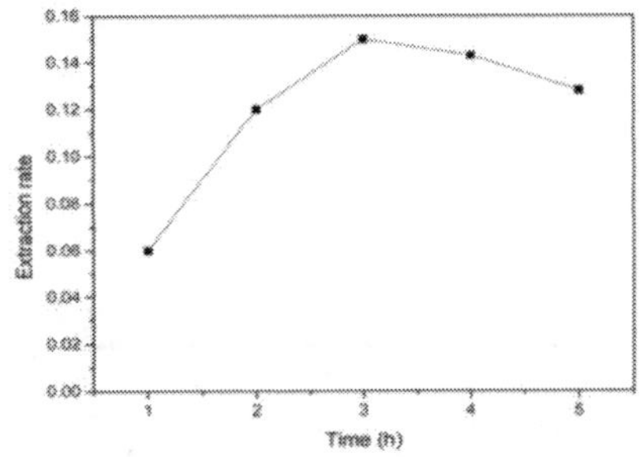

Figure 4. Effect of ultrasonic time on flavonoids extraction.

50 °C was set as the extraction temperature, ultrasonic power was set as 400 W, 70% was set as the ethanol concentration. Figure 4 showed that before 3 hours with the extension of ultrasonic time, the flavonoids extraction rate was increased, and after 3 hours, the extraction rate showed a downward trend, the reason might be that with the incerase of the extraction time, the solvent was evaporated.

3.2 Experimental results of response surface

The Box Behnken model by software Design Expert 8.0.6 was used to design the response surface experiment. The design parameter contained the ratio of liquid to solid, ultrasonic time and ethanol concentration[10].

Table 1. Response surface factor and level table.

Code	Ratio of liquid to solid（A）	Ethanol concentration（B）	Time（C）
-1	10:1	50	1
0	20:1	70	2
1	30:1	90	3

Table 2. Design and the results of the response surface.

No	Ratio of liquid to solid（A）	Ethanol concentration（B）	Time（C）	Extraction rate
1	-1	-1	0	0.06
2	1	-1	0	0.14
3	-1	1	0	0.19
4	1	1	0	0.31
5	-1	0	-1	0.07
6	1	0	-1	0.21
7	-1	0	1	0.08
8	1	0	1	0.24
9	0	-1	-1	0.10
10	0	1	-1	0.29
11	0	-1	1	0.10
12	0	1	1	0.27
13	0	0	0	0.12
14	0	0	0	0.12
15	0	0	0	0.13
16	0	0	0	0.14
17	0	0	0	0.13

The multiple regression equation model was obtained: $Y=0.13+0.063A+0.083B+2.500\times10^{-3}C+1.000\times10^{-2}AB+5.000\times10^{-3}AC-5.000\times10^{-3}BC+3.500\times10^{-3}A^2+0.043B^2+0.018C^2$。

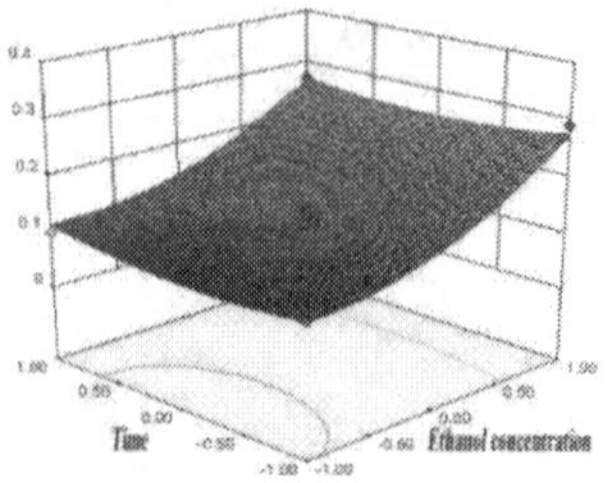

Figure 5. Interaction between ultrasonic time and ethanol concentration on flavonoids extraction rate.

According to the results of response surface, the optimized process condition was that: when ultrasonic time was 1.95 h, the ratio of liquid to solid was 20:1 (ml/g), and the ethanol concentration was 83.26%, the predicted extraction rate of flavonoids was 0.363%. Considering the convenience of the actual operation, the condition in the optimized process was modified to: the ultrasonic time 2 h, the ratio of liquid to solid 20:1 (ml/g) and the ethanol concentration 90%. Three parallel repeated validation experiments were conducted, and the average flavonoids extraction rate under the modified conditions was 0.378%. The experimental value was only 0.015% different from the predicted value of the model equation, indicating that the optimized extraction condition by response surface method was accurate.

Table 3. Variance analysis of the response surface results.

Source of variance	Sum of squares	mean square	F value	P value
Model	0.095	0.011	30.84	<0.0001
A	0.031	0.031	90.02	<0.0001
C	5.000×10^{-5}	5.000×10^{-3}	0.14	0.7156
AB	4.000×10^{-4}	4.000×10^{-4}	1.15	0.3187
AC	1.000×10^{-4}	1.000×10^{-4}	0.29	0.6081
BC	1.000×10^{-4}	1.000×10^{-4}	0.29	0.6081
A^2	5.158×10^{-5}	5.158×10^{-5}	0.15	0.7113
B^2	7.967×10^{-3}	7.967×10^{-3}	22.95	0.0020
C^2	1.441×10^{-3}	1.441×10^{-3}	4.15	0.0310
Residual	2.430×10^{-3}	3.471×10^{-4}		
Unintended term	2.150×10^{-3}	7.167×10^{-4}	10.24	0.0239
Pure error	2.800×10^{-4}	7.000×10^{-5}		
Total deviation	0.099			

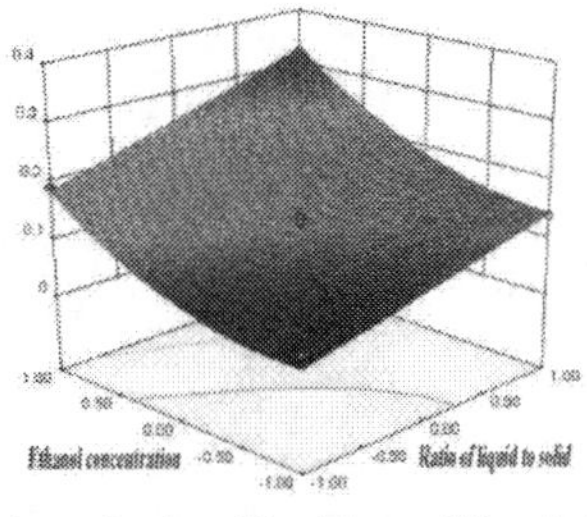

Figure 6. Interaction of ratio of liquid to solid and ethanol concentration on flavonoids extraction rate.

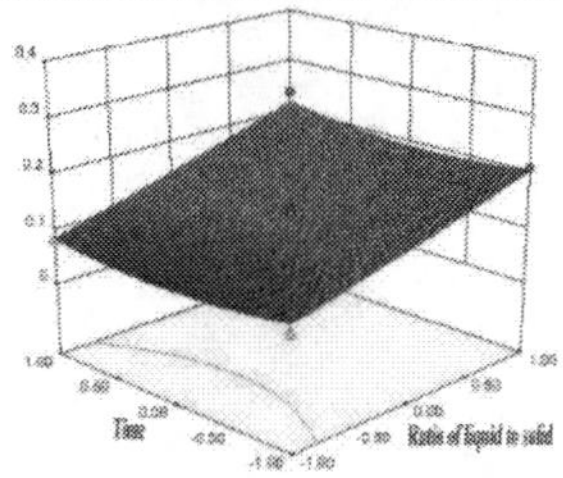

Figure 7. Interaction between ultrasonic time and ratio of liquid to solid on flavonoids extraction rate.

It could be seen in Figure 5, 6 and 7 that ethanol concentration had the greatest influence on the extraction rate, followed by ratio of liquid to solid and ultrasonic time.

3.3 Infrared spectral characteristics and analysis

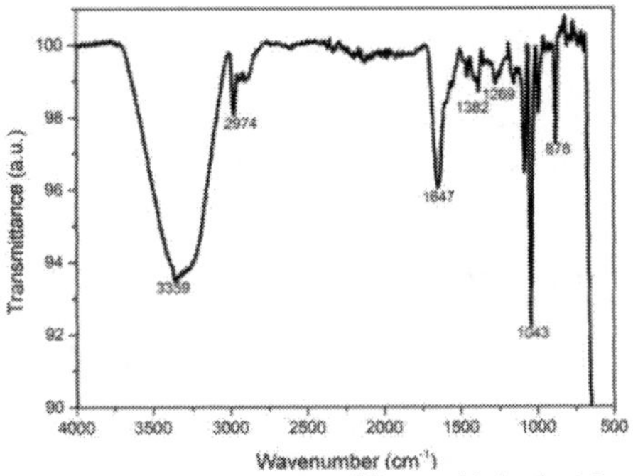

Figure 8. Spectrogram of flavonoids in the Momordica cochinchinensis.

It could be seen in Figure 8, there was a strong absorption peak at 3359 cm^{-1}, indicating that there was a large amount of phenolic or hydroxyl groups. The -CH stretching vibration was found at 2974 cm^{-1}. C=C double bond was found at 1647 cm^{-1}.-CH3 was appeared at 1382 cm^{-1}. The peak 1269 cm^{-1} was C-C skeleton vibration absorption peak. The absorption peak at 1043 cm^{-1} was the stretching vibration peak of C-O. The peak 878 cm^{-1} might be hexagon double oxygen ring.

3.4 Analysis of HPLC results

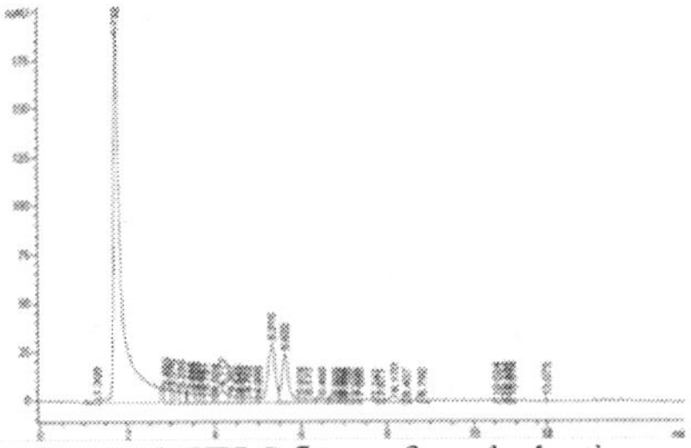

Figure 9. HPLC figure of standard rutin.

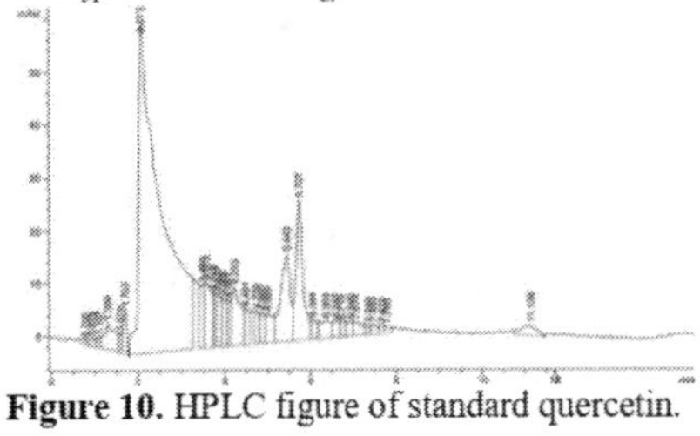

Figure 10. HPLC figure of standard quercetin.

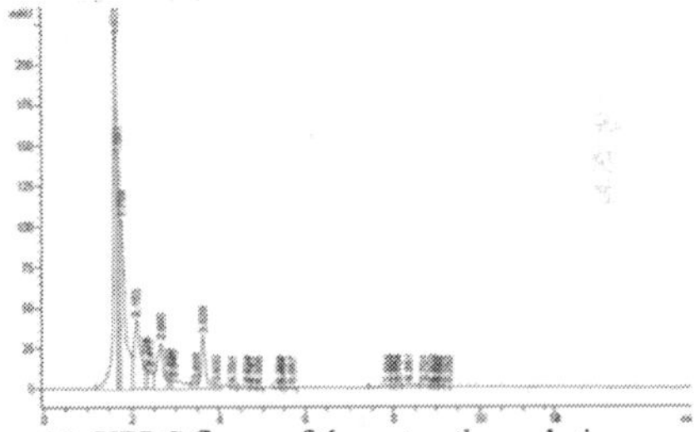

Figure 11. HPLC figure of the extraction solution.

As shown in the figures above, the peak of rutin was appeared at 1.732 min, the peak of quercetin was appeared at 2.073 min, the peak of the extraction solution was appeared at 1.631 min, followed by 1.691 min ,1.769 min, and 2.103 min, which indicating that the compound might contain substances similar to rutin and quercetin.

3.5 Antioxidant analysis

3.5.1 Scavenging of superoxide anion free radical

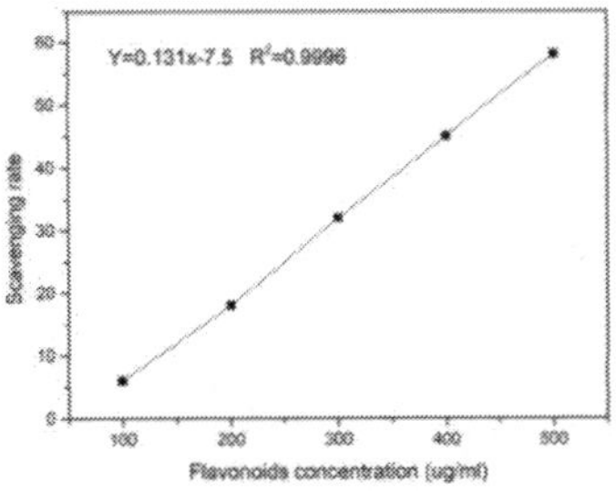

Figure 12. Scavenging ability of flavonoids in the kernel of chiton sinensis to superoxide anion free radicals.

As shown in Figure 12, the flavonoids solution had a good scavenging ability to superoxide anion free radicals, and the scavenging rate was increased with the increase of flavonoids concentration[11].

3.5.2 Scavenging hydroxyl radical

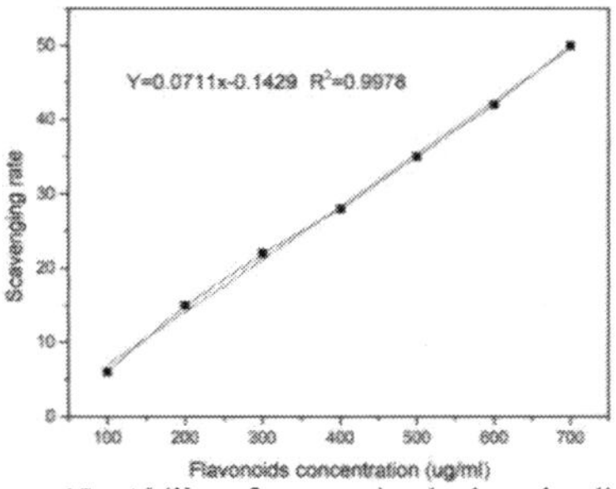

Figure 13. Ability of scavenging hydroxyl radical.

Figure 13 showed that flavonoids solution had the higher scavenging ability for hydroxyl radicals. Moreover, its antioxidant capacity was increased with the increase of flavonoids in the extraction solution. There was a good linear relation in the range of 0-700 ug/ml.

Conclusions

According to the optimization of response surface, under the condition of 400 W ultrasonic power, the optimal extraction condition of Momordia cochinchinensis seeds flavonoids was: the ratio of liquid to solid was 20:1, the ethanol concentration was 90% and the extraction time was 2 h. Under this condition, the actual average yield of flavonoids could reach 0.363%, which was close to the predicted value of 0.378%.

It is a good research direction to extract functional compounds from natural products and use them as natural antioxidants. The purpose of this study is to provide references for the extraction of natural products by studying the extraction conditions, we hope more and more natural functional compounds could be developed to new pharmaceutical preparations and benefited human health.

Acknowledge

This work was financially supported by Liaoning Natural Science Foundation Project (project number: 20170520093, 20170540466), Liaoning Provincial Education Office Service Local Project (project number: 2017FWDF05).

References

1. Adina Elena Segneanu, Daniel Damian, Iosif Hulka, Ioan Grozescu, Athanasios Salifoglou. A simple and rapid method for calixarene based selective extraction of bioactive molecules from natural products[J]. Amino Acids 48:849-858 (2016)
2. Ramarathnam, N.Ochi, H.; Takeuchi, M. Antioxidant defense system in vegetable extracts. In Natural Antioxidants: Chemistry, Health Effects, and Applications; Shahidi, F., Ed.; AOCS Press: Champaign, IL, 76-87 (1997)
3. Larson, R. A. The antioxidants of higher plants. Phytochemistry, 27, 969-978 (1988)
4. S. Velioglu, G. Mazza, L. Gao, and B. D. Oomah.Antioxidant Activity and Total Phenolics in Selected Fruits, Vegetables, and Grain Products[J]. Agric. Food Chem, 46, 4113-4117 (1998)
5. Marja P. Kähkönen, Anu I. Hopia, Heikki J. Vuorela, Jussi-Pekka Rauha, Kalevi Pihlaja,Tytti S. Kujala, and Marina Heinonen. Antioxidant Activity of Plant Extracts Containing Phenolic Compounds[J]. Agric. Food Chem, 47, 3954-3962 (1999)
6. Serafini, M.; Maiani, G.; Ferro-Luzzi, A. Alcohol-free red wine enhances plasma antioxidant capacity in humans[J]. Nutr, 128 (6), 1003-1007 (1998)
7. Carbonneau, M.-A.; Le´ger, C. L.; Descomps, B.; Michel, F.; Monnier, L. Improvement in the antioxidant status of plasma and low-density lipoprotein in subjects receiving a red wine phenolics mixture[J]. Am. Oil Chem. Soc,75, 235-240 (1998)
8. Chen J, Chen SK. Spermatophyta. In: Wu ZY, editor. Flora Yunnanica. 1st ed. Beijing: Science Press; 329-30 (1995)
9. D. Esclapez, J. V. Garcı´a-Pe´rez, A. Mulet, J. A. Ca´rcel.Ultrasound-Assisted Extraction of Natural Products[J]. Food Eng Rev, 3:108-120 (2011)
10. Sibel Tunali Akar, Fatih Sayin, Serpil Turkyilmaz, Tamer Akar. Multivariate optimization of the decolorization process by surface modified biomaterial: Box-Behnken design and mechanism analysis[J]. Environ Sci Pollut Res, 21:13055-13068 (2014)
11. Baudouin C, Pisella PJ, Ettaiche M, Goldschild M, Becquet F, Gastaud P, Droy-Lefaix MT.Effects of EGb761 and superoxide dismutase in an experimental model of retinopathy generated by intravitreal production of superoxide anion radical[J]. Graefes Arch Clin Exp Ophthalmol, 237 (1): 58-66 (1999)

Response surface optimization of ultrasonic assisted extraction of flavonoids from Brucea javanica

Aiping Fei, Junyu Jiang, Yong Wang, Zhigang Fang, Yingxue Teng, Xiaoliang Hao*

University of Science and Technology Liaoning, Anshan China, 114000

Abstract. This text was mainly studied the response surface optimization of ultrasonic assisted extraction of flavonoids from Brucea javanica. The optimum extraction process conditions were obtained by color reaction to prove the presence of flavonoids and by response surface simulation and quantification, the optimum condition was: ethanol concentration was 90%, the ratio of liquid to solid was 10:1, and the extraction time was 2h, the total extraction rate was the largest, up to about 1.43%.

1 INTRODUCTION

Brucea javanica is the mature seed of Brucea javanica (Simaroubaceae), also known as Hazelnut, Sophora flavescens and so on. It is mainly in China's Guangdong and Guangxi provinces, and is rich in resources. The main chemical constituents from Brucea javanica are bitter lignans, flavonoids, triterpenoids and alkaloids. These ingredients are great of using in the field of medicine and human health. Many researchers have studied the chemical constituents of Brucea Javanica[1-4]. With the continuous development of plant extraction technology, various chemical components in Brucea Javanica are extracted by people for scientific research[5-7]. Especially, the anti-cancer effect of Brucea javanica oil has been paid more attention[8-13], and the anti-cancer effect of Brucea javanica oil has been extensively studied. It is reported that flavonoids have anti-inflammatory and analgesic effects. In this study, flavonoids were extracted from Brucea javanica, hoping to provide references for further processing and utilization of traditional Chinese medicine[14].

2 MATERIALS AND METHODS

2.1 Materials

Brucea javanica was bought from the local supermarket. Anhydrous ethanol, sodium nitrite, aluminum sulfate, sodium hydroxide were bought from Beijing North Chemical Fine Chemicals Co., Ltd. Rutin, saffron red, o-phenol, disodium hydrogen phosphate, sodium dihydrogen phosphate, and sodium dihydrogen phosphate were bought from National drug group chemical reagents Co., Ltd. All the reagents were analytically pure.

[a] Corresponding author: Haoxiaoliang1980@163.com

2.2 Instruments

Dfy-400 swing high-speed universal mill (Wenling Linda machinery manufacturing co., LTD); 00000792-1 electronic balance (Zhuji chaoze equalizer equipment co., LTD); Kq-250b ultrasonic cleaner (Kunshan ultrasonic instrument co., LTD); ReactlR 15 type infrared analyzer (German mettler company); 1100 chromatographic analyzer (Agilent company).

2.3 Experimental

2.3.1 Extraction procedure of flavonoids

Brucea javanica was smashed with mill grinding, the powder weight was 1.00 g, a certain volume fraction of ethanol solution was added, and adjusting the ratio of liquid to solid, put triangular flask which contained the extraction sloution into ultrasonic cleaner, set temperature as 50 ℃, set power as 400 W. After extraction, the extraction liquid was filtered by filter paper.

2.3.2 Drawing of standard curve of rutin

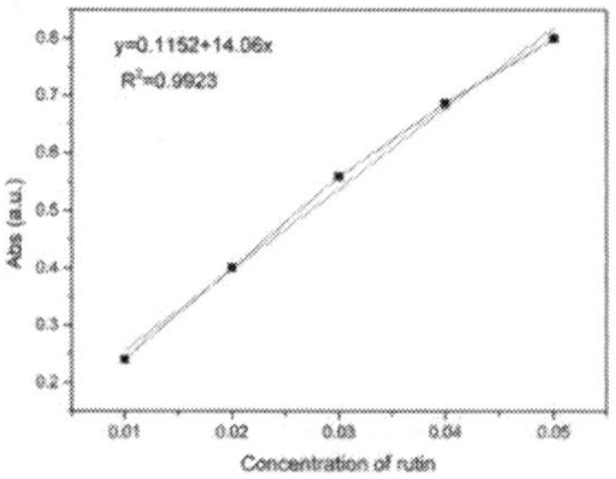

Figure 1. Standard curve of rutin.

2.3.3 Calculation of flavonoids content

The flavonoids content C% was calculated according to the formula (1).

$$C(\%)=\frac{V\times X}{100\times A\times W}\times 100\% \quad (1)$$

X:Content of flavonoids (mg/ml)
V:Extract volume (ml)
A:Sampling volume (ml)
W:Sample weight(g)

2.3.4 Color analysis identification

The above extracted liquid of Brucea javanica was taken as sample, and color analysis was carried out by UV lamp irradiation, ammonia reaction, and sodium hydroxide reaction.

3 RESULTS AND DISSCUSSION

3.1 Single factor experiment

3.1.1 Effect of ratio of liquid to solid on the extraction effect of flavonoids

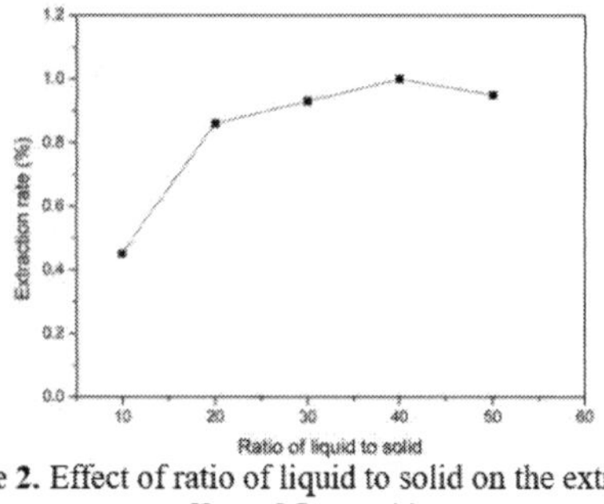

Figure 2. Effect of ratio of liquid to solid on the extraction effect of flavonoids.

As shown in Figure 2, with the increase of ratio of liquid to solid, flavonoids extraction rate in Brucea javanica was first increased and then showed a gentle trend. When the ratio of liquid to solid was too low, the sample could not be completely immersed in the solvent, the extraction was incomplete, and the extraction rate was low. When the ratio was higher than 20:1, the curve showed a gentle trend, indicating that this volume of solvent had reached the extraction effect, which led to waste of solvent. So when the extraction temperature was 50 ℃, the ultrasonic power was 400 W, the ethanol concentration was 70%, extraction time was 2 h, flavonoids extraction rate was better.

3.1.2 Effect of ultrasonic time on the extraction effect of flavonoids

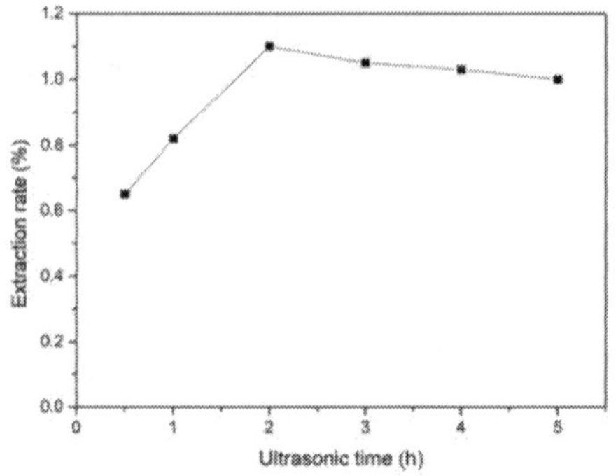

Figure 3. Effect of ultrasonic time on the extraction rate of flavonoids.

As shown in Figure 3, with the increase of ultrasonic time, the extraction rate of flavonoids became higher. After 2h, the extraction rate of flavonoids decreased slightly, the reason was that the ultrasonic time was too long, caused the instrument temperature more than 50 ℃, depressed flavonoids material degeneration. After 2h, the flavonoids in the sample had been fully extracted, and the further time would only increase the energy consumption. So when the temperature was 50 ℃, ethanol concentration was 70%, ultrasonic time was 2 h, flavonoids extraction rate was better.

3.1.3 Effect of ethanol concentration on the extraction effect of flavonoids

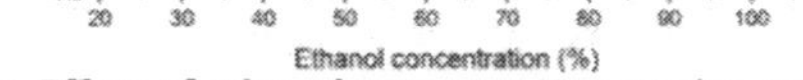

Figure 4. Effect of ethanol concentration on the extraction rate of flavonoids.

With the increase of ethanol concentration, the extraction rate of flavonoids increased first and then decreased (see Figure 4). It might be that if the ethanolconcentration was different and the polarity was different. After the ethanol concentration was over 70%, the extraction rate began to decline greatly, and the dissolution of flavonoids was reduced, while some impurities were increased, leading to the reduction of the extraction rate. So when extraction temperature was 50 ℃, ultrasonic time was 2 h, the ethanol concentration was 70%, flavonoids extraction rate was better.

3.2 Experimental results of response surface

The Box Behnken model by software Design Expert 8.0.6 was used to design the response surface experiment. The design parameter contained the ratio of liquid to solid, ultrasonic time and ethanol concentration.

Table 1. Response surface factor and level table.

Code	Ratio of liquid to solid (A)	Ethanol concentration (B)	Time (C)
-1	10:1	50	1
0	20:1	70	2
1	30:1	90	3

Table 2. Design and the results of the response surface.

No	Ratio of liquid to solid (A)	Ethanol concentration (B)	Time (C)	Extraction rate
1	-1	-1	0	1.32
2	1	-1	0	0.99
3	-1	1	0	1.35
4	1	1	0	1.01
5	-1	0	-1	1.10
6	1	0	-1	0.45
7	-1	0	1	0.95
8	1	0	1	0.56
9	0	-1	-1	0.41
10	0	1	-1	0.83
11	0	-1	1	1.11
12	0	1	1	0.82
13	0	0	0	0.57
14	0	0	0	1.10
15	0	0	0	1.02
16	0	0	0	1.01
17	0	0	0	1.01

The multiple regression equation model was obtained:

$Y=0.94-0.21A+0.023B+0.081C-2.5\times10^{-3}\times AB+0.065AC-0.18BC+0.099A^2+0.13B^2-0.28C^2$.

By the Design Expert 8.0.6 software analysis, the optimum condition for flavonoids was: the extraction temperature was 50 ℃, ultrasonic power was 400 W, ratio of liquid to solid was 10:1, ethanol concentration was 90%, ultrasonic time was 2 h, under the condition, the flavonoid extraction rate could reach to 1.43%.

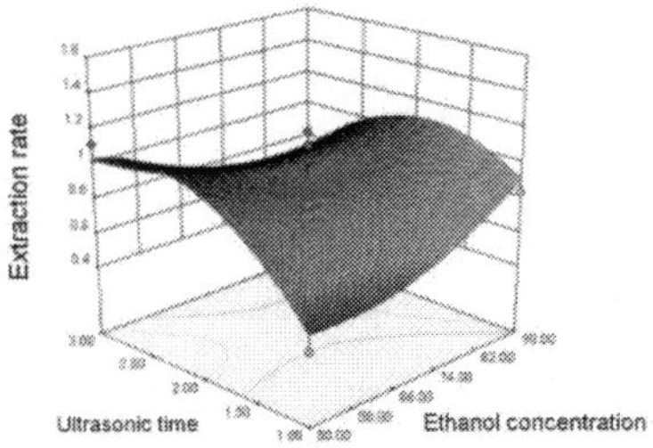

Figure 5. Interaction between ultrasonic time and ethanol concentration on flavonoids extraction rate.

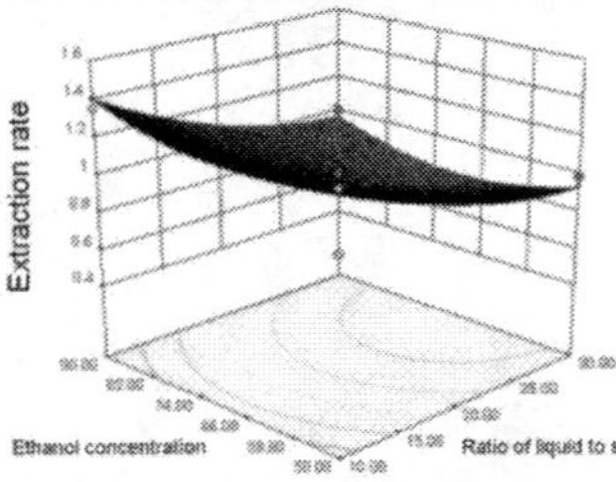

Figure 6. Interaction of ratio of liquid to solid and ethanol concentration on flavonoids extraction rate.

Table 3. Variance analysis of the response surface results.

Source of variance	Sum of squares	of Freedom	Mean square	F value	P value
Model	0.97	9	0.11	2.88	0.0887
A	0.37	1	0.37	9.72	0.0169
C	4.050×10^{-3}	1	4.050×10^{-3}	0.11	0.7524
AB	0.053	1	0.053	1.40	0.2746
AC	2.500×10^{-5}	1	2.500×10^{-5}	6.649×10^{-4}	0.9801
BC	0.017	1	0.017	0.45	0.5241
A^2	0.13	1	0.13	3.35	0.1098
B^2	0.041	1	0.041	1.10	0.3296
C^2	0.067	1	0.067	1.79	0.2225
Residual	0.32	1	0.32	8.53	0.0223
Unintended term	0.26	7	0.038		
Pure error	0.085	3	0.028	0.63	0.6327
Total deviation	0.18	4			

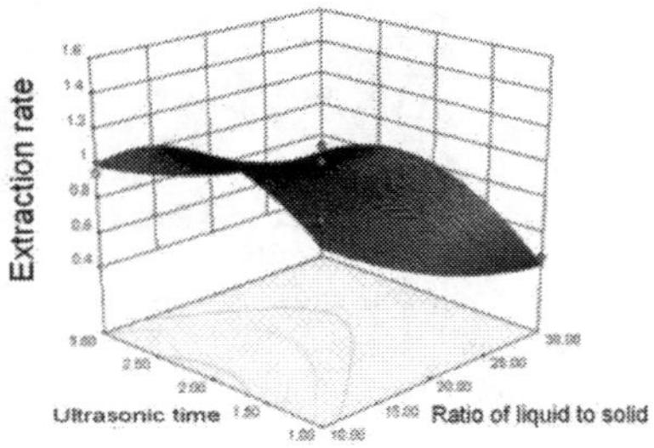

Figure 7. Interaction between ultrasonic time and ratio of liquid to solid on flavonoids extraction rate.

It could be seen from Figure 5, 6 and 7 that three factors ethanol concentration, ultrasonic time, ratio of liquid to solid had the influence on the extraction rate. The best extraction rate could be achieved by seeking the optimal experimental conditions.

3.3 Color reaction result

Figure 8. Color reaction under ultraviolet light.

A few drops of sample solution were put on the filter paper and irradiated by ultraviolet lamp. yellow-green fluorescence could be seen, it appeared that the extraction solution contained flavonoids.

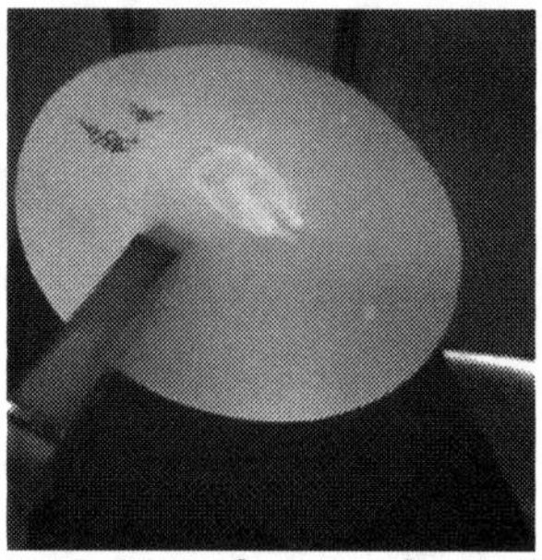

Figure 9. Ammonia water color reaction.

A few drops of sample solution were taken and dripped on the filter paper for air drying. Ammonia solution was poured into the small beaker. The filter paper was smoked on the small beaker for 1min, yellow-green fluorescence appeared.

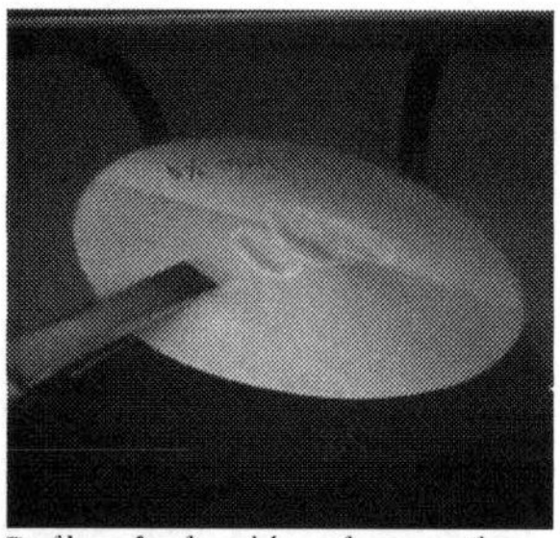

Figure 10. Sodium hydroxide color reaction.

2ml sample solution and 2 ml 5% sodium hydroxide solution were added to observe the color change. Flavonoids would be yellow, orange or red when dissolved in alkaline solution. The result suggested that it might contain both a small amount of flavonoids and a small amount of flavonols.

Conclusions

This experiment mainly studied the effects of various factors on the extraction rate of flavonoids from Brucea javanica, providing some theoretical basis for its extraction in the food and drug industry. The results showed that for Brucea javanica extraction, when extraction temperature was 50 ℃, the ultrasonic power was 400 W, the ratio of liquid to solid was 10:1, ethanol concentration was 90%, ultrasonic time was 2 h, the extraction rate could reach to 1.43%. By color reaction it could be known that the flavonoids extracted from the Brucea javanica were mainly flavonols.

Acknowledge

This work was financially supported by National Natural Science Foundation of China (project number:U1860112), Liaoning Natural Science Foundation Project (project number: 20170520093, 20170540466), Liaoning Provincial Education Office Service Local Project (project number: 2017FWDF05).

References

1. Shihui Dong, Jia Liu, Yingzi Ge, Lei Dong, Chenghui Xu, Jian Ding and Jianmin Yue. Chemical constituents from Brucea javanica [J]. Phytochemistry, 85 (1):175-184 (2013)
2. LIxia Fu, Conggang Huang, Minbao Lin and Qi Hou. The research progress of quassionoids components in Brucea javanica and their pharmacological activities [J]. Chinese Pharmacological Bulletin,32(11): 1481-1485 (2016)
3. Mohd-Al-Faisal Nordin, Wan Himratul Aznita Wan Harun, Fathilah Abdul Razak. An in vitro study on the anti-adherence effect of Brucea javanica and piper betle extracts towards oral Candida [J]. Archives of Oral Biology, 7(001):1335-1342 (2013)
4. Seung Mok Ryu, Jaeyoung Kwon, Young Hye Seo, Eun Gyeong Song, Seong Su Hong, Beom Seok Kim, Jin Sung Hong, Ki Hyun Ryu and Dongho Lee. Quassinoids isolated from Brucea javanica inhibit pepper mottle virus in pepper [J]. Virus Research 09(013): 49-56 (2016)
5. Salprima Yudha S, Doni Notriawan, Eka Angasa, Totok Eka Suharto, John Hendri and Yuta Nishina. Green synthesis of silver nanoparticles using aqueous rinds extract of Bruces javanica (L.) Merr at ambient temperature [J]. Materials Letters, 1(114): 181-183 (2013)
6. Chumkaew, Parinuch, Srisawat, Theera. Phytochemical and antimalarial screening of Brucea javanica plant extracts [J]. Journal of Chemical and Pharmaceutical Research, 12: 253-256 (2014)

7. SRIWILAIJAROEN Nongluk, KONDO Sumalee, NANTHASRI Prance, AUPARAKKITANON Saranya, SUZUKI Yasuo and WILAIRAT Prapon. Antiplasmodial effects of Brucea javanica (L.) Merr. and Eurycoma longifolia Jack extracts and their combination with chloroquine and quinine on plasmodium falciparum in culture[J]. Japanese Journal of Tropical Medicine and Hygiene, 2:61-68 (2010)
8. Anna P. Roswiem, Bambang Kiranadi, Sanjaya Putra Bachtiar and Raafqi Ranasasmita. Antihypertensive Effect of Brucea javanica(L.) Merr. Fruit Extract[J]. Makara Seri Sains, 2: 71-76 (2012)
9. Jiupeng Zhou, Haixia Yang. Meta-analysis on efficacy and safety of Brucea javanica oil emulsion injection combined with chemotherapy for patients with advanced gastric carcinoma [J]. China Journal of Chinese Materia Medica, 41(2): 326-332 (2016)
10. Khalil Hasni, Zul Liham, Surya Dharma, Mahendra Varman. Optimization of biodiesel production from Brucea javanica seeds oil as novel nonedible feedstock using response surface methodology [J]. Energy Conversion and Management, 07(037): 392-400 (2017)
11. Haohuan Liu, Bing Zhang, Wu Zhou, Xiangmin Li, Liwen Hu. Anti-tumor effect of Brucea javanica oil [J]. Journal of NanChang University, 1: 22-26 (2017)
12. Fengjuan Han, Dongyan Cai, Xiaoke Wu, Xiaolong Zhai, Long Wang. Research progress of Brucea javanica antitumor mechanism [J]. Journal of Modern Oncology, 3: 669-671 (2013)
13. Min He, Yu Zhou, Jun kuang, Jian-yong Zhu, Li-bo Zhang. Effect of Brucea javanica oil emulsion combined with Aidinjection on non-small cell lung cancer patient's immune function [J]. China Journal of Modern Medicine, 7: 47-50 (2015)
14. ALEX P,DARREN B.Ultrasonic innovations in the food industry:from the laboratory to commercial production[J].Innovative Food Science and Emerging technologies,9(2):147-154 (2008)

Research on the parameters of producing filamentous textured soybean protein with soybean protein isolated and soybean protein concentrate

Aiping Fei, Xiaoliang Hao, Junyu Jiang, Yong Wang, Yingxue Teng, Yongfei Wang, Yun Gao*

University of Science and Technology Liaoning, Anshan China, 114000

Abstract. Textured soybean protein (TSP) is a product made from cooking and extrusion of soybean protein, which has been widely used in food, feed and other industries. This text made soybean protein isolated (SPI) and soybean protein concentrate (SPC) as the raw materials to produce filamentous protein production. By experiment, the influence of puffing temperature, screw speed and feed rate on the quality of the protein products was studied. Finally it was concluded that when the temperature of the barrel was 152 ℃, the screw rotation speed was 119 rpm, the feed rate was 0.426 kg/min, the TSP product had the biggest expansion degree.

1 INTRODUCTION

Protein is one of the important nutrients in food and plays an important role in improving human immunity. Soybean is rich in soybean protein, and more efficient proteins are developed from soybean, which is beneficial to the improvement of human life quality.

Soybean is rich in protein, and has been an important cash crop which has been widely cultivated all around the world. As the raw material for soybean oil, soy-based products, such as soymilk and tofu, soybean is widely consumed in China[1]. At present, the processing degree of soybean is low, which is based on primary processing. How to further process the soybean and improve the added value of soybean products is one of the key problems to be solved in soybean production.

As a high-grade product, textured soybean protein (TSP) is more and more popular in the market because of its high protein content, low cost, simple production process and tastes like meat. At present, TSP products in the market have the bean smell and the structure are in a bad state. One main reason is partly because the raw materials are improperly selected, and the other reason is that the processing technology is improperly selected[2-4].

So far, there have been few studies on textured soybean protein production. In fact, many parameters such as the temperature, feed rate, screw rotation speed would affect the quality of TSP products[5-7].

In this study, soybean protein isolated (SPI) and soybean protein concentrate (SPC) were used as raw materials, flour and corn flour were used as auxiliary materials to produce TSP by improving processing technology and controlling processing parameters. The protein content of the product was high, the expansion degree was good, and the taste of the product was better. We hope our research might provide more reference for actual production.

2 MATERIALS AND METHODS

2.1 Materials

Soybean protein isolated powder (Harbin Gongda Protein Factory), soybean protein concentrate powder (Qinhuangdao Jinhai Cereals and Oils Co., Ltd.), other raw materials are commercially available.

2.2 Instruments

Soxhlet extractor (Tianjin Tianke Glass Instrument Manufacturing Co., Ltd.), micro-Kjeldahl nitrogen plant (Tianjin Tianke Glass Instrument Manufacturing Co., Ltd.), DS56-III twin-screw extruder (Ji Nan Saixin Extrusion Machinery Co., Ltd. the company).

2.3 Experimental

2.3.1 Determination of main raw material composition

The raw material included SPI, SPC, flour, corn flour. Because the usage amount of flour and corn flour was very few, their nutrients were refer to market averages. The test items included protein, fat, water, ash and nitrogen solubility index.

2.3.2 The production process

[a] Corresponding author: Yuner9111@163.com

The raw and auxiliary materials were weighed in a certain proportion, and mixed well. The twin-screw extruder needed to warm up half an hour in advance. After the machine parameters were set, the feed production test was carried out. Pay attention to the changes of materials and processing parameters in the production process at any time, and turned off the machine switch after the experiment. The expansion degree of the product was tested to analyze the quality difference of the product under different experimental conditions.

3 RESULTS AND DISSCUSSION

3.1 Raw material composition determination results

Table 1. Results of determination of main raw material components in extrusion test.

Raw material	Protein/ %	Fat/%	Water/ %	Ash/%	Nitrogen solubility index/%
SPC	71.3	0.85	5.3	5.5	71.6
SPI	91.5	0.65	3.4	3.7	89.7

It could be seen from Table 1 that the protein contents of both main raw materials were greater than 50%, and the nitrogen solubility index was greater than 70%, which was consistent with the requirements of the production of tissue proteins for the raw materials used. In this research, the ratio of raw materials was: SPC (40%), SPI (40%), flour (10%), and corn flour (10%). Such ratio not only ensured normal production, but also met the requirements of high protein and low cost[8].

3.2 Effect of single factor effect on the expansion degree (Y)

In order to better study the effect of production conditions on TSP, the mathematical model was used to optimize the condition. Three main factors, such as the temperature, screw speed, and feed rate were chosen (Table 2).

Table 2. Experimental factors and their range of variation.

No	Factors	Range of variation
Z1	Temperature (°C)	130~160
Z2	Screw speed (rpm)	60~120
Z3	Feed rate (kg/min)	0.2~0.6

3.2.1 Effect of temperature (X_1) on the expansion degree (Y)

In order to intuitively find out the influence of the barrel temperature (X_1) on the index, set $X_2=X_3=-1$; $X_2=X_3=0$; $X_2=X_3=1$, and the following three equations could be obtained:

$$Y_1=3.20+0.39X_1-0.23X_1^2 \quad (1)$$

$$Y_2=3.58+0.39X_1-0.23X_1^2 \quad (2)$$

$$Y_3=3.36+0.39X_1-0.23X_1^2 \quad (3)$$

The curves of the equations were shown in Figure 1:

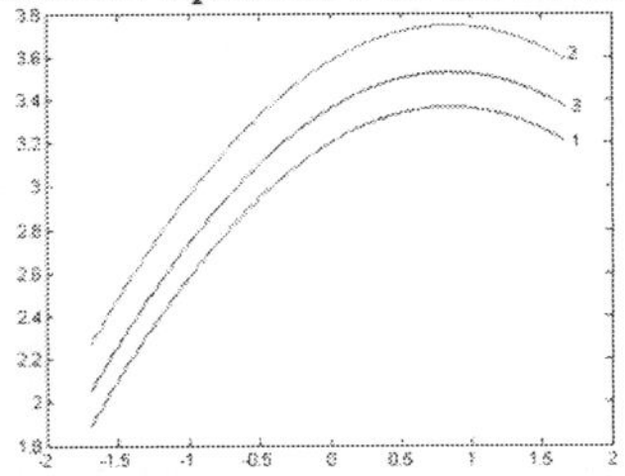

Figure 1. Effect of temperature on the expansion degree.

As shown in Figure 1, the curve of the processing temperature on the expansion degree was parabolic. When the processing temperature was increased, the expansion degree of the product increased, but under a certain extent, the expansion degree decreased. The reasion was that within a certain temperature range, increasing the temperature made the material more favorable for gelatinization and increased the expansion degree of the product.

3.2.2 Effect of screw speed (X_2) on the expansion degree (Y)

X_1 and X_3 were fixed and the following equation could be obtained:

$$Y_1=2.56-0.02X_2^2 \quad (4)$$

$$Y_2=3.58+0.27X_2^2 \quad (5)$$

$$Y_3=2.96+0.56X_2^2 \quad (6)$$

The curves of the equations were shown in Figure 2:

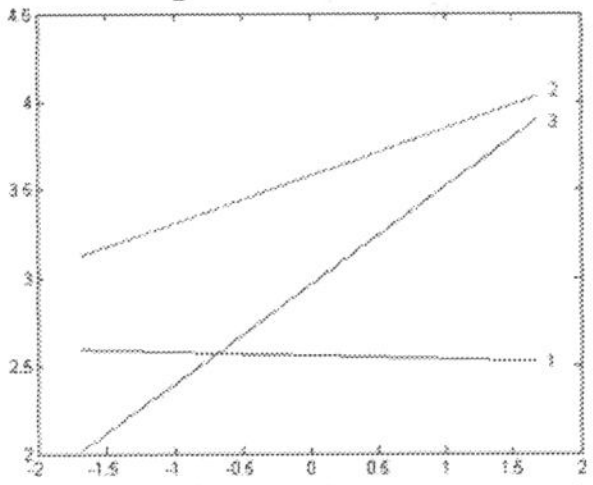

Figure 2. Effect of screw speed on the expansion degree.

As shown in Figure 2, curve 1 was relatively flat, indicating that when the barrel temperature and the feed rate were very slow, it was difficult to expand the material, and the material had to be expanded to have a certain temperature to change the internal structure. For curve 2 and curve 3, it could be seen that at a certain temperature and feed rate, as the screw speed increased, the internal action of the material increased, and the expansion degree increased. For the actual production, the screw speed could be increased within a certain range to increase the product expansion degree and output, but the screw speed could not be too fast, otherwise it was easy to block the material and the product was not expanded[8-13].

https://doi.org/10.1051/matecconf/201823804006

3.2.3 Effect of feed rate (X_3) on the expansion degree (Y)

Fixing X_1 and X_2, the following equations could be obtained:

$Y_1=2.69-0.48X_3-0.59X_3^2$ (7)

$Y_2=3.58-0.19X_3-0.59X_3^2$ (8)

$Y_3=4.01+0.10X_3-0.59X_3^2$ (9)

The curves they presented were shown in Figure 3.

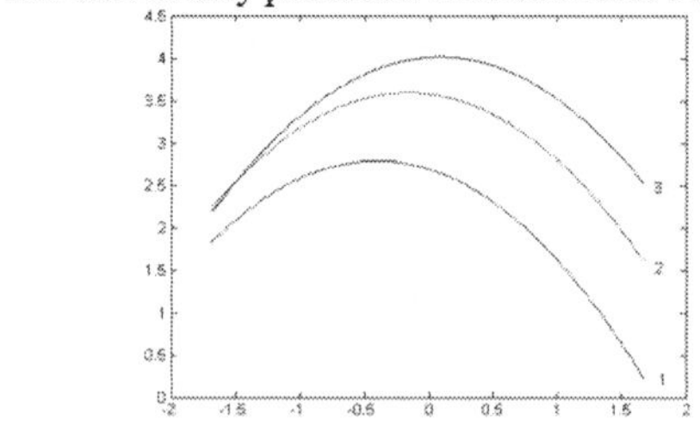

Figure 3. Effect of feed rate on the expansion degree.

As shown in Figure 3, within a certain range, as the temperature of the barrel rised, the internal gelatinization of the product was enhanced, and the expansion degree of the product was increased, but when the temperature exceeded a certain temperature, the product would be burnt, the structure became hard, the surface was browned, and the expansion degree was lower. For actual production, the product could be expanded within a certain temperature range, and the temperature was slightly higher, which was more conducive to product expansion.

3.3 Two-factor effect analysis

3.3.1 Effect of screw speed (X_2) and feed rate (X_3) on the expansion degree

By fixing the barrel temperature (X_1) at the 0 level, a model of the screw speed (X_2) and the feed rate (X_3) versus the expansion degree could be obtained: $Y=3.58+0.27X_2-0.19X_3+0.29X_2X_3-0.59X_3^2$ (10)

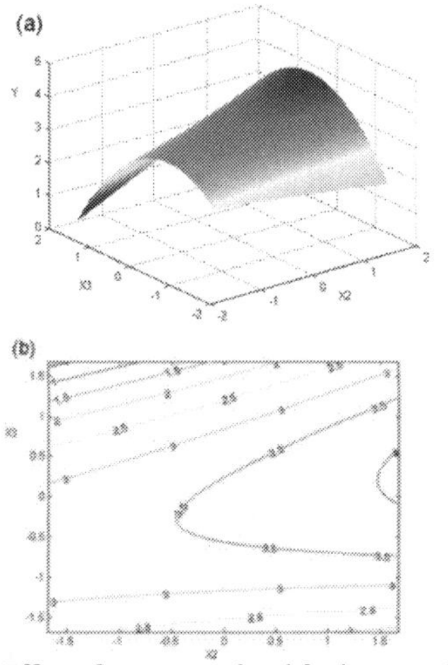

Figure 4. Effect of screw speed and feed rate on the expansion degree.

It could be seen from the Figure 4 that the screw speed X_2 = 1.48 ~ 1.68, and the feed rate X_3 = -0.2 ~ 0.45 had the largest expansion degree. This range could be used as the scope of the subsequent verification experiment. The mechanism was that when the screw speed and feed rate were larger, the processing ability began to be stronger. The extreme points were selected as: X_1 = 0, X_2 = 1.68, X_3=0.45, and the subsequent experiments were verified.

3.3.2 Effect of barrel temperature (X_1) and feed rate (X_3) on the expansion degree

By fixing the screw speed (X_2) at the 0 level, a model of the barrel temperature (X_1) and the feed rate (X_3) versus the expansion degree could be obtained:

$$Y=3.58+0.39X_1-0.19X_3-0.23X_1^2-0.59X_3^2 \quad (11)$$

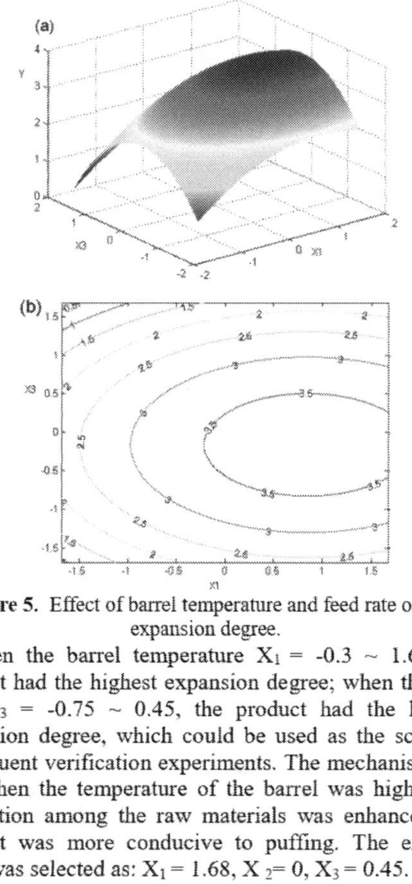

Figure 5. Effect of barrel temperature and feed rate on the expansion degree.

When the barrel temperature X_1 = -0.3 ~ 1.68, the product had the highest expansion degree; when the feed rate X_3 = -0.75 ~ 0.45, the product had the highest expansion degree, which could be used as the scope of subsequent verification experiments. The mechanism was that when the temperature of the barrel was higher, the interaction among the raw materials was enhanced, the product was more conducive to puffing. The extreme point was selected as: X_1 = 1.68, X_2 = 0, X_3 = 0.45.

3.3.3 Influence of barrel temperature (X_1) and screw speed (X_2) on the expansion degree

$$Y=3.58+0.39X_1+0.27X_2-0.23X_1^2 \quad (12)$$

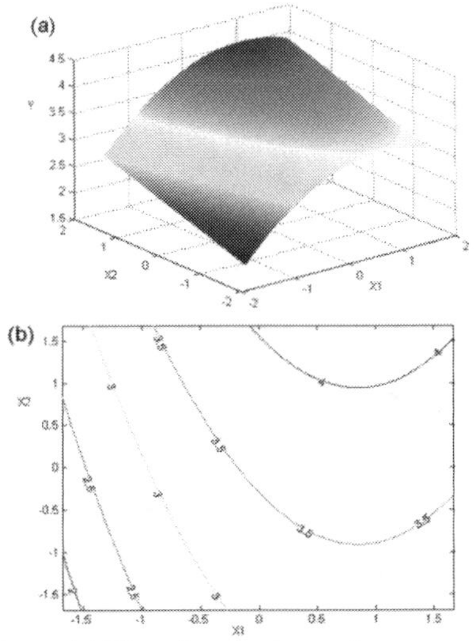

Figure 6. Effect of temperature and screw speed on the expansion degree.

When the barrel temperature X_1= -0.1 ~ 1.68, the product had the highest expansion degree; when the screw speed X_2 = 0.95 ~ 1.68, the product had the highest expansion degree, which could be used as the scope of subsequent verification experiments. The mechanism was that when the temperature of the barrel was higher, the interaction among the raw materials was enhanced, which was more conducive to puffing; when the screw was higher, the product feed began to be smoother. The extreme point was selected as: X_1 = 1.68, X_2 = 1.68, X_3 = 0.

3.4 Optimization of process parameters

Regression equation for the expansion degree: $Y=3.58+0.39X_1+0.27X_2-0.19X_3+0.29X_2X_3-0.23X_1^2-0.59X_3^2$

The optimized condition was that: when X_1=0.85, X_2=-3.13, X_3=-0.93, the Y value was equal to 2.57. Because the range of X was between -1.682 and 1.682, X was out of range, therefore, it had no meaning for this equation and should be discarded.

The three extreme points were obtained by the two factors and the extreme points were obtained by the optimization program. Some random datas were also selected as the experimental parameters, The extrusion experiment optimized parameters were shown in Table 3.

Table 3. Extrusion experiment optimized parameters.

No	X_1	X_2	X_3	Y(By equation)	Y(By experiment)	Error	Fractional error(%)
1	0	1.68	0.45	4.04	3.85	-0.19	4.70
2	1.68	0	0.45	3.70	3.92	0.20	5.41
3	1.68	1.68	0	4.04	4.19	0.15	3.71
4	0.82	1.62	0.22	4.15	4.29	0.14	3.37
5	0	0	0	3.58	3.73	0.15	4.19
6	-0.5	0	0.5	2.94	3.21	0.27	9.18

As shown in Table 3, No 4 had the highest Y value, no matter by equation and by experiment. The X_1, X_2, X_3 were all in the range of -1.68 and 1.68, it was seen that the values obtained were shown to be valid. The error and fractional error were relatively low, indicating a good reproducibility of the experiment, we could use equations to simulate actual production. Changed the code of the equation into the actual operating parameters, and we got the optimal condition was that: when the temperature of the main barrel was 152 °C; the screw speed was 119 rpm; the feed rate was 0.426 kg/min, the expansion degree was highest.

Conclusions

In our experiment, the optimal condition for the production of filamentous TSP was obtained by equation modeling and experimental operation. It was proved that the equation modeling was a better method to optimize the optimal condition. The error between the equation modeling and experimental operation was lower (<10%), it illustrated the feasibility of the method. We hope our research might provide more reference for actual production.

Acknowledge

This work was financially supported by Liaoning Natural Science Foundation Project (project number: 20170520093, 20170540466), Liaoning Provincial Education Office Service Local Project (project number: 2017FWDF05).

References

1. Li, Y.H., Guan, R.X., Liu, Z.X., Ma, Y.S., Wang, L.X., Li, L.H., et al., Genetic structure and diversity of cultivated soybean (Glycine max (L.) Merr.) landraces in China. Theor. Appl. Genet, 117, 857-871(2008)
2. de Kruif, C.G., Weinbreck, F., de Vries, R., Complex coacervation of proteins and anionic polysaccharides. Curr. Opin. Colloid Interface Sci, 9 (5), 340-349 (2004)
3. Lemetter, C.Y.G., Meeuse, F.M., Zuidam, N.J., Control of the morphology and the size of complex coacervate microcapsules during scale-up. AIChE, 55 (6), 1487-1496 (2009)
4. Yan Li Du, Guo Qing Huang, Hai Ou Wang, Jun Xia Xiao. Effect of high coacervation temperature on the physicochemical properties of resultant microcapsules through induction of Maillard reaction between soybean protein isolate and chitosan. Journal of Food Engineering, 234, 91-97(2018)
5. K.E. Preecea, N. Hooshyarb, N.J. Zuidam. Whole soybean protein extraction processes: A review. Innovative Food Science and Emerging Technologies, 43, 163-172(2017)

6. Zhang Qiuhui, Yue Xiqing, Li Miaoyun, et al. Application of Soybean tissue protein in Recombinant Meat products. Food and fermentation Industry, (1), 101-105(2015)
7. Liu Fusheng. Study on the production of tissue protein by medium-sized twin-screw extruder. The Master thesis of Northeast Agricultural University of China, 11-10, 35-39(2006)
8. Yang Qingyu. High moisture cooking extruded noodle food and its application in spicy food. The Master thesis of Northeast Agricultural University, 10-20(2008)
9. Zhu Feng, Zhu Yinyue, Li Tianjiao, et al. Study on the production of instant sausages with soybean meal systematized products. Food and feed industry, (10), 20-25(2016)
10. Hao Xiaoliang, Liu Yi. Comparison of fibrous Soybean tissue protein with General tissue protein. Food Industry Technology, (9), 113-114(2007)
11. Hao Xiaoliang, Sun Yuemei, Jiang Lianzhou, et al. Preparation of Soybean tissue protein Ham sausage. Soybean Bulletin, (3), 23-25(2007)
12. Badrie N, Mellowes W A. Effect of extrusion variables on cassar extrudate. Food Science, 56(5), 1443(1991)
13. Camire M L. Protein functionality modification by extrusion cooking. JAOCS, 68(3), 200-204(1991)

Experimental Study on Water Erosion Resistance Characteristics to Steam Turbine Blade Material

Zheyuan Zhang [1], *Yonghui* Xie [1,*], and *Di* Zhang[2]

[1] School of Energy and Power Engineering, Xi'an Jiaotong University, Xi'an, Shaanxi Province, 710049, P. R. China.
[2] Key Laboratory of Thermal Fluid Science and Engineering of Ministry of Education, School of Energy and Power Engineering, Xi'an Jiaotong University, Xi'an, Shaanxi Province, 710049, P. R. China.

Abstract. An experimental system to assess the water erosion resistance of the steam turbine blade material was built. It can conduct the test of the blade material water erosion resistance characteristic with different water jet diameter and different impact velocity. The experimental investigation of the water erosion resistance characteristics of the blade materials with different surface treatment processes are carried out. By weighing the specimens of each stage, the curves of mass loss with time are obtained. The 2D metal micrograph, 3D surface topography and cross section pit curves are achieved with a laser 3D microscopy. The results reveal the characteristics of the specimens at all stages during the experiment. A method was proposed that the water erosion resistance characteristic can be evaluated by the volume loss based on the pit cross section size, which solves the problem that it's difficult to compare the masses with different material densities or coating densities. The experimental study in this paper has great value for the turbine blade material selection in the steam stages and surface protection technology.

1 Introduction

The last-stage blades of modern high-power ultra-supercritical and nuclear turbine are longer than before, which leads to a great increase in the circumferential velocity of the blade shroud. Water erosion areas are basically located on the blade inlet edge top, and the impact velocity of water drops greatly increase as a result of the high circumferential velocity, which leads to the larger impact stress and reduction of the fatigue life. Consequently, the operation of turbine plant is seriously threatened.

Much experimental study on water erosion has been conducted since 1920s, and several researchers tried to build accurate and stable experimental analysis systems in order to establish the standard test process. Scholars agree that water erosion experimental evaluation is important in the process of blade design and material selection. Water erosion resistance test on real turbine blade is expensive, so the key parameters are selected to simplify the experimental system in order to complete the analysis system. The serious influence caused by the high-velocity drop impact was proposed by Honegger [1], and later an experiment was carried out on account of the turbine blade water erosion accidents. During the experiment, the turbine blade material was impacted by the water jet with a velocity of 200m/s. A large-scale analysis system for rain erosion was established by German Dornier Company [2], and many different working conditions could be analysed. Ripken [3] also established a simple vertically arranged water erosion analysis system. With the increase of turbine power, the last-stage blade is longer, and a number of analysis systems were established, such as the experimental systems in British Electric Company, CA Parsons and British Central Electricity Authority [4]. In 1970, a small analysis system with high-velocity jet was established by NASA to conduct experimental analysis on aluminium 1100-0 and stainless steel 316[5]. In addition to the rotary experimental systems, the experimental system on single high-velocity jet impact was built as well. The first single jet impact device was built by Bowden [6]. It was founded that the erosion area consisted of pits and it was surrounded by some broken area caused by the high shear force. In addition, a measuring device was set up to measure the pressure distribution inside the water drop by Rochester and Brunton [7]. Hoff [2] developed a small-scale experimental facility to study energy conversion problem during the impact of the pendulum. D.C.Jenkins [8] established an experimental system for the study of the impact between a liquid drop and a surface moving at high speed. Toshiyuki [9] et al developed a device to study the impact of high-speed steam-droplet spray on solid surface.

These experimental systems show that water erosion depends on various experimental conditions and material characteristic parameters. For the majority of the materials with good resistance to water erosion, the water erosion begins from an incubation period with little mass loss, then the damage expands until it reaches a maximum, after which the water erosion rate decreases until it is stable.

* Corresponding author: yhxie@mail.xjtu.edu.cn

A rotary water erosion experimental system for turbine wet steam stage blade material was designed and built based on the actual operating conditions of turbine blade and the research on water erosion in the recent years. Since the average experiment time reaches to at least dozens of hours, it is very difficult to guarantee the same parameters completely every time. The system can analyse at most 24 specimens simultaneously, so the variables can be controlled as much as possible. The experimental analysis for specimens of base material and those protected by two different surface protection treatments was conducted in this paper. Some significant results were obtained including the water erosion characteristics of all the specimens, the sort of water erosion resistance ability of different specimens and two important parameters influencing water erosion.

2 Experimental system

An experiment platform is designed for the research on water erosion resistance characteristics of wet steam stage blades, and the water erosion resistance performances of base material and surface-treated material are obtained directly. It provides evaluation criteria for the water erosion of turbine wet steam stage blade and basic data for water erosion resistance treatments.

The shaft velocity ranges from 0 to 9000r / min, and the drop diameter ranges from 0.1mm to 0.3mm while the maximum impact velocity is above 650m/s. The experiment platform consists of a DC motor, a gear increaser, an ultra-high pressure pump, a vacuum pump, an experiment box, a lube oil system and a control system. The schematic diagram of the experiment platform is shown in Fig. 1.

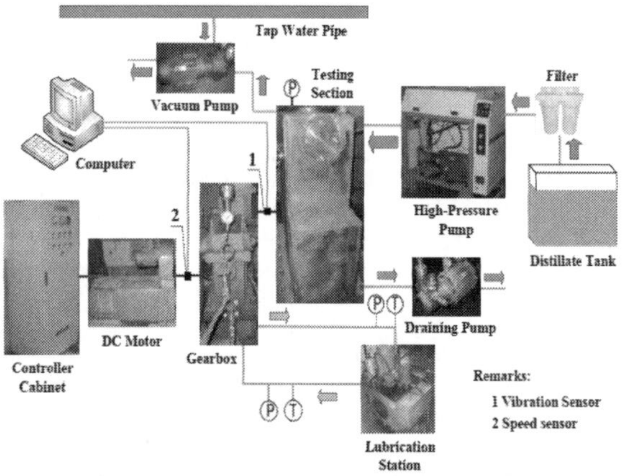

Fig. 1. Schematic Diagram of the Experimental System

The DC motor is controlled by the motor control cabinet and the motor gets connected to the gearbox. The coupling is applied to connect the output end of gearbox and shaft, and there is roulette on the shaft, which can rotate in a high velocity with specimens. There is also a sealing shell and a nozzle outside. The nozzle is used to lead the water from the pump outside the shell, and its position greatly agrees with the specimen position. There is a connection between the shell and the vacuum pump, and a drain pump is located at the bottom of the cylinder.

The high-pressure pump nozzle, shaft and roulette are shown in Fig. 2.

Fig. 2 High-pressure pump nozzle, shaft and roulette

The ultra-high pressure pump and generator are shown in Fig. 3 and Fig. 4. The main parameters of the ultra-high pressure pump are described as follows: the maximum of jet outlet pressure is 300MPa and the maximum working pressure can reach 240 ~ 260MPa. Because the pressure is adjustable and measurable, the jet velocity can fully simulate the real velocity of wet steam drops.

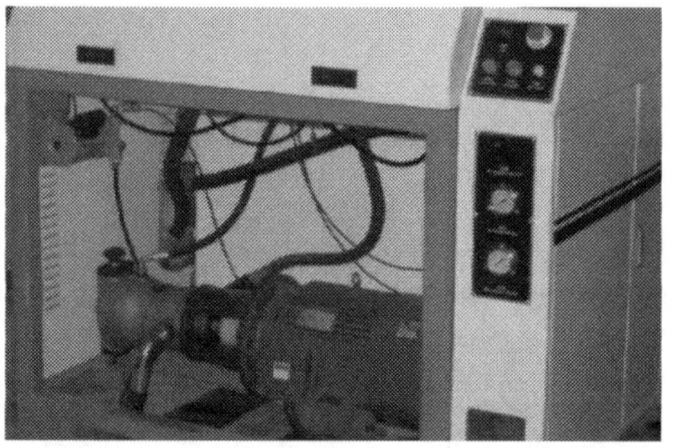

Fig. 3 Ultra-high pressure pump

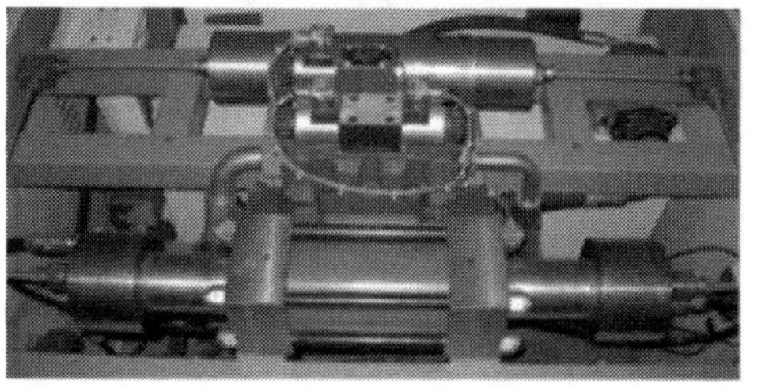

Fig. 4 Ultra-high pressure generator

The diameter of the throttle nozzle is a key part and it must be adjustable. The jet injector and throttle nozzles with different diameters are shown in Fig. 5. The jet injector is fixed on the shell and the nozzle extending into the shell inside faces the specimen surfaces.

It is necessary to stop the machine and observe the specimen surface at intervals during the experiment. Initially the interval time is short, and once the water erosion phenomenon occurs, the surface morphology characteristics should be recorded immediately. The weighing is conducted after the preheating of the precise balance and the cleaning of the specimen surface. The data record is carried out when it remains stable.

Fig. 5 Jet injector and nozzles with different diameters

3 Results and discussion

The water erosion analysis for three kinds of specimens is carried out on the rotary experiment platform. The specimens are the base material specimen with simple heat treatment, high frequency quenching material specimen with hardened surface and stellite-welded specimen respectively. The water erosion photos at different moments for testing material specimen are shown in Fig. 6

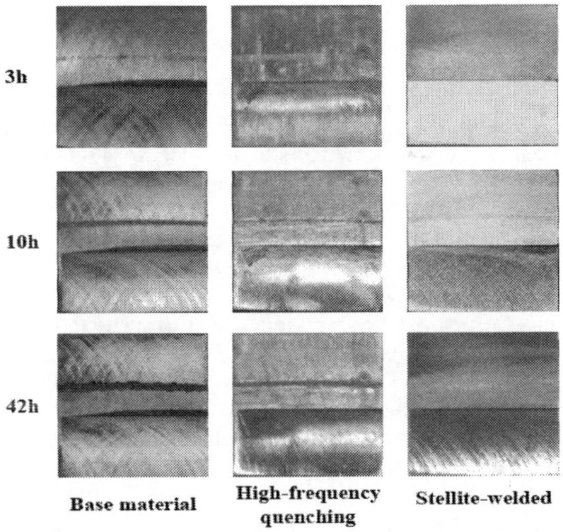

Fig. 6 Water erosion photos at different moments for testing material specimen

As shown in Fig. 6, in the early stage, the water erosion traces consist of a series of discrete micro pits. Then these pits are connected to each other and the erosion grooves form and expand gradually. As can be seen above, the size of water erosion grooves is the largest. At the same time, the erosion traces for stellite-welded specimen are less obvious compared with those of other two specimens, which shows that stellite-welded specimen has the best resistance of water erosion the present study.

The mass loss per unit area can be used to measure the extent of water erosion for the testing material. During the experiment, mass weighing is conducted 6 times and the mass loss curves are shown in Fig. 7 to Fig. 9.

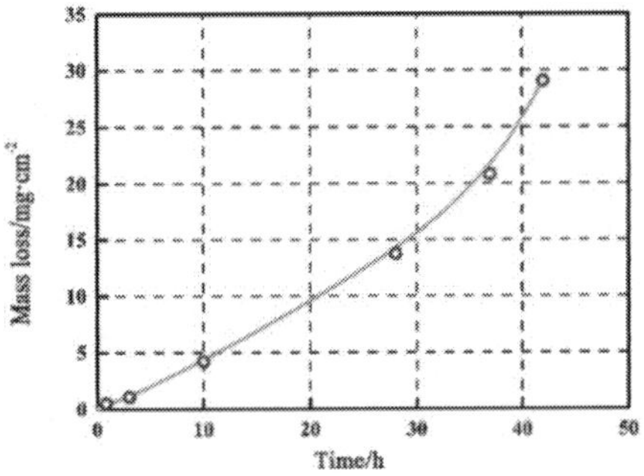

Fig. 7 Mass loss curve of water erosion with base material specimen

It can be seen that the water erosion occurs when the experiment starts. The water erosion rate is relatively constant, which should be regarded as a straight line theoretically. However, after 30 hours the slope of the curve begins to rise, and it is found that the pit depth is larger than 400μm. This may be because that the heat treatment for the base material specimen is not enough, which results in the stress focus and other defects.

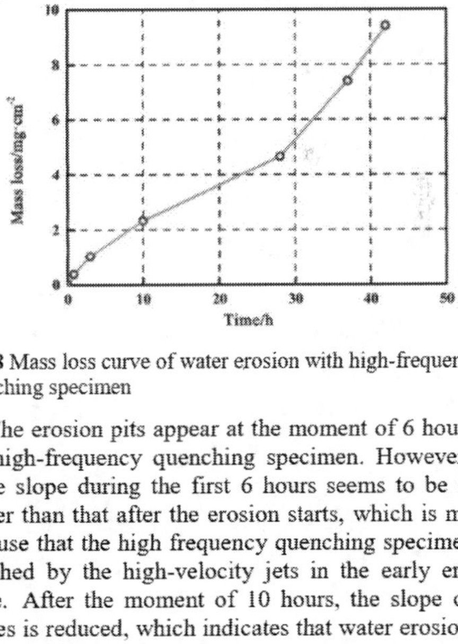

Fig. 8 Mass loss curve of water erosion with high-frequency quenching specimen

The erosion pits appear at the moment of 6 hours for the high-frequency quenching specimen. However, the curve slope during the first 6 hours seems to be much higher than that after the erosion starts, which is mainly because that the high frequency quenching specimen are polished by the high-velocity jets in the early erosion stage. After the moment of 10 hours, the slope of the curves is reduced, which indicates that water erosion pits have reached the hardened layer. The time when the erosion reaches the base material can be estimated as 28 hours in the experiment process from the Fig. 8 because of the rising water erosion rate.

https://doi.org/10.1051/matecconf/201823804007

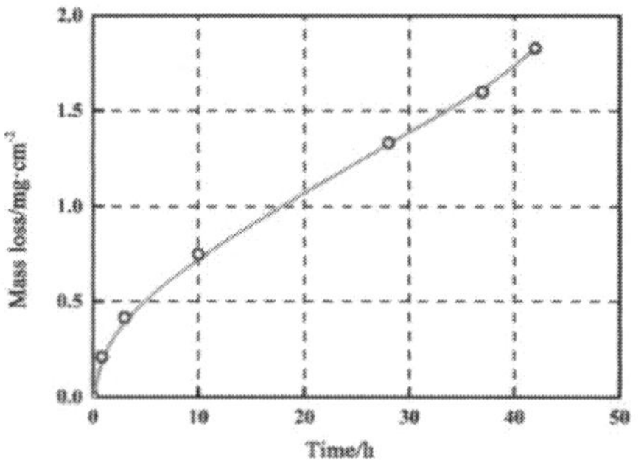

Fig. 9 Mass loss curve of water erosion with stellite-welded specimen

As shown in Fig. 9, the initial slope of the mass loss curve for stellite-welded material specimen is quite large, which is the same with that of the high-frequency quenching specimen. Because the specimen surface is rough and the lateral water jet polishes the target surface, which leads to a great quantity of mass loss. In addition, another excellent characteristic of the stellite-welded specimen is that the mass loss is still little after obvious water erosion. The water erosion rate nearly keeps at a constant value during the whole experiment with the total mass loss of only 1.83mg/cm^2.

In order to observe the changes of water erosion pits, the 2D microscopic water erosion regions at different moments are shown in Fig. 10.

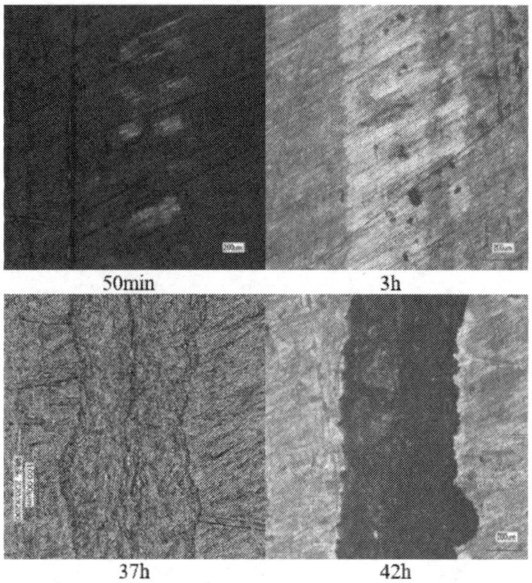

Fig. 10 2D microscopic water erosion regions at different moments with high-frequency quenching specimen

It is shown from Fig. 6 that the water erosion is not obvious and there are only some friction traces in the first 3 hours. When it reaches to the moment of 50 minutes, the friction traces appear as discrete micro pits. After the moment of 6 hours, water erosion develops faster until the experiment is over. It can be observed within 2D photos that the pit width nearly remains the same value during the later erosion period.

In order to analyse the water erosion morphology and specific values of erosion pit width and depth, the 3D microscopic morphologies are obtained, shown in Fig. 11.

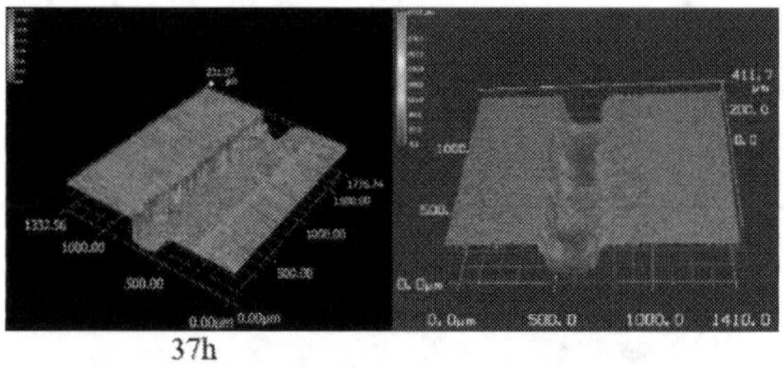

Fig.11 3D water erosion area morphologies at different moments with high frequency quenching specimen

Because no significant water erosion phenomenon occurs within the incipient 6 hours, the 3D pit topography cannot be obtained. Then with the accumulation of impact effect, the water erosion accelerate gradually and pits start to appear. The maximum height differences for the moment of 37 hours and 42 hours are 231μm and 411.70μm respectively.

In order to illustrate the expansion law of water erosion pits, the basic dimensions of the specimen pits are listed in Table 1.

Table 1 Pit sizes at different moments with base material specimen

Items	3hours	37hours	42hours
D/μm	42.00	382.01	445.32
W/μm	175.90	560.00	578.00

The pit depth (***D***) of the specimen keeps increasing from the moment of 3 hours to 42 hours. In addition, pit width (***W***) also develops to 578μm, which grows much faster than the pit depth. After some time since water erosion occurs, it can be approximately believed that the pit width stops increasing, but the depth gets larger. So it can be easily seen that there is a gradual increase in the depth. However, the pit width nearly remains the same value during the erosion later period.

On the other hand, the density and the surface coating are different for materials treated by different technologies, so it cannot meet the requirements just by comparing the mass loss. Therefore, the volume per unit length should be obtained to make comparisons according to the section dimensions.

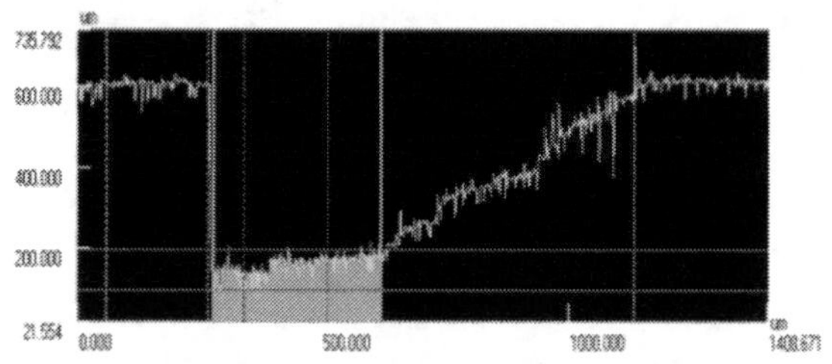

(a) Base material

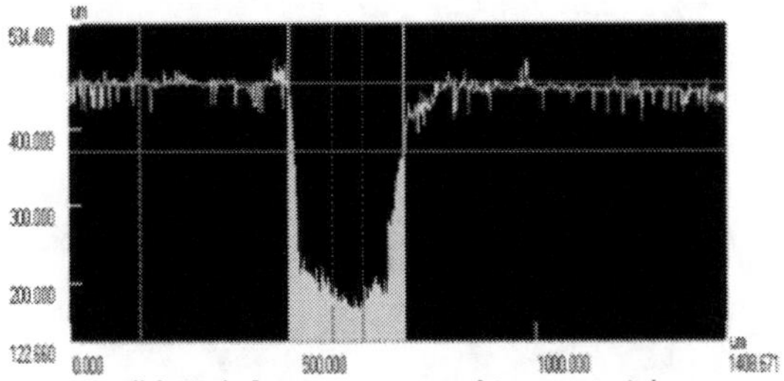
(b) High frequency quenching material

Fig. 12 Section height curves with the testing material

As shown in Fig. 12, the position where the curve trends to go down and the position where the pit area is over are recorded. The integration is conducted between the two regions to obtain the area between the curve and the coordinate axis. The total original area among the section can be obtained according to the average height of the region without water erosion. Then the difference of the two areas is the pit area. The contrast histograms for the pit section dimension of all the specimens are shown in Fig. 13.

The histograms can reflect the huge difference of water erosion resistance performances for different treatment processes. The most serious water erosion appears on the base material specimen, and the pit section area is $0.26313mm^2$ while it is $0.06628mm^2$ for the high-frequency quenching material specimen, which indicates that after high-frequency quenching, the water erosion resistance performance is strengthened 3.97 times relative to the base material. Compared with the base material specimen, the surface treatment technology of high-frequency quenching improves the surface hardness by forming a hardened layer, so that the water erosion resistance performance gets significantly enhanced. The pit section area of stellite-welded specimen is about $0.00239mm^2$, which indicates that the water erosion resistance of stellite-welded specimen is 110 times and 28 times stronger than the base material specimen and the high frequency quenching material specimen respectively.

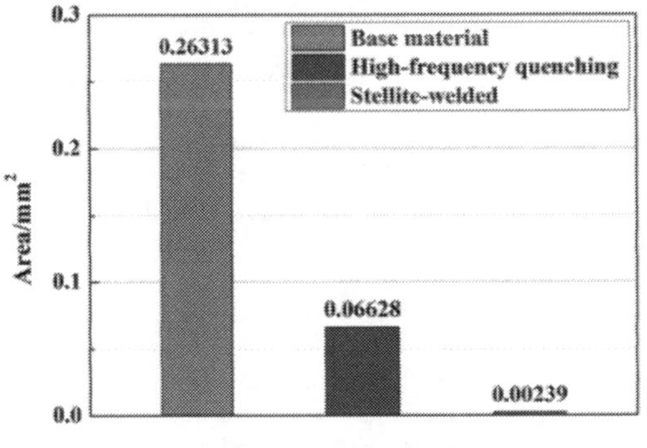

Fig. 13 Water erosion pit area of the testing material specimens

In the present study, the hardened layer of the high frequency quenching material specimen is probably equal to the thickness of the stellite alloy. It is also found that stellite-welded specimen has not only high hardness, but also strong toughness. The time when water erosion pits for stellite-welded specimen appear is just 4 hours later than that of the high-frequency quenching material specimen, however the total mass loss for stellite-welded specimen is extremely small in the 38 hours left. At the same time, the hardened layer of the high-frequency quenching material specimen has been penetrated and water jet has reached the base material, which results in a large of mass and volume loss.

4 Conclusions

The rotary water erosion experiment platform, which is designed and established independently, is applied for the water erosion experimental analysis of the base material specimen, high-frequency quenching material specimen and stellite-welded specimen. The main conclusions are as follows:

1) The water erosion experiment results for three kinds of specimens during 42 hours are obtained and analysed in detail. A detailed analysis of the water erosion characteristics of each material of each stage and the curves of mass loss are obtained. The 2D and 3D morphologies of water erosion pits are acquired.

2) The depth and width data are gained according to erosion pit section curves. It is important to find that the width develops faster than the depth, which indicates that the cutting and scouring effect of lateral jet caused by impact is the main cause of the water erosion phenomenon.

3) It is found that the water erosion resistance performance of stellite-welded specimen is 110 times and 28 times stronger than that of the base material specimen and the high frequency quenching material specimen respectively by contrast.

References

1. Honegger, E. Brown Boveri Rev., 12: 263 (1924)
2. Hoff G, Langbein G, Rieger H. Erosion by Cavitation or Impingement, **408**, 42-69 (1967)
3. Ripken J F. *University of Minnesota, St. Anthony Falls Hydraulic Laboratory* (1962)
4. Elliott D E, Marriott J B, Smith A. ASTM Spec. Tech. Publ, **474**, 127-161 (1970)
5. Rudy S L, Thiruvengadam A. NASA CR-1288 (1971)
6. Bowden F P, Brunton JH. *Proceedings of the Royal Society Series A* (1961)
7. Brunton J H. *Proceedings of the Royal Society Series A* (1966)
8. Jenkins D C, Booker J D, Sweed J W. *Aeronautical Research Council* R&M No.3203 (1961)
9. Toshiyuki S, Masao W, Minori S, et al. Fluid Dynamics Research, **40(7-8)**: 627-636.(2008)

Comparisons of Inter-tablet Coating Variability under Different Tablet Shapes and Filling Levels by Using DEM Simulations

Qiushuang Zhang[1,*], *Hui* Wan[1], *Bin* Jian[1], and *Guoming* Hu[1]

[1]School of Power and Mechanical Engineering, Wuhan University, Hubei District, Wuhan 430072, PR China

Abstract: The purpose of this paper is to investigate inter-tablet coating thickness variability based on different tablet shapes and filling levels. To achieve this goal, the Discrete Element Method (DEM) is applied, and efforts are made as following: For the first step, four different tablet shapes, referred to as Sphere, SRC (Standard Round Concave), Oval and Triangular, are built by importing three (except Sphere) CAD templates to create one or more spherical surfaces. Furthermore, a spray zone in the coating pan is presented to monitor the change of the "coating mass", by virtue of which we calculate the inter-tablet coating variability. And to complete this work, custom contact models and properties are created to describe coating. Then, three different filling levels are added to the simulation. Eventually, the compared results are analyzed between three diverse filling degrees among tablet particles with four discrepant shapes.

1 Introduction

Coating serves a wide variety of functions in pharmaceutical industry, such as avoiding the influences of external factors (moisture, air, light) and gastric acidity to guarantee the bioavailability effectiveness of drugs, masking drugs color, flavors and odors to improve the physical attribution and esthetics level of medicines and controlling the release profile of the drug to regulate its absorption degree. Consequently, controlling inter-tablet coating variability is of great significance for film coated tablets to work well. To succeed in obtaining a qualified batch of tablets, a prescribed coating with a thin film of polymer onto each tablet's surface which means low inter-tablet variability is needed.

Coating variability is divided into two types [1]: the inter-tablet coating variability which represents the coating variability from tablet to tablet, and the intra-tablet coating variability which indicates the variability in the coating mass over the surface of a single tablet. In this work, the inter-tablet variability is of particular notice to research the influential factors in the coating to minimize the inter-tablet variability.

In this paper, the Discrete Element Method (DEM) is used to simulate the coating process so as to track the inter-tablet variability. A spray zone and the spray droplets are modeled to get the coating mass in order to calculate the inter-tablet variability. Besides, an analytical work and a comparative study have been done between tablet particles in various shapes, and between several kinds of filling levels.

2 Background

It is a public consensus that the intra- and inter-tablet variability is two major standards that weigh coating performance. Tobiska and Kleinebudde [2] have done a research about both the intra- and inter-tablet homogeneity by using analysis of variance (ANOVA). While Mueller and Kleinebudde [3] investigated the particle motion in which the tablet velocity and surface time were included in rotating pan coaters by using imaging techniques. In recent years, DEM has been on the center stage as a popular instrument of coating process analysis. As a deluxe simulation tool with faster computation capability and stronger analytical function, it has been used frequently by many researchers in mathematical modeling of the coating effect [4, 5], most of whom having achieved excellent outcomes.

A multitude of scholars have taken tablet shape and pan filling into consideration to get a relatively minor numerical value of inter-tablet variability by means of experimental technique. Pandey and Turton [6] obtained a conclusion by using novel video-imaging techniques: Compared to the spherical-particles, the shaped tablets possessed faster velocity, higher dynamic angle of repose and wavier bed surface. Lower pan loads were required for minor inter-tablet coating uniformity according to Ketterhagen [7]. He also pointed out that the pan loading played a big part in the inter-tablet coating uniformity, while the tablet shape played a small one through a DEM model. Suzzi et al. [8] analyzed the influence of tablet shapes (bi-convex, oval and spherical) and filling degrees on the inter-tablet coating variability in a continuous drum.

Previous works researching inter-tablet coating variability mostly focused on tablet residence time.

* *Corresponding author*: 2016202080018@whu.edu.cn

Kalbag and Wassgren [9] introduced a variable quantity of coating trials to investigate the tablet residence time with a thought that there was a positive correlation between the tablet residence time and coating mass. Kumar and Wassgren [10] used a Peclet number which was defined as the ratio of axial advection rate to the axial diffusive rate to describe the inter-tablet coating variability at the premise of the residence time with a finite variance, showing a strong effect of the residence time variability on the inter-tablet coating variability.

The focus of the current work is the obtainment of the coating mass captured through an algorithm by the aid of API. So DEM simulations for four tablet shapes and three filling levels have been carried out to observe the inter-tablet coating variability in the coating process within a coating pan. A static spray zone and uniform spray droplets are applied to the DEM simulations to seek a relatively objective simulation result.

3 Discrete element method (DEM) simulations

3.1. Geometry model

The simulated tablet coating procedure in this work was operated in a Thomas Engineering Accela-Cota pan with a diameter of 24 inch (61.0 cm), as shown in Fig. 1. The modeled coating drum contained two baffles serving to gather the tablets to the center of the span and four anti-slip bars helping to avoid sliding which arrayed symmetrically in the cylindrical drum. Kalbag et al. [4] have given a full account of the geometry model.

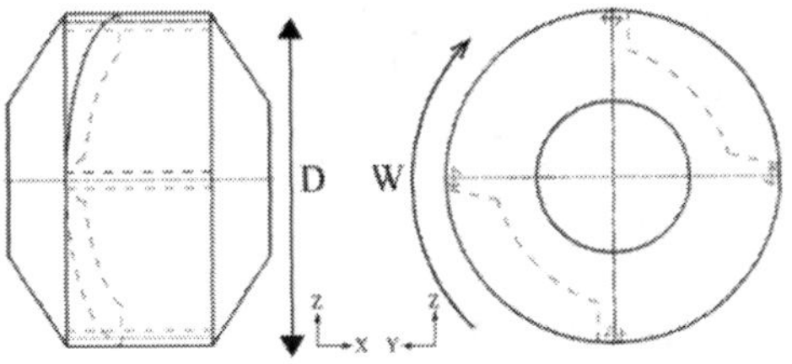

Fig.1. Diagram of the modeled coating drum. Front view (left) and side view (right)

3.2 Tablet model

Four different tablet shapes, including Sphere, SRC, Oval, and Triangular, were created to make out what roles they played in coating homogeneity, as shown in Fig.2. To take on a clear and all-round display, the top views of three CAD templates and a sketch of a sphere particle were presented on the upper side, and the tablet models developed by the glued spheres were displayed on the lower side for every kind of tablet shape.

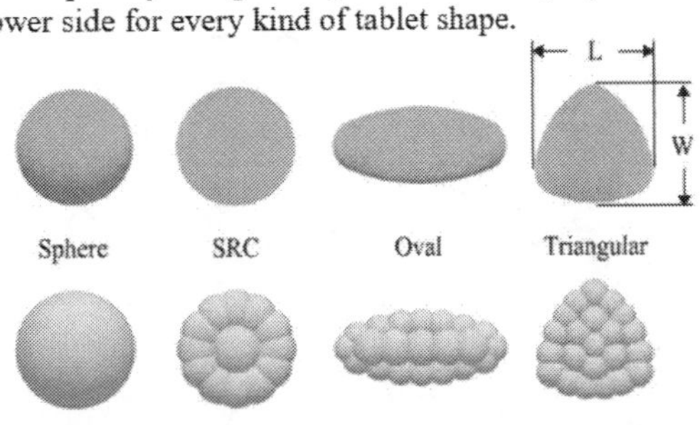

Fig. 2. Four tablet shapes (including Sphere, SRC, Oval, and Triangular) presented in the current work

The glued sphere method was applied in the current work to approach tablets' original shapes to the greatest extent by light of a clump of rigidly bonded spheres [11]. Hence, several CAD templates of tablets were established to provide references for determining the size and position of particles filling into them. The number of the constituent particles in the composite tablet shapes was chosen moderately in consideration of over-damping phenomenon (if too many) and distortion of original shapes (if too few). Otherwise, the theory that the rounder the objective particle is, the fewer spheres are demanded for tablets expression [12] was also demonstrated according to Fig.2.

The detailed information of tablets randomly selected as representatives of differentiated shapes is illustrated in Table 1, in which several measurable parameters are given. Subsequently, the material features of the geometry as well as simulation settings indispensable for mutual contact between particle-steel and particle-particle are shown in Table 2.

3.3 Contact model

To reproduce the coating process, a static plane which is set to be practically vertical to the normal of the tablet bed surface is introduced in, acting as the hypothetical spray zone to produce spray droplets. For convenience, it is assumed that the spray droplets are evenly distributed within the spray zone.

To realize the spraying function, a user defined contact model is created to closely model the interactions between particles to a more accurate extent which the traditional one can not do. The approach primarily considers three contact conditions:

Table 1. Physical properties of four kinds of tablets

Tablet shape	Mass (g)	Density (kg/m3)	Length (cm)	Width (cm)	Thickness (cm)
Triangular	0.520	1300	1.59	1.57	0.47
Oval	0.520	1300	1.61	0.70	0.61
SRC	0.520	1300	1.08	—	0.70
Sphere	0.520	1300	0.85	—	—

Table 2. Primary parameters set for the DEM simulation

Parameter	Value
Physical properties-geometry(steel)	
Pan speed	22 rpm
Poisson's ratio	0.3
Shear modulus	7×10^{10} Pa
Density	7800 kg/m3
Material properties-particle(tablet)	
Poisson's ratio	0.25
Shear modulus	1×10^{6} Pa
Contact interaction(particle-steel/particle-particle)	
Coefficient of restitution	0.5
Coefficient of static friction	0.5
Coefficient of rolling friction	0.01

Spray droplet-spray droplet: The contact is ignored in this simulation, and the program handles with nothing and goes on next step,

Spray droplet-tablet: The mass of spray droplet is transferred to added mass of the tablet once they meet, and the spray droplet is removed at the same time,

Spray droplet-geometry: The collision is addressed in the pattern of the removal of the spray particles in this contact situation.

4 Simulations under different filling levels and tablet shapes

In this work, the inter-tablet coating variability is gauged according to its definition of the coefficient of variation of the coating mass between tablets [13], namely:

$$CoV_{\text{inter}} = \frac{\sigma}{\overline{m}} \tag{1}$$

where $\overline{m}$, the mean coating mass over the total time, is directly obtained from the results data exported from the EDEM software depending on the contact algorithm between tablet particles and spray particles, and σ, the standard deviation, can be obtained using the following formula:

$$\sigma = \sqrt{\frac{1}{N}\sum_{i=1}^{N}(m_i-\overline{m})^2} \tag{2}$$

Where m_i represents the added mass that every individual tablet particle gets resulting from passing through the spray zone, and N represents the total number of tablets generated in the coating drum.

As illustrated in Fig. 3, three different tablets simulations (a) Triangular (14% filling level), (b) Oval (22% filling level) and (c) SRC (28% filling level) are presented. The rectangular section in green color is the hypothetical spray zone, in which spray particles are uniformly generated in a dynamic manner. Triangular, Oval and SRC tablets are presented in a way of template models created by a three-dimensional modeling software according to Fig. 3, for a particle template method is employed in this paper. The colors of tablets indicate a reflection of the amount of coating mass (red, blue and white colors stand for much, medium and little coating mass respectively) monitored by a variable quantity "mass added".

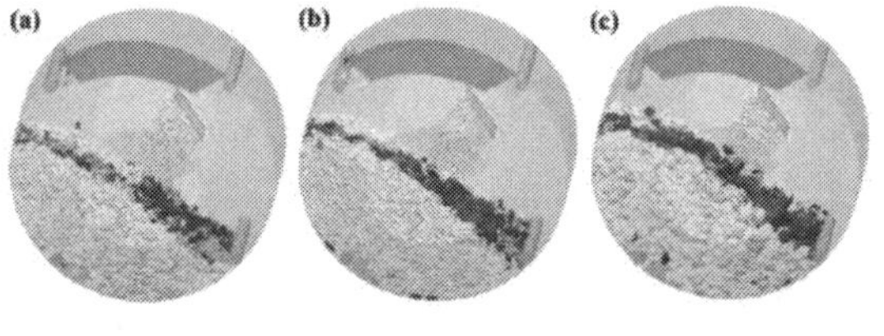

Fig. 3. Three irregular-shaping tablet simulation: (a) Triangular (14% filling level), (b) Oval (22% filling level) and (c) SRC (28% filling level)

5 Results and comparison

Fig. 4 plots the average coating mass for each kind of tablet (Sphere, SRC, Oval and Triangular) under three different filling levels (14%, 22% and 28%). It can be observed that although under different filling levels, tablets of different shapes take on similar features from Fig. 4 (a)-(c). Four kinds of tablets have an approximately linear increasing average coating mass, among which Triangular tablets possess greatest average coating mass due to the bigger surface area comparative to other different-shaping tablets, and the rest three kinds of tablets (Sphere, SRC and Oval) have close values in mean coating mass, but Oval tablets bear the least average coating mass resulting from the smaller surface area.

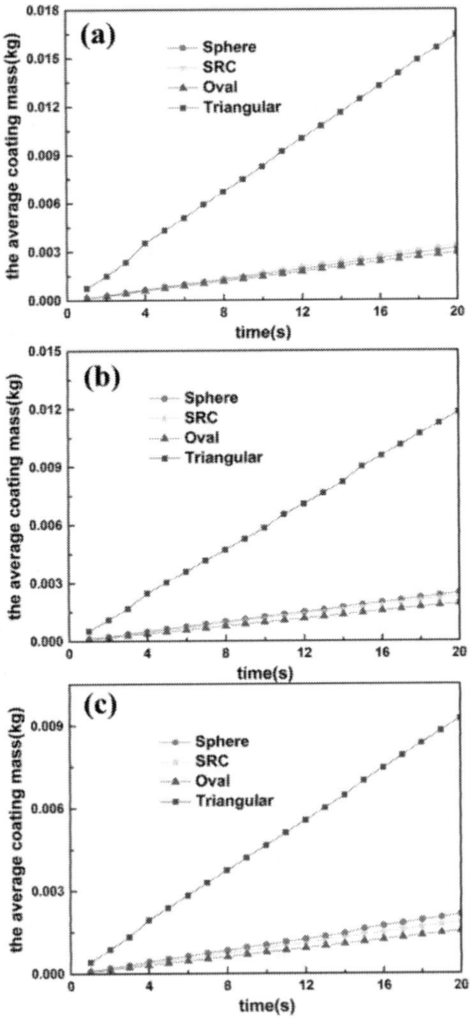

Fig. 4. Average coating mass of Sphere, and another three kinds of particles irregular in shapes, SRC, Oval, and Triangular, under three different filling levels: (a) 14% filling level; (b) 22% filling level; (c) 28% filling level

Fig. 5 indicates the average coating mass under 14%, 22% and 28% filling levels for four kinds of irregular-shaping tablets. Fig. 5 (a)-(d) illustrates that the higher the filling level is, the less the mean coating mass attributing to the added numbers of tablets, and the bigger the digital gap is, the faster the mean coating mass decreases. It also shows that the filling ratio has the greatest influence on Triangular tablets which undulate greatly when filling level changes and the least influence on Sphere tablets which fluctuate relatively evenly on the condition of different filling ratios, and SRC and Oval tablets are impacted at the almost same level.

Fig. 6 depicts the inter-tablet coating variability (CoV_{inter}) for Sphere, SRC, Oval and Triangular tablets versus time curve during simulation process under three different filling levels (14%, 22% and 28%). Fig. 6 (a)-(c) indicates that CoV_{inter} presents a downward trend with the simulation time getting longer no matter what kind of tablets, and finally stabilizes at a value when the time is at

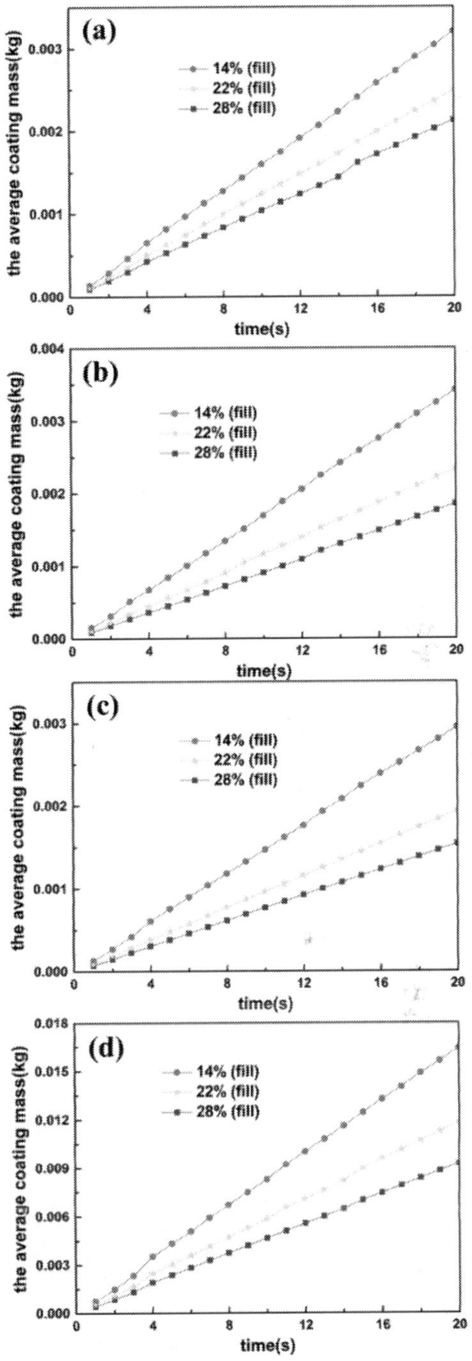

Fig. 5. Average coating mass of particles in four different shapes, (a) Sphere, (b) SRC, (c) Oval and (d) Triangular, under 14%, 22%, and 28% filling levels

18 s. Oval tablets get a greater CoV_{inter} at a low filling level as seen from Fig. 6 (a), and a smaller CoV_{inter} at high filling levels as shown in Fig. 6 (b) and (c) compared with other kinds of tablets, while Sphere tablets display contrary regularities, with the implication that Oval tablets will be a better choice when high filling levels are needed while Sphere tablets may be a reasonable option

https://doi.org/10.1051/matecconf/201823804008

when low filling levels are demanded. Gaining a minimum CoV_{inter} under three filling levels as shown in Fig. 6 (a)-(c), Triangular tablets exhibit a good adaptability to filling ratio, indicating that they have a good comprehensive performance.

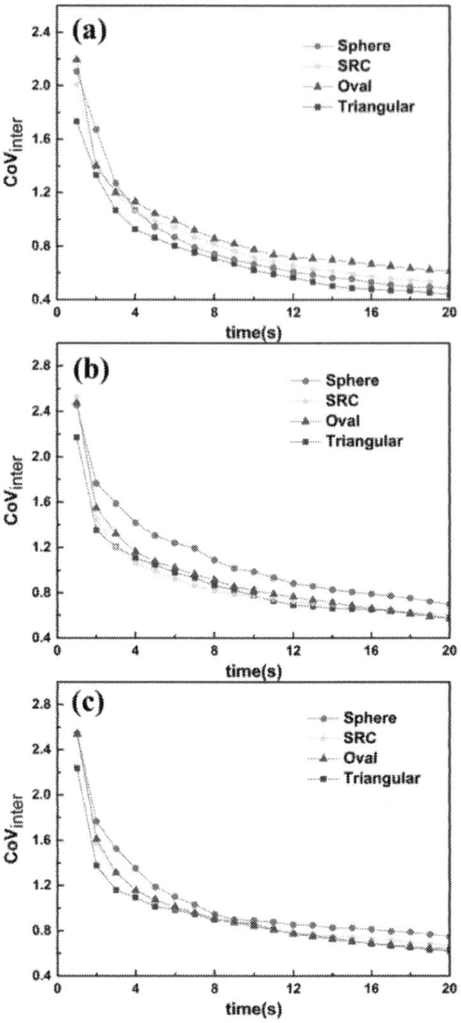

Fig. 6. Inter-tablet coating variability of Sphere, and another three kinds of particles irregular in shapes SRC, Oval, and Triangular, under three different filling levels, (a) 14% filling level, (b) 22% filling level and (c) 28% filling level

Fig. 7 describes the CoV_{inter} versus time curve under 14%, 22% and 28% filling levels for each kind of tablets respectively during the whole simulation course. Contrary to the average coating mass, the higher the filling degree fraction is, the greater the CoV_{inter} becomes, indicating a worse case coating uniformity. All tablets are compliant to this law, except Oval tablets, which obtain the greatest CoV_{inter} at 28% filling ratio, the smallest at 22%, and a medium at 14%, while CoV_{inter} figures are very close as illustrated in Fig. 7 (c), suggesting that Oval tablets are not sensitive to filling levels but filling levels chosen for them should be prudent, because too high or too low filling levels are not fit for them either. The CoV_{inter} values get a small oscillation in terms of Sphere tablets at high filling ratios, an obvious drop emerges at a low filling ratio, and a close parallel situation happens with Triangular tablets, showing that they are both propitious to low filling levels. SRC tablets demonstrate a relatively even change with the filling degree fraction getting higher.

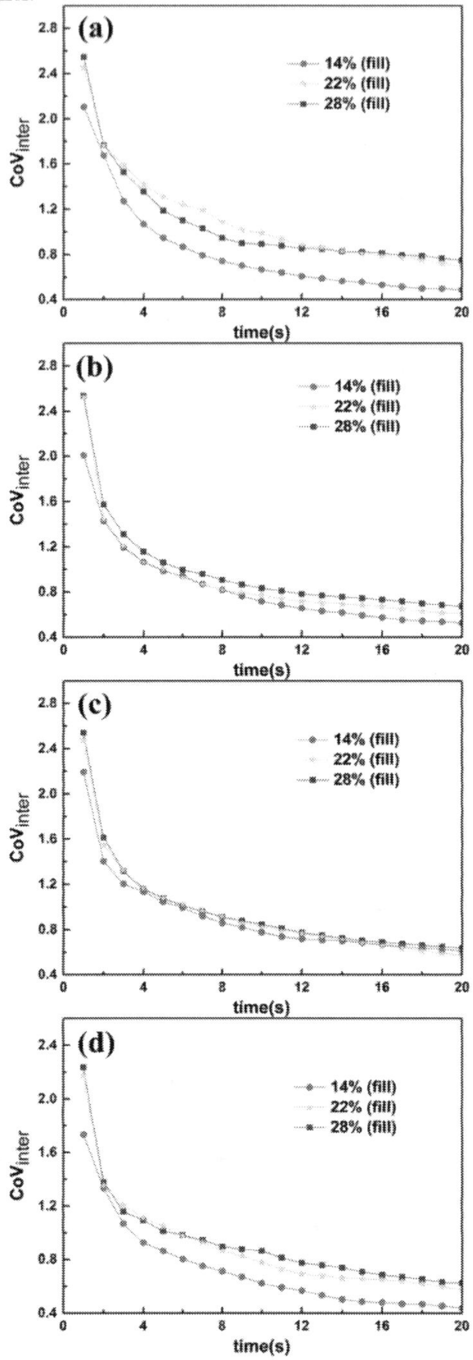

Fig. 7. Inter-tablet coating variability of particles in four different shapes, (a) Sphere, (b) SRC, (c) Oval and (d) Triangular, under 14%, 22%, and 28% filling levels

6 Conclusions

A research of inter-tablet coating thickness variability inspected from two respects, four tablet shapes (Sphere, SRC, Oval and Triangular) and three filling degrees (14%, 22% and 28%), in a revolving coating drum is made in this paper. Based on the special user defined contact model, the average coating mass is obtained, and finally the inter-tablet coating variability is evaluated. Through an analysis from Figs. 4-7, we come to the following conclusions:

The tablet shape acts as a decisive role in inter-tablet homogeneity. The following opinions are all concluded on the premise of comparing with the rest kinds of tablets. Triangular tablets exhibit excellent comprehensive properties, for they not only obtain the greatest average coating mass, but also the smallest CoV_{inter}, which means the highest homogeneity. Sphere tablets have a bad homogeneity in high filling height condition, and a better uniformity at a low filling degree. On the contrary, in terms of Oval tablets, heterogeneity is more evident in situation of low filling levels, and a better uniformity appears in high filling degree fraction.

The filling level also has a crucial influence on the ultimate quality of the coating results. Higher filling height receives a worse coating performance in return, a conclusion in accordance with Suzzi *et al.* [8], in condition, it results in lower mean coating mass. Filling ratio has a greater impact on Sphere and Triangular tablets which gain a noticeably smaller homogeneity in terms of low filling height, and a tiny influence on Oval tablets, whose uniformity changes little with it.

Based on the above, we can draw a conclusion that tablet shape and filling level investigated in this paper are both indispensable considering factors to the inter-tablet coating variability. This work could serve as a reference for a better decision, choosing what kind of tablet and which filling degree. The different coater speeds and other measures investigating the coating effectiveness will be considered in further studies.

Acknowledgments

This research is supported by the National Natural Science Foundation of China (Grant No.51575404).

References

1 B. Freireich, R. Kumar, W. Ketterhagen, Chem. Eng. Sci. **131**, 197-212 (2015)

2 S. Tobiska, P. Kleinebudde, Eur. J. Pharm. Biopharm. **56**, 3–9 (2003)

3 R. Mueller, P. Kleinebudde, Powder Technol. **173**, 51–58 (2007)

4 A. Kalbag, C. Wassgren, S.S. Penumetcha, J.D. Pérez-Ramos, Chem. Eng. Sci. **63**, 2881–2894 (2008)

5 G. Toschkoff, J.G. Khinast, Eur. J. Pharm. Biopharm. **457**, 407-422 (2013)

6 P. Pandey, R. Turton, AAPS PharmSciTech. **6**, E237-E244 (2005)

7 W.R. Ketterhagen, Int. J. Pharm. **409**, 137–149 (2011)

8 D. Suzzi, G. Toschkoff, S. Radl, D. Machold, S.D. Fraser, B.J. Glasser, J.G. Khinast, Chem. Eng. Sci. **69**, 107–121 (2012)

9 R. Mueller, P. Kleinebudde, Powder Technol. **173**, 51–58 (2007)

10 R. Kumar, B. Freireich, C. Wassgren, Chem. Eng. Sci. 125, 144–157 (2014)

11 M. Kodam, R. Bharadwaj, J. Curtis, B. Hancock, C. Wassgren, Chem. Eng. Sci. **64**, 3466–3475 (2009)

12 M. Price, V. Murariu, G. Morrison, *Fourth International Conference on Discrete Element Methods* (Brisbane, Australia 2007)

Preparation and Properties of Polylactic Acid/Modified Starch/POE Degradation Materials

Xiaobo Zhang[1,*]
[1]Yanching Institute of Technology, 065201 Sanhe, China

Abstract. Polylactic acid (PLA) / polyolefin elastomer (POE) / modified starch composites were prepared by melt blending. The composites were characterized by softening point, polarizing microscope, melt flow rate, tensile strength and degradation properties. The results showed that with the increase of modified starch and POE content, phase separation occurred between modified starch and PLA, and the tensile strength decreased compared with pure PLA. Modified starch could increase the crystallinity of PLA to increase the softening point and improve the thermal stability of PLA. When the modified starch and POE content were 15 and 20 parts respectively; the composite material has the best performance. At this time, the softening point was 75°C and the tensile strength could reach 47.7 MPa. The composite materials in different media showed the quality decreased with the prolongation of degradation time, especially in acidic and alkaline conditions, because the degradation rate of hydrolysis was accelerated.

1 Introduction

With the progress and development of society, the environmental pollution problems caused by the wide application of plastics are becoming more and more serious. Polylactic acid (PLA), as a kind of material made from starch and biodegradable, has attracted people's attention. The polyester bond in the polylactic acid molecule has the advantages of high modulus and high strength. It can be formed by blow molding, molding and extrusion processing, but its poor impact resistance, heat resistance and hydrophilicity limit its wide application. [1-3]At present, the main method to improve the disadvantages of PLA is to toughen PLA, including copolymerization, blending and processing modification. Blending modification is not only simple, feasible and low cost, but also the blend system can balance the performance of each component, achieving the effect of complementing each other and obtaining a modified material with good comprehensive performance. The commonly used materials for modified PLA are: polybutylene succinate (PBS), poly (adipic acid)/butylene terephthalate (PBAT), poly-ε-caprolactone and so on. [2, 3]

POE has good toughness, good plasticity, excellent dispersion and weather resistance, [4-6] high transparency. Starch has the advantages of a wide range of sources, low cost and short degradation cycle. [7, 8] In this paper, starch and POE were used as toughening modifiers to modify PLA. The effects of various components on the properties of PLA were studied. The application range of PLA was extended.

2 Experimental

2.1 Materials

Polylactic acid (PLA) (industrial pure, injection molding grade)was purchased from Zhejiang Hai Zheng biological material Limited by Share Ltd.; PEO(industrial pure, injection molding grade) was purchased from Changchun Dadi Fine Chemical Co., Ltd.; Modified starch, (analytically pure) was purchased from homemade; Carboxymethyl cellulose(analytically pure) was purchased from Tianjin Yongda Chemical Reagent Co., Ltd.; Cyclodextrin(analytically pure) and Polyvinylpyrrolidone(analytically pure) was purchased from Tianjin Fuchen Chemical Reagent Co., Ltd.; talcum powder(analytically pure) and glycerin(analytically pure) was purchased from Tianjin Dingsheng Chemical Co., Ltd..

2.2 Experiment process

2.2.1 Preparation of modified starch

The starch is dissolved in water and gelatinized at 85-90 degrees Celsius for 15 minutes. After cooling, acrylic acid, polyvinyl alcohol and initiator are added. The reaction was carried out in a microwave reactor for a certain time, and then silica is added and stirred to obtain a modified starch.

* Corresponding author: zxb_1015@163.com

https://doi.org/10.1051/matecconf/201823804009

2.2.2 Preparation of composite PLA materials

The weighted PLA, POE, modified starch and CMC were blended evenly and then add other additives such as glycerin, and then the co-rotating twin screw extruder (SHJ-20, length-diameter ratio 40:1, Nanjing Jeente Electromechanical Co., Ltd.) was used for granulation. The extrusion temperatures are 135 ° C in zone I, 150 ° C in zone II, 170 ° C in zone III, 170 ° C in zone IV and 135 ° C in zone V (from feed inlet to outlet). Finally, the obtained pellets were subjected to hot pressing in the range of 150-230 ° C to prepare a PLA modified material.

2.3 Experimental instrument and test

Softening Point: According to GB/T 1633-2000 "Determination of softening point temperature of thermoplastic Vicat", the temperature of thermal deformation and Vicat softening point was measured by RHWK-300B type thermal deformation and Vicat softening point tester of Beijing Crown Testing Instrument Co., Ltd.; Polarization microscope test: the sample was melted at 300℃ for 3 min, pressed into 10 μm thin sheets, and then quenched, the crystal morphology of the material was observed under Chongqing UP203i polarization microscope; Tensile performance test: the sample size is 3cm×1cm, and the mechanical tensile properties are tested by Shenzhen New Sansi E44 electronic universal testing machine (speed 50mm/min); Degradation performance test: The composite material was prepared into a film by casting method, then the film was placed in the solution, the sampling interval was 5 days, and the weight loss rate of the sample was measured.

3 Results and discussion

3.1 Softening point of composite

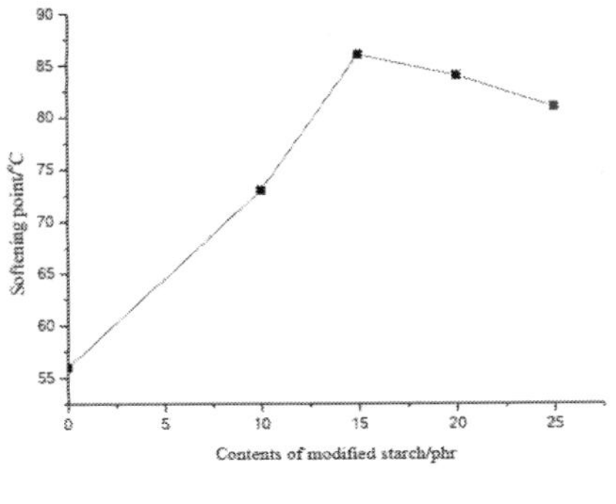

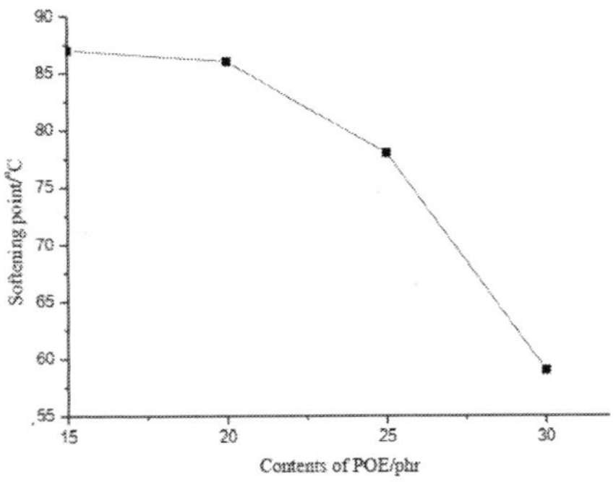

Fig. 1. Different starch and PEO ratios softening point diagram

It can be seen from Fig. 1 that when the quality of the modified starch is increased from 10 parts to 15 parts, the softening point of the blending system increases, but from 15 parts to 20 parts, the softening point temperature begins to decrease. It is known that when the starch is 15 parts, the softening point reaches the best condition of this experiment. With the addition of starch, the crystallinity of PLA was promoted, but after POE is excessive, the heat resistance of the system is poor due to its poor heat resistance. PEO itself plays a toughening effect, therefore its addition leads to The degree of crystallization decreased. When the amount of POE increased to 30 parts, the softening point of the system dropped drastically to 58 degrees. Therefore, in order to improve the heat resistance of the system, the amount of PEO must be strictly controlled.

3.2 Microstructure of composite materials

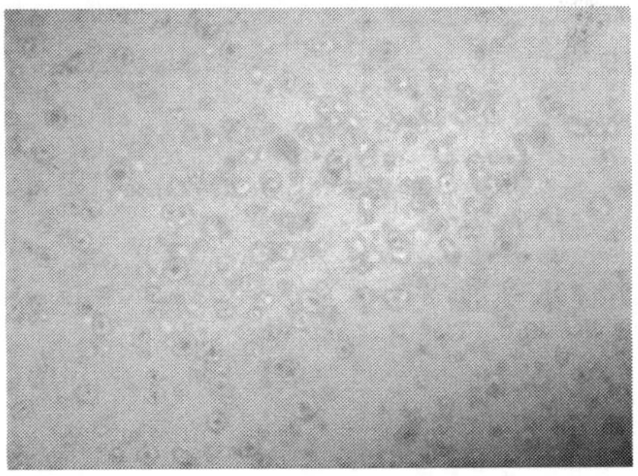

PLA: starch=20:10

https://doi.org/10.1051/matecconf/201823804009

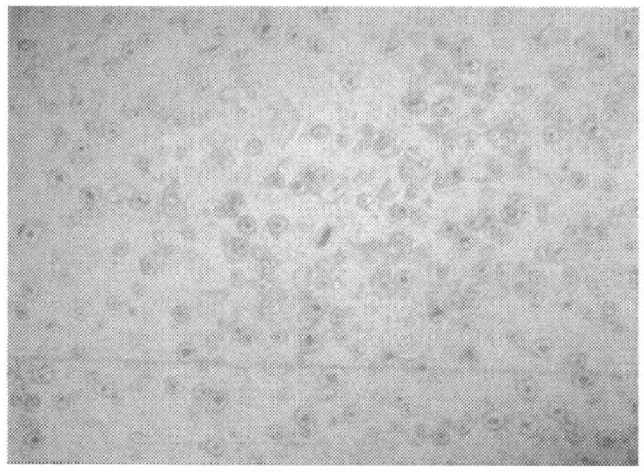

PLA: starch=20:15

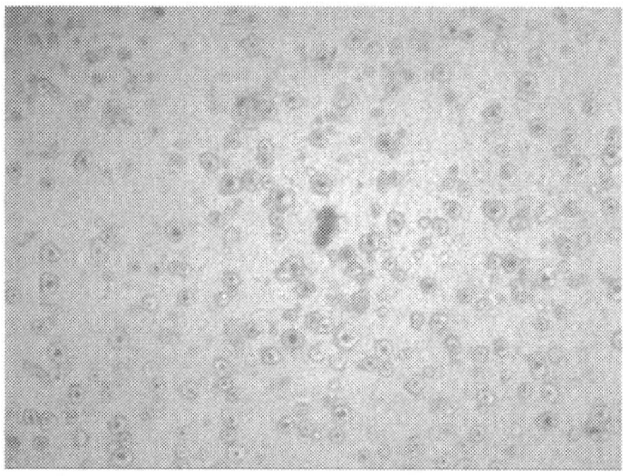

PLA: starch=25:15

Fig. 2. Polarizing micrographs of starch modified PLA composites with different proportions

Fig. 2 is a polarizing microscope image of the composite material with POE content of 25 parts and different modified starch contents. PLA is a semi-crystalline polymer. As can be seen from Fig. 2, there is obvious particle distribution on the surface of the composites, and the PLA is the continuous phase and the starch is the packed phase in the composites. At 140 ° C crystallization conditions, the composite material has crystallized completely, because the interaction force with the matrix PLA restricts the movement of the molecular chain, and forms a rigid interface between them. With the starch content increases, the phenomenon becomes more obvious. As a result, the interfacial compatibility of the composite material is worse, the number of crystals and the size increases. However, the particles are closely distributed and uniformly dispersed, showing higher melt dispersibility with PLA, and the initial and final temperatures of crystallization are much higher than pure PLA. At this time, PLA modified with POE and modified starch has been transformed from the brittle material to the ductile material.

3.3 Analysis of melt flow rate (MFR) of blend system

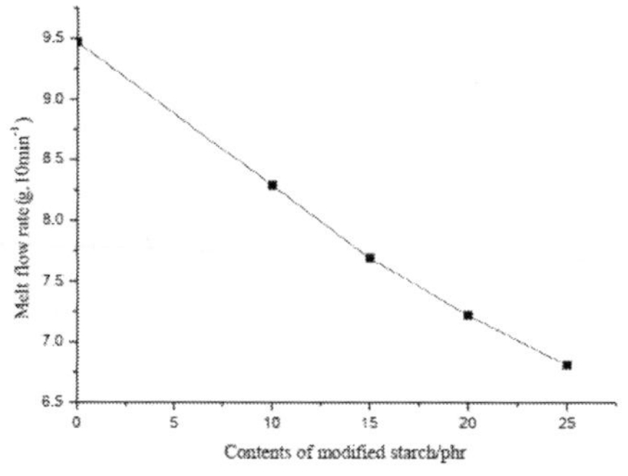

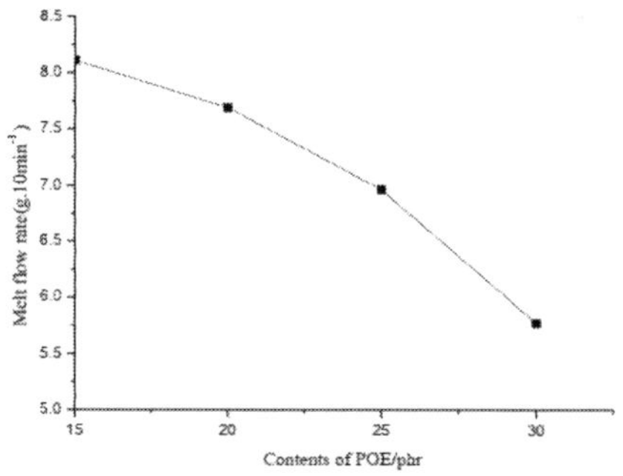

Fig. 3. The melt flow rate of different systems

It can be seen from Fig. 3 that with the increase of POE content in starch and elastomer MFR shows a downward trend, and the effect of POE was greater than that of starch. It shows that the processing performance of the system decreases with the increase of elastomer content. This is because the melt flow rate of the elastomer itself is low. In the blending, the elastomer is dispersed in the PLA melt with tiny droplets. The addition of elastomer increases the internal friction resistance of the blend system, which makes the relative movement of polymer chains difficult and dissipates additional energy, thus reducing the fluidity of the blend system. It is manifested the viscosity increased and the MFR value decreased. Therefore, the proper proportion should be chosen to facilitate the processing of the blends.

3.4 Mechanical properties of composite materials

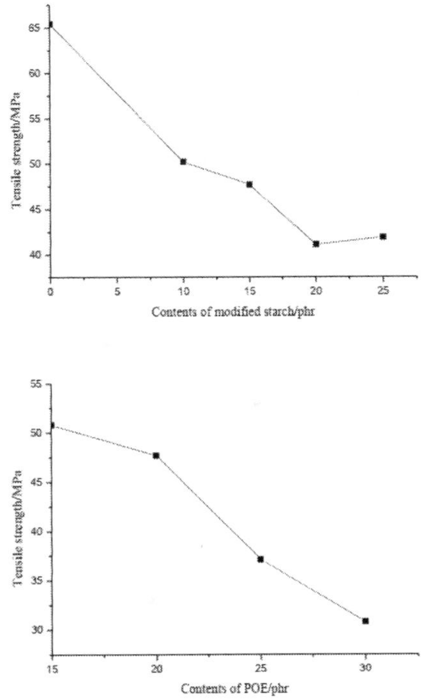

Fig. 4. The mechanical properties of PLA and PLA blends

Fig. 4 shows the mechanical properties of PLA blend with different contents of modified starch. The tensile strength of pure PLA is 65.4 MPa. The tensile strength of the composites decreased first and then increased with the increase of modified starch content. When the starch content is small, the tensile strength of the material is mainly derived from polylactic acid. But because both PLA and starch are brittle materials, phase separation occurs between the modified starch and PLA, the interfacial adhesion is deteriorated, and the tensile strength decreases. With the increase of starch content, glycerol makes PLA-COOH and starch-OH cross-linked excessively, restricting the movement of the interface between PLA and starch, thus increasing the small molecule chain forging of PLA, accelerating the crystallization and improving the crystallinity, so the tensile strength of PLA composites increases.

With the increase of POE content, the tensile strength remains basically unchanged at first and then decreases. At higher POE content, the tensile strength remains at 30 MPa. The addition of POE elastomers has a toughening effect on PLA blends. The addition of POE elastomer weakened the interaction force between PLA macromolecules. However, due to the low rigidity of POE itself, when the amount of POE was too much, the POE forming the elastic toughening phase microspheres in the PLA matrix increases the POE phase microspheres with the increase in content, which can not further strengthen the toughness, and the tensile strength decreases significantly.

3.5 Degradation properties of composite materials

The degradation performance of the composites is shown in Fig. 5. It can be seen from Fig. 5 that in the acid, alkali and neutral medium, the quality of the composite material tends to decrease with the extension of time. The modified polylactic acid can be degraded in acid, alkali and neutral medium, and its degradation rate is slowed down in turn, especially in alkaline media. This is because the side chain of polylactic acid can be subjected to deesterification in alkaline, which is more easily hydrolyzed than under acidic conditions. Moreover, the modified starch has certain hydrophilicity, and its addition can form a phase-separated blend with PLA, which can accelerate the hydrolysis of the sample.

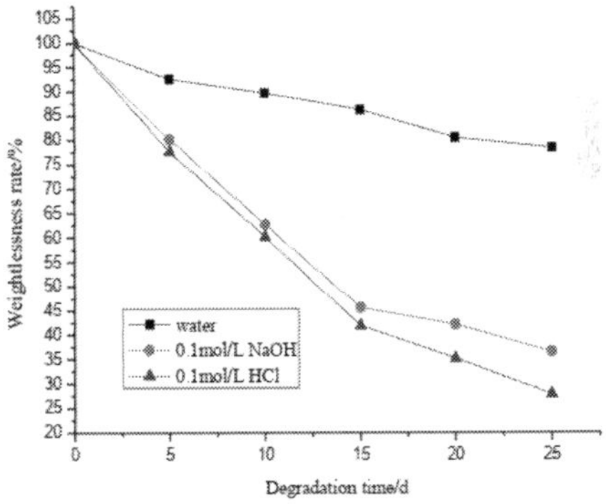

Fig. 5. Degradation curve of composite materials in different medium

4 Conclusions

The composite materials of modified starch and POE modified PLA were prepared by melt blending method. Modified starch and POE can improve the crystallization of PLA to some extent. Modified starch, POE and PLA can be effectively bonded, which improves the tensile strength and heat resistance of the composite, but the flow performance decline slightly. When the content of modified starch and POE is increased at 15 and 20 parts respectively, the tensile strength can reach 47.7 MPa and softening point is 75℃. The weight loss of the composite materials in different media is obviously different, and the degradation rate in alkaline condition is much faster than that in acidic and neutral conditions.

References

1. K. W. Kim, S. I. Woo, Macromol Chem Phys. **203**,2245-2250(2002)
2. K. Das, D. Ray, I. Banerjee, J Appl Polym Sci. **118**,143-151(2010)

3. H. L. Wang, M. G. Sheng, T. J. Shi, Polym. Mater.Sci. & Eng. **20(6)**, 20-23(2004)
4. L. Fang, J. Li, R.R. Qi, Engineering Plastics Application., **37(6)**,5-8(2009) (In Chinese)
5. F. Du, China plastics. **25(11)**,39-41(2011) (In Chinese)
6. H. T .Oyama, Polymer. **50**, 745-751 (2009)
7. D. Garlotta, W. Doane, R. Shagren, J.Appl.Polym.Sci. **88(7)**, 1775-1786(2003)
8. O. Martin, L. Averous, Polymer. **42(14)**, 6209-6219(2001)

Research on the Steel for Oil and Gas Pipelines in Sour Environment

Xiaodong Shao [1, 2*]

[1]State Key Laboratory for Performance and Structure Safety of Petroleum Tubular Goods and Equipment Materials, No. 89 Jinyeer Road, Xi'an 710077, China
[2]CNPC Tubular Goods Research Institute, No. 89 Jinyeer Road, Xi'an 710077, China

Abstract. In recent years, there are more and more oil and gas fields containing H_2S corrosive media, and it is imperative to develop oil and gas transmission pipes in acidic environment. The research progress of steel for oil and gas pipelines in acidic environment is reviewed from the aspects of H_2S corrosion mechanism and influencing factors in pipeline service under acidic service conditions. At the same time, the development status of sour-resistance corrosion oil and gas transmission steel tubes at home and abroad is introduced. Pipeline steel is further explored in terms of SSCC-resistance research and improvement of corresponding standards.

1 Introduction

Oil and gas pipeline transportation is characterized by high efficiency, economy and safety, and is the main form of long-distance transportation of oil and gas [1]. The development of pipelines tends to be large-caliber and high-pressure, the service conditions of pipelines are becoming more and more demanding, the transportation medium is complex, and many pipelines need to cross densely populated areas or deserts, cold regions, etc., which puts forward the performance of pipelines. High technical requirements, especially the requirements of corrosion resistance, Academician Li Helin pointed out that anti-hydrogen induced cracking (HIC) pipeline steel and steel pipe is one of the development directions of high performance oil and gas transmission steel pipe. Corrosion is a key factor affecting the reliability and service life of pipeline transportation systems. It can not only cause perforation, but also cause oil and gas leakage. It may also cause an explosion, threaten personal safety, pollute the environment, and the consequences are extremely serious. According to the National Transportation Safety Administration, 74% of the leaks in the US gas pipelines and gas gathering pipelines are caused by corrosion, and H_2S corrosion is one of the main forms of pipeline corrosion. According to the American Society of Corrosion Engineers (NACE), in natural gas containing water and H_2S, when the partial pressure of H_2S in the gas is equal to or greater than 0.0003 MPa, it is called an acidic environment.

Since the 1970s, the development conditions of oil and natural gas have changed significantly in various countries. At present, although the purification process has been carried out before the natural gas transmission, the natural gas to be transported into the gas pipeline does not meet the requirements of dry and purified natural gas. H_2S in the pipeline. In-pipe corrosion caused by the presence of water is still inevitable. In recent years, accidents caused by pipe H_2S corrosion have occurred, resulting in huge economic losses and environmental pollution. Therefore, solving or mitigating H_2S corrosion of oil and gas pipelines has very important economic and social benefits. According to the development needs of Chinese oil and gas industry, it is imperative to develop Chinese own acidic service conditions to transport steel pipes. Foreign countries have started research in this area earlier, and there are specialized agencies to study the transportation of steel pipes in an acidic environment. Although domestic work in this area started late, it also carried out a series of research work. The author mainly reviews the research progress of steel for acid and oil pipelines from the corrosion mechanism and influencing factors of H_2S in acidic environment pipelines, and introduces the development of oil and gas pipelines in acidic environment at home and abroad. Research direction.

2 H_2S corrosion mechanism of pipeline steel

The corrosion of pipeline steel in acidic environment mainly includes hydrogen induced cracking and sulfide stress corrosion cracking (SSCC). At present, the research on H_2S corrosion of pipeline steel is mainly focused on HIC. Pipeline steel in the oil-rich environment rich in H_2S in the state of no stress or non-tensile stress, the hydrogen generated by the corrosion into the steel to form a crack called HIC, generally refers to hydrogen-induced bubbling (surface crack) and hydrogen-induced step cracking (Internal crack). SSCC

*Corresponding author: shaoxd@163.com

means that the hydrogen atoms generated by H_2S penetrate into the interior of the steel, dissolve in the crystal lattice, cause brittleness, and form cracks under the action of external tensile stress or residual stress. SSCC is in sensitive materials, acidic environment and stretching. The three conditions of stress occur together [2]. When transporting acidic oil and gas, the inside of the pipeline is exposed to H_2S. Under the joint action of stress and H_2S and other corrosive media, SSCC often occurs in the pipeline.

3 Research on the HIC of pipeline steel

3.1 Influence of the environment on HIC

The results show that when the partial pressure of H_2S is 0.1~0.5MPa, the corresponding hydrogen permeability is higher, which means that the sensitivity of HIC is greater [3]. As the pH decreases, the HIC sensitivity of the pipeline steel also increases. When CO_2 is dissolved in water to form carbonic acid, hydrogen ions are released, which lowers the pH of the environment and increases the sensitivity of HIC. Cl^- accelerates the corrosion rate in the range of pH 3.5~4.5, and the HIC sensitivity increases [4]. When the temperature is around 30°C, the additive effect of hydrogen adsorption and diffusion is the largest, that is, the HIC sensitivity is the strongest, below 30°C, the hydrogen diffusion rate and activity decrease, and it is difficult to aggregate active hydrogen above 30°C.

3.2 Influence of composition and structure on HIC

The influence of pipeline steel composition and structure on HIC is the most in-depth study, and the results are also the most, which can be summarized into three aspects.

(1) Chemical composition. The elements that are more sensitive to HIC are carbon, manganese, sulfur, phosphorus, calcium, copper, and molybdenum. As the carbon content increases, the HIC sensitivity increases, and the increase in carbon content and carbon equivalent in the pipeline steel causes the steel to form the martensite structure most sensitive to hydrogen bubbling during hot rolling, thus reducing the carbon content and carbon equivalent. It can improve the H_2S-resistance corrosion performance of pipeline steel. Sulfur can promote the occurrence of HIC, which is the most easily nucleated site of HIC with manganese-generated MnS. Baosteel's research indicates that the addition of calcium can change the form of sulfide inclusions, making it a dispersed spheroid, thereby improving the HIC-resistance ability of steel. The test found that the best effect is when the calcium-sulfur content ratio is greater than 1.5. Adding an appropriate amount of manganese to the pipeline steel can improve the hardenability of the steel, cause solid solution strengthening, and compensate for the decrease in strength caused by low carbon. The segregation of manganese and phosphorus can cause the formation of band-like structures sensitive to HIC, thus increasing the manganese content., will lead to more banded tissue generation, which increases HIC sensitivity. The effect of copper on HIC is obvious. In NACE B solution, HIC sensitivity decreases significantly with increasing copper content. However, in the H_2S environment with pH<4.5, the passivation film of copper is no longer formed, and at this time, the copper prevents the effect of HIC from disappearing. The addition of molybdenum can lower the phase transition temperature, inhibit the formation of massive ferrite, promote the transformation of acicular ferrite, improve the strength of steel, reduce the ductile-brittle transition temperature, and improve the HIC-resistance ability. The addition of microalloying elements such as bismuth, vanadium and titanium to pipeline steel can effectively prevent the growth of austenite grains, refine grains and enhance the corrosion resistance of pipeline steels [5,6]. Due to the influence of chemical composition on the HIC resistance of pipeline steel, the chemical composition of different steel grade HIC-resistance oil and gas transmission steel pipes has clear technical requirements in ISO 3183.

(2) Microstructure. Sodium thiocyanate was selected as the chromogenic reagent in the proposed method. The effect of the chromogenic reagent was examined by measuring the absorbance of solution containing certain amounts of molybdenum and variable amounts of sodium thiocyanate. It was found that 10ml of $100.0mg \cdot mL^{-1}$ sodium thiocyanate solution sufficed to complex the amounts of molybdenum taken, with higher concentrations the absorbance was essentially constant. Ten millilitres of $100.0mg \cdot mL^{-1}$ sodium thiocyanate solution were recommended as a suitable amount of chromogenic reagent. The thermodynamically balanced and stable fine grain structure is the ideal structure for HIC-resistance. For medium and low-strength pipeline steels, HIC is prone to low temperature conversion of manganese, phosphorus and other elements in the banded pearlite structure and the center of the plate thickness. In particular, in order to improve the strength of pipeline steel, the addition of many alloying elements promotes the formation of low temperature conversion microstructure, which provides a convenient place for the occurrence of HIC [7]. With the increase of the microstructure of steel, the HIC-resistance performance decreases gradually, and the HIC-resistance performance decreases with the increase of martensite structure. With the increase of the band structure level, the HIC-resistance performance of pipeline steel is gradually reduced. Yin Guanghong *et al* found that the HIC-resistance performance of pipeline steel will be significantly enhanced after normalizing heat treatment.

(3) Non-metallic inclusions. The morphology and distribution of non-metallic inclusions affect the HIC resistance of pipeline steel. The non-metallic inclusion interface is a strong hydrogen trap [8]. On the interface between the pipeline steel and H_2S, the hydrogen ions generated by the electrochemical reaction are trapped in the hydrogen trap after the electrons are converted into hydrogen atoms at the anode. When the hydrogen pressure rises to a certain value, it is in the non-metal. Cracks occur at the interface of the inclusions, and

plastic deformation occurs at the crack tip. In pipeline steel, the higher the content of non-metallic inclusions, the smaller the threshold value of hydrogen concentration required for HIC, that is, the stronger the HIC sensitivity [9]. In addition, in the same pipeline steel, the coarser the grain, the worse the HIC-resistance performance [10]. Some scholars have pointed out that the hardness of the material also has a great influence on HIC. In the medium and low strength steel, the higher the strength and the greater the hardness, the greater the HIC sensitivity.

4 Research on the SSCC of pipeline steel

4.1 Influence of the environment on SSCC

The lower the pH of the environment, the greater the hydrogen concentration and the stronger the sensitivity of the SSCC. At the same time, the amount of hydrogen permeation in the pipeline steel increases with the increase of H_2S concentration, and the SSCC sensitivity of the pipeline steel is greater [11], the high-strength steel SSCC has a most sensitive temperature, near this temperature, occurs The SSCC is the most likely.

4.2 Influence of composition and structure on SSCC

The composition and organization of pipeline steel is another major factor affecting pipeline steel SSCC.

(1) Chemical composition. Regarding the influence of carbon, different scholars have different opinions, but regardless of the explanation, the lower the carbon content, the lower the sensitivity of the SSCC of pipeline steel. Du Shiyu and other research through the domestic X70 pipeline steel, pointed out that the main chemical elements affecting its SSCC are manganese and phosphorus content, not carbon content. Molybdenum is a beneficial element, especially in the case where high strength of the material is required, a small amount of molybdenum is added to form carbides, so that the amount of solid solution carbon is reduced, and thus the performance against SSCC is improved. The role of chromium has two different perspectives, but it is beneficial for improving general corrosion resistance. The addition of titanium and boron improves the hardenability and also has a certain effect on improving the resistance to SSCC [12,13]. Aluminum is a beneficial element, especially for materials that add both chromium and aluminum.

(2) Microstructure. The microstructure of pipeline steel has a great influence on the sensitivity of SSCC. Martensite is generally considered to be more sensitive than pearlite and austenite. Xie Guangyu *et al* studied the SSCC behavior of X70 grade pipeline steel with different microstructures by three-point bending method. The results show that the uniform fine acicular ferrite microstructure has excellent SSCC resistance. Zhao Mingchun *et al* studied the SSCC-resistance behavior of different microstructure pipeline steels. The results showed that the acicular ferrite microstructure had the best performance against SSCC, the ultrafine ferrite structure was the second, and the microstructure of ferrite pearlite. Worst. Mansour *et al.* [14] studied the SSCC behavior of X100 pipeline steel and pointed out that with the increase of the microstructure of steel, the performance of SSCC decreased. In general, the higher the strength level of pipeline steel, the greater the sensitivity to SSCC, the highest hardness value of SSCC does not occur between 20~27HRC, the higher the hardness, the lower the critical stress value and the fracture time. In addition, due to a series of physical and chemical changes in the weld during the welding process and some weld defects, the ability of the welded joint to resist SSCC is reduced relative to the base metal. The SSCC resistance of the pipeline steel can be improved by eliminating the residual stress annealing treatment. Tao Yongqi *et al.* showed that different weld matching in H_2S-containing media resulted in different corrosion resistance of pipeline steels, with the increase of cold deformation, the resistance of pipeline steel to SSCC was reduced.

5 Development of H_2S-resistance corrosion pipeline steel

Some scholars believe that there is a difference and a relationship between the HIC and SSCC of pipeline steel. It can be considered that the SSCC crack of pipeline steel is a hydrogen-induced cracking stress corrosion. If the HIC of the pipeline steel can be controlled, the necessary conditions for the occurrence of SSCC H_2S corrosion or material sensitivity are not available, so that the SSCC crack of the pipeline steel can be controlled. Therefore, the development and evaluation of H_2S-resistance corrosion transport steel pipes at home and abroad mainly focus on the HIC-resistance performance of steel. Pipeline steels used in acidic environments are characterized by high toughness and resistance to HIC properties that depend on the high purity of the steel.

In order to meet the strength, weldability and manufacturability, it is very important to control the process parameters of the HIC pipeline steel, such as composition design, smelting, continuous casting, rolling, etc., in order to ensure a uniform microstructure. Through the research on the corrosion mechanism and influencing factors of H_2S, the measures to improve the HIC resistance of pipeline steel have formed a consensus, using concentrate and high-efficiency iron pretreatment and re-refining outside the furnace to improve the purity of steel. Calcium treatment is carried out while reducing sulfur content, electromagnetic stirring is carried out during steel smelting and continuous casting, light reduction technology is adopted in continuous casting process, multi-stage controlled rolling and forced accelerated cooling process are used to limit band structure, reduce carbon content, control manganese content, add copper and nickel, fully refine the grains by microalloying and controlled rolling processes.

At present, the general method adopted for the evaluation of steel HIC-resistance ability is the standard test method NACE TM0284 issued by NACE, including three evaluations of crack sensitivity rate (CSR), crack length ratio (CLR) and crack thickness ratio (CTR). parameter. Since the 1990s, the HIC-resistance steel pipes supplied by foreign countries are mainly concentrated in the X65 steel grade. The HIC-resistance oil and gas transmission steel pipes produced by the European steel pipe companies have accounted for more than 30% of their sales, the X70 steel pipes with HIC-resistance have also been successfully developed. It is used on a pipeline in Mexico. Some countries have developed X80 steel pipes resistant to HIC, but they have not been supplied in bulk and are still under further study. Sumitomo Metal Industries Co., Ltd. has developed X80 steel grade pipe steel with excellent HIC resistance and steel pipe made of this steel. For X70 and below steel grade HIC-resistance steel pipes, there are clear requirements for HIC resistance evaluation in the world: CSR≤2%, CLR≤15%, CTR≤5%. Although the research and development of domestic HIC-resistance pipeline steel has just started, it has also achieved some results. According to the above design principles, China has developed HIC-resistance pipeline steel including X80 steel grade. Pangang Group Chengdu Iron and Steel Co., Ltd. is at the domestic leading level in the research and development of anti-acid corrosion seamless steel tubes, with mass production capacity. The submarine pipelines developed by the company are resistant to acid corrosion X56, X60 and X65 steel grade series seamless steel tubes. The microalloyed component design and the corresponding process technology route solve the technical problems of smelting, rolling and heat treatment processes, and the product indicators meet the requirements of the corresponding standards. Baoji Petroleum Pipe Co., Ltd. also developed a spiral submerged arc welded pipe for X52 H_2S-resistance corrosion gas gathering pipeline. Laiwu Iron & Steel Co., Ltd. has developed a manufacturing method for HIC-resistance pipeline steel. The physical and chemical properties of the product have reached X46 and X52 steel grades, which can realize multi-purpose steel. The X52, X60 and X65 grade HIC pipeline steels developed by Baosteel have high strength and high toughness and good HIC resistance. The main properties are listed in Table 1. X65 grade pipeline steel also has better resistance to dynamic tearing. Performance and higher impact toughness, which helps to ensure the safety of high pressure gas pipeline operation. Wugang developed a large-thickness and high-performance HIC-resistance steel for the Sichuan Puguang gas field, and formed mass production, which successfully replaced imported products. Baosteel is at the domestic leading level in the development of X80 steel grade HIC-resistance pipeline steel. It adopts lower carbon, phosphorus and sulfur content in the composition design, and at the same time, in order to make up for the strength reduction caused by the decrease of carbon content, proper addition is beneficial to the HIC-resistance performance. Alloying elements such as copper and nickel. In order to obtain a microstructure with less segregation, increase the niobium content and low carbon content, NbC particles are obtained, the microstructure is refined to increase the recrystallization termination temperature, and the tissue segregation is reduced, thereby ensuring strength and toughness and ensuring HIC-resistance performance. The method also adopts the pipeline steel composition and process design with "ultra-low carbon acicular ferrite" as the structural feature to eliminate the banded pearlite structure in the steel to ensure the HIC-resistance performance of the pipeline steel. The HIC-resistance performance of the product was evaluated as CSR≤0.01%, CLR≤0.09%, and CTR≤0.02%.

6 Conclusions

With the advancement of metallurgical technology, after years of development and production, the output of pipeline steel in China has increased year by year.

Table 1. Properties of HIC-resistance pipeline steel produced by Baosteel

Grade	Specification mm	$R_{t0.5}$ MPa	R_m MPa	A %	A_{KV} (-30℃) / J			NACE TM0284 A Solution
					Tube	Heat affected zone	Weld	
X52	Φ508×11.9	425	546	44	375	239	198	No crack
X60	Φ711×8.0	476	598	41	208	180	178	No crack
X65	Φ508×8.7	479	599	39	388	196	192	No crack

The products have been applied to a series of major long-distance oil and gas pipeline projects such as the West-East Gas Pipeline Project. It has been able to mass produce X65 pipeline steel resistant to HIC, X80 steel grade HIC-resistance pipeline steel has also been developed. However, it must be recognized that the research and development of Chinese HIC-resistance pipeline steel has just started, and there is still a certain gap with the advanced level of foreign countries.

With Chinese energy structure adjustment and environmental protection demand, the demand for natural gas has been increasing, and natural gas development conditions have also undergone significant changes. Acidic oil and gas fields have been rapidly developed. Therefore, in the future, it is necessary to further strengthen the research and development of steel for oil and gas pipelines in acidic environments.First of all, although there is a relationship between HIC and SSCC of pipeline steel, the research on pipeline steel HIC can not completely replace the research on SSCC performance of pipeline steel. At present, there are some researches on the influencing factors of SSCC, but the lack of SSCC is unique. The study of the key factors affecting tensile stress should strengthen the research work in this area and further develop the development of targeted SSCC-resistance pipeline steel. Secondly, in order to better study the HIC generated by pipeline steel in the actual acidic environment, the standard evaluation test of NACE TM0284 should be extended and improved through relevant experimental research. NACE has published the corresponding standard test method NACE TM0177 for evaluation. The material's ability to resist SSCC, but there is no clear technical parameter requirement in the world to measure the anti-SSCC ability of steel, and should be further explored and improved. Finally, there are few studies on the influence of H_2S corrosion products on the stress corrosion rate of hydrogen sulfide, and the H_2S corrosion products have some influence on the HIC crack and SSCC crack growth rate, and then the crack growth rate model can be predicted. And the establishment of difficulties.

Acknowledgments

The authors gratefully acknowledge the *State Key Laboratory for Performance and Structure Safety of Petroleum Tubular Goods and Equipment Materials, CNPC Tubular Goods Research Institute*, and *China National Quality Supervision, Testing and Inspection Center of Oil Tubular Goods*.

References

1. E.J. Omonbude, The transit oil and gas pipeline and the role of bargaining: a non-technical discussion, Energy Policy, **35**, 6188 (2007).
2. L. Zhang, X. Li, C. Du, Effect of applied potentials on stress corrosion cracking of X70 pipeline steel in alkali solution, Materials & Design, **30**, 2259 (2009).
3. W.K. Kim, S.U. Koh, B.Y. Yang, Effect of environmental and metallurgical factors on hydrogen induced cracking of HSLA steels, Corrosion Science, **50**, 3336 (2008).
4. J.L. Gonzalez, R. Ramirez, J.M. Hallen, Hydrogen-induced crack growth rate in steel plates exposed to sour environments, Corrosion, **53**, 935 (1997).
5. B. Beidokhti, A.H. Koukabi, A. Dolati, Influences of titanium and manganese on high strength low alloy SAW weld metal properties, Materials Characterization, **60**, 225 (2009).
6. S.S. Nayak, R.D.K. Misra, J. Hartmann, Microstructure and properties of low manganese and niobium containing HIC pipeline steel, Materials Science and Engineering: A, **494**, 456 (2008).
7. G.T. Park, S.U. Koh, H.G. Jung, Effect of microstructure on the hydrogen trapping efficiency and hydrogen induced cracking of linepipe steel, Corrosion Science, **50**, 1865 (2008).
8. X. Ren, W. Chu, J. Li, The effects of inclusions and second phase particles on hydrogen-induced blistering in iron, Materials Chemistry and Physics, **107**, 231 (2008).
9. M. Elboujdaini, V.S. Sastri, J.R. Perumareddi, Studies on inhibition of hydrogen-induced cracking of linepipe steels, Corrosion, **62**, 29 (2006).
10. V. Venegas, F. Caleyo, T. Baudin, Role of microtexture in the interaction and coalescence of hydrogen-induced cracks, Corrosion Science, **51**, 1140 (2009).
11. C. Natividad, M. Salazar, A. Contreras, Sulfide stress cracking susceptibility of welded X-60 and X-65 pipeline steels, Corrosion, **62**, 375 (2006).
12. B. Beidokhti, A.H. Koukabi, A. Dolati, Effect of titanium addition on the microstructure and inclusion formation in submerged arc welded HSLA pipeline steel, Journal of Materials Processing Technology, **209**, 4027 (2009).
13. B. Beidokhti, A. Dolati, A.H. Koukabi, Effects of alloying elements and microstructure on the susceptibility of the welded HSLA steel to hydrogen-induced cracking and sulfide stress cracking, Materials Science and Engineering: A, **507**, 167 (2009).
14. M.A. Mansour, A.M. Alfantazi, M. El-boujdaini, Sulfide stress cracking resistance of API-X100 high strength low alloy steel, Materials & Design, **30**, 4088 (2009).

Study on the effect of electrochemical treatment of nicosulfuron wastewater

Rui Zhao and Wenqiang Jiang*

College of environmental science and engineering, Qilu University of Technology (Shandong Academy of Sciences), Jinan 250353, People's Republic of China

Abstract. Nicosulfuron, a kind of sulfonylurea herbicide, because of its high activity and low dosage, has been widely used in corn planting weeding. Electrochemical method was chosen to treat its simulated waste water and the influence factors such as current intensity, plate spacing and electrolyte pH on electrolysis effect were investigated. A double-coated titanium dioxide titanium plate was chosen as the anode accompanied with a single-coated titanium plate as the cathode. The results showed that the removal efficiency of nicosulfuron was the highest when the electrolyte pH=3, the plate spacing is 1.5cm and the current intensity is 0.8A. Under the optimal condition, the removal rate of nicosulfuron reached 86.6% after 7 hours electrolysis. Thus, electrochemical technology might provide a significant method to remove the synthetic organic compounds.

1 Introduction

Nicosulfuron has been widely used in corn planting weeding because of its high activity and low dosage. There are many kinds of methods to treat its waste water . Sabadie indicated that hydrolysis of nicosulfuron principally involves the breakdown of the urea part of the molecule at 30°C as pH ranging from 4 to 11. Its rate constants decreased from 0.50 d^{-1} to 0.002 d^{-1} as pH increased from 4 to 8, then remained stable under alkaline conditions[1]. Chen adopted advanced oxidation technology to increase the biodegradability of nicosulfuron, and then activated sludge process was used to improve the quality of effluent. The optimal removal rate of 83.2% was obtained through the optimization of the reaction process [2]. The toxic and harmful substances in the wastewater can not be completely removed only by biological treatment.

Advanced oxidation processes (AOPs) have been emerged as potentially powerful methods to remove organic pollutants in water through in situ generation of strong oxygen-based oxidizers, among which the formation of hydroxyl radical (·OH) is especially desired, since it is the strongest oxidizer and can react non-selectively with various types of organics [3]. Electrochemical oxidation is one of the AOPs and has drawn considerable research attention due to its robustness and simplicity in structure and operation [4].

In this study, the high performance liquid chromatographic method (HPLC) was estabilished to determine the concentration of nicosulfuron. The following parameters such as electrolyte pH, plate spacing, and current intensity were studied in order to find the optimal electrolysis condition for nicosulfuron. The results might provide a theoretical basis for the electorchemical treatment of nicosulfuron wastewater.

2 Experimental materials and methods

2.1 Experimental materials

Nicosulfuron was selected as the model pollutant of electrolysis. It was a kind of industrial product obtained from Jingbo agricultural Polytron Technologies Inc. Its structure formula and general characteristics were given in Table 1.

Table 1. General characteristics of nicosulfuron.

Name	Nicosulfuron
Chemical structure	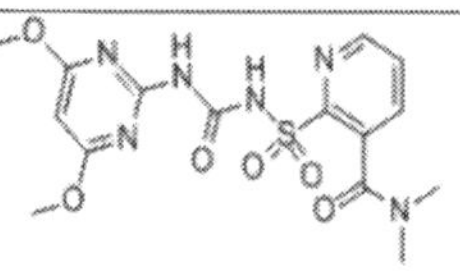
Molecular Formula	$C_{15}H_{18}N_6O_6S$
CAS Number	111991-09-4
Molecular mass	410.41g/mol
Solubility	acetonitrile 2.3%, acetone 1.8%, ethanol 0.45%, water 12%

The other chemicals were of analytical reagent grade and used as received without further purification.

[a] Corresponding author: spgc@163.com

2.2 Experimental methods

2.2.1 Electrolysis

A double-coated titanium dioxide titanium plate was chosen as the anode accompanied with a single-coated titanium plate of the same size as the cathode at a certain space between electrodes. To examine the influencing parameters, a volume of 400 mL nicosulfuron solution was used to conduct the electrocatalytic degradation experiments.

Electrolysis was performed with galvanostat to study the electrocatalytic performance of the electrodes. 3g of anhydrous sodium sulfate was dissolve in the solution. During the experiments, samples were drawn from the reactor at certain time intervals and the concentration of nicosulfuron was quantitatively analyzed by a high performance liquid chromatography (LC-10ATvp, SHIMADZU Corporation, Japan), equipped with a N2000 data station. The column was chose to use an inertSustain C18 (4.6mm×150mm×5μm), which was filled with octadecylsilane bonded silicao and its temperature was controlled at room temperature. The mobile phase was methanol, water and acetic acid (60:40:2, V/V/V). The eluent was delivered at a flow rate of 0.88mL/min and the absorbance at 254 nm was measured. The injection volume was 20 μL.

2.2.2 Linear relationship between peak area and the concentration of nicosulfuron solution

0.2000g nicosulfuron was accurately weighed and diluted to 2000ml with distilled water to obtain a standard solution of nicosulfuron at a concentration of 100mg/L. The standard solution was diluted to different concentrations, such as 1 mg/L, 5 mg/L, 10 mg/L, 25 mg/L, 50 mg/L and 100 mg/L. 20 μL solution was injected into the HPLC to determined the concentration of nicosulfuron. The linear relationship between the concentration of nicosulfuron and the peak area was analyzed and shown in Figure 1.

It could be seen from Figure 1 that there was a good linear relationship between the peak area and the concentration of nicosulfuron solution in the range of 0-100 mg/L, and the R^2 value was 0.9996.

In the next expeirments, the concentration of nicosulfuron could be calculated by equation 1.

$$C = \frac{A - 16797.75}{18001.43167} \quad (1)$$

2.2.3 simple stirring control experiment method

400mL of nicosulfuron solution was placed in the electrolytic cell and 3g of anhydrous sodium sulfate was added into the solution. The electrolyte was continuously stirred for 9h using a magnetic stirrer. Samples were taken at intervals of 3h to be analyzed by using HPLC. The peak area of nicosulfuron was determined by chromatography.

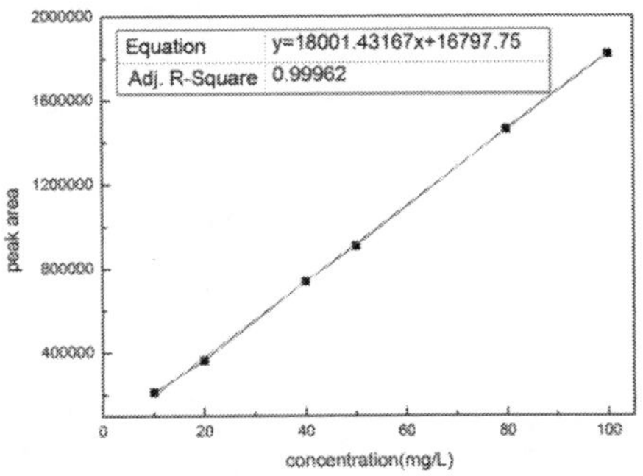

Figure 1. Linear relationship between peak area and the concentration of nicosulfuron solution

3 Results and discussion

3.1 Simply stirring control experiment

The degradation of nicosulfuron in stirring process was shown in Table 1. There was little change in peak area during 9 hours stirring process, which indicated that nicosulfuron was stable under the condition of simple agitation. In all the subsequent electrolysis experiments, the reaction time was within 9 hours, therefore, the degradation of nicosulfuron due to stirring process could considered to be negligible.

Table 1. Change of peak area for nicosulfuron.

Time (h)	0	3	6	9
Peak area ($\times10^6$)	1.820	1.819	1.821	1.819

3.2 Effect of current intensity

The current intensity directly is a key experimental parameter that affects the electrochemical oxidation process as it regulates the capability of hydroxyl radical generation on the electrode, which gonverns the degradation efficiency of organic matter in aqueous solution[5]. However, the higher the current intensity is, the greater the power consumption is. Therefore, it is extremely necessary to choose a suitable current intensity to obtain a high electrolysis efficiency with low energy consumption.

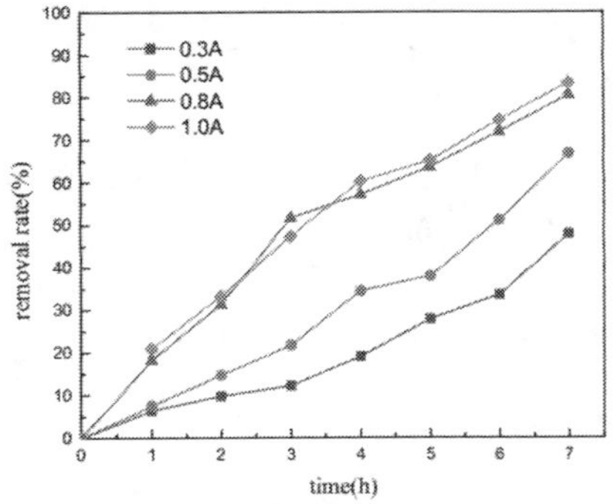

Figure 1. Effect of current intensity on electrolysis of nicosulfuron

Electrolysis experiments were performed under different current density and results were shown in Figure 1. Increasing current intensity had a positive influence on the degradation of nicosulfuron in the range of 0.3-1.0A. When the current intensity was 1A, its removal rate of nicosulfuron was almost the same as that of 0.8A. The removal rate of nicosulfuron reached and 83.3%, respectively, after electrolyzing for 7h. The increase of current intensity has dual effect on the effect of electrolysis. On the one hand, the high current intensity can accelerate the oxidation rate of organic matter on the surface of the electrode, while it also accelerates the generation of indirect oxidant hydroxyl radicals (·OH) [6]. When the applied current intensity is too high, the formation of side products may hamper the generation and availability of ·OH radicals. If the current intensity is too low to meet the energy demand for the chemical bond breakage of nicosulfuron, the degradation process could not occur. Therefore, combining a variety of factors, the optimal current intensity selected in this experiment was 0.8A.

3.3 Effect of electrode plate spacing

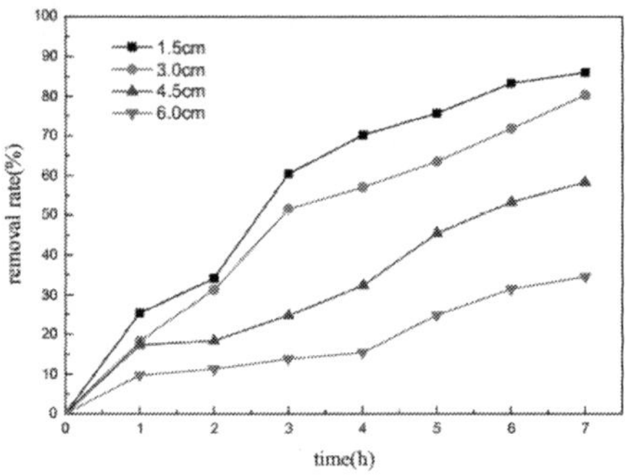

Figure 2. Effect of plate spacing on the electrolysis of nicosulfuron.

Electrolytic experiments were performed at different spacings to study the influence on the degradation of nicosulfuron and results were shown in Figure 2. It could be seen that when the distance between the plates was 1.5cm, the nicosulfuron solution had the highest electrolytic removal rate. In the same electrolyte system, the smaller the electrode spacing was, the shorter distance the electrons were demanded to transfer in the electrolyte, which is advantageous to increase the transmission of energy transfer during the electrolysis process. Therefore, the short distance is help to increase the electrolysis efficiency [7]. However, the excessively small plate spacing would lead to a low treating volume between the electrodes, which was not benefit the improvement of the processing efficiency. Herein, the plate distance of 3cm was chosen for the rest of the experiment.

3.4 Influence of electrolyte pH

Solution pH is one of the important factors that affects the performance of electrochemical process. For incosulfuron, the acidity and alkalinity of the solution have a great influence on its hydrolysis process. In neutral solution, its hydrolysis rate is faster than in acidic solution or in alkaline solution and the half-life of hydrolysis is 5.2 d. In acidic solution and alkaline solution, the half-life of hydrolysis is 7.7 d and 20.5 d, respectively [8].

The effect of pH on the degradation of nicosulfuron in the electrolysis process may be different from that of natural hydrolysis. pH As a consequence, the effect of pH value ranging from 1 to 9 on the nicosulfuron degradation was investigated, and the results were presented in Figure 3. From Figure 3, it was found that the maximum removal rates of nicosulfuron was observed at pH 3, which was 86.6% at 7 hrs. The minimum removal rate reached at pH 6, suggesting that it was unfavorable for the degradation process of nicosulfuron in neutral condition. It might be due to the highest resistance of the solution and the lowest conductivity in the neutral solution, which led to the lowest removal rate of nicosulfuron [9]. The optimum reaction pH was 3.0 for the experiment.

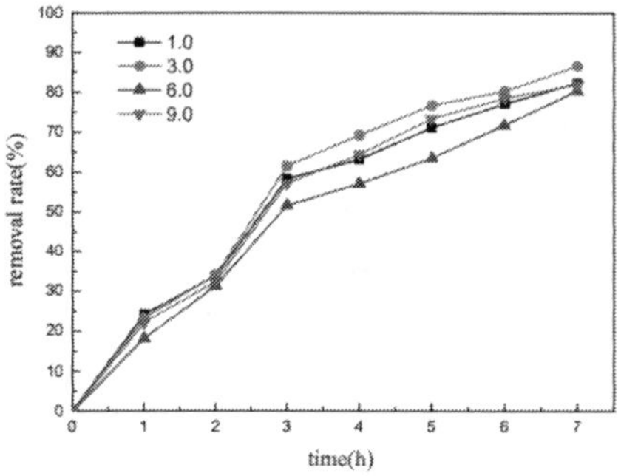

Figure 3. Effect of pH on the degradation of nicosulfuron

4 Conclusion

A method for the determination of nicosulfuron concentration by HPLC was established. The optimal condition of electrolysis was investigated to obtain highest removal rate for nicosulfuron. The results were shown as follows :

(1) HPLC, equipped with an inertSustain C18 column (4.6mm×150mm×5μm) could be used to determine the concentration of nicosulfuron. There was a good linear relationship between the peak area and the concentration of nicosulfuron solution in the range of 0-100 mg/L.

(2) No significant change of nicosulfuron was achieved in the electrolysis experiment after 7h of agitation. The degradation of nicosulfuron caused by continuous stirring can be ignored.

(3) A double-coated titanium dioxide titanium plate was used as the anode accompanied with a single-coated titanium plate as the cathode. The highest removal efficiency of nicosulfuron was obtained when the electrolyte pH = 3, the plate spacing is 1.5 cm and the current intensity is 0.8A. Under the optimal condition, the

highest removal rate could reach 86.6% after 7 hours electrolysis.

Acknowledgment

This work was supported by the Natural Science Foundation of Shandong Province, China (Grant No. ZR2014CM032).

References

1. Sabadie J. *J AGR FOOD CHEM*, **50,** 3. (2002)
2. F. L. Chen, W.Q.Jiang., *ICMSE*, **30**, 11 (2016)
3. Z. Q. Liu, J. Ma, Y. H. Cui, L. Zhao, B. P. Zhang, *APPL CATAL B-ENVIRON.* **101**,74 (2010)
4. T. Velegraki, G. Balayiannis, Chem. Eng. J. **160**,538(2010)
5. J. M. Chen, Y. J. Xia, Q. Z. Dai, *Electrochimica Acta*, **16**,5 (2015)
6. W.J.Zhi. *Harbin Institute of Technology*, (2011)
7. L. B.Qi, S. Q. Zhong , Meng. W. *Contemporary Chemical Industry*. **20**. 9. (2015)
8. L. Wan, Z. G. Hou, L. C. Li. *Journal of Anhui Agricultural Sciences*. **39,** 25. (2011)
9. Y. Li, *Shandong Institute of Light Industry*, **61**. 9. (2010)

Conversion of multi-layered Mg-based carrier to hydrogen storage performance

Ningning Zhou[a, b] and Dongying Ju[b*]

[a]*Petrochemical Research Institute, Liaoning Petrocchemical Vocational and Technology, No. 4, District 2 Beijing Road, Jinzhou, Liaoning, 121001, China*
[b]*Department of Electronic Engineering, Saitama Institute of Technology, Fukaya, Saitama 369-0293, Japan*

Abstract The mixed powder method and covering method were adopted to prepare a composite that has hydrogen storage properties. In the experimental process, the alloy phases Mg_2Ni and NiTi generated regularly; the penetration of the hydrogen atoms to the material was considered as the adsorption of the material to the hydrogen atoms; the original thermodynamic equation was edited; the crystal phases were optimized by adjusting the parameters about the material composition and gas concentration, so that providing advantageous channels of the hydrogen atoms in and out to improve the hydrogen storage properties. The results showed that integer multiples adjustment to trace element Ti could make hydrogen storage properties maximizing optimize and have controllability for the capacity and rate of hydrogen storage according to the newly editing equation. The innovation was to edit out equation that can improve the hydrogen storage properties through adjusting three parameters of trace element Ti, adsorption and penetration to find the most optimal condition improving the hydrogen storage properties.

1 Introduction

In recent years, the fossil fuel reserves are rapidly reduced lead to the existing reserves cant not meet the growing demand. The majority of the world energy currently comes from the fossil fuels, but the use of the fossil fuels can produce large amounts of the harmful substances, which causes a great threat to the environment. Therefore, accelerating the conversion of the energy system to renewable energy meets the energy needs of the current and future world has become urgent problem to be solved.[1-4] In particular, how to adopt more efficient functional materials and to effectively control hydrogen absorption and desorption under a certain condition have been main research directions. Expanding applied range of hydrogen storage material, developing various potential functional, particularly, the using of the hydrogen storage media have became concern issues. In the present work, low price, abundant resources and light quality Mg based material was adopted for two-way development of hydrogen storage material. A new type prepared techniques included that synthesis-forming-sintering-alloying achieved crystalline phase restructuring to generate the composite of clad Mg_2NiX-NiTiX (watchcase)- pure Mg (inner casing). To distinguish conventional methods, the advantages from the powder and alloy were synthesized by covered method to achieve the adding of the powder to the metal and the alloy phases Mg_2Ni and NiTi generated continuously from the surface to the inner. Through activated treatment on the surface, crystalline phase characteristic and organizational structure were improved and optimized.

2 Experimental results and discussion

2.1 Actual results before adjusting Ti content

As shown in Figure 1a, the composite at the molar ratios of Ni： Ti : Mg =β： γ: η absorbed hydrogen relatively slowly under the constant pressure and constant temperature of 340°C, with the increasing of hydrogen pressure, hydrogen absorption tended to increase slowly. The saturation produced when the hydrogen absorption reached 2.6%. Hydrogen desorption started with the changes of hydrogen pressure for 6 hours after reaching saturation, the actual capacity of hydrogen desorption was 2%. As shown Figure 1b, the proportion of the materials had greater effect on the hydrogen storage performance based on the directions of present experiment. There were no obvious changes about the capacity of hydrogen absorption (in unit of /mg) within the 35 hours area, the initial activation and the concentration of the alloy was relatively lower. It indicated that the effective area was relatively small due to the insufficient reaction of the powder with pure Mg ingot or the insufficient unit content of a certain element. It is considered that the addition of powder can effectively control the surface area. The addition of powder Ni was definite purpose in this experiment. The

[a] Corresponding author: zhouningning_yes@163.com

addition of Ni powder at a certain ratio enable the reaction with pure Mg enhanced to generate Mg_2Ni alloy phase. If any deviation produced, the $MgNi_2$ alloy phase will be generated and it will prevent hydrogen absorption and adsorption. Therefore, the added amount of powder Ti is controllable. Figure 1c showed that the penetrated capacity of the hydrogen to the material is the key factor in the process of the hydrogen permeated to the composite under the temperature of 340°C (the adsorption of the material to H_2 was considered as the penetration of H_2 to the material). If the activation time did not considered, the formula as shown in

$$\Pi = C \times R \times T \qquad (1)$$

Πis the permeability of hydrogen; C is a penetrated concentration; R is a gas constant; T is an absolute temperature. According to a amount of H atoms, it was considered as the absorbed capacity of the material after penetrating to the material, the maximum permeate pressure reached 6500mOsm/Kg.

The rate of hydrogen absorption and desorption showed the materials used in this experiment was hard to activate. In terms of materials, Mg_2Ni alloy phase can absorb sufficiently hydrogen atoms but its hydrogen absorption rate is relatively slow, NiTi alloy phase can effectively provide H atoms to enter channel. The above data suggested that the generation of NiTi was insufficiently. Therefore, the adjustment of Ti as a key can improve the hydrogen absorption and adsorption properties.

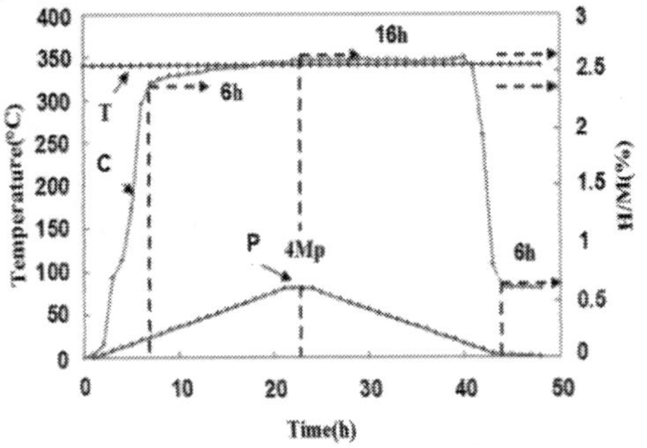

a

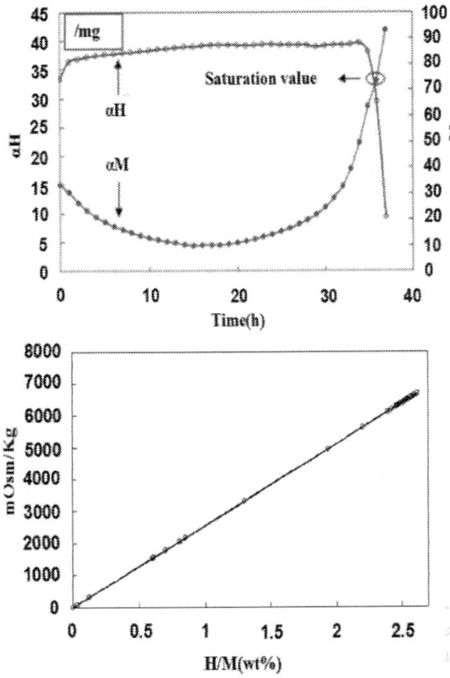

b

c

Figure 1. Analysis and calculation of P-C-T curve before adjusting Ti. 1. P-C-T curve under constant temperature 340°C. 2. Relative concentrations between H atoms and alloy. 3. Actual relation between the permeate pressure of H_2 to the material and the concentration of the material absorbed H_2 before adjusting Ti content.

2.2 Simulated result after adjusting Ti content

The research purpose was to improved the hydrogen absorption and desorption ability under a lower temperature 220°C by adjusting the concentration of the hydrogen. In order to confirm the consistency of the simulated value and actual value after the absorbed capacity, the hydrogen absorption ability of the material to the hydrogen was considered as the penetration of hydrogen to the material. According to Ti atom could enter into face centered cubic position, two Ni-Ti compound bonds might be form in an aeromonas with the addition of Ti at molar amount. NiTi alloy phase provided the channel of H atoms entered into the material. Therefore, the concentration of H would be expand 2 times (analog value) [C_H (Fig. 1a) ×2] according to $\Pi = C \times R \times T$, the simulated result showed that the permeate ability of hydrogen to the material could be improved obviously with the increasing of the hydrogen concentration, so that the permeate pressure reached 10656.89mOsm/Kg in Figure 2.

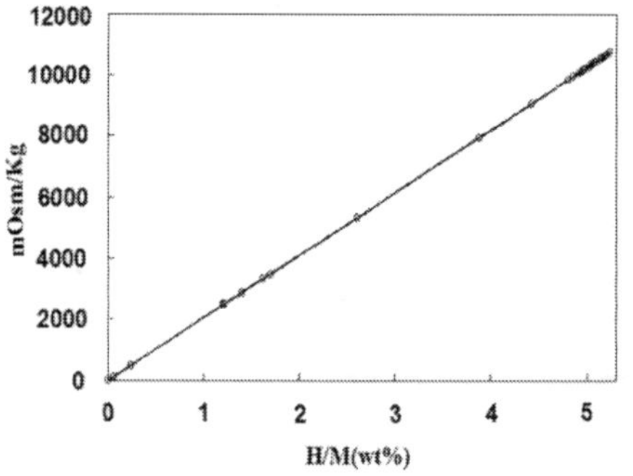

Figure 2. Simulated relation between the permeation of H_2 to the material and the concentration of the material absorbed H_2 after adjusting Ti content.

2.3 Actual results on substituted into revised formulas after adjusting Ti content

The study conducted structure-recombination in sintered process, the materials were constituted at the molar ratio of Ni： Ti : Mg =β： (γ+1): η under the constant temperature of 220°C,which is far below 300°C. The actual verified result was confirmed by adjusting the component and revising the formulas. The original formula and the specific revised process as shown in follow

$$\ln p = \frac{\Delta H}{RT} - \frac{\Delta S}{R} \quad (2)$$

Where P is equilibrium pressure; △H is the endothermic enthalpy; △S is the endothermic entropy; T is the absolute temperature; R is the gas constant， through extending the original formula, the relationship formula between formed the metal hydrides and the metal is discussed

$$K_p = \frac{\alpha_{MH_X}}{\alpha_M \left(f_{H_2}\right)^{\frac{X}{2}}} \quad (2.1)$$

Where K is the equilibrium constant; α is substance activity; M, MH is the standard state; αMH and αM are activity; f is gas fugacity. In which the equilibrium constant Kp equal to

$$K_p = P_{H_2} \quad (2.2)$$

The formulas could be changed to

$$P_{H_2} = \frac{\alpha_{MH}}{\alpha_M \left(f_{H_2}\right)^{\frac{X}{2}}} \quad (f_{H_2} \approx 1) \quad (2.3)$$

P_{H2} is equilibrium hydrogen pressure. In the formulas, f_{H2} is considered as 1, therefore, the equilibrium H_2 pressure is related to the concentration of metal hydride and the pure metal, as shown in below

$$P_{H_2} = \frac{\alpha_{MH}}{\alpha_M} \quad (2.4)$$

In order to macroscopic analysis the effect of introducing components on the hydrogen storage, the capacity of the metal hydride after the cycle of hydrogen absorption and desorption became seldom and uncertain. αM was approximately equal to the concentration of the pure metals, the metal can be divided to three kinds of pure metal αMg, αNi and αTi, respectively.

$$\alpha_M \approx \alpha Mg_{\frac{\eta}{24}} + \alpha Ni_{\frac{\beta}{59}} + \alpha Ti_{\frac{r}{48}} \quad (2.5)$$

Substituting into formula (2.4)

$$P_{H_2} = \frac{\alpha_H}{\alpha Mg_{\frac{\eta}{24}} + \alpha Ni_{\frac{\beta}{59}} + \alpha Ti_{\frac{r}{48}}} \quad (2.6)$$

η is the quality of Mg; β is the quality of Ni; γ is the quality of Ti. Therefore, the adding of trace element Ti can be adjusted according to the formulas

$$\alpha_M \approx \alpha Mg_{\frac{\eta}{24}} + \alpha Ni_{\frac{\beta}{59}} + (\alpha Ti_{\frac{r}{48}} + \alpha Ti_{\frac{\varepsilon}{48}}) \quad (2.7)$$

So the below formula is obtained

$$P_{H_2} = \frac{\alpha_{MH}}{n\alpha_M} \quad (2.8)$$

n is a alloy phase concentration factor. It is considered that the hydrogen pressure had relationship with the component of adding the element, the performance of hydrogen absorption and desorption can be improved by adjusting the component of the raw material. Accordance to above export formula, the absorption capacity of hydrogen atoms was viewed as the gas permeation. α_{MH} is deformed into α_H as shown in follow

$$P_{H_2} = \frac{\alpha_H}{n\alpha_M} \quad (2.9)$$

In order to avoid the deviation from traditional calculated method, the absorbed hydrogen concentration is regarded as the permeability ability of gas H_2. The below formulas can be obtained

$$\alpha_H = C_H = \frac{\Pi}{RT} \approx m\frac{\Pi}{RT} \quad (3.0)$$

$\frac{\Pi}{RT}$ is the gas permeability. Through adjusting the adding of trace element Ti, H_2 permeability had been validated according to the below formulas

$$P_{H_2} = \frac{m\frac{\Pi}{RT}}{n\alpha_M}(n \approx m) \quad (3.1)$$

n and m are concentration factors and the differ are small, the value of n and m approximately equal 2. It is indicated that effectively control the content of Ti can provides a powerful channel for hydrogen absorption and desorption. The formula (3.1) is substituted into (2.1) and (2.2)

$$P_{H_2} = \frac{m\frac{\Pi}{RT}}{n\alpha_M \left(f_{H_2}\right)^{\frac{X}{2}}}(m \approx n) \quad (3.2)$$

$$\Pi = \frac{P_{H_2} \times n\alpha_M \left(f_{H_2}\right)^{\frac{X}{2}} \times RT}{m} \left(m \approx n\right) \quad (3.3)$$

As shown Figure 3a, the hydrogen absorption speed was improved; hydrogen absorption amounted to 3% within 3 minutes and 5% within 16 minutes. Not only the temperatures for hydrogen absorption had been reduced significantly, but also the amount and the speed of hydrogen absorption had been also improved significantly. As shown Fig. 3b, the hydrogen absorption reached quickly the saturation value and the concentration of the alloy (in unit of /g) was relatively lower, it indicated the concentration of hydrogen had enhanced within the relatively short time. As shown in Fig. 3c, the concentration of Ti was adjusted to verify and simulate the permeation of the hydrogen to the material due to the concentration of hydrogen could not be controlled through the simulation. According to the hydrogen atom enter into the composite in form of tetrahedral or octahedral gap, the molar ratio was adjusted to Ni： Ti : Mg =β： (γ+1): η, it is considered that the effective increasing of Ti content could form fully NiTi alloy phase in the sintered process and reduce the loss of the reaction from Ni and Mg, which provided more channels for hydrogen absorption/desorption. To increase the molar amount of Ti, the ideal state was each Ti atom and Ni atom reacted to generate NiTi, so that effectively increasing the surface of the composite material and improving the performance of hydrogen absorption/desorption. According to this theory, the formulas would be revised and obtained verification, the maximum permeated pressure reached to 11005.28mOsm/Kg by the revised formula (3.3). This result was consistent with above simulated result of 10656.89mOsm/Kg; it was showed that adjusting Ti content could achieve controllability of the absorbed capacity of the material considered as the penetration of the hydrogen according to revise formulas.

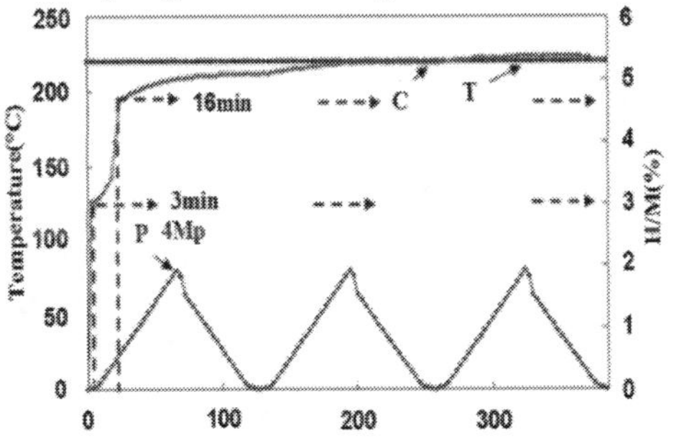

a

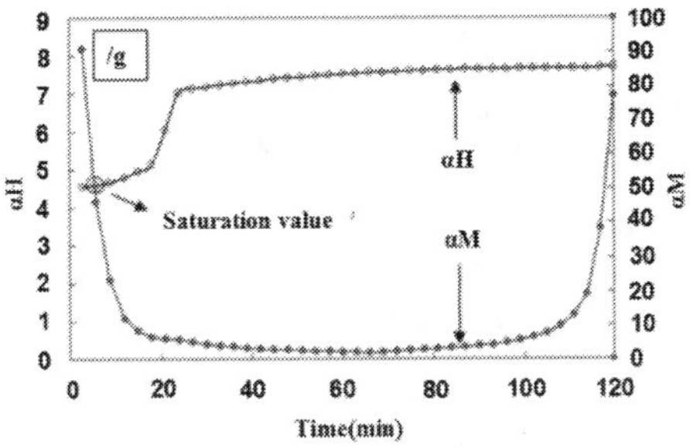

b

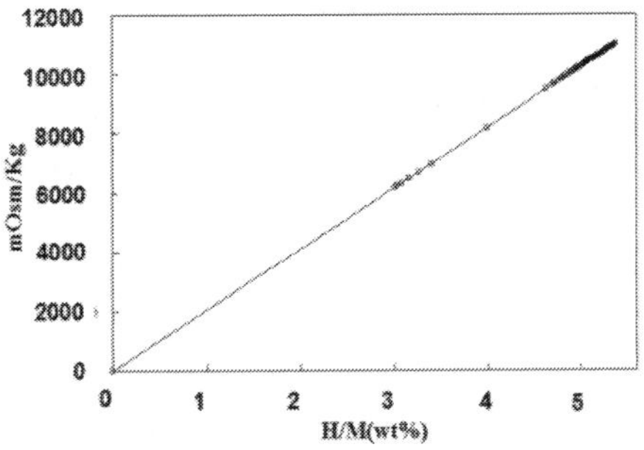

c

Figure 3. Analysis and calculation of P-C-T curve after adjusting Ti. 1. P-C-T curve under constant temperature 220°C. 2. Relative concentrations between H atoms and alloy. 3. Actual relation between the permeate pressure of H_2 for the material and the concentration of the material absorbed H_2 after adjusting Ti content.

The results from the actual operation and simulation were consistence,the absorbed channel of the hydrogen atom could be increased effectively through adjusting the concentration of Ti, and original formula would be revised by the new method that adjusting the concentration of the original material. This actual result was closed to that after simulating. Adjusting the concentration of Ti and simulating concentration of hydrogen was basically consistent.

3 Conclusion

In this work, the composite effectively overcame inconvenient from the powder on storage aspect and avoided negative factors of small surface area due to an increasing of the grain boundary between the crystal grains and the channels of H entered into the material increased. For the composite, hydrogen absorption and desorption rate was obviously improved and even superior to the effects from a powder by traditional method. The composite had improved greatly the capacity and rate of hydrogen absorption/desorption. The capacity of hydrogen absorption reached 3% within 3min and the highest amount reached 5.3% within 16min. In this study, the adjusting of Ti had a certain range, what effects will produce if the proportion will become excessive need to be further researched.

4 References

[1] D. Kyoi, T. Sakai, N. Kitamura, A. Ueda and S. Tanase, Synthesis of FCC Mg–Ta Hydrides Using GPa Hydro-gen Pressure Method and Their Hydrogen-

Desorption Properties. ALLOYS COMPOUNDS. J. **463**, 306-310 (2008).

[2] A. Singh, B.K. Singh, D.J. Davidson and O.N. Srivastava, Studies on improvement of hydrogenstorage capacity of AB5 type: =MmNi4.6Fe0.4 alloy. HYDOGEN ENERG. J. **29**, 1151-1156 (2004).

[3] L. Schlapbach, A. Züttel, Hydogen-Storage Matcrials for Mobile Application. Journal of Nature. J. **414**,353-358 (2001)

[4] M. V. Simičić, M. Zdujić, R. Dimitrijević, L. Nikolić-Bujanović, N. H. Popović, Hydrogen Absorption and Electrochemical Properties of Mg_2Ni-Type Alloys Synthesized by Mechanical Alloying. J POWER SOURCES. J. **158**, 730-734 (2006).

Reinforcing high-density polyethylene by phenolic spheres

Lien Zhu[1], Di Wu[1], Baolong Wang[1], Jing Zhao[1], Meihua Liu[1],Zheng Jin[1,a*], Kai Zhao[2,b*]

[1] *Key Laboratory of Chemical Engineering Process & Technology for High-efficiency Conversion, College of Heilongjiang Province, College of Chemistry Engineering and Materials, Heilongjiang University, Harbin, 150080, PR China.*

[2] *Laboratory of Microbiology, School of Biological Science and Technology, Heilongjiang University, Harbin, 150080, PR China.*

Abstract. Phenolic spheres are synthesized through resorcinol and formaldehyde. The phenolic spheres were blended with HDPE to prepare binary composites. The rheological properties and mechanical properties of the composites were studied. The composite materials have higher tensile strength and impact strength than pure HDPE, which extends the application of the material.

1 Introduction

High-density polyethylene (HDPE), as a semi-crystalline polyolefin, is one of the most common plastics with a wide range of applications such as bottles, pipes and containers [1]. It is a chemically resistant, flexible, easy to process, tough plastic, and most significantly it can be processed at low temperature [2]. However, HDPE cannot approach the typical properties of engineering or high performance polymer due to its low mechanical properties [3,4] which limited its applications in a broader field.

In order to increase the tensile strength of the material, the blending of the polymer with the fiber is an effective way [5]. The principle is that the polymer matrix can transfer stress to the fiber [6]. However, due to the poor interfacial strength between the fiber and the matrix, the tensile strength of the material tends to damage its toughness at the same time.

In this study, phenolic spheres were first synthesized. The prepared phenolic spheres were melt-blended with HDPE to form binary composites, of which rheological properties and mechanical properties of composite materials were analyzed. Due to the well compatibility of phenolic spheres with HDPE, phenolic spheres have successfully acted as reinforcements, resulting in a significant increase in the tensile strength of the composites. Also, due to the proper size of the phenolic spheres, the impact strength of the composite material is improved. These expand the application of materials.

2 Experiment

Under room temperature conditions, resorcinol and formaldehyde were mixed in a ratio of 1:2. Anhydrous sodium carbonate was used as a catalyst, and a certain amount of distilled water was added. After stirring, the mixture was added to a three-necked flask and the three-necked flask was placed in an oil bath. In the pot, the set temperature was 85°C. and the reaction was carried out at a constant temperature for 8 hours. After that, it was washed with anhydrous ethanol, centrifuged, and dried at a constant temperature of 60°C for 48 hours to obtain phenolic spheres that were relatively dispersed and uniform in particle size. The experimental reaction device is shown in Fig. 1 and prepared phenolic spherical is shown in Fig. 5(a). The resulting phenolic spheres and HDPE were melt-blended through a twin-screw extruder. After extrusion and granulation, the dumbbell-shaped samples were made using an injection molding machine.

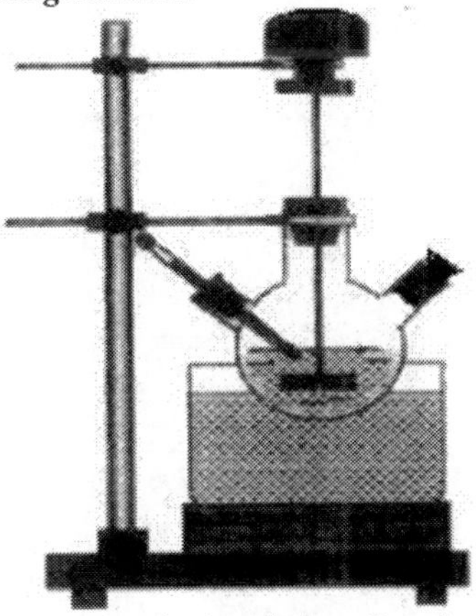

Fig. 1 The reaction device of phenolic spheres

*Corresponding author : [a] jinzheng@hlju.edu.cn, [b]zybin395@126.com

The morphology of all the blends was characterized by scanning electron microscopy (SEM). The samples were fractured after immersing in liquid nitrogen and then covered with gold. Then the fractured samples were observed in a SEM instrument (Model: FEI-INSPECT F, USA) at an acceleration voltage of 5 kV.

Tensile propertie of samples was examined according to GB/T 1040.1-06 standard using a tensile testing machine (YHS-229WJ) and the tests were carried out at a displacement rate of 50 mm/min at room temperature. The notched izod impact strength of the specimen was determined as per ASTM D256-10 standard using an impact strength testing machine (XJU-2.75J).

Rheological characterization of the samples was carried out at 170 °C on a rheometer (TA-AR2000ex USA) with a 25 millimeter parallel plate. Strain sweep was carried out to determine the limit on the strain for linear viscoelastic response and the strain range of 0.01~100% was used. The testing samples with a volume of 20×20×4 mm3 were prepared by cutting the dumbbell-shaped standard sample. The gap between the two parallel plates in rheology test was set as 2mm.

The elasticity of samples was measured using a DMS6100 dynamic thermomechanical analyser with alternating stress of the frequency of 1 Hz at a temperature rise rate of 2°C/min. The specimen was prepared by cutting the dumbbell-shaped standard sample with a volume of 35×10×2.5mm3.

3 Results and discussions

Fig. 2 and Fig. 3 show the dependence of the storage modulus and the loss modulus of the composites with different phenolic sphere content on the frequency. As can be seen, the storage modulus and the loss modulus all show an increasing trend with the increase of the shear rate. The change in storage modulus is due to the greater shear force, the greater the short-term recovery ability of the composite, which showing a significant increase in elasticity. The change in the loss modulus is due to the increased movement of the molecular chain under external forces, which manifests as an increase in system viscosity. In the low frequency region, the storage modulus of the composite material is less than the loss modulus, indicating that the polymer melt is dominated by viscous flow. After the frequency reaches a certain value, the material's storage modulus is greater than the loss modulus, indicating that the polymer melt is dominated by viscous flow.

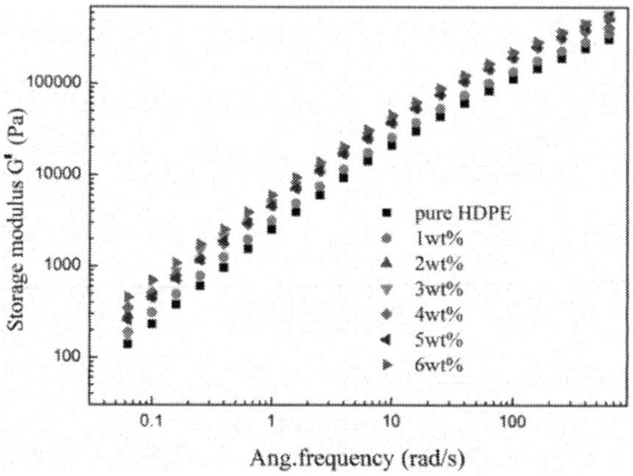

Fig. 2 The plots of storage modulus versus ang.frequency of composites with different content of phenolic spheres

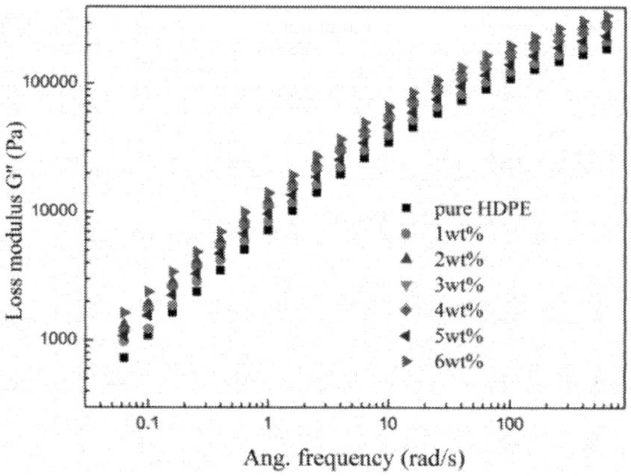

Fig. 3 The plots of loss modulus versus ang.frequency of composites with different content of phenolic spheres

Fig. 4 shows the tensile strength and elongation at break of composites with different phenolic sphere contents. It can be seen from the figure that the tensile strength of the composite increases first and then decreases, reaching a maximum of 25 MPa when the phenolic sphere content is 4 wt%, which is a 17.8% increase over pure high density polyethylene resin. However, the elongation at break of the composite material decreased significantly with the increase of the phenolic sphere content. When the phenolic sphere content was 4 wt%, the elongation at break decreased by 70%.

There are two main reasons to explain the trend of the tensile strength curve. One is that the heterogeneous nucleation of the phenolic spheres in the composite material increases the crystallinity of the high density polyethylene, which leads to an increase in the tensile strength of the material. The other is related to the degree of dispersion of the phenolic spheres in the composite material, which can be visualized by cross-sectional scanning electron microscopy of composites with different phenolic sphere contents in Fig. 5. As can be seen from the figure, when the phenolic sphere content is low, it is independently and homogeneously dispersed in the high-density polyethylene resin matrix, and the particle diameter is about 1 to 1.5 μm. The phenolic sphere will be pulled out from the resin matrix under the tensile stress, which promotes the transfer of stress and leads to an increase in tensile strength. When the phenolic sphere content exceeds 4 wt%, a part of the phenolic spheres in the composite material stacks to form agglomeration. Since the aggregate is formed by intermolecular forces, a stress concentration will occurs under the tensile stress, causing a decrease in tensile strength. The decrease in elongation at break is due to the fact that the phenolic spheres are organic rigid particles. The addition of the phenolic spheres to the high-density polyethylene resin destroys the flexibility of the matrix molecular chains and causes them to change from flexible to rigid.

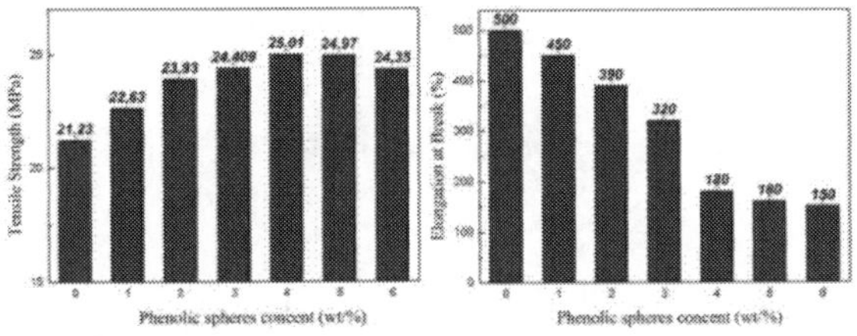

Fig. 4 Tensile strength and elongation at break curves of composites with different content of phenolic spheres

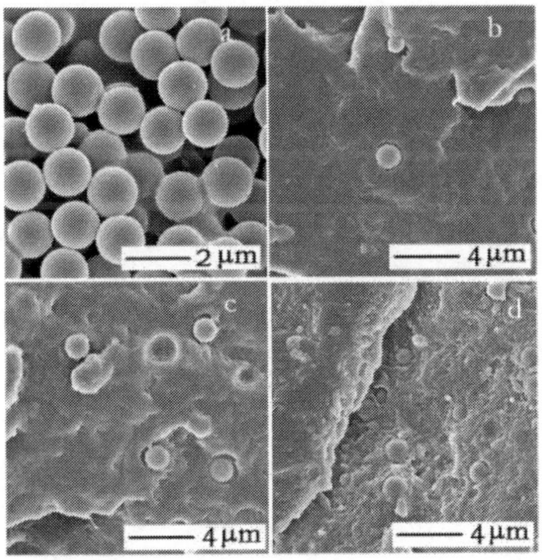

Fig. 5 SEM images of composites with different content of phenolic spheres: (a) phenolic spheres, (b) 2wt%, (c) 4wt%, (d) 6wt%

Fig. 6 shows the impact strength curves of composites with different phenolic sphere contents. It can be seen from the figure that the impact strength increases first and then decreases, reaching a maximum of 32.93 at 4 wt% which is an increase of 65.98% compared with pure high-density polyethylene resin. This can be explained by the "silver shear band" theory. When the substrate is subjected to impact stress, the presence of phenolic spheres causes local stress concentration, which induces a large amount of silver lines and shear bands, absorbing a large amount of fracture energy and hindering crack propagation. Also, since the interface strength between the phenolic spheres and the matrix is high, a large amount of fracture energy is absorbed when phenolic spheres are separated from the matrix. All of these lead to an increase in the impact strength of the material. However, when the content of phenolic spheres is too high, more and more silver striations are produced in the composite material. The silver striation is a precursor to submicroscopic fracture of the material. Under the influence of internal and external stresses, the craze gradually develops into a macroscopic fracture, resulting in the decrease of impact strength.

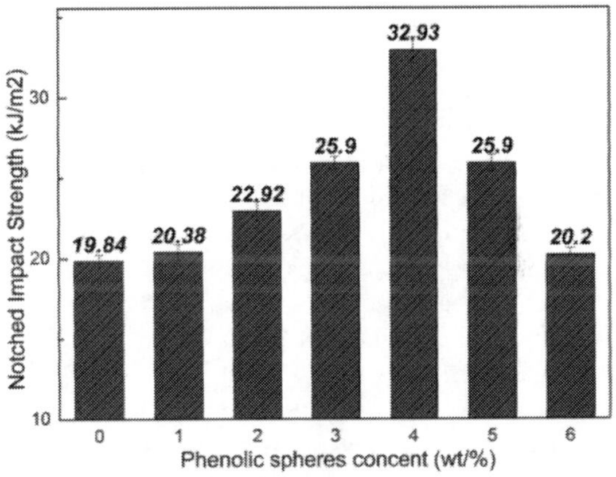

Fig. 6 Notched impact strength of composites with different content of phenolic spheres

Dynamic thermomechanical analysis was used to study molecule movement. In this work, polymer chain movement behavior will be influenced with the addition of fibers. Storage modulus is a parameter characterising molecular chain segment motion. The higher storage modulus, the harder to elastic deform. In other words, storage modulus increasement could enhance resistance capacity to deformation. As displayed in Fig.7, the storage modulus of all composites increases first and then decreases with the increase of phenolic spheres content, reaching the maximum at 4 wt%. This is because when the content of phenolic spheres is small, the microspheres form a good dispersion effect in the matrix, and the independent microspheres can effectively prevent the movement of the molecular chains. The deformation of the segments can not keep up with the change of stress, so that the deformation elastic energy storage is gradually generated. When the content of phenolic aldehydes exceeds 4%, the microspheres begin to aggregate together by intermolecular forces. The chain motion makes it easier for the deformation to keep up with the change in stress and the storage modulus is reduced. This phenomenon signified that the resistance capacity to deformation of composites enhanced with the addition of phenolic spheres.

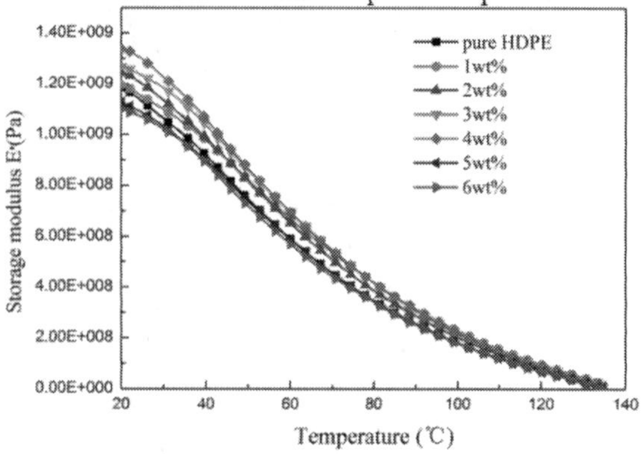

Fig.7 Storage modulus of composites with different content of phenolic spheres

Conclusion

With the addition of the phenolic spheres, the storage modulus, loss modulus, tensile strength and impact strength of the materials have all improved significantly, which extends the application of the material.The best mechanical properties are obtained when the phenolic sphere content is 4%.

Acknowledgements

The authors gratefully acknowledge the National Key Research and Development Program of China (Grant Numbers 2017YFD0500706 and 2017YFD0500603), National Natural Science Foundation of China (Grant Numbers 31570929 and 31771000), Natural Science Foundation of Heilongjiang Province (Grant Number C2017058), Innovation Foundation of Harbin (Grant Number 2017RAXXJ001), Graduate Student Innovation Research Project Funding of Heilongjiang University (Grant Number YJSCX2018-061HLJU), Students Innovation and Entrepreneurship Training Project of China (Grant Number 201810212020) and Key Scientific Technological Planning Project of Harbin (Grant Number 2016AB3BN036).

References

1. J. Zhang, G. M. Rizvi, C. B. Park. J. Mater. Sci-mater. 46,11(2011)
2. K. Sewda, S. N. Maiti. Polym. Bull. 70,10(2013)
3. A. V. Ortiza, J. G. Teixeiraa, M. G. Gomesa, R. R. Oliveiraa, F.R.V. Díazb, E.A.B.Moura, Appl. Surf. Sci. 310,8(2014)
4. C. L. Simões， LMC Pinto， C.A Bernardo, C. L. Pinto, L. M. Bernardo, C. A. Waste. Manage. Res.32,5(2014)
5. L. Xia, T. Zhang, Z. Chai, X. Hu, F.Jin, G. Wen. J. Eur. Ceram. Soc.36,14(2016)
6. L. Zhu, D. Wu, B. Wang, J. Zhao, Z. Jin. Pigm. Resin. Technol. 47, 1(2017)

MATEC Web of Conferences **238**, 05004 (2018) https://doi.org/10.1051/matecconf/201823805004
ICNFM2018

Effect of transition metal ions on luminescence of MOFs

Bo Ruan[1,+], Huan-Li Liu[1,+], Xue-Qing Zhan[1], Hui Ding[1], Lei Xie[1], Fang-Chang Tsai[1,*]

[1] *School of Materials Science and Engineering, Hubei University, Wuhan 430062, China*
[+] *Both authors contributed equally*

Abstract. Three kinds of metal-organic framework materials (MOF-5, MLI-101(Cr), UiO-66) are synthesized, which are based on transition metal (Zn, Cr, Zr) as the center ions, and terephthalic acid (H_2BDC) as organic ligand. At the same time, three kinds of MOFs were characterized and analyzed with X-ray diffraction (XRD), scanning electron microscopy (SEM), UV-visible absorption spectrometer (UV-Vis), and fluorescence spectrometer (PL). This article analyzes the fluorescence effects of these three MOFs and investigates the role of different transition metal ions in MOF luminescence. The results show that the ionic radii of Zn^{2+} and Zr^{4+} are similar, which are 0.074 nm and 0.072 nm, respectively. MOF-5 and UiO-66 have similar fluorescence effects, both of which emit blue-violet light. However, MIL-101 (Cr) fluorescence centers on Cr^{3+} with an ionic radius of 0.0651 nm emits yellow-green light, which is different from the foregoing MOFs. Therefore, the ionic radius of the transition metal has an important effect on the fluorescence effect of the MOF. For larger ionic radii, the fluorescence of MOF is generally in the blue-violet square region.

1. Introduction

Luminescence is a very rare phenomenon, which generally refers to the luminescence ability of the molecule when the ground state electrons in the molecule absorb energy to transition to the excited state and then return to the ground state by means of radiative transition[1, 2]. Therefore, various luminescent substances have received extensive attention. As people's demand for luminescent materials continues to increase, many new functional materials have been explored and discovered. Luminescent metal-organic framework materials, due to their design of the main structure of the fusion of the two major properties of porosity and light, can show a relatively rich host-guest interaction, and thus get a wide range of attention[3-6]. The metal-organic framework material is a porous material composed of a metal ion or a metal oxygen cluster as a central node and an organic chain-like molecule ligand, which simultaneously possesses an inorganic block and an organic block, which may have luminescent properties, and the combination of inorganic blocks and organic blocks can also form unique metal-ligand intramolecular charge transfer to induce new luminescence properties[7-10]. Therefore, the metal organic framework material is a very good luminescent material.

Most of the metal elements in the periodic table of elements have been used to build MOFs, of which transition metals are very common. Therefore, it is very meaningful to study the effect of transition metals on the luminescence properties of MOFs. Here we select transition metal ions of different valences as the central ions. Zinc ions are the most common among the divalent ions which have suitable hardness and have good binding with common donor atoms such as oxygen and nitrogen[11-13]. Trivalent transition metal ions such as Cr^{3+} have a smaller radius and a higher charge, which can form strong coordination bonds with oxygen-containing ligands, and for tetravalent transition metal ions such as Zr^{4+}, the coordination bond is more strong[14]. Metal-organic framework materials generally use π-conjugated organic molecules as organic ligands, which have very good rigidity, and most of them are based on functionalization with polycarboxylate groups or heterocyclic groups[12, 13]. One of the most common and simplest is terephthalic acid (H_2BDC)[15-17]. Therefore terephthalic acid is often used as an organic ligand for MOFs. Therefore, this article selects terephthalic acid as the organic ligand and uses Zn^{2+}, Cr^{3+}, and Zr^{4+} as the central ions to synthesize MOF-5, MIL-101(Cr) and UiO-66 and study the luminescence effects of these MOFs.

2 Experimental

2.1 Materials and synthesis

All reagents were purchased commercially and used without further purification.

The synthesis of MOFs are shown in Figure 1.

*Corresponding author: tfc0323@gmail.con

Taking MIL-101(Cr) as an example, $Cr(NO_3)_3{\cdot}9H_2O$ (2 mmol, 0.8054 g), H_2BDC (2 mmol, 0.3326 g), 0.1 mL hydrofluoric acid (wt.%, 35%) and 9.6 mL deionized water were placed in a 50 ml Teflon lining. Then, the inner liner was placed in a stainless steel reactor, and the lid of the reactor was placed in an oven. The temperature of the oven was increased from room temperature to 200°C within 3 hours and maintained at 200°C for 8 hours. After that, it was cooled to room temperature then filtered through a fritted funnel to give a green precipitate. The product was washed with DMF and EtOH, respectively. Then it was centrifuged for 5 minutes to obtain precipitate (revolution speed 8000 rpm) and repeatedly washed twice more. Finally, the suction filtration was used. The solid obtained by suction filtration was dried in a vacuum oven at 150°C for 12 h. In this way, MIL-101 (Cr) used in the following experiment was prepared.

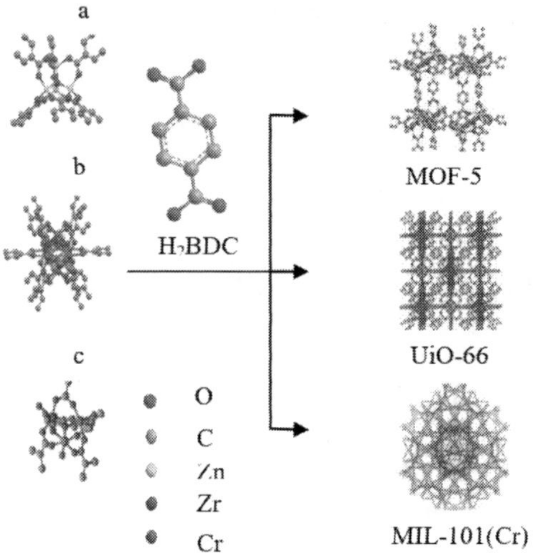

Figure 1. The concept of syntheses of MOF-5, MIL-101(Cr) and UiO-66; a, b and c are SBU of Zn^{2+}, Zr^{4+} and Cr^{3+}, respectively.

2.2 Characterization

The sample powder was ground uniformly and then compacted in a mold and placed on a Bruker D8A25 X-ray powder diffraction (PXRD). With a Cu target, $\lambda K\alpha$ = 0.154 nm, I = 40 mA, and U = 4 kV, the test was performed in a continuous scanning mode with a scanning range of 2–40°, a scanning speed of 4 deg/min, and a step length of 0.02 step size. The crystal morphology of MOFs was observed on a scanning electron microscope (SEM) (JSM7100F).

UV-visible absorption spectra of the prepared metal-organic framework materials were obtained with integrating sphere test mode of UV3600. And the wavelength coverage is 200 nm to 700 nm. The digital photo was taken mainly under white light and 365 nm UV light. The fluorescence emission spectra (PL) of the prepared materials were obtained using an LS-55 fluorescence spectrophotometer from Perkin Elmer, USA. The entire test was performed at room temperature with an applied voltage of 650 V and a test step of 0.5 nm.

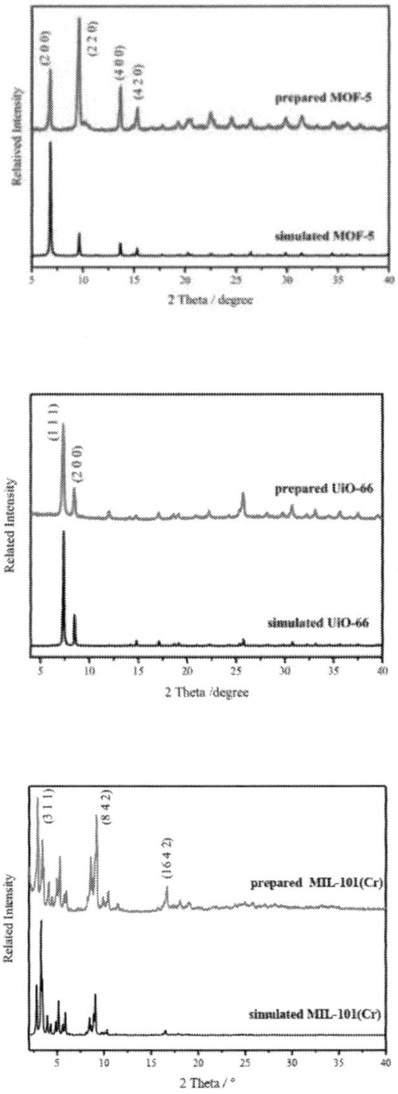

Figure 2. PXRD of MOF-5, UiO-66, and MIL-101(Cr)

3. Results and discussion

3.1 Powder X-ray diffraction (PXRD)

MOFs have the characteristics of both inorganic and organic materials. The specific crystal structures of the MOFs are generally determined by powder X-ray diffraction analyses. Decrystallization of the crystal CIF file can then lead to an X-ray diffraction pattern that can be used to evaluate whether or not the same structural MOFs were prepared. The polycrystalline powder X-ray diffraction of MOFs herein are shown in Figure 2. Through the analyses of the maps, we can find that the location and relative intensity of as-synthesized MOFs diffraction peaks are consistent with those of the simulation. These show that MOFs were successfully prepared.

Figure 3. SEM images of MOF-5, UiO-66 and MIL-101(Cr)

3.2 Scanning electron microscope (SEM)

Scanning electron microscope images of metal-organic frameworks are shown in Figure 3. MIL-101 (Cr) has a cage topology with MTN molecular sieves with a regular octahedral crystal morphology. The crystal morphology of UiO-66 is a particularly small granular morphology. The morphology of MOF-5 prepared by the mechanical method is a stack structure of various block structures. Although the MOFs have the same ligand, the crystal morphology of the MOFs is different due to the different transition metal ions. Therefore, we can speculate that the central ion plays an important role in the formation of crystals in MOFs, which further affects the luminescence properties of MOFs.

3.3 Luminescence properties

The UV absorption and fluorescence emission spectra of the MOFs are shown in Figure 4. The UV absorption peaks and fluorescence emission peaks of the MOFs are shown in Table 1. UiO-66 was excited by light with the excitation wavelength of 330 nm. The complex produces a single-peak emission of 350-500 nm. The main emission band was at 398.5 nm, which was attributed to the internal π-π^* charge transition (ILCT) of the ligand. Also UV absorption peak of UiO-66 is a broad singlet of 200-350 nm. With it excited at an excitation wavelength of 338 nm, MOF-5 produces a single emission of 400-500 nm. The most important emission band is at 440 nm, which is attributed to the π-π^* charge transition (ILCT) inside the ligand. UV absorption peak of MOF-5 is a broad singlet of 200-400 nm. The MIL-101 (Cr) was excited by light at a wavelength of 395 nm. The complex produces three emission peaks at 485 nm, 525 nm, and 580 nm. The first peak is attributed to the ligand's internal π-π^* charge transition (ILCT), and the latter two peaks are attributed to electronic transitions between the internal energy levels of the metal ion. UV absorption of MIL-101(Cr) is 200-700 nm full-spectrum absorption. In Table 2, we can see that the original white color of MOF-5 is changed to blue-violet under the irradiation of ultraviolet light. This means that MOF-5 emits blue-violet light. At the same time, under the irradiation of ultraviolet light, white color of UiO-66 becomes blue-purple, which shows that UiO-66 emits blue-violet light. MIL-101 (Cr) changes from blue-green to yellow-green under irradiation of ultraviolet light, which indicates that it emits yellow-green light.

For MOF-5, it has been studied that the luminescence of MOF-5 is derived from carboxylic acid to zinc ion charge transfer (LMCT), but theoretical calculations and spectroscopic studies have shown that MOF-5 luminescence is more likely to be based on internal ligand transitions[18, 19]. In Figure 4 we can see that terephthalic acid itself is not fluorescent. And why is the emission of MOF-5 caused by terephthalic acid when terephthalic acid is combined with zinc ions to form MOF-5? This may be because the combination of terephthalic acid and metal ions to form MOF enhances the intermolecular force, which makes the excited state molecules more susceptible to optical radiation and thus enhances fluorescence. At the same time, the outermost electron of zinc ion is the structure of d^{10}, which is in a state of full filling. This means that the outer electrons of the zinc ion are very stable and do not easily undergo electron transitions. As a result, the terephthalic acid ligand can emit light.

Table 1. UV absorption peak and fluorescence emission peak of H_3BDC, MOF-5,UiO-66 and MIL-101(Cr)

	Ultraviolet absorption peak	Flurescent emission peak
Terephthalic acid	200-350nm	※
MOF-5	200-400nm	440nm
UiO-66	200-400nm	398nm
MIL-101(Cr)	200-700nm	485nm,525nm, 580nm

※ means no fluorescence

Table 2. Photograph of three kinds of MOF under white light and ultraviolet light

	Nature light	Ultraviolet light/365nm
MOF-5		
UiO-66		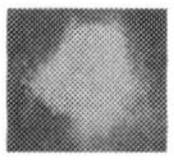
MIL-101(Cr)		

Similarly, we can find that the outermost electron of zirconium ion is the structure of d0, and it is in the empty state. This means that the outer electrons of the zirconium ion are also very stable and electron transitions are unlikely to occur. UiO-66, like MOF-5, has an organic linker ligand that is also terephthalic acid. Therefore, we can speculate that UiO-66, like MOF-5, emits light because it is based on the internal π-π^* electron transition of the organic ligand. Meanwhile, we can find that UiO-66 and MOF-5 have similar UV absorption and fluorescence emission in Figure 4. This further shows that UiO-66 emits light based on the π-π^* charge transition inside the ligand. For MIL-101(Cr), although the organic ligand of MIL-101(Cr) is also terephthalic acid, its luminescence mechanism is different from that of MOF-5. Since the outermost electron of the chromium ion is the

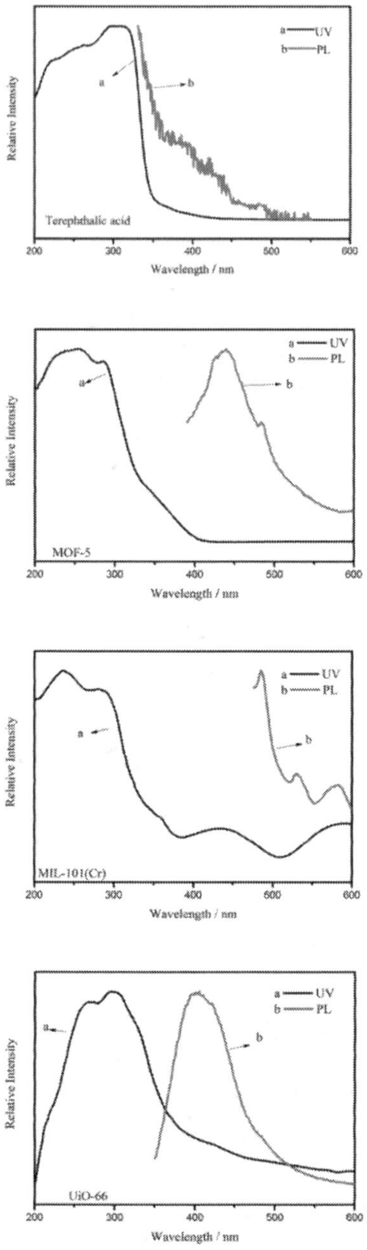

Figure 4. UV absorption and fluorescence emission spectra of terephthalic acid (H_2BDC), MOF-5, UiO-66 and MIL-101(Cr)

structure of d^3, this means that the outer electrons of the chromium ion are very active and are prone to electronic transitions. So we can speculate that chromium ions play an important role in the luminescence of MOL-101(Cr).

From Figure 5, we can see that the ground state of Cr^{3+} is ^4F. In the octahedral symmetric crystal field, ^4F splits into $^4A_2$ (ground state), $^4T_2$ and $^4T_1$; Excited ^4P splits into $^4T_1$. ^2G splits into ^2E, $^2T_1$ and $^2T_2$. ^2F splits into $^2A_2$. In Figure 4 we can see that the ultraviolet absorption spectrum of MIL-101(Cr) is composed of four main absorption peaks. According to the spin selection rule of the optical transition $\Delta S = 0$, transitions from the ground state $^4A_2$ to $^4T_2$, $^4T_1(^4F)$ and $^4T_1(^4P)$ are allowed. The absorption peak at 273 nm comes from the absorption of terephthalic acid. The absorption peaks at 237 nm, 431 nm, and 588 nm are all generated by Cr^{3+}, which are respectively caused by the energy level transitions of $^4A_2$-$^4T_1(^4P)$ and $^4A_2$-$^4T_1(^4F)$ and $^4A_2$-$^4T_2$ in Cr^{3+}. In Figure 4 we can see that there is a large overlap between the UV absorption spectrum of MIL-101(Cr) and the UV absorption spectrum of terephthalic acid, which provides plenty of possibilities that energy is transferred from terephthalic acid to chromium ions. In Figure 4 we can see that the fluorescence emission spectrum of MIL-101(Cr) consists of three emission peaks. We can speculate that the emission peak at 485 nm is caused by the charge transition of terephthalic acid to Cr^{3+}. The emission peak at 525 nm is generated by the ^2E-$^4A_2$ transition of Cr^{3+}. The emission peak at 580 nm is generated by the $^4T_2$-$^4A_2$ transition of Cr^{3+}.

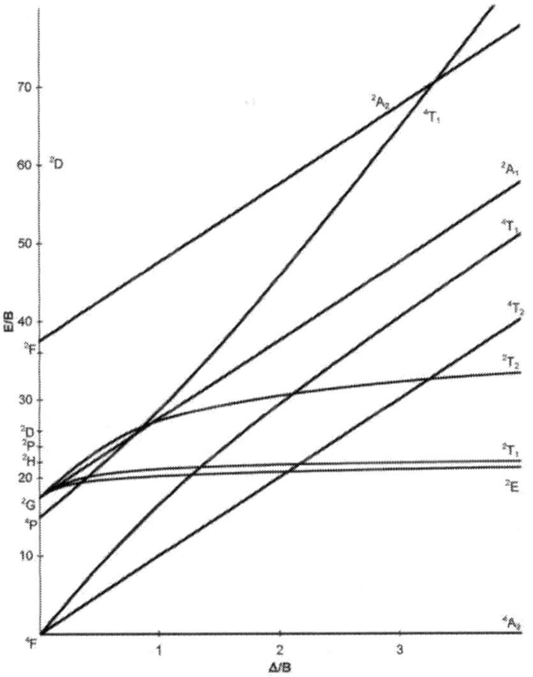

Figure 5. Crystal field splitting Tanabe-Sugano diagram of transition metal Cr^{3+}. [20]

Although the organic ligands are terephthalic acid, the three MOFs have different fluorescence effects due to their different transition metal ions. After consulting the literature, the radius of zirconium ions is 0.072 nm, the radius of chromium ions is 0.0615 nm, and the radius of zinc ions is 0.074 nm. We can find that the radius of zirconium ion is similar to the radius of zinc ion. At the same time, it can be seen from Figure 6 that the fluorescence emitted by MOF-5 and UiO-66 is mainly

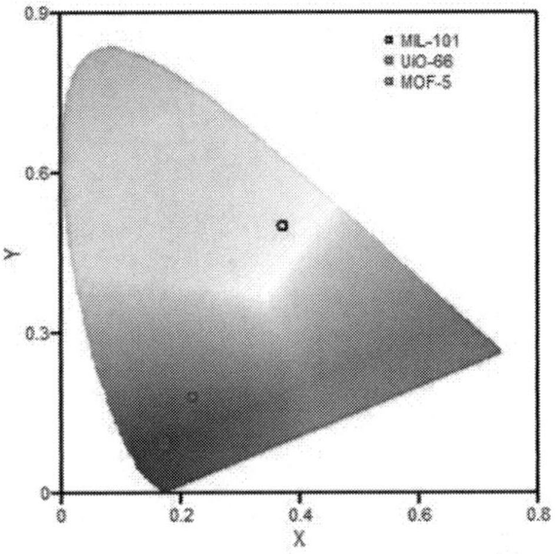

Figure 6. Three kinds of MOF chromaticity diagram

distributed in the blue-violet region of the chromaticity diagram, while the fluorescence emitted by MIL-101 (Cr) is mainly distributed in the yellow-green region of the chromaticity diagram, which is consistent with the previous analysis. We can find that MOF-5 with similar ionic radius has a similar fluorescence effect as UiO-66, which is different from the fluorescence effect of MIL-101 (Cr) with a smaller ion radius. Similarly, Wang and colleagues found that the observation of LMCT or linker luminescence depends on the metal ion radius in transition metal-containing MOFs[21]. Therefore, it is speculated that when the organic ligands are the same, the radius of the transition metal ions plays an important role in the luminescence of the MOF.

4. Conclusions

MOF-5, UiO-66, and MIL-101 (Cr) have been successfully synthesized. All three MOFs have a strong fluorescence effect, in which the fluorescence emission spectra of both MOF-5 and UIO-66 are relatively broad and emit blue-violet light, while the fluorescence emission spectrum of MIL-101 (Cr) is a multimodal fluorescence emission with multiple excited states. In addition, MOF-5 and UiO-66 with the similar transition metal ion radii have similar fluorescence effects, while the fluorescence of MIL-101(Cr) with smaller transition metal ion radius is very different from the former two MOFs. Therefore, the luminescence of the MOF is closely related to the transition metal ion radius in the MOF.

Acknowledgement

This study has been supported by the National High Technology Research and Development Program (863 program) of China No. 2012AA06A111; Hubei Provincial Natural Science Foundation of China No. 2014CFB552; Hubei Collaborative Innovation Center for Advanced Organic Chemical Materials of China No. 000-01647909.

References

1. M. D. Allendorf, C. A. Bauer, R. K. Bhakta and R. J. Houk, Chemical Society reviews, **38**, 1330-1352 (2009).

2. V. Amendola, L. Fabbrizzi, F. Foti, M. Licchelli, C. Mangano, P. Pallavicini, A. Poggi, D. Sacchi and A. Taglietti, Coordination Chemistry Reviews, **250**, 273-299 (2006).

3. R. H. Kyriakos C. Stylianou, Samantha Y. Chong, John Bacsa,, Y. Z. K. James T. A. Jones, Darren Bradshaw,* and and M. J. Rosseinsky, J. AM. CHEM. SOC., **132**, 4119–4130 (2010).

4. Banglin Chen, † Liangbo Wang,‡ Fatima Zapata,† Guodong Qian,*,‡ and Emil B. Lobkovsky§, J. AM. CHEM. SOC, **130**, 6718–6719 (2008).

5. Z. Hu, B. J. Deibert and J. Li, Chemical Society reviews, **43**, 5815-5840 (2014).

6. L. E. Kreno, K. Leong, O. K. Farha, M. Allendorf, R. P. Van Duyne and J. T. Hupp, Chemical reviews, **112**, 1105-1125 (2012).

7. Y. Cui, B. Li, H. He, W. Zhou, B. Chen and G. Qian, Accounts of chemical research, **49**, 483-493 (2016).

8. R. J. Kuppler, D. J. Timmons, Q.-R. Fang, J.-R. Li, T. A. Makal, M. D. Young, D. Yuan, D. Zhao, W. Zhuang and H.-C. Zhou, Coordination Chemistry Reviews, **253**, 3042-3066 (2009).

9. N. Stock and S. Biswas, Chemical reviews, **112**, 933-969 (2012).

10. H. C. Zhou, J. R. Long and O. M. Yaghi, Chemical reviews, **112**, 673-674 (2012).

11. P. Silva, S. M. Vilela, J. P. Tome and F. A. Almeida Paz, Chemical Society reviews, **44**, 6774-6803 (2015).

12. M. Yu, M. Hu and Z. Wu, RSC Advances, **3**, 25175 (2013).

13. Y. Cui, Y. Yue, G. Qian and B. Chen, Chemical reviews, **112**, 1126-1162 (2012).

14. T. Devic and C. Serre, Chemical Society reviews, **43**, 6097-6115 (2014).

15. D. Jiang, A. D. Burrows and K. J. Edler, CrystEngComm, **13**, 6916 (2011).

16. M. Kandiah, M. H. Nilsen, S. Usseglio, S. Jakobsen, U. Olsbye, M. Tilset, C. Larabi, E. A. Quadrelli, F. Bonino and K. P. Lillerud, Chemistry of Materials, **22**, 6632-6640 (2010).

17. W. J. Son, J. Kim, J. Kim and W. S. Ahn, Chemical communications, 6336-6338 (2008).

18. M. Ji, X. Lan, Z. Han, C. Hao and J. Qiu, Inorganic chemistry, **51**, 12389-12394 (2012).

19. P. Feng, J. Perry Iv, S. Nikodemski, B. Jacobs, S. Meek and M. Allendorf, Journal of the American Chemical Society,**44**,2010.

20. T. U. o. C. Davis, University of California Davis, 2010.

21. Y.-T. Wang, G.-M. Tang and Z.-W. Qiang, Polyhedron, **26**, 4542-4550 (2007).

An Electroactive Polymer Modified Separator with A Potential Clamp Function

Kaiyue Zhang[1], Wei Xiao[1,a] , Jianguo Liu[1] and Chuanwei Yan[1]

[1] *Institute of Metal Research, Chinese Academy of Sciences, Shenyang 110016, China*

Abstract. An inexpensive electroactive polymer polytriphenylamine (PTPAn) was synthesized through chemical oxidative polymerization of TPA. The composite separator was simply prepared by impregnating a commercial PP separator with PTPAn polymer solution. PTPAn was observed on the surface of the composite separator and formed a uniform coating. In cyclic voltammetry test, PTPAn polymer exhibited a pair of well-defined redox peaks between 3.0 and 4.5 V, which could be regarded as a reversible electrochemical p-doping and dedoping process. When incorporated in a blocking cell, the composite separator was able to hold a constant potential for current densities up to 5 mA/cm^2 to protect the cell from overcharging.

1 Introduction

Overcharge is a condition in which charging current is applied to a battery that has reached its end-of-charge potential, raising cell potential to undesirably high values. Overcharge may cause electrolyte breakdown, electrode degradation, current collector corrosion, gas evolution and so on [1]. Meanwhile, increased internal pressure and generated heat increase the risk of explosions and fires.

The common external methods for overcharge protection add weight, cost, and complexity to the battery pack. Redox shuttles, as a reversible internal protection mechanism, are widely used to allow for continued operation in the event of overcharge [2-4]. However, the overcharge performance is fundamentally limited due to redox shuttle solubility, diffusion coefficient, and chemical stability [5]. In recent years, an alternative approach using electroactive polymers for overcharge protection has aroused the attention of researchers. When incorporated into a battery separator, electroactive polymers can provide a reversible internal short upon overcharging to maintain the cell potential within a safe range. According to physical properties of electroactive polymers, various film-forming methods have been discussed, such as roll pressing [6], impregnating [7], electrospinning [8] and so on. The film-forming method seriously affects utilization and stability of the polymers. Therefore, the film-forming methods of electroactive polymers hold considerable future promise.

In this study, we tried to synthesize an inexpensive electroactive polymer polytriphenylamine (PTPAn), which can be oxidized to a conductive state at a higher potential. The composite separator was prepared simply by impregnating a commercial porous separator with PTPAn polymer solution. The microstructure and electrochemical property of the composite separator were evaluated. The overcharge property of blocking cells assembled with the composite separator and the commercial PP separator was also investigated..

2 Experimental

2.1 Preparation of polytriphenylamine

The triphenylamine (TPAn) of 0.025 mol was dissolved in 100 mL chloroform to form a mixed solution, with brief ultrasonic treatment to speed dissolution. Then 0.1 mol $FeCl_3$ oxidant (4 times than TPAn) was added into the reaction solution within an hour. The oxidative polymerization of TPA usually continued for 12 h in an inert gas condition. After that, the reaction mixture was poured into methanol to deposit the polymer product, which was then filtered and washed with methanol several times. In order to purify the coarse PTPAn, the collected polymer powders were dissolved again in chloroform to remove the insoluble residues by filtration. The filtrate was re-deposited with methanol. Finally, the polymer product was filtered and dried in vacuum at 60 °C for 24 h.

2.2 Preparation of the composite separator

The PTPAn-modified separator was prepared using a dip-coating process. Firstly, the electroactive polymer of PTPAn was dissolved in chloroform to form a 2 wt.% polymer solution. Then the PTPAn-modified separator was prepared simply by immersing a commercial PP separator (Celgard 2500, thick=25μm, porosity=55%) into the polymer solution for 10 min under ultrasonic

[a] Corresponding author: wxiao@imr.ac.cn

condition, evaporating the solvents off to leave a modified separator. The resulting composite separator was 27 μm thick and the mass gain is about 0.2 mg / cm^2 after the PTPAn coating progress.

2.3 Characterization of the composite separator

The surface morphology of the separators was examined using a scanning electron microscopy (JEOL, JSM-6300). The phase structure of prepared PTPAn was studied using FT-IR (IRTracer-100) and XRD (D/max2200VPC). The thermal shrinkage of the separators was determined by measuring the dimensional change before and after being exposed at 140 °C for 0.5 h. The electrochemical behavior of prepared PTPAn was examined by cyclic voltammetry (CV). The working electrode used for CV was a PTPAn-coated Al foil electrode, which was prepared by impregnating Al foil with a chloroform solution of PTPAn, and then taking out for evaporating the solvents off to leave a PTPAn coating layer on the Al foil electrode. The CV curves were recorded on an electrochemical workstation (Princeton Applied Research 273, USA) using a two-electrode CR2032-type cell with PTPAn-coated Al foil electrode as working electrode and a Li foil as both counter electrode and reference electrode in the electrolyte (1.0 M $LiPF_6$ in EC/DEC). The overcharge performance of the composite separator and the commercial PP separator was evaluated in a blocking cell with the composite separator sandwiched between a stainless steel current collector and a Li foil in the electrolyte. After a 30 min rest period, increasing currents were applied to the cell, each holding for 10 min before stepping to the next current level and the cell potential was recorded.

3 Results and Discussion

It is well known that PTPAn can be synthesized by chemical and electrochemical polymerization of TPA. In contrast, PTPAn produced from chemical polymerization has a higher degree of polymerization and structural homogeneousness. Therefore, we adopted chemical polymerization to prepare PTPAn.

The FTIR spectra of TPAn and PTPAn are shown in Fig. 1. The main vibration bands characteristic of TPAn can also be found in PTPAn, such as C–C stretching at 1491 cm^{-1}, C–H bending at 1322 cm^{-1} and C–N stretching at 1274 cm^{-1}.Compared with TPAn, a new band at 819 cm^{-1} appears in PTPAn, which belongs to the C–C bending between the benzene rings of TPA. In addition, the peak at 1275 cm^{-1} is broader than that in the TPAn and the peak at 1596 cm^{-1} is shifted from its position in the monomer to higher wave numbers. In the region 900 to 1200 cm^{-1}, the peak profile has changed due to higher concentration of para substituted phenylene rings in PTPAn. These results suggest a highly crosslinked structure of the PTPAn polymer, which is consistent with previous study [9,10].

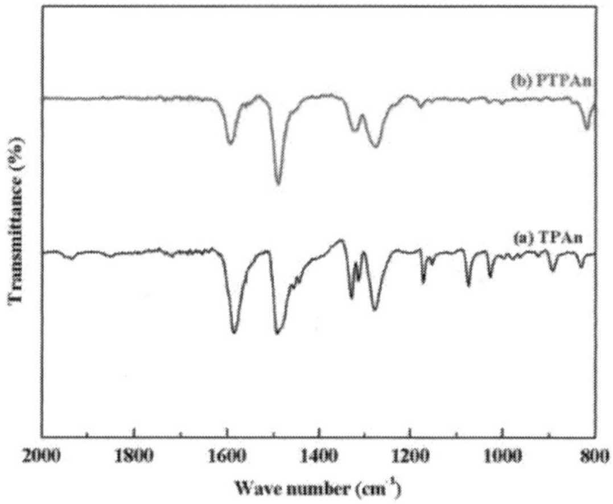

Figure 1. The FTIR spectra of TPAn and PTPAn.

Fig. 2 illustrates the XRD patterns of TPAn and PTPAn. It is obvious that TPAn exists as a crystal phase. After oxidation polymerization, the crystallographic structure has been significantly changed. And a single broad diffraction peak recorded at 20° can be found, indicating that TPAn can entirely change into PTPAn (amorphous phase).

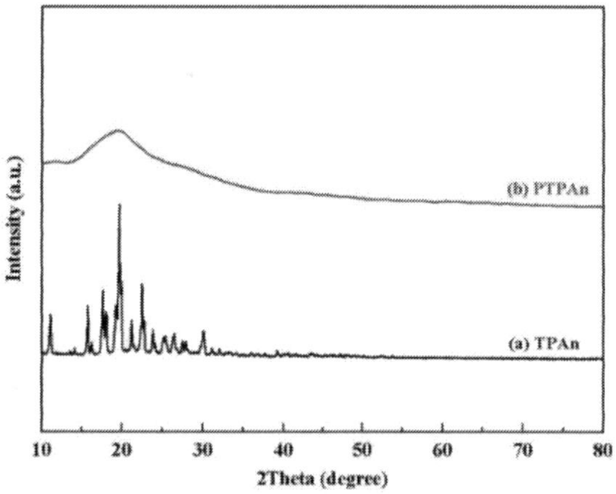

Figure 2. The XRD patterns of TPAn and PTPAn.

Fig. 3 shows the CV curves of PTPAn.The PTPAn polymer exhibits a pair of overlapped redox peaks between 3.0 and 4.5 V, which can be regarded as a reversible electrochemical p-doping and dedoping process. The anodic and cathodic branches of PTPAn are very symmetric in the peak height and peak areas, which contributes to reversible and rapid electrochemical processes of the polymer. In the doping process (forward scan), the anodic current of PTPAn appears at the onset potential of 3.7 V. In the dedoping process (reverse scan), the cathodic current of PTPAn starts at the reversed potential of 3.8 V and reaches its maximum current at 3.5 V. These results indicate that the PTPAn can be used to provide reversible overcharge protection.

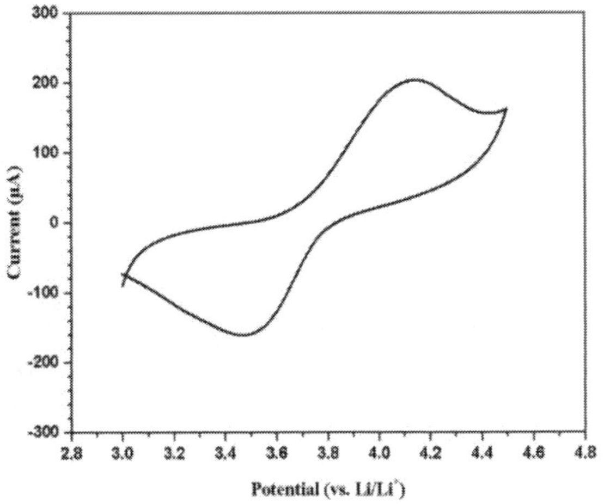

Figure 3. The CV curve of PTPAn. Scan rate: $10mVs^{-1}$.

The surface morphologies of the PP separator and the composite separator are shown in Fig. 4. The PP separator exhibits a uniform porous structure. After modification, PTPAn is observed on the surface of the composite separator and forms a uniform coating. Also PTPAn polymer exists within the PP separator pores and does not completely block the pores.

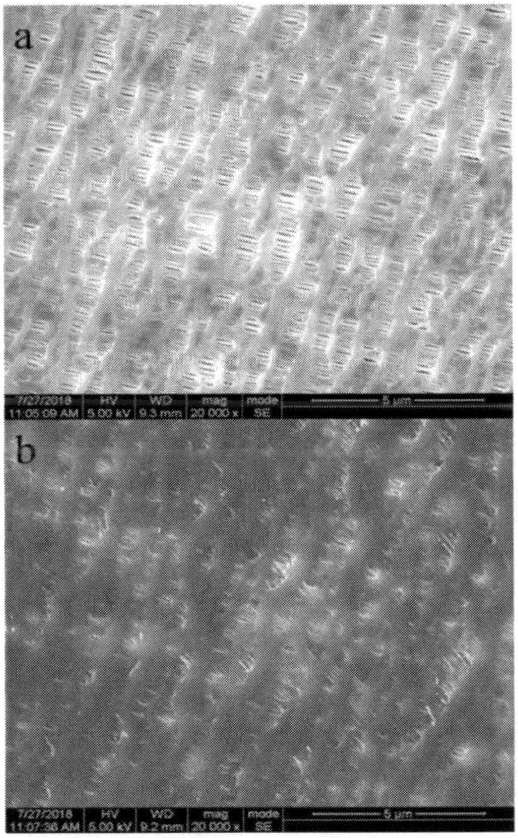

Figure 4. The surface morphologies of (a) the PP separator and (b) the composite separator

Fig. 5 shows the photographs of the PP separator and composite separator before/after being exposed to 140 °C for 0.5 h. Since the PP separator has a melting point of about 160 °C, it easily loses dimensional stability upon 140 °C. While the composite separator distinctly suppress the thermal shrinkage compared with the PP separator owing to the existence of the electroactive polymer PTPAn. As with other polymer materials, the heat resistant PTPAn is believed to effectively prevent the composite separator from being thermally shrunk.

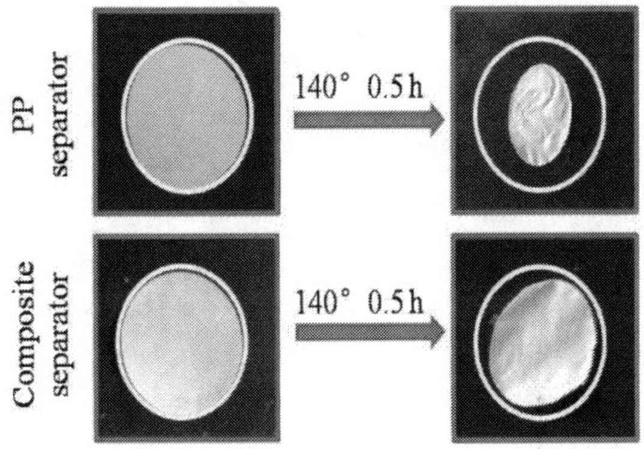

Figure 5. The photographs of the PP separator and composite separator before/after being exposed to 140 °C for 0.5 h.

Fig. 6 shows the voltage profile of the blocking cells with the PP separator and composite separator at various applied current densities. As shown in Fig. 6, the cell with PP separator can not remain on a constant potential. And the potential increases sharply with the increase of applied current density. While the cell with the composite separator is able to hold a constant potential for current densities up to 5 mA/cm^2, which is ascribed as the electrochemical p-doping process of PTPAn. The p-doping process results in cell short. Therefore, the composite separator can maintain the cell potential within a safe range in case of overcharging based on its reversible internal short.

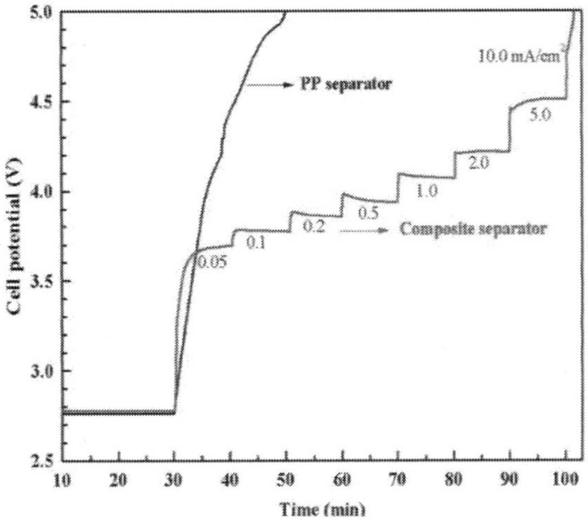

Figure 6. The voltage profiles of the blocking cells with the PP separator and composite separator at various current densities.

4 Conclusion

The composite separator was successfully prepared by impregnating a commercial PP separator with PTPAn polymer solution. PTPAn was observed on the surface of the composite separator and formed a uniform coating. The PTPAn polymer exhibited a pair of well-defined redox peaks between 3.0 and 4.5 V, which could be regarded as a reversible electrochemical p-doping and dedoping process. When incorporated in a blocking cell, the composite separator was able to hold a constant potential for current densities up to 5 mA/cm^2 to protect the cell from overcharging.

Acknowledgements

This work was supported by the National Natural Science Foundation of China (No. 21676282), Natural Science Foundation of Liaoning (No. 20180510039) and Shenyang Science and Technology Project (No. 18013051).

References

1. S. A. Odom, S. Ergun, P. P. Poudel, R. P. Sean, Energy Environ. Sci. **7**, 760 (2014)
2. Z. Zhang, L. Hu, H. Wu, W. Weng, M. Koh, P. C. Redfern, L.A. Curtissb, K. Aminead, Energy Environ. Sci. **6**, 1806 (2013)
3. W. Weng, Z. Zhang, P.C. Redfern, L.A. Curtiss, K. Amine, J. Power Sources **196**, 1530 (2011)
4. B. Wang, T. J. Richardson, G. Chen, Phys. Chem. Chem.Phys. **15**, 6849 (2013)
5. L. Zhang, Z. Zhang, H. Wu, K. Amine, Energy Environ. Sci. **4**, 2858 (2011)
6. X.M. Feng, J.Y. Zheng, J.J. Zhang, R.F. Li, Z.J. Li, Electrochim. Acta **54**, 4036 (2009)
7. S.L. Li, L. Xia, H.Y. Zhang, X.P. Ai, H.X. Yang, Y.L. Cao, J. Power Sources **196**, 7021 (2011)
8. B. Wang, T.J. Richardson, G. Chen, J. Electrochem. Soc. **161**, A1039 (2014)
9. J.K. Feng, Y.L. Cao, X.P. Ai, H.X. Yang, J. Power Sources **177**, 199 (2008)
10. A. Petr, C. Kvarnstrom, L. Dunsch, A. Ivaska, Synthetic Met. **108**, 245 (2000)

Study on aluminum foam as a filler material for impact limiter of spent fuel transport cask

Zhongfang Li, Siyi Yang*, Haile Xu and Yukun An

School of Mechanical Engineer, Shandong University of Technology, Zibo, China

Abstract. Spent fuel transport cask is a significant carrier of spent fuel transport. The main function of impact limiters installed at both ends of the container is to absorb energy and limit overload to ensure the integrity of the structure. The quasi-static compression process of aluminum foam was simulated on the platform of ANSYS Workbench. Foam aluminum was prepared by melt foaming method and quasi static compression test was carried out. The experimental results show that the deformation process of aluminum foam is basically the same as that of experiment, and the aluminum foam has good compressive and energy absorption properties. The yield stress (σ_{ys}) and plateau stress (σ_{pl}) of aluminum foam with density of 0.64 g/cm^3 can reach 8.26 MPa and 11.11 MPa respectively, and the energy absorption capacity (W_{EA}) and unit energy absorption capacity (W_{SEA}) can reach 6.31×10^3KJ/m^3 and 9.87 KJ/Kg respectively, and the difference between the foam with density of 0.61g/cm^3 and its various properties is very small. It can be concluded that the aluminum foam in a certain density range has roughly the same performance, and it also reflected the stability of aluminum foam's performance. Additionally, aluminum foam is an isotropic material, which can overcome directional limitation when used as shock absorber filler material for spent fuel transport cask.

1 Introduction

Spent fuel transport is the link between nuclear power plant and reprocessing plant or final disposal plant. It controls the continuous operation of both plants to a certain extent, and spent fuel transport cask is a significant carrier to achieving this process [1, 2]. To ensure the integrity of the transport cask under normal transport and accident conditions, the International Atomic Energy Agency (IAEA) issued the "Regulations for the Safe Transport of Radioactive Materials"[3]. China had made GB11806 standards on this basis as well [4]. The transport cask should be tested through 9 m drop test, 1 m penetration test, 30 min fire (800 °C) and 15 m flooding test. For the first test, the cask dropped freely onto an unyielding surface from 9 m. Pair of impact limiters installed on the top and the bottom of a transport cask is the crucial safety component to absorb most of the impact energy and limit overload, which can ensure the integrity of the structure.

Over the years, many scholars have studied several of impact limiter filler materials. Zhou Y, You M and Zhang Y [2] designed and studied China's RY-I type spent fuel transport cask, in which paulownia wood was used for the first time to replace light wood as impact limiter filler material. Silva and Kyriakides [5] evaluated the compressive response and failure behavior of balsa wood based on its microstructure and found that the internal structure of balsa wood was composed of slender cells with similar hexagonal cross-sections that arranged axially. Choi and Seo [6] conducted dynamic impact experiments on polyurethane foam, honeycomb, and balsa wood respectively. The experimental results showed that balsa wood with the axial wood grain had the most significant absorbed energy.

As the drop mode of spent fuel transport cask under accident conditions was summarized as three types (Fig. 1): vertical drop, horizontal drop, and corner drop [6]. Wood, the main filler material at present, is restricted by directivity because it is an anisotropic material. In addition, due to the water absorption of wood and low ignition point, the influence of environmental factors would reduce the life of the impact limiter. Aluminum foam is an isotropic material with good compressive and energy absorption properties, which can ensure the integrity of the cask when three possible dropping modes occur. In this paper, the compression process of aluminum foam was simulated on the platform of ANSYS Workbench, and quasi-static compression experiment was designed on the basis of which the compression performance and energy absorption performance were verified.

* Corresponding author: yangsiyichina@yeah.net (Siyi Yang)

https://doi.org/10.1051/matecconf/201823805006

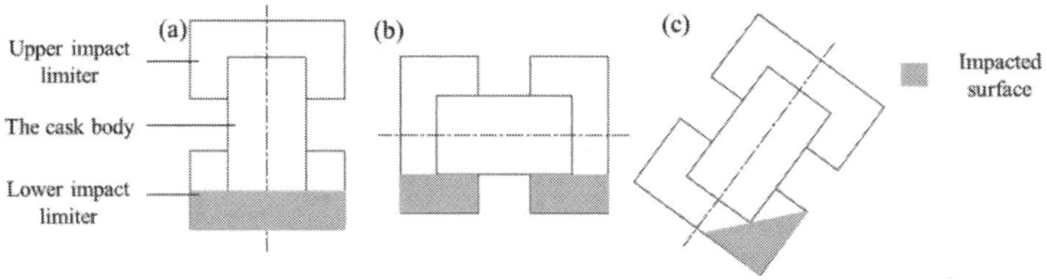

Figure 1. Dropping conditions of spent fuel transport container: (a) end drop; (b) corner drop; (c) side drop.

2 Modeling and simulation

2.1 Preparation of aluminum foam model

There are three main modeling methods for aluminum foam: repeated fixed pore structure (spatial lattice solid and beam structure) [7], randomly generated pore (2D circular pore and 3D Voronoi) [8] and physical scanning technology (CT scanning) [9]. Simulating foam materials with repeated fixed structure models have been common in the study of various foam materials [10]. The results of the large deformation compression simulation of the tetrahedron model of aluminum foam show that the stress-strain relationship of [10] begins to converge obviously after using 27 elements of 3 × 3 × 3, so it is unnecessary to use more repetitive structures to obtain more accurate results. From reference [11], the mechanical properties of aluminum foams with high porosity (Pr = 85%~90%) are similar, because their Young's modulus is basically the same, and the stress of yield platform is less than 1 MPa. Therefore, FCC pore structure model with porosity of 90% (Fig. 2) was used in this paper.

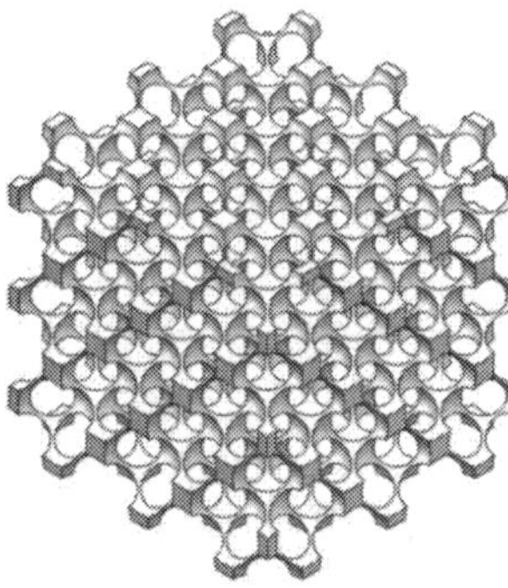

Figure 2. FCC pore structure model.

2.2 Quasi-static compression simulation and results discussion

2.2.1 Setting of simulation parameters

The quasi-static compression behavior of aluminum foam was simulated by using the Static Structure module of ANSYS Workbench. In the simulation, aluminum (ALUMINUM ALLOY) in the ANSYS material library was selected as the basic material, and the specific parameters of aluminum material were determined by referring to the parameters of ANSYS and other data. The specific parameters were as follows: Young's modulus was 71GPa, Poisson's ratio was 0.33, bulk modulus was 69.61GPa, and shear modulus was 26.69 GPa. The basic stress-strain relationship of aluminum is bilinear isotropic hardening (BISO). The yield point and the tangent modulus were set as 2.8 MPa and 500 MPa respectively [10].

2.2.2 Meshing and setting of initial boundary conditions

The Aggressive Mechanical option for large deformation simulation was used for the whole mesh, and Mesh Morphing was activated, which allows geometric adjustment of meshes under large deformation. The mesh type was the default ten-node tetrahedron of the system platform, and the cell size was defined as 0.1mm. Fig. 3 (a) shows the overall mesh generation effect of the model.

Fig. 3 (b) shows the setting of initial boundary conditions. Fixed Support was set at the lower end face of the lower die, and displacement was set at the upper end face of the upper die with a value of -24 mm.

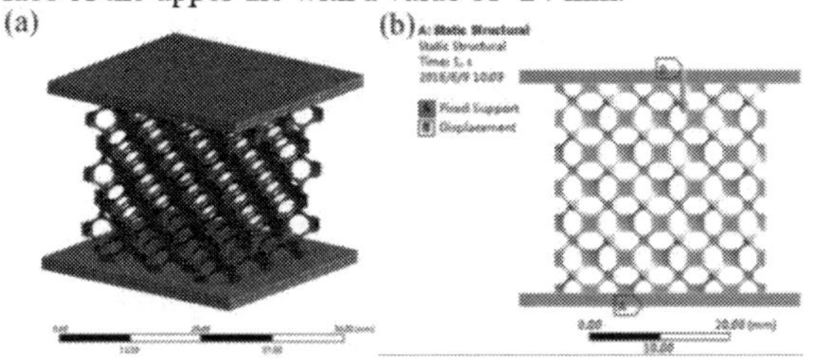

Figure 3. (a) mesh of the FCC model; (b) boundary and initial conditions.

2.2.3 Discussion of simulation results

Fig. 4 shows that there is no obvious nonlinear buckling deformation in the quasi-static compression process, and the forces acting on the wall and the beam of the face-centered cubic cavity structure are basically the same. There is a small stress concentration phenomenon on where the foam aluminum model contacts with the upper and lower pressure plates. It shows that the foam aluminum model with face-centered cubic cavity structure had uniform stress distribution during quasi-static compression, and the stress of the die can be well dispersed, which indicates that the foam aluminum model has good performance characteristics.

https://doi.org/10.1051/matecconf/201823805006

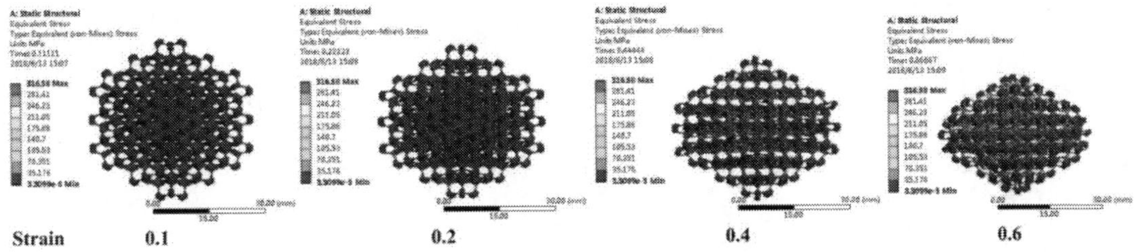

Figure 4. Von-Mises stress cloud of Al-foam model (FCC) with different strain under.

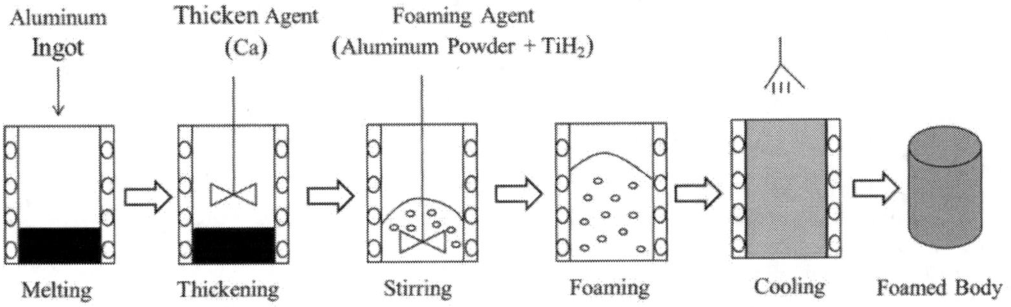

Figure 5. Production flow chart of aluminum foam.

3 Materials and method

3.1 Preparation of aluminum foam body

In this paper, closed-cell aluminum foam body was prepared by melt foaming method. After melting, thickening, stirring, foaming and cooling, the foamed body was produced finally. Among them, Ca particles with mass fraction of 2% were used as thicken agent, and mixed foaming agent, which was composed of 800g aluminum powder and 8g TiH2 and milled for 5 hours by planetary ball mill (QM-3SP2), was used as foaming agent. The specific preparation process [12] is shown in Fig. 5.

3.2 Method

To avoid size effect, samples for compression test were cut into the dimension of 25 × 25 × 25 mm3 using a wire electro-discharging machine (WEDM). As the density of foam increases from top to bottom, aluminum foam with high density at the bottom was chosen as the sample. the location illustration of sample selection is shown in Fig. 6. Quasi-static compression tests were implemented using a universal testing machine (WDW-3100) under a pressure head speed of 2 mm/min up to a nominal strain of 0.8. In order to avoid the uncertainty and anisotropy during the tests, at least three samples were tested and the average data was adopted for each parameter.

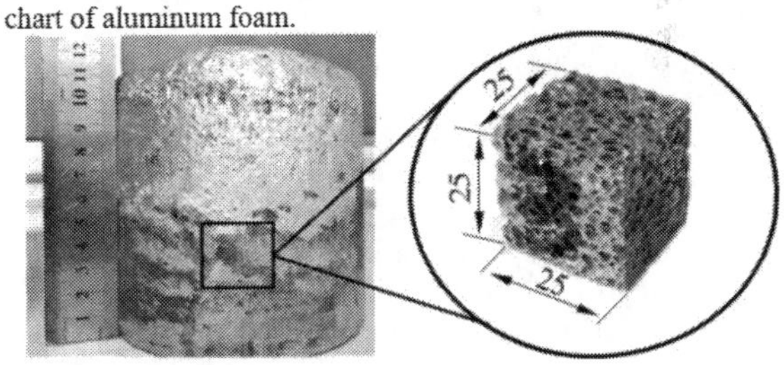

Figure 6. Location illustration of sample selection and sample size.

4 Results and discussion

4.1 Compressive property

Fig. 7 shows the stress-strain curves of aluminum foams with different densities after quasi-static compression tests. With the increase of pressure, aluminum foams have experienced three typical compression deformation stages. Under the compressive load, the aluminum foam first entered the elastic deformation stage. After the yield stress had been reached, the foamed aluminum began to enter the stress plateau stage where the deformation process was relatively stable and it stopped until the strain was 0.55. Subsequently, the aluminum foam began to enter the densification stage, during which the compressive stress increased rapidly [13]. The whole compression deformation process is basically consistent with the finite element simulation of the compression process.

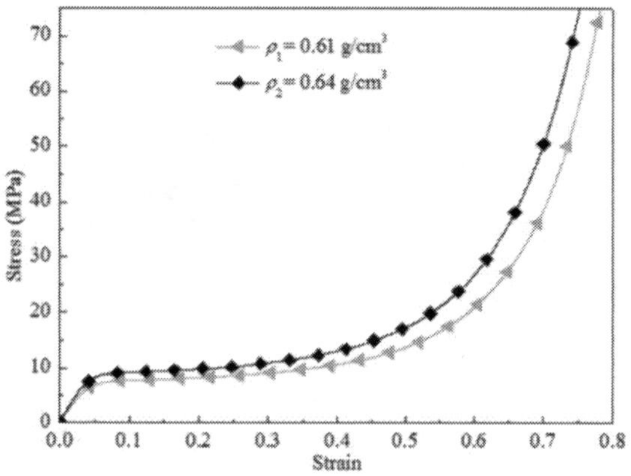

Figure 7. Stress-strain curves of aluminum foam compression with different strain rate.

In order to compare and characterize the test data of the aluminum foam with different densities, the peak stress (σ_{ys}) and the plateau stress (σ_{pl}) were selected as the main parameters of the compression property comparison. σ_{pe} defined as the highest stress point in the elastic phase, σ_{pl} defined as the average stress of strain between 0.2~0.4 [14] respectively. The above parameters were statistically calculated, and the obtained data were shown in Table 1. From the data of Table 1, we can see that the two kinds of aluminum foam have good compression properties. However, the yielding stress (σ_{ys}) and plateau stress (σ_{pl}) of aluminum foam with density of 0.64 g/cm^3 are 1.45 MPa and 1.89 MPa higher than those of aluminum foam with density of 0.61 g/cm^3, respectively. It can be seen that the compressive properties of aluminum foam improve with the increase of density.

Table 1. Formatting sections, subsections and subsubsections.

Density (g/cm^3)	Stress (MPa)	
	σ_{ys}	σ_{pl}
0.61	6.81	9.22
0.64	8.26	11.11

4.2 Energy Absorption Property

Energy absorption capacity (W_{EA}), and the specific energy absorption (W_{SEA}) for weight sensitive applications were applied to characterize the energy absorption property of the materials [15], defined as:

$$W_{EA} = \int_0^{\varepsilon} \sigma(\varepsilon) d\varepsilon \tag{1}$$

$$W_{SEA} = \frac{W_{EA}}{\rho} \tag{2}$$

Where ε is the strain, $\sigma(\varepsilon)$ is the stress at the strain, ρ is the density of the sample. W_{EA} and W_{SEA} versus compressive strain curves were plotted in Fig. 8.

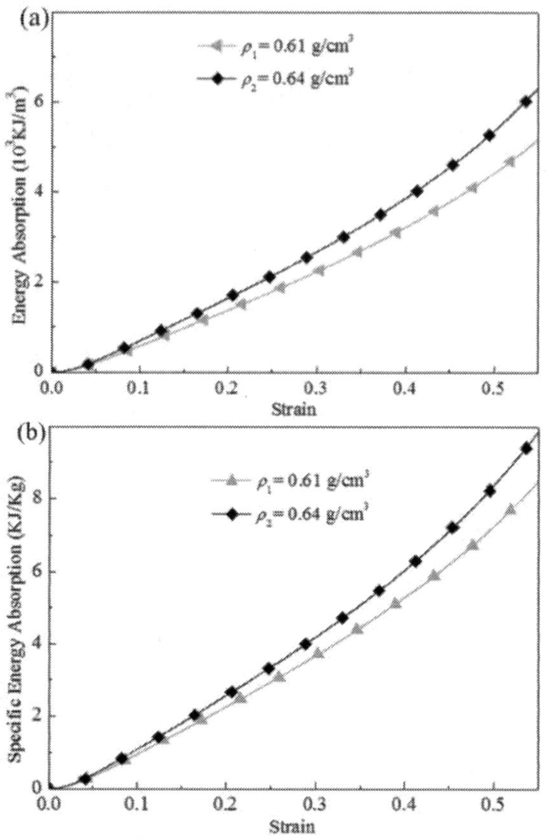

Figure 8. (a) Energy absorption, and (b) specific energy absorption curves of aluminum foam.

As can be seen from Fig. 8, parameters WEA and WSEA increase with the increase of compressive strain, and the slope of the curves increase gradually during the densification stage. The energy absorption capacity (WEA) of aluminum foams with densities of 0.61 g/cm^3 and 0.64 g/cm^3 were 5.18 × 10^{3}KJ/m^3 and 6.31 × 10^{3}KJ/m^3 respectively, and the W_{SEA} values were 8.49 KJ/Kg and 9.87 KJ/Kg respectively. For these two values, the differences between aluminum foams with densities of 0.61 g/cm^3 and 0.64 g/cm^3 are about 1.5. Hence, the energy absorption performance of aluminum foam is stable in a density range. When aluminum foam is applied to the shock absorber of spent fuel transport cask, the material can be filled into the impact limiter according to the density to obtain the required energy absorption performance requirements.

5. Conclusion

According to the mechanism of filler material for impact limiter of spent fuel transport cask, aluminum foam was selected and prepared by melt foaming method as filler material. The results of finite element simulation and quasi-static compression experiments show that aluminum foam has better compressive performance and overall energy absorption than traditional filler materials. The yield stress (σ_{ys}) and plateau stress (σ_{pl}) of aluminum foam with density of 0.64g/cm^3 can reach 8.26 MPa and 11.11 MPa respectively, and the energy absorption capacity (W_{EA}) and unit energy absorption capacity (W_{SEA}) can reach 6.31 × 10^{3}KJ/m^3 and 9.87 KJ/Kg respectively, and the difference between the foam with density of 0.61g/cm^3 and its various properties is very small.

It can be concluded that the aluminum foam in a certain density range has roughly the same performance, which reflects the stability of its performance more easily, and can be used for impact limiter installed on spent fuel transport cask more convenient.

Acknowledgment

This work was supported by the Natural Science Foundation of Shandong Province under Grant [number ZR2018MEM006]

References

1. G-S You, W-M Choung, J-H Ku, et.al. NUCL ENG TECHNOL, **41**, 859-866 (2009)
2. Y Zhou, M You and Y Zhang, China Nuclear Science and Technology Report, 315-325 (1993)
3. IAEA, Regulations for the Safe Transport of Radioactive Material, Standard No. TS-R-1 (2009)
4. AQSIQ and SAC, Regulations for the Safe Transport of Radioactive Material, Standard No. GB11806-2004 (2014)
5. Silva AD, Kyriakides S. INT J SOLIDS STRUCT.; **44**, 685-717 (2007)
6. Choi W-S, Seo K-S. NUCL ENG DES.**240**, 925-32 (2010)
7. Zhang X, Jia G, and Huang H, Chinese J Aeronautics, **24**, 734 (2011)
8. Siegkas P, Tagariell V, and Petrinic N, Procedia Mater Sci, **4**, 212 (2014)
9. X Zhu, S Ai, D Fang, B Liu, and X Lu, Computational Materials Science, **82**, 451-456 (2014)
10. Z Hu, A Ma, D Yang, J Chen, and J Jiang, Materials Review, **29**, 124-131 (2015)
11. T Mukai, T Miyoshi, S Nakano, H Somekawa, and K Higashi, **54**, 533-537 (2006)
12. Y An, S Yang, Wu H, E Zhao, Z Wang, MATER DESIGN, **134**, 44-53 (2017)
13. Y An, S Yang, E Zhao, Z Wang, COMPOS STRUCT, **178**, 288-296 (2017)
14. Institution BS. Mechanical testing of metals-Ductility testing-Compression test for porous and cellular metals. 2011
15. LL Yan, ZY Zhao, B Han, TJ Lu and BH Lu, Materials Letters, 70-73 (2018).

Preliminary study of vibration characteristics of intelligent beam with the influence of SMA's restoring force

Chuanliang Shen[1], Wei Su[1,a], Guochao Sun[1], Binyu Zhou[1], Tairong Sun[1]and Yi Liu[1]
[1]*State Key Laboratory of Automotive Simulation and Control, Jilin University, Changchun 130025, China*

Abstract. In order to solve the problem of noise and comfort caused by vibration during automobile driving, this paper proposes to use shape memory alloy wire as driving element and simplified beam structure to form intelligent beam composite structure, and use the restoring force generated by shape memory alloy wire to realize vibration control of the structure. several macroscopic phenomenological constitutive models of NiTi alloy were summarized and analyzed. On the basis of the Liang-Rogers one-dimensional constitutive model, Simulating the response force changing process of the SMA wire by useing MATLAB. According to the oretical analysis, we can obtain the system stiffness matrix and mass matrix, establish the dynamic model of smart beam composite structure, and analyze the interaction between shape memory composite wire prestress and natural frequency. The numerical simulation method is used to establish the simulation model and study the influence of the size and position parameters of the intelligent beam structure and the restoring force on its vibration characteristics. The results show that the natural frequency of the beam decreases and the natural frequency of the SMA wire increases with the increasing restoring force of SMA wire. As the distance from the SMA wire to the neutral axis increases, the natural frequency of the system increases, revealing the mechanism of vibration control of the shape memory alloy smart beam composite structure.

1. Introduction

Automobile vibration and noise are the important indexes that affect the ride comfort of the occupant. Serious car vibration and interior noise will also affect the safety of the vehicle. The automobile body is mainly composed of thin-walled panel and beam structure, which are connected to each other and constitute the overall frame structure of the automobile body. It is very necessary to restrain the vibration of body beam structure because the vibration of body beam will affect the ride comfort. The strength and stiffness of the structure may be adaptive if the shape memory alloy be used as the driving force of execution device compound or embedded into the structure. There are various deformation under the external excitation, elastic modulus can be changed along with the phase change, and great restoring force can be generated when limited recovery is achieved, so as to effectively suppress the vibration of the body structure.

When SMA plays the shape memory effect, the material can generate recovery stress of up to 700MPa and strain of about 8%[1]. The huge recovery force generated when limited recovery is used can change the stiffness characteristics of the structure, make the structure's own frequency avoid the external excitation frequency, and reduce the vibration response of the structure. At the same time, the elastic modulus of SMA wire is increased by more than 3 times after the phase transformation under heat, which can increase the local or overall stiffness of structure [1,2].In 1989, Rogers [3] first made theoretical analysis of SMA applied in vibration control of composite structures. According to different modes of vibration suppression, SMA was used for structural vibration control into active control, passive control, semi-active control and mixed control[4].Using the shape memory effect of SMA wire, McGavin[5] applied it to the vibration suppression of frame structure and carried out experimental verification. The test results showed that the restoration force generated by SMA wire under heat can effectively change the natural frequency of frame structure, and the fundamental frequency of the driven system increased by 32%.Jaakko Heinomen et al. explained the principle of SMA for semi-active control in a single-degree of freedom system in detail, activated SMA when the system frequency was 2.4hz, and increased the stiffness property of the structure[6]. By stopping heating SMA at 7.5hz, the stiffness of the structure is reduced, and the amplitude of the system is effectively reduced. Then, carried out the simulation of semi-active control of the intelligent system by using ABAQUS, and obtained the same damping effect. Barbara Tiseo et al. made an in-depth study on the change of stiffness caused by the change of the elastic modulus of SMA wire, and designed an ATDVA device [7]. The working mechanism was: when the SMA wire was heated, the NiTi alloy wire changed from martensite to austenite, the elastic modulus increased, and the stress inside the wire in the limited recovery increased. The structure is applied to the design of aviation industry, and aluminium alloy plate and glass fibre can effectively suppress vibration . In existing research for restoring force effect on the vibration of the beam structure study is infrequently.In existing research for restoring force effect

[a] Corresponding author: 3339177157@qq.com

on the vibration of the beam structure is infrequent.In this paper, the finite element model of the composite structure is established ,this structure contains intelligent beam with shape memory alloy wire and simple support beam. The influence law between the restoring force of shape memory alloy wire and the vibration characteristics of intelligent beam is analyzed through numerical analysis and simulation, to reveal the mechanism of vibration control.

2. Modeling of intelligent beam structure

2.1Simplification of intelligent beam structure model

The simplification of the shape memory alloy intelligent beam composite structure and the discretization of the structure can make it convenient to describe the engineering problem and use time variable to describe the mathematical model of the discrete system in the dynamic problem. The motion equation can also be expressed in the form of differential. In this paper, the beam -SMA structure system is discretized into a finite number of n elements. The main mechanical behaviours of each element (such as strain, stress, displacement, etc.) can be expressed by the displacement of the nodes on the element. The mechanical behaviours of the system can be described by the displacement of all nodes. It is assumed that both ends of each unit are connected with pre-stressed SMA filaments, as shown in figureure 1.The length of the element is L_i, and the nodes at both ends are i and j. V_i and V_j, M_i and M_j respectively represent the external force and torque applied on node i and j,M_N is the control torque generated by SMA wire.

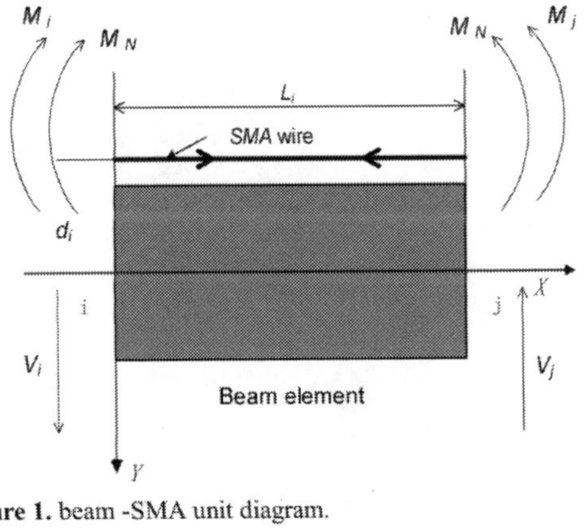

figure 1. beam -SMA unit diagram.

2.1. Natural frequency of the system

The dynamic equation of the beam -SMA system can be described as[8] :

$$M\ddot{\delta} + C\dot{\delta} + K\delta = F \quad (1)$$

M、C、K are the Mass matrix, damping matrix and stiffness matrix respectively.δ is the displacement vector of the whole system.

If the effect of damping and external load is ignored, the above equation of vibration differential equation of linear time-invariant system can be described as:

$$[M]\{\ddot{\delta}\} + [K]\{\delta\} = 0 \quad (2)$$

$$\{\delta\} = \{\varphi\}_i \cos\omega_i \quad (3)$$

Then

$$\left([K] - \omega_i^2 [M]\right)\{\varphi\}_i = \{0\} \quad (4)$$

$$\{\varphi\}_i \neq 0 \quad (5)$$

$$[K] - \omega_i^2 [M] = 0 \quad (6)$$

$$f_i = \frac{\omega_i^2}{2\pi} \quad (i = 1,2,3\cdots) \quad (7)$$

In the above equation, ω_i is the natural circular frequency (radian/second) of the beam-SMA structure, f_i is the natural frequency, and the mass matrix and stiffness matrix of the structure determine the natural frequency of the system. Considering that small deformation of the structure will not change the stiffness characteristics of the structure, the structure mass matrix [M] is not affected by the temperature, and the material properties and restoring force of the stiffness matrix [K] are all functions of the temperature. It can be seen that the variation of SMA wire temperature causes the change of structural stiffness and changes the vibration characteristics of the structure.

3. Analytical analysis of intelligent beam structure

3.1. Relationship between pressure and frequency

The restoring force generated by SMA wire can be regarded as the axial force exerted on both ends of the simple beam structure, regardless of whether the axial force is eccentric or not. Based on ignoring shear and rotational inertia, the bending vibration balance equation of the simple beam is as follows[9] :

$$\frac{\partial^2}{\partial x^2}\left(EI\frac{\partial^2 y}{\partial x^2}\right) + N\frac{\partial^2 y}{\partial x^2} + \overline{m}\frac{\partial^2 y}{\partial t^2} = 0 \quad (8)$$

EI is the combined stiffness of simply supported steel beam and prestressed SMA wire. In other words, $EI = E_C I_C + E_P I_P$, *N* is the axial pressure generated by SMA wire,m is the quality of simply supported steel.

The solution is assumed to be:

$$y(x,t) = \varphi(x)q(t) \tag{9}$$

$\varphi(x)$ is a function of x, q(t) is a function of t.

Then:

$$\frac{\mathrm{EI}\Phi''''(x)}{\Phi(x)} + \frac{N\Phi''(x)}{\Phi(x)} = -\overline{m}\frac{\ddot{q}(t)}{q(t)} = \mu^4 \tag{10}$$

According to the above equation, the following two equations can be obtained:

$$\ddot{q}(t) + \omega^2 q(t) = 0 \tag{11}$$

$$I\Phi''(x) + N\Phi''(x) - \overline{m}\omega^2\Phi(x) = 0 \tag{12}$$

It can be solved:

$$\omega_n = \frac{n^2\pi^2}{l^2}\sqrt{\left(1 - \frac{Nl^2}{n^2\pi^2 EI}\right)}\sqrt{\frac{EI}{m}} \tag{13}$$

The above formula is the calculation formula of beam frequency when the beam is subjected to axial force. EI is the combined stiffness of SMA wire and beam. It can be seen from equation (13), the vibration frequency of the beam decreases with the increasing pre-stress.

3.2. Relationship between SMA wire tension and frequency

The tension force generated by the heat recovery of the SMA wire is applied to the beam, causing the whole structure of the beam to be compressed.For the whole structure of simple supported beams with prestress, the tension of restoring force on SMA itself should be considered at the same time. The restoring force is regarded as the axial tension on both ends of SMA wire, and its motion differential equation is as follows[10]:

$$\frac{N\partial^2 y}{\partial x^2} + \rho_P A_P \frac{\partial^2 y}{\partial t^2} = 0 \tag{14}$$

$\rho_P A_P$ is the mass of the pre-stressed SMA wire of unit length, this value is small relative to the steel beam structure. The calculation formula of pre-stressed SMA wire frequency under axial tension is as follows:

$$\omega_n^{P^2} = \left(\frac{n\pi}{l}\right)^2 \frac{N}{\rho_P A_P} \tag{15}$$

According to the above equation, as the axial restoring force of SMA filament increases, its natural frequency increases.

4. Numerical analysis and results

4.1. Influence of shape memory alloy wire regeneration force on the natural frequency of the system

In this section, a numerical model of simple support beam and SMA wire is established by using MATLAB software, as shown in figureure 2.The size of the beam is 500mm x 50mm x 0.7mm, the diameter of SMA wire is 0.5mm and the number of SMA wires is 3, and the vertical distance d of SMA wire to the neutral axis of the beam is 1.35mm.

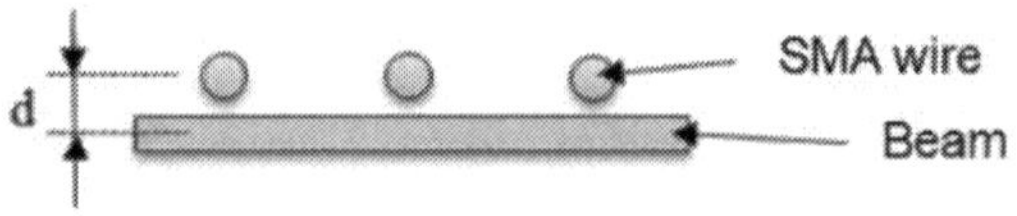

figure2. Cross section diagram of simple beam -SMA intelligent structure.

In the process of meshing, a lot of irregular units will be generated if you use a circular SMA wire. Leading the calculation time increased. In this paper, the length, arrangement and restoring force of SMA materials are considered as the main factors that affect the inhibition performance of SMA, ignoring the influence of section shape . Therefore, the square section is adopted to simulate the alloy wire, the section side length is 0.5mm, which shortens the calculation time and improves the calculation accuracy. Every initial strain of the shape memory alloy wire is 6.31%, with the method of node coupling, bonding together five discrete coupling collection and the metal plate , create a contact between SMA wire and metal plate. The finite element model is shown as figure 3.

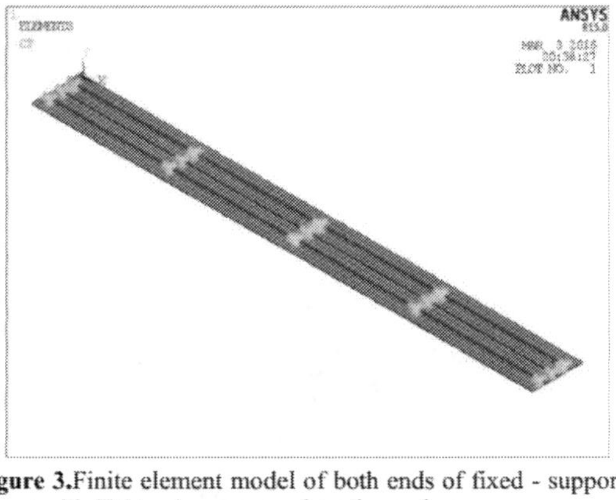

figure 3. Finite element model of both ends of fixed - supported beam with SMA wires arranged on the surface

In practice, each SMA wire is connected to the beam through 5 bolts and uniformly distributed, the arrangement is shown in the figure 4.

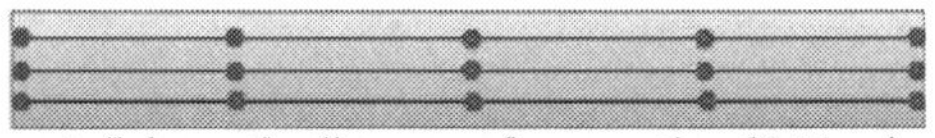

figure 4. Schematic diagram of connecting SMA wire and plate with bolts

As can be seen from figure 5, with the increasing restoring force of SMA wire, the first third order natural frequency of the beam decreases, and the first order frequency decreases obviously, and the natural frequency of SMA wire increases gradually, and the frequency of S-MA wire increases significantly compared with beam.

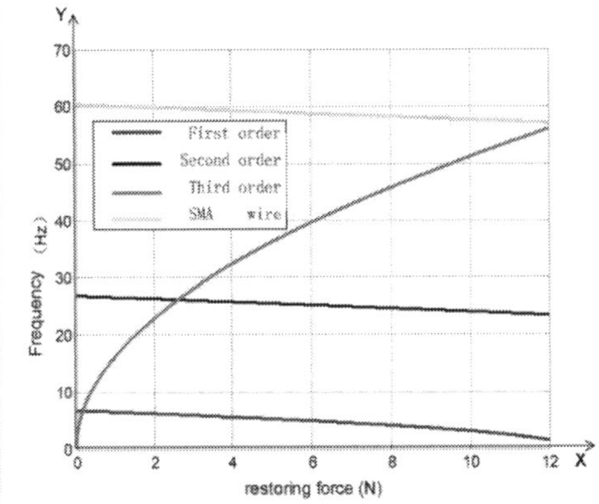

figure5. Relation of beam and SMA wire frequency with restoring force

4.2. Influence of distance location parameters on the natural frequency of the system

In order to study the effect of SMA wire placement on the natural vibration frequency of steel beam, numerical simulation was conducted for SMA wire at different .

As can be seen from figureure 6, as the distance from the SMA wire to the neutral axis of the beam increases, the natural frequency of the system increases significantly. Therefore, in order to enhance the SMA control effect on the variable stiffness of the body beam structure, it should be positioned away from the neutral layer of the beam. In the following simulation and experiment; the vertical distance from the SMA wire to the neutral axis of the beam is 2mm.eccentric position d, and the analysis results are shown in figure 6.

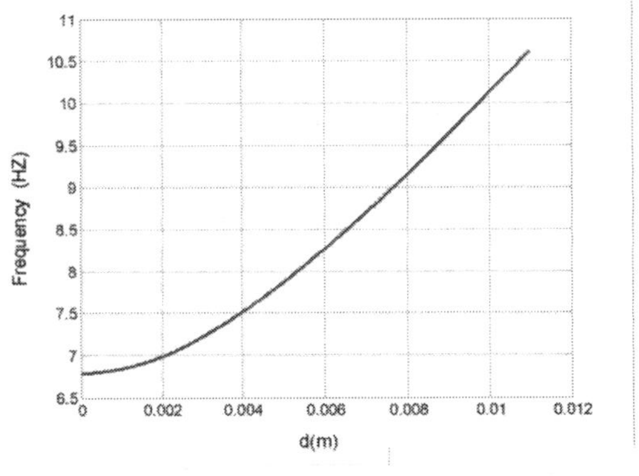

figure 6. Relation of system frequency with the neutral wheelbase distance from SMA wire to beam

5. Conclusion

The above content analyzed the relationship between the restoring force of SMA wire and the beam frequency theoretically, and the system dynamics model is established. Numerical simulation was carried out by using the numerical simulation software. the results showed that with the increasing restoring force of SMA wire, the first third order natural frequency of the beam decreased, and the first order frequency decreased significantly. As the distance from the SMA wire to the beam's neutral axis increases, the natural frequency of the system increases, and the SMA wire's control effect on the variable stiffness of the body beam structure

increases. In the application, SMA wire should be positioned as far as possible away from the beam's neutral layer. By analyzing the effect of the restoring force of shape memory alloy on structural vibration characteristics, the feasibility of using shape memory alloy wire to control the semi-active vibration of beam structure with variable stiffness was verified.

Acknowledgments

This project is supported by National Natural Science Foundation of China (Grant No. 51875237).The authors are grateful for the financial support from National Natural Science Foundation of China (Grant No. 51875237), the Science and Technology Development Project of Jilin Province (Project No. 20160101283JC) and the "13th Five-Year" Science and Technology Research Project of Education Department of Jilin Province (Project No. 2016-417).

References

1. S.Gao,YP.Shen,D.G.Xu.Shape memory alloy and its mechanical analysis in engineering applications[FJ].Mechanical progress,3:301-312(1997)
2. T. Kobori, M. Takahashi, T. Nasu, Seismic response controlled structure with Active Variable Stiffness system[J]. Earthquake Engineering&Structural Dynamics, 22(11):925-941(1993)
3.C. Rogers. Behavior of shape memory alloy reinforced composite plates. I — Modelformulations and control concepts[C]// 30th Structures. Structural Dynamics andMaterials Conference (1989)
4. H.Li, Application of shape memory alloy in structural vibration control[D}.Beijing jiaotong university(2007)
5. G. McGavin, G.Guerin. Real-time seismic damping and frequency structures using nitinol wire[C]. Proceedings of SPIE,4696:175-184(2002)
6. J. Heinonen, I. Vessonen, P. Klinge. Controlling stiffness of a frame spring by changing the boundary condition with an SMA actuator[J]. Computers&Structures,86:398-406(2008)
7. B. Tiseo, A. Concilio, S. Ameduri. A shape memory alloys based tuneable dynamic vibration absorber for vibration tonal control[J]. Journal of Theoretical&Applied Mechanics,48:135-153(2010)
8. M. Paz, H. Saunders. Structural Dynamics Theory and Computations[J].Journal of Vibration&Acoustics, 107(4):319-347.(1985)
9. Y.G.Tang,Advanced structural mechanics[M].Tianjin university press(2002)
10.X.Yang,J.Hong,Y.H.Ma.Experimental study on vibration control of SMA smart beam

MATEC Web of Conferences **238**, 05008 (2017) https://doi.org/10.1051/matecconf/201823805008
ICNFM2018

Effect Comparison of Rotated Tiles and Symmetrical Tiles on Binary Counter in DNA Self-Assembly

Aijun Zhu[1,2,3], *Chunxia* Zhao[1], *Duanyong* Chen[1], *Chuanpei* Xu[1], *Cong* Hu[1] and *Junhao* Niu[1,*]

[1]School of Electronic Engineering and Automation, Guilin University of Electronic Technology, Guilin 541004, P.R.China
[2]Department of Computer Science and Engineering, Texas A&M University, College Station, TX, USA
[3]Guangxi Key Laboratory of Automatic Detecting Technology and Instruments, Guilin 541004, P.R.China

Abstract. With the development of DNA Self-Assembly, several phenomena have happened and been discovered, which maybe lead to faulty crystals or patterns with faults. Among those unexpected phenomena, rotated tiles and symmetrical tiles have still not been discussed in detail. There is also no such research that compares the effect of rotated tiles and symmetrical tiles on Binary Counter in DNA Self-Assembly This paper compares and analyses the effect of rotated tiles and symmetrical tiles on binary counter in DNA Self-Assembly. The precise relationship between rotated tiles and symmetrical tiles is achieved taking into account the type of rotated tiles and symmetrical tiles. Experimental simulation results through Xgrow prove that the comparison and analysis is effective and correct.

1 Introduction

With the shrinking feature size of nano-fabrication, Self-Assembly is generally treated as an emerging subject and technology for its wide use and application in nano-fabrication or computation [1,7-15]. In nanomanufacture system and algorithmic system, DNA is a basic element in DNA Self-Assembly. From a seed tile to final pattern, DNA is treated as a small and standard part and is used to set up a complex system or a unique pattern. Generally speaking, DNA Self-Assembly starts from a seed tile and goes on its growing by appending of several boundary tiles and rule tiles in one specific orientation. As we all know, it starts from the southeast corner to northwest. The attachment and appending of tiles is on the basis of a rigorous rule as to the growing corner and bonding feature. With the development of DNA Self-Assembly, several phenomena have happened and been discovered, which maybe lead to faulty crystals or patterns with faults [5]. Among those unexpected phenomena, rotated tiles and symmetrical tiles have still not been discussed in detail [3]. There is also no such research that compares the effect of rotated tiles and symmetrical tiles on Binary Counter in DNA Self-Assembly. In the case of rotated tiles and symmetrical tiles, several bonding action and behaviours are completely different from the original growth. In the case of rotated tiles and symmetrical tiles, the final patterns always differ from the reserved patterns; the reason is that the growth with rotated tiles and symmetrical tiles would conduct with uncertain ultimate patterns.

Generally speaking, the binary counter is a basic element and key unit in manufacture system of self-assembly and algorithmic system. There is a special tile set and its corresponding graphic pattern in the binary counter. For instance, literature [2] uses the binary counter for the component of a de-multiplexer. Literature [6] uses the binary counter for estimation framework. Literature [4] uses demultiplexers by self-assembly for memory design, while Literature [5] uses binary counter for assembly of a square. Literature [7] discusses the effect of rotated tile on binary counter. This paper mainly discusses the effect comparison of rotated tiles and symmetrical tiles on binary counter.

Organization of the rest of this paper is as follows. The mathematical model is briefly introduced in Section 2. The effect comparison of rotated tiles and symmetrical tiles on binary counter are presented in Section 3. The effect comparison of counting rules after single tile symmetry is introduced in Section 4. Finally, concluding remarks is given in Section 5.

2 Mathematical model

In order to establish the model of the growth behaviour, it is appropriate to apply a transition graph for ensuring the attachment in an established growth corner in DNA self-assembly. The transition graph could be defined as a directed graph P(D,B). Let D be a vertex set and Let B be a directed edge set. Because the growth direction begins from south-east to north-west, without any doubt, the south-east corner should be defined as the growth corner. The directed edge set B is all the possible and feasible attachments in all growth corners. Therefore, every rule tile composes of two nodes and a directed edge.

* Corresponding author: *Junhao* Niu, Email: 45953000@qq.com

Definitely, there is special transition graph P in the binary counter tile set. The tile set of binary counter pattern is shown in Fig.1.

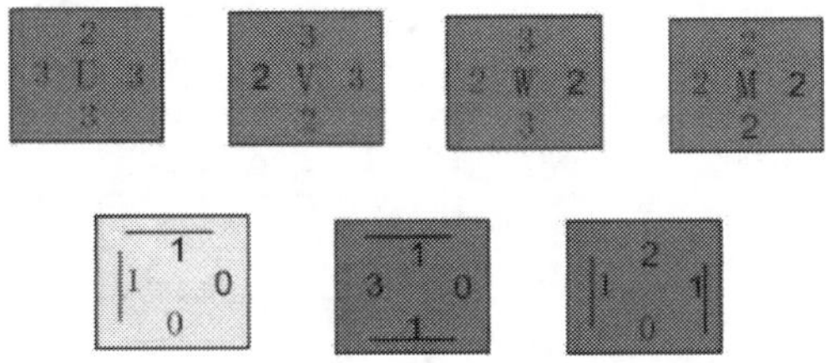

Fig. 1. All the tiles in the binary counter tile set.

There are three types of tile in binary counter which is rule tile, seed tile and boundary tile. However, growth is mostly obtained through rule tiles; there is the transition graph applied only for the rule tiles of binary counter in Fig.2. In binary counter, there is a seed tile [(N,E,S,W)] [(1,0,0,1)] and two boundary tiles [(2,1,0,1), (1,0,1,3)]. Rule tiles [(2,3,3,3), (3,3,2,2), (3,2,3,2), (2,2,2,2)] are defined as [U ,V, W, M], respectively. In the case of counting, V and W are defined as digit "1", and the corresponding U and M are defined as digit "0". Under the assumption of fault-free attachment in the tile set, the counter number will increase one by each line from zero.

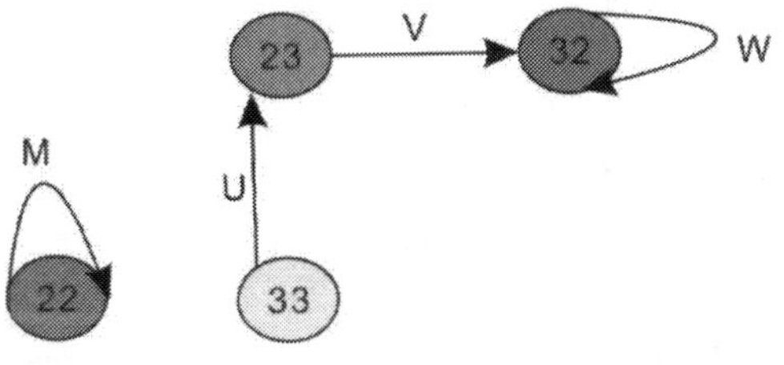

Fig. 2. Transition graph of binary counter tile set.

Let P'(D',B') be symmetrical transition graph or rotated transition graph which is achieved by the symmetrical tiles or rotated tiles of the original rule tiles. Let P(D,B) is the original transition graph. X symmetry of rule tile is the exchange of South with North. Y symmetry of rule tile is the exchange of East with West. XY symmetry of rule tile is both the exchange of North with South and the exchange of West with East. The rotation operations are defined as being in the clockwise direction and could be identified by the rotation angle(i.e., 90, 180, and 270 degrees). Therefore, the following situations should be taken into account as to the relationship between P and P'.

(1)P'=P, which also means D'=D and B'=B. In this case, symmetrical tile or rotated tile is exactly the same as the original tile. There are rule tiles W and M as to symmetrical tile, and there is rule tiles M as to rotated tile.

(2) P'! =P, to be exact, D'=D and $B^{'} = B \cup \{\forall b \mid b \notin B\}$. In such occasion, there are new edges without new vertexes which means no new growth corner. There are two subclasses. The first subclass is that the new edge connects two existing vertexes which are connected too in original P. For instance, X symmetry of rule tile U will lead to a new edge between nodes [2, 3] and [3, 3] in Fig.3 ; 90 degrees rotation of rule tile U will lead to a new edge between nodes [3, 3] and [3, 2] in Fig.4. Therefore, transitions between some nodes possess several different choices. Generally, the final pattern would be not the same as what we wanted. The second subclass is that the new edge would connect two existing vertexes in P' while no connection in P. For instance, X symmetry of rule tile V will lead to a new edge between nodes [3, 3] and [2, 2] in Fig.3; 90 degrees rotation of rule tile U will lead to a new edge between nodes [3, 2] and [3, 3] in Fig.4. Therefore, transitions between such nodes are not sole. As a result, the final pattern would be not the same as what we wanted.

(3) P'!=P , to be exact, $D^{'} = D \cup \{\forall d \mid d \notin D\}$ and $B^{'} = B$. The above occasion is out of question in theory. Generally speaking, the new node must posses an out-edge. As a result, $B^{'} = B$ is out of question and null.

(4) P'!=P , to be exact, $D^{'} = D \cup \{\forall d \mid d \notin D\}$ and $B^{'} = B \cup \{\forall b \mid b \notin B\}$. Such occasion most likely appears. There are also two subclasses. The first subclass is that the new nodes in the symmetrical transaction graph or rotated transaction graph perhaps be linked to the other nodes in P. Therefore, new edges are coming into being and the transitions between some nodes are not sole. As a result, the final pattern would be not the same as what we wanted. The second subclass is that the new nodes in the symmetrical transaction graph or rotated transaction graph are not connected to the other nodes in P. In such occasion, it is unlikely to get a new growth corner. Naturally, the final assembly must be what we want.

3 Effects comparison on counting

To compare the effects on counting, we should create symmetrical transition graph and rotated transition graph of binary counter tile set first. Symmetrical transition graph of the binary counter tile set is shown in Fig.3 and rotated transition graph of binary counter tile set of the binary counter tile set is shown in Fig.4.

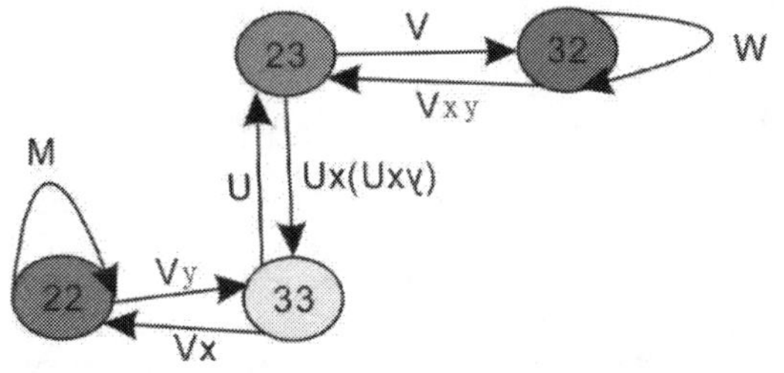

Fig. 3. Symmetrical transition graph of binary counter tile set.

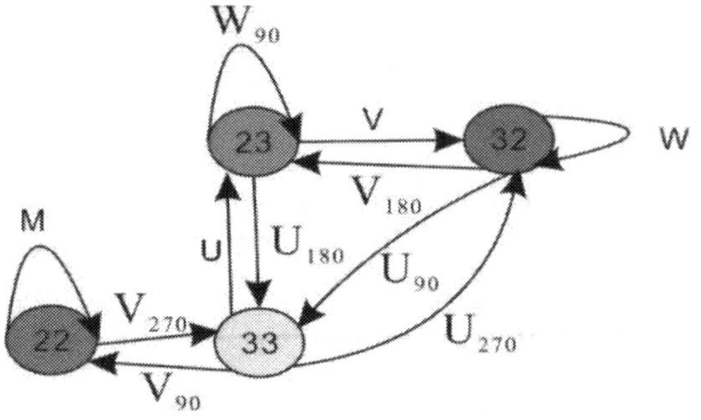

Fig. 4. Rotated transition graph of binary counter tile set.

As shown in Fig.3, Ux, Uxy, Vx, Vy and Vxy will result in new edges. While in Fig.4, V90, V180, V270, U90, U180 , U 270 and W90 will result in new edges.

Compared the effects of rotated transition and symmetrical transition, we could conclude that rotation of tile U are the same as symmetry of tile U and rotation of tile V are the same as symmetry of tile V. As to tile M, no matter what operation, it will not change at all. As to tile W, it will not change at all for any symmetry. However, as to tile W, it maybe change for some rotation. Generally speaking, the binary counter starts to increase by one each row from zero. Every row in the binary counter pattern denotes a count value. Let a tile be expressed by a four-tuple (N, E, S, W), which means the values of the bonds with east, south and west, respectively. Definitely, we can find exact relationship between each tile sequence and the counting value. Let the tile V(3,3,2,2) and the tile W(3,2,3,2) equal a binary value "1", and let the tile (2,2,2,2) and the tile (2,3,3,3) equal a binary value "0". As a result, symmetrical or rotated tile would cause a tile meaning "1" substituted by a tile meaning "0". Of course, we can find other transaction which may be happen. For example, an X symmetry or 90 degrees rotation in tile (3, 3, 2, 2) (a digital "1") would both produce a new tile (2, 3, 3, 2).

We would give a detailed example to demonstrate an exact occasion where a rotated tile or symmetrical tile could change the original final pattern.

3.1 Both a rotated tile and symmetrical tile: leading to lower or smaller number

In order to obtain the exact effect impact on counting, we can suppose current count value be 3 ("MMMMMMWV"or "00000011"). Naturally, the next fault-free count value should be 4("MMMMMVUU" or "00000100"). Fig.5 (a) demonstrate a segment of its self-assembly without any unwanted fault. Next step suppose a rotated tile and symmetrical tile occur. For example, a tile V (3, 3, 2, 2) with x symmetry or 90 degrees rotation would produce a new tile (2, 3, 3, 2) with the same growth corner as the tile U. As a result, Vx or V90 is sure to attach in the place of the tile U. This phenomenal will occur for the second U in the fourth row in the Fig.5 (a) in the fault free occasion, the tile sequence in the fourth row is "MMMMMVUU" or "00000100" in the Fig.5(a). With a rotated tile or symmetrical tile, the tile sequence is changed to be "MMMMMMVxU" or "MMMMMMV90U". Fig.5 (b) and Fig.5(c) show part of the final pattern under the condition of the rotated tile or symmetrical tile.

Count Value : 6	M	M	M	M	M	W	V	U
Count Value : 5	M	M	M	M	M	W	M	V
Count Value : 4	M	M	M	M	M	V	U	U
Count Value : 3	M	M	M	M	M	M	W	V
Count Value : 2	M	M	M	M	M	M	V	U
Count Value : 1	M	M	M	M	M	M	M	V

(a)

Count Value : 2	M	M	M	M	M	M	V	U
Count Value : 1	M	M	M	M	M	M	M	V
Count Value : ..	M	M	M	M	M	M	Vx	U
Count Value : 3	M	M	M	M	M	M	W	V
Count Value : 2	M	M	M	M	M	M	V	U
Count Value : 1	M	M	M	M	M	M	M	V

(b)

Count Value : 2	M	M	M	M	M	M	V	U
Count Value : 1	M	M	M	M	M	M	M	V
Count Value : ..	M	M	M	M	M	M	V_{90}	U
Count Value : 3	M	M	M	M	M	M	W	V
Count Value : 2	M	M	M	M	M	M	V	U
Count Value : 1	M	M	M	M	M	M	M	V

(c)

Fig. 5. Part of Self-assembled pattern of binary counter (a) fault-free and (b) with a symmetric tile and (c) with a rotated tile

Compared Fig.5 (b) with Fig.5(c), we could find that a single Vx has the same effect with a single V90 on final pattern. They both lead to a lower number.

4 Counting rules after one single symmetrical or rotated tile

After the above consideration, we would give a discussion about counting rule in detail for binary counter with one single symmetrical and rotated tile. For simplification, suppose only o one single symmetrical or rotated tile on every row. Generally, every row in the binary counter pattern is divided into four parts. The value of the counter is shown in the following part. Because of one single symmetrical or rotated tile on every row, the counting value of is not N+1. Suppose j is the exact position of symmetrical or rotated tile on every row. Such string "M..MW..WM..MW..WVU..U" in a row denotes a row in a binary counter. Let the right be the least significant digit and the left be the most significant digit. There is either zero number of "U" or more, exactly one "V", either zero number of "W" or more, and either zero number of "M" or more. Above occasions contain all possible circumstances that maybe happen.

The difference between the counting value before (demonstrated as N) and after one single symmetrical or rotated tile (demonstrated as Nnext) is shown in the following part. With each circumstance of single symmetrical or rotated tile, we will give a strict and detailed proof. There are four different circumstances.

1) Any V90 degrees rotation or Vx symmetrical tile: Nnext is in the range of 1 and N.

Proof: first, let us take into account the case which V90 degrees rotation or Vx symmetrical tile would impact the "U" region that are seated in the least significant zeros and on the right side of the least significant ones. Such Nnext is at the right side. Let j equal 0 then Nnext=N.

Second, let the V90 degrees rotation or Vx symmetrical tile affect the most significant zero of the "U" region. Suppose the "U" regions are t bits long, and we get

$$\mathbf{N} = \mathbf{2}^t - 1 + constant \ \ and \ \ j = t - 1 \tag{1}$$

$$\mathbf{N}_{next} = \mathbf{2}^t - 1 + constant - \mathbf{2}^{t-1+1} + 2 = constant + 1 \tag{2}$$

Where $constant = \sum b_i \times 2^i; b_i \in \{0,1\}$ and $i > t$.

It is evident that $\mathbf{N}_{next} = constant + 1 \geq 1$.

2) Any U270 degrees rotation or Ux (Uxy) symmetrical tile: Nnext is in the range of N+4 and 3*N+4.

Proof: U270 degrees rotation and Ux or Uxy symmetrical tile would impact "V" region clearly as their growth corners are exactly the same as that of "V" region. First, as "V" region may be the least significant bit in the case of j equal zero, and $\mathbf{N}_{next} = N + \mathbf{2}^{0+1} + 2 = N + 4$.

Second, let "V" region is the most significant bit o. Suppose the binary counter are t bits long, and we get

$$\mathbf{N} = \mathbf{2}^t - 1 + constant \ \ and \ \ j = t \tag{3}$$

$$\mathbf{N}_{next} = N + \mathbf{2}^{t+1} + 2 \leq 3 \times N + 4 \tag{4}$$

It is evident $N + 4 \leq \mathbf{N}_{next} \leq 3 \times N + 4$.

3) Any V270 degrees rotation or Vxy symmetrical tile: Nnext is in the range of N+4 and 2*(N+1).

Proof: such V270 degrees rotation or Vxy symmetrical tile would impact "W" region clearly as its growth corner is the same as that of "W" region. First, as "W" region may be the least significant bit in the case of j equal one ("…WV"), and $\mathbf{N}_{next} = N + \mathbf{2}^1 + 2 = N + 4$.

Second, let "W" region be the most significant bit that is followed by "M", "V", and "U". Suppose the binary counter is t bits long, and we get

$$\mathbf{N} = \mathbf{2}^t - 1 + constant \ \ and \ \ j = t - 1 \tag{5}$$

$$\mathbf{N}_{next} = N + \mathbf{2}^{t-1} + 2 \leq 2 \times (N + 1) \tag{6}$$

It is evident $N + 4 \leq \mathbf{N}_{next} \leq 2 \times (N + 1)$

4) Any V90 degrees rotation or Vy symmetrical tile: Nnext is greater than or equal N+8.

Proof: such V90 degrees rotation or Vy symmetrical tile would impact "M" region clearly as its growth corner is the same as that of "M" region. First, as "M" region may be the least significant bit in the case of j equal one ("…MU"), and $\mathbf{N}_{next} = N + \mathbf{2}^{1+1} + \mathbf{2}^1 + 2 = N + 8$.

Second, as "W" region does not have the most significant bit of binary counter. Let the binary counter be t bits long, and we get

$$\mathbf{N} = \mathbf{2}^t - 1 + constant \ \ and \ \ j = t \tag{7}$$

$$\mathbf{N}_{next} = N + \mathbf{2}^{t+1} + \mathbf{2}^t + 2, t \in [2, +\infty], \ \mathbf{N}_{next} \to +\infty \tag{8}$$

It is evident $\mathbf{N}_{next} \geq N + 8$.

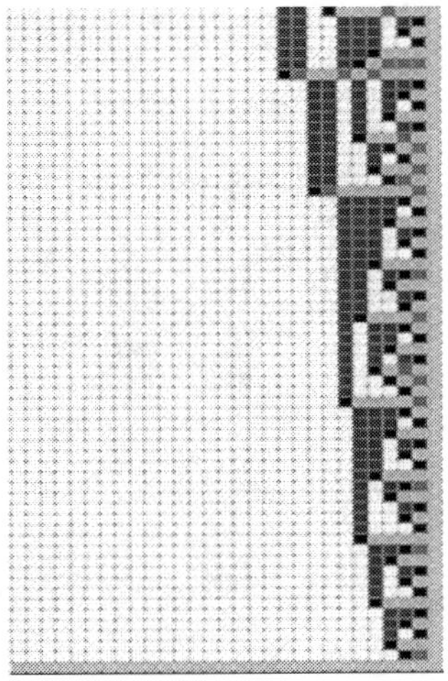

Fig. 6. U_{90} degrees rotated tile or Ux (Uxy) symmetrical tile and its impact on binary counter.

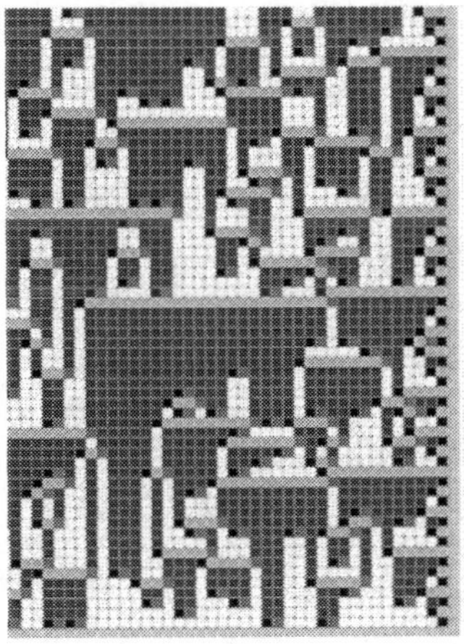

Fig. 7. V_{180} degrees rotated tile or Vy symmetrical tile and its impact on binary counter.

Simulation results obtained by Xgrow prove that the related analysis is effective and errorless. Some sections of the simulation results are shown in Figs 6-7.

5 Conclusions

We give detailed analyses for the effect comparison of rotated tile and symmetrical tile on binary counter. The detailed comparison of rotated tile and symmetrical tile on binary counter is obtained after taking into account the type of rotated tile and symmetrical tile. The analysis demonstrates that the final results are partly dissimilar with what we originally designed in the case of rotated tile and symmetrical tile because of the altered assembly mode. Simulation results from Xgrow show that the analysis is effective and correct.

Acknowledgment

This research has been partially funded by National Natural Science Foundation of China (61861012, 61561012, 61562013), Guangxi Natural Science Foundation Joint Funding Project (2018GXNSFAA138115), Guangxi Key Laboratory of Automatic Detecting Technology and Instruments (YQ16110), Talent Project of Guilin university of electronic technology (UF15008Y), Guangxi Natural Science Foundation (2017GXNSFAA198021, 2015GXNSFDA139030), China Scholarship Council (201508455020). Thanks Professor Fabrizio Lombardi, who is with Northeastern University, USA, for his valuable suggestions.

References

1. S. Zhang, D. M. Marini, W. Huang, and S. Santoro, "Design of nanostruc-Tured biological materials through self-assembly of peptides and proteins," Curr. Opin. Chem. Biol, vol. **6**, 865–871(2002).
2. M. Cook, P. W. K. Rothemund, and E. Winfree, "Self-assembled circuitpatterns," DNA Computers 9, (Lecture Notes in Computing Science,vol. **2943**). New York: Springer-Verlag, 91–107(2004).
3. F. Fujibayashi and S. Murata, "Precise simulation model for DNA tile self-assembly," IEEE Trans. Nanotechnol, vol. **8**, no. 3, 361–368(2009).
4. M. Hashempour, Z. M. Arani, and F. Lombardi, "A graph model for tile sets in DNA self-assembly," in Proc. *IEEE Int. Workshop Design Test Nano Devices, Circuits Syst*, Boston, MA, 77–81(2008).
5. M.Hashempour,Z.M.Arani,andF.Lombardi,"Healing assessmentoftile sets for error tolerance in DNA self-assembly," IET Trans. Nanobiotech-nol, vol. **2**, no. 4, 81–92(2008).
6. M. Y. Kao and R. Schweller, "Randomized self-assembly for approxi-mate shapes," in *International Colloqium on Automata, Languages, and Programming (ICALP)*, NewYork:Springer-Verlag, 370–384(2008).
7. M. Hashempour, Z. M. Arani, and F. Lombardi, "Counting by DNA Self-Assembly in the Presence of Rotated Tiles," IEEE Trans. Nanotechnol, vol. **10**, no. 3, 77–81 (2011).
8. J. Tan, X. Zhang and D. Liu, "Facile Preparation of CO2-Responsive Polymer Nano-Objects via Aqueous Photoinitiated Polymerization-Induced Self-Assembly (Photo-PISA)", Macromol Rapid Comm, vol. **38**, no. 13, 15759-15765(2017).
9. K.T. Geun, P. Nokyoung, and K.U. Jeong, "Electrical Modulation of Graphene by the Self-Assembly of DNA-Functionalized Gold Nanoparticles," J Nanosci Nanotechno, vol. **17**, no. 11, 8007-8011(2011).
10. E. Ducrot, M. He, G. Yi, "Colloidal alloys with preassembled clusters and spheres," Nat Mater, vol.**16**, no. 6, 652-657(2017).
11. S. Jiang, F. Hong, H. Hu, "Understanding the Elementary Steps in DNA Tile-Based Self-Assembly," ACS Nano, vol. **11**, no. 9, 9370-9381(2017).
12. S. Dong, X. Ding, T. Guo, "Self-assembled hollow sphere shaped Bi2WO6/RGO composites for efficient sunlight-driven photocatalytic degradation of organic pollutants," Chem Eng J, vol. **316**, no. 17, 778-789(2017).
13. D. Sun, T. Guo, B. Guan, "Counting by DNA Self-Assembly in the Presence of Rotated Tiles," J Lightwave Technol, vol.**35**, no. 16, 3354-3359(2017).
14. S. Xu, G. Ng, J. Xu, P. Rhiannon , Y. Jonathan and B.Cyrille,2-(Methylthio)ethyl Methacrylate: A Versatile Monomer for Stimuli Responsiveness and Polymerization-Induced Self-Assembly in the Presence of Air, ACS Macro Lett, vol. **6**, no.11, 1237-1244(2017).
15. C. Zhu, S. Fu, J. Song, "Self-Assembled Fe-N-Doped Carbon Nanotube Aerogels with Single-Atom Catalyst Feature as High-Efficiency Oxygen Reduction Electrocatalysts," Small, vol. **13**, no. 15, 157-167(2017).

MATEC Web of Conferences 238, 05009 (2017) https://doi.org/10.1051/matecconf/201823805009
ICNFM2018

Proportion Design Method and Water Stability of Cement-Emulsified Asphalt Cold Recycling Mixtures

Junxiao Li [1,*], *Wei* Fu [2] and *Xiaobo* Yin [3]

[1] Institute of Oceanographic Instrmentation, Qilu University of Technology (Shandong Academy of Sciences),Shandong Provincial Key Laboratory of Marine monitoring instrument equipment technology, National Engineering and Technological Research Center of Marine Monitoring Equipment, Qingdao 266001, China
[2] Dept. of No.2 sanatorium, Navy Qingdao No.1 Sanatorium of PLA, Qingdao 266001, China
[3] Zhuhai Dacheng Engineering Testing Co. Ltd, Zhuhai 519000, China

Abstract. Nowadays, a large amount of work has been done in the field of emulsified asphalt cold regeneration. Three different proportions of emulsified asphalt cold recycling mixtures were designed by mixing different ratios of new aggregates and RAP, the effect of cement content and RAP content on the mixtures's water stability was studied. The results show that the addition of cement can improve the mixtures's water stability, but there is an optimum content; the mixtures's water stability reduces with the increase of RAP content.

1 Introduction

In china, the total number of highways was 4,773,500 kilometers by the end of 2017. Asphalt road occupies 76% in high-grade road construction, so there will be a large amount of recycled asphalt mixture in the road pavement renovation, renovation and conservation every year. If calculate by 10% of highway pavement need repair, there will be 2.2 million tons recycled asphalt mixture every year. It is easy to be predicted that more and more recycled asphalt mixture will be generated in the future. The recycled asphalt mixture was treated as waste in some engineering projects, which is a waste of resources. In addition, they will occupy a lot of land, and pollute the environment. Asphalt cold recycling can be a good solution, which can make the old pavement materials reused and save 40%-50% of the total investment [1,2,3] .In recent years, the cold regeneration technique of recycled waste road materials by foam asphalt has been famous for its remarkable advantages in energy saving, environmental protection and economy. Nowadays, a large amount of work has been done in the field of emulsified asphalt cold regeneration [4,5] .

The effect of emulsified asphalt cold regeneration is influenced by many factors, such as the type and amount of emulsified asphalt, the gradation of the mixture, the dosage of the new aggregate, and the dosage of the cement. Based on the different proportions of RAP and new aggregate expected to cooperate, three kind of emulsified asphalt cold recycled mixture were designed. The optimum water consumption of the mixture was determined by the compaction test. By designing the proportion of emulsified asphalt cold recycled mixture combined with the water stability experiment, for this mixture the best proportion are determined. The results of this study can provide some reference for the design of practical emulsified asphalt cold regeneration project.

2 Raw materials

Table 1. Gradation of RAP and new mixture.

Mesh size /mm	Pass rate /%			
	RAP	1# new mixture	2# new mixture	3# new mixture
26.5	100.00	100.00	100	100
19.00	98.87	98.55	100	100
16.00	96.82	93.33	100	100
13.20	92.75	69.27	100	100
9.50	86.69	21.64	100	100
4.75	73.00	0.07	99.82	100
2.36	51.49	0.02	74.92	100
1.18	36.02	0.00	59.78	100
0.60	19.94	0.00	46.11	100
0.30	4.67	0.00	28.43	100
0.15	1.38	0.00	19.45	100
0.075	0.36	0.00	13.24	95.54

The recycled asphalt pavement (RAP) is from highway milling material in Wuhan, China. As for the role of RAP in cold regenerated mixture, there is still no consensus in academia. In this study, RAP is regarded as the black aggregates, and the mechanism of aging

* Corresponding author: caike0501@163.com

asphalt in the process of cold regeneration is not considered. Simply using RAP as the regenerated mixture aggregates can't meet the design requirements. So, the new materials used in this study are of three kinds, which are limestone and their technical indexes meet the specification requirements. Table 1 is Gradation of RAP and new mixture. Emulsified asphalt is homemade cationic emulsified asphalt. Table 2. is its technical indicators. Cement is P.O. 425 ordinary Portland cement from Huaxin cement factory, and table 3. is its chemical composition. Water is just general water.

Table 2. Technical indicators of emulsified asphalt.

Test item		Quality requirements	Testing result
Demulsification velocity		Slow crack or Medium speedcrack	Slow crack
Particle charge		Cationic(+)	Cationic
1.18mmsieving residue(%)		≤0.1	0.027
His viscosityE25		2~30	10.5
Evaporation residue	Residual content of points(%)	≥62	63.4
	solubility(%)	≥97.5	99.2
	25℃penetration(0.1mm)	50~300	69.5
	15℃ductility(cm)	≥40	＞100
Wrapped with coarse aggregate area		≥2/3	＞3/4
Coarse aggregate and Fine aggregate mixing experiments		uniform	uniform
Room temperature storage stability(%)	1d 5d	≤1 ≤5	0.50 1.0

Table 3. Technical indicators of emulsified asphalt.

Chemical composition	Content /%
SiO_2	21.73
Al_2O_3	4.25
Fe_2O_3	2.67
CaO	63.59
MgO	2.21
Na_2O	0.55
SO_3	2.39
Ignition loss	1.75

3 Design for proportion of cement-emulsified asphalt cold recycled mixture

3.1 Gradation design

There are a lot of specifications for the gradation design of cold recycled mixture, but there is no uniform standard. The grading design of this study is based on the <Technical Specification for Road Asphalt Pavement Regeneration> (JTG F41-2008). Mixture 1, Mixture 2, and Mixture 3 were three different proportions of emulsified asphalt cold recycling mixtures that were designed by mixing different ratios of new aggregates and RAP.

Mixture 1=25%RAP+40%1#new mixture+35%2# new mixture;

Mixture 2=50%RAP+25%1#new mixture+25%2# new mixture;

Mixture 3=75%RAP+15%1#new mixture+5%2# new mixture +5% 3# new mixture.

Each of the three gradations meets the requirements of the medium-grain grading of <Technical Specification for Road Asphalt Pavement Regeneration> (JTG F41-2008). Fig.1. is cold recycling mixtures gradation curves which compared with the standard grading.

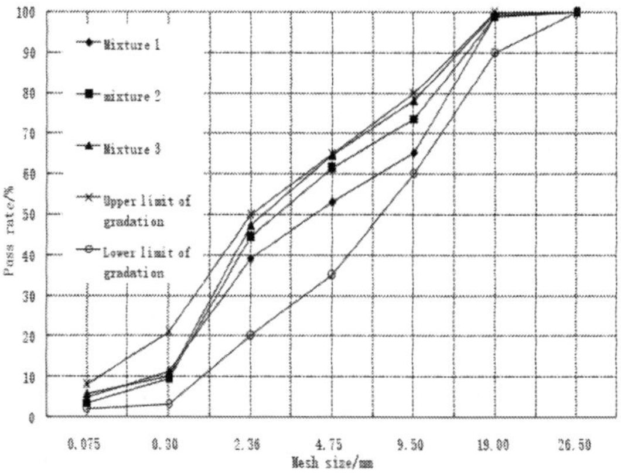

Fig. 1. Cold recycling mixtures gradation curves.

3.2. The amount of cement and emulsified asphalt

Cement is only used as a stabilizer and it is not considered as active filler, regardless of its effect on the mixture grading. In order to investigate the effect of cement content change on the stability of mixed material water, this study set five kinds of cement content, which are 0%,1%,2%,3% and 4% of the RAP and new mixture total quality. The amount of emulsified asphalt is 4.2% of the RAP and new mixture total quality.

3.3. The amount of cement and emulsified asphalt

Mix and mix well and need a moderate amount of water. If the water is too little, the emulsified asphalt is difficult to disperse, and the lubrication between the aggregate particles is not enough. It is difficult to compaction the mixture and the cement cannot be fully hydrated. If there is too much water, the pressure of moving water will increase, and the mixture will be hard to compaction. At the same time, emulsified asphalt and cement slurry loss. Hydrate crystal development is not sufficient, and health time is long, and material strength is low. Therefore, it is necessary to determine the optimum mixing water consumption through experiments.

The amount of emulsified asphalt is 4.2% of the RAP and new mixture total quality. The optimized amount of water (OWC) used and the maximum dry density were determined by the compaction test, and the mixing water consumption is calculated. The impaction test was referring to the method in <Highway Engineering stable inorganic binder materials testing procedures> (JTG E 51-2009). The RAP and new aggregates used in this experiment were air-dried, which could be considered as 0% water content. Therefore, the added amount of water is equal to the difference value between the OWC of the mixture and the water content of the asphalt emulsion. Table 4 shows the total amount of mixed materials and the maximum dry density obtained by the compaction test, and calculates the actual amount of water needed in the experiment.

Table 4. Compaction test results.

Type of mixture	Cement /%	OWC/ %	Maximum dry density /g·cm^{-3}	Water/%
1	0	6.02	2.04	4.42
	1	6.48	2.14	4.88
	2	6.69	2.09	5.09
	3	6.90	2.11	5.30
	4	7.01	2.19	5.41
2	0	6.14	2.19	4.54
	1	6.47	2.08	4.87
	2	6.68	2.21	5.08
	3	6.97	2.14	5.15
	4	7.09	2.16	5.49
3	0	6.09	2.09	4.49
	1	6.39	2.07	4.79
	2	6.58	2.18	4.98
	3	6.74	2.12	5.14
	4	6.95	2.20	5.35

4 Evaluation of water stability of cold recycling mixtures

4.1. Test method

This experiment adopts wet splitting strength and dry splitting strength ratio at 15 ℃ to evaluate the water stability of mixture. The impaction test was referring to the method in <Highway Engineering stable inorganic binder materials testing procedures> (JTG E 51-2009) the preparation of mixture standard Marshall Specimens was refer to the methods in < Road engineering asphalt and asphalt mixture test procedure > (JTG 052-2000). After demoulding, specimens kept in 60 ℃ ventilation oven by 72h curing. This curing method can simulate the condition after 10d actual pavement construction. Splitting test: Specimens at 15℃ environment after 1.5 h do dry splitting test; After specimens soak in 25 ℃ constant temperature water bath 23 h, then in 15 ℃ constant temperature water bath for 1 h, do wet splitting test at 15 ℃ [6] .

4.2. The effect of cement dosage on the water stability of mixture

Fig. 2 shows the variation of the dry splitting strength and wet splitting strength of the three mixtures respectively under different cement content.

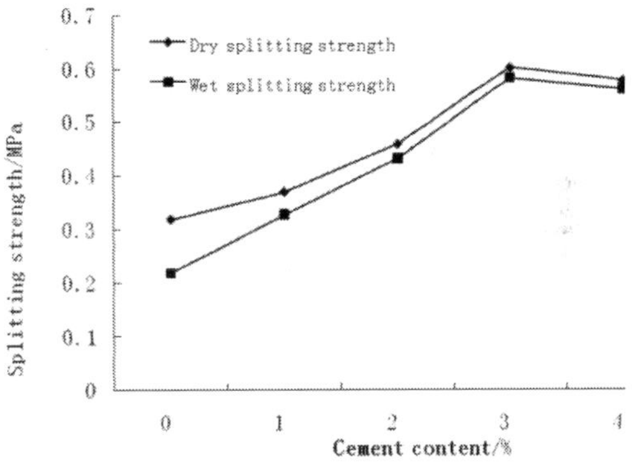

(a)Mixture 1

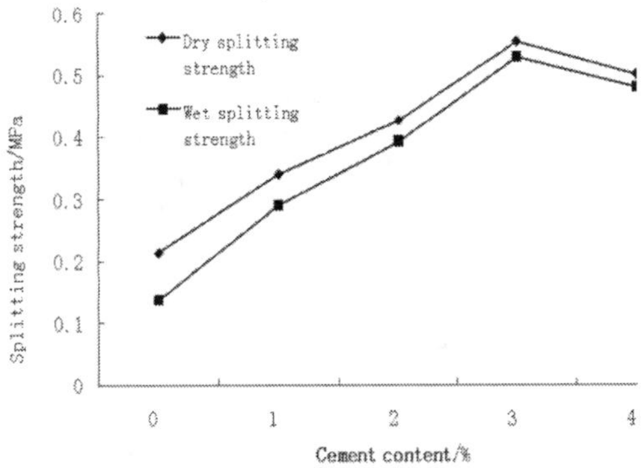

(b)Mixture 2

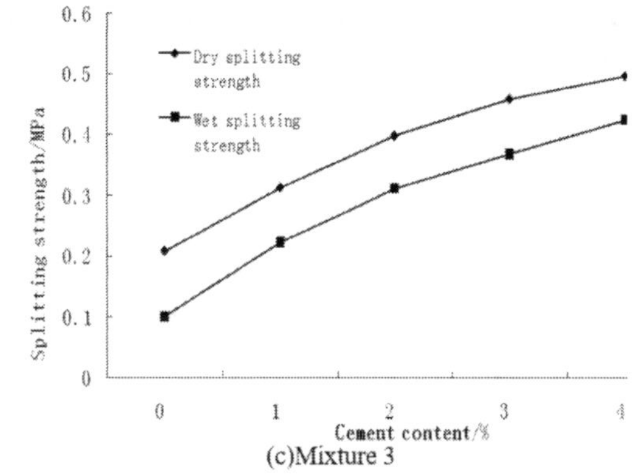

(c)Mixture 3

Fig. 2. The relationship between splitting strength and cement content.

It can be seen from Fig. 2. that the dry splitting strength and wet splitting strength of the mixture 3 are increased with the increase of cement content. When the cement content is less than 3%, the dry splitting strength and wet splitting strength of A and B increase with the increase of cement content.

Table 5. is the water stability of different cement content mixtures. As shown in table 5, for certain types of mixtures, the water stability increases with the increase of cement content. When comparing 1% cement content and 0% cement content, the increase of water stability of the three mixed materials was 20.3%, 20.8% and 22.6% respectively. Therefore, the addition of cement to the water stability of the mixture is very obvious.

Table 5. Water stability of different cement content mixtures

Mixture type	Cement content /%	0	1	2	3	4
Mixture 1	Dry and wet splitting strength ratio.	0.684	0.887	0.941	0.967	0.970
Mixture 2		0.646	0.854	0.921	0.955	0.961
Mixture 3		0.488	0.714	0.785	0.804	0.855

After the hydration of the cement in cold cycling mixture, some of the hydration products C_4AH_{13} react with gypsum: C_4AH_{13}+ 3C-S-H_2+14H=AFt+CH. The needle bar AFt will be filled space within the mix and form a network structure that increased mixture strength. Others hydration products C-S-H coagulation forms a fibrous structure that further increased mixture strength[7]. Hydration products develop to the surrounding space of the cement particles, criss-cross, and gradually fill the mix of all the capillary space. Thereby it forms a uniform, dense, pore closed the whole structure, and enhances the overall strength of the mixture of water and anti-damaging. On the other hand, the crystal structure of cement hydration products can be formed and developed in the water phase of asphalt emulsion. The cement hydration products and the released asphalt after emulsion breaking are intertwined, and the mixture is tightly integrated into a whole. Under the combined action of these two aspects, the splitting strength and water stability of the mixture are improved with the increase of cement content. However, cement content is not the higher the better[8]. Because the increase of cement dosage will lead to the increase of the dry shrinkage of the mixture, resulting in a severe shrinkage crack to reduce the strength. This is the reason that when the cement content is 4%, the strength of Mixture 2 and Mixture 3 is slightly lower than that of 3% cement. Therefore, the optimal cement dosage for this test was 3%.

4.3. The effect of RAP dosage on the water stability of mixture

Fig.3. Cold recycling mixtures gradation curves.

When the cement content was selected as 3%, the dry and wet splitting strength values of the three mixed materials were compared, and the results were shown in Fig. 3.

The RAP dosage of mixture 1, mixture 2 and mixture 3 was 25%, 50% and 75% respectively. As shown in Fig. 3, with the increase of RAP proportion in the mixture, the dry and wet splitting strength of the mixture decreased gradually. The dry and wet splitting strength of the three kinds of mixtures was 0.967, 0.955 and 0.804 respectively. But 25%RAP content is only 1.2% higher than the 50% RAP content that is 16.3 percent larger than the 75% RAP mixture. This is due to the fact that as the amount of RAP increases, the content of stone debris in the mixture will be reduced. Thus, the adhesion of aggregates and emulsified asphalt is weakened, and the dry, wet splitting strength and water stability of the mixture are reduced. However, due to the cementation effect of cement, the amount of RAP content increased from 25% to 50%, the reduction of the mixture splitting strength and water stability were minor. The cementation of cement is not sufficient to compensate the strength of the loss due to the attenuation of aggregates and emulsified asphalt, when the amount of RAP in the mixture is further increased. The results show that the mixture drying, wet splitting strength and water stability are greatly reduced.

In the actual cold regeneration construction, the lower the new aggregate amount in the mixture, the lower the material cost. But the higher the RAP dosage, the worse the performance of the mixture. Considering the factors of these two aspects, this experiment determines that the Mixture 2 is the optimal gradation.

5 Conclusions

By adding new aggregates of certain amount and granule, it can be combined with recycled asphalt pavement materials (RAP) to meet the design requirements. The design of the emulsion asphalt cold regenerated mixture gradation meets <Technical specification for road asphalt pavement regeneration>(JTG F41-2008).

The addition of cement can significantly improve the splitting strength and water stability of emulsified asphalt cold regeneration mixture. However, the high cement content will cause serious shrinkage crack in the mixture, which will reduce the strength of the mixture.

With the increase of RAP dosage, the splitting strength and water stability of the mixture were reduced. Compared with the mixture of 25% RAP content, the mixture of 50% RAP, mixture of 25% RAP has a small reduction in splitting strength and water stability. Compared with the mixture of 25% RAP content, the mixture of 75% RAP, mixture of 75% RAP has a big reduction in splitting strength and water stability.

The best mixture ratio of this study is the Mixture 2 (50%RAP+25%1#new mixture+25%2# new mixture), and the cement content is 3%, and the emulsified asphalt content is 4.2%, and the water content is 5.15%. Best of proportion of mixture of 15℃ dry splitting strength is 0.554 MPa, and wet splitting strength is 0.529 MPa, and dry wet splitting strength ratio 0.955. All these meet the requirements of the design of the emulsified asphalt cold regeneration mixture.

References

1. Wang Guo-ke, Zhou Zhi-li, Foamed Asphalt Cold in-plant Recycling Applied in Pavement, Reconstruction Transport Standardization, **219** (2010)
2. Guidelines for Cold In-Place Recycling, Asphalt Recycling and Reclaiming Association, Annapolis, MD, (1991)
3. Murphy DT, Emery JJ. Modified cold in-place asphalt recycling. Transportation Research Record, Washington, D.C., (1996)
4. Wang Dengzhou, Wang Ler , Zhao Yongbo, Study on the design method of emulsified asphalt cold regeneration mixture,Journal of China & Foreign Highway, **29**,2(2009),
5. Zhong Mengwu, Wu Chaofan, Yu Yongsheng, Study on the design method of emulsified asphalt mixture with cement.Road, **1** (2008)
6. Fang zhi SHI, Xiu jun LI, Da quan SUN, Wei min LV. Discussion About Design Methods for Recycling Cold Mix Asphalt.Highway, **11** (2004)
7. Yuan Yunzhang, *Gelling Material Science* (Wuhan University of Technology Press, 2008)
8. Xiaobo Yin: Performance of Emulsified Asphalt Full Thickness Cold Recycling Subgrade Materials, Advanced Materials Research, (2014)

Finite Element Simulation and Construction Technology Research of Cement-Emulsified Asphalt Cold Recycling System

Junxiao LI [1,*], *Wei* FU [2] and *Xiaobo* YIN [3]

[1] Institute of 0ceanographic Instrmentation, Qilu University of Technology (Shandong Academy of Sciences), Shandong Provincial Key Laboratory of Marine monitoring instrument equipment technology, National Engineering and Technological Research Center of Marine Monitoring Equipment, Qingdao 266001, China
[2] Dept. of No.2 sanatorium, Navy Qingdao No.1 Sanatorium of PLA, Qingdao 266001, China
[3] Zhuhai Dacheng Engineering Testing Co. Ltd, Zhuhai 519000, China

Abstract. The force model of pavement structure is established. In order to investigate the force on pavement when the ratio and thickness of cold recycled pavement vary, three concrete pavement structures, material parameters and vehicle load parameters are designed. By finite element analysis, type 2 pavement structure make the cold-regenerative pavement structure anti-XY surface shear stress, anti-Z compressive stress and anti-Z deformation three aspects are kept in a reasonable range, and it is considered that the cold regenerative pavement structure type 2 is the best. Type 2 is specifically composed of 8 cm new asphalt surface layer +20 cm no stabilized soil regeneration base +20 cm stabilized soil regeneration base + original road soil base. The results of this paper can provide some reference for the actual design of emulsified asphalt cold regeneration construction process.

1 Introduction

In china, the total number of highways was 4,773,500 km by the end of 2017. About 2.2 million tonnes of recycled asphalt pavement material (RAP) is generated annually, based on 12% of high-grade road surface maintenance [1,2,3] . If RAP is abandoned directly, it will not only cause pollution to the environment, but also be a serious waste of resources. Therefore, it is very necessary to study RAP regeneration technology. Emulsified asphalt cold regeneration technology is a kind of asphalt mixture regeneration technology which can reuse RAP to make the recycled mixture have good workability and can be constructed at room temperature [4] . This is a cost-effective way to maintain and upgrade asphalt pavement. At present, the research progress on emulsified asphalt cold regeneration is very fast at home and abroad[5,6].

The cold regeneration of emulsified asphalt is affected by many factors. In this paper, the force model of pavement structure is established. In order to investigate the force on pavement when the ratio and thickness of cold recycled pavement vary, three concrete pavement structures, material parameters and vehicle load parameters are designed. By finite element analysis, type 2 pavement structure make the cold-regenerative pavement structure anti-XY surface shear stress, anti-Z compressive stress and anti-Z deformation three aspects are kept in a reasonable range, and it is considered that the cold regenerative pavement structure type 2 is the best. Type 2 is specifically composed of 8 cm new asphalt surface layer +20 cm no stabilized soil regeneration base +20 cm Stabilized soil regeneration base + original road soil base. The results of this paper can provide some reference for the actual design of emulsified asphalt cold regeneration construction process.

2 Parameter setting and finite element model

2.1. The basic assumptions

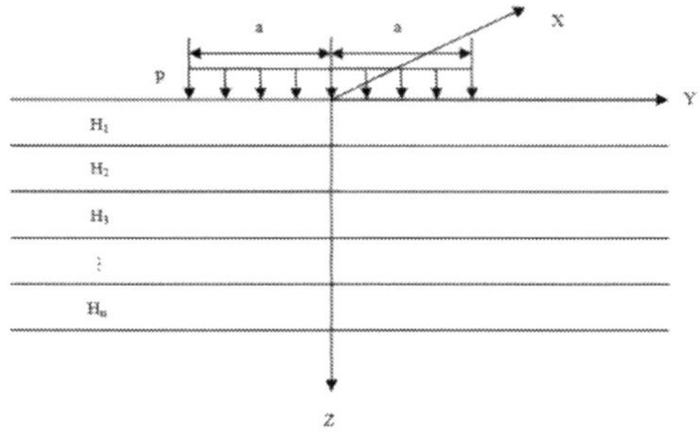

Fig. 1. Pavement stress model

(1) All layers are homogeneous linear elastic materials, and the elastic modulus E and Poisson's ratio are used to characterize.
(2) The contact between the layers is completely continuous.

* Corresponding author: caike0501@163.com

(3) The foundation is a elastic half-space body, and It is infinite in both horizontal and vertical downward directions.
(4) The thickness of each layer on the foundation is limited, but the horizontal direction is infinite.
(5) The load on the road meter is only vertical static load, and the load axis is uniformly distributed. The force model of road surface is shown in Fig. 1.

2.2 Cold recycling mixtures system structure layer and combination design

There are many types of pavement structure. As shown in Figure 2, after cold recycling simplify the pavement structure to New asphalt pavement + No stabilized soil regeneration base + Stabilized soil regeneration base + Original roadbed.

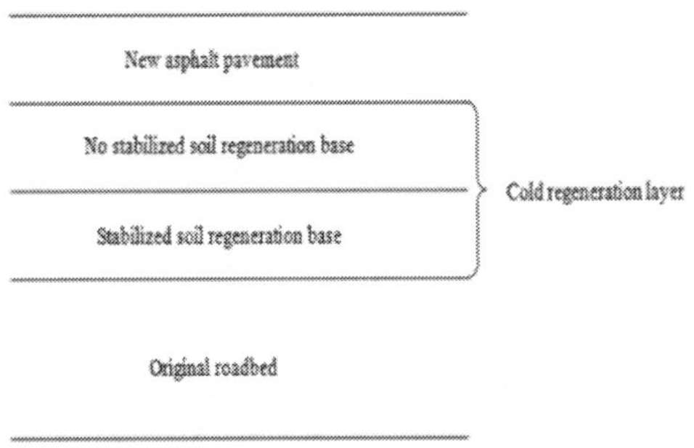

Fig. 2. Pavement stress model.

As shown in Table 1, three types of cold recycled pavement structures are designed in order to investigate the force on pavement when the ratio and thickness of cold recycled pavement vary.

Table 1. The thickness of each layer under the structure of three types of cold regenerative pavement

Structural Layer	Type 1	Type 2	Type 3
	Thickness /cm		
New asphalt pavement	8	8	8
No stabilized soil regeneration base	10	20	30
Stabilized soil regeneration base	30	20	10
Original roadbed	-	-	-

2.3 Parameter setting

2.3.1 Formatting author names and author affiliations

As shown in Fig. 3, under single wheel loading, the ground area of the vehicle tire can be approximately a rectangle and two semicirons with radius of 0.3L. The tire pressure P is regarded as uniform load. L is the load acting length. A is the loading area, and it can be calculated by dividing the load on each tire by the tire pressure.

$$L=\sqrt{\frac{A}{0.5227}} \tag{1}$$

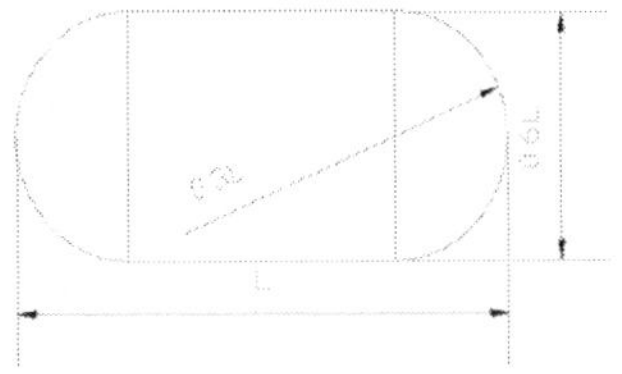

Fig. 3. Actual operating area

The effective area is reduced to a rectangular equivalent area. As shown in Figure 4, length is 0.8712 L and width is 0.6L.

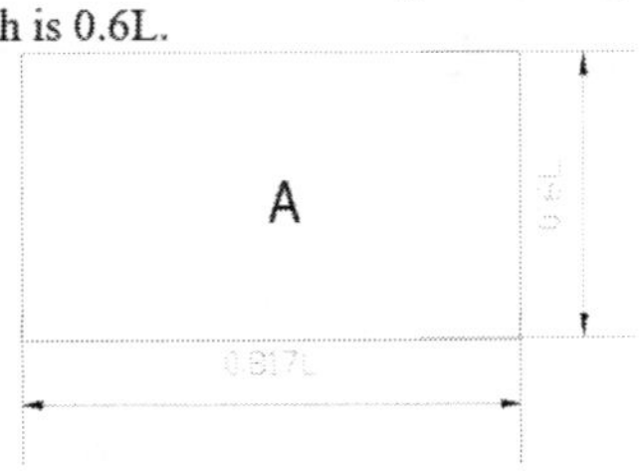

Fig. 4. Equivalent area

When two wheel loads are applied, the tire area can be equivalent to a rectangle. As shown in Figure 5, its length is the same as the rectangular length of single wheel, and its width is the center distance of two wheels. A* is equivalent area of two wheels. L* is equivalent rectangular length. D is center distance of two wheels.

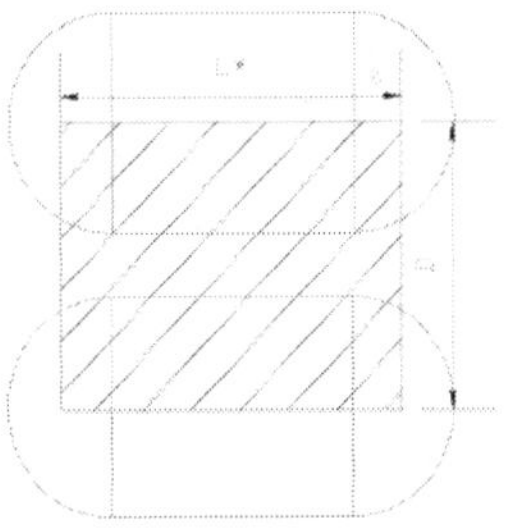

Fig. 5. Equivalent diagram of double wheel load

$$A^{*}=L^{*}\times D \tag{2}$$

The vehicle load is specified in the < Specifications of asphalt pavement design > (JTG D50-2006). The standard axle load is BZZ-100. Double wheel axle load is 100kN. The tyre ground pressure P is 0.7mpa. The tyre ground pressure p is 0.7mpa. One wheel pressure surface equivalent circle diameter d is 21.3cm. The center distance between the two wheels is 1.5d. The load ACTS on the center of the road surface, and the parameters related to standard load are calculated as follows.

Single wheel area:

$$A = \frac{F}{P} = \frac{100/4}{700} = 357cm^2 \quad (3)$$

Length of single wheel load:

$$A = \frac{F}{P} = \frac{100/4}{700} = 357cm^2 \quad (4)$$

Equivalent acting length of two wheels:

$$L^* = \frac{A}{0.6L} = \frac{357}{0.6 \times 26.13} = 22.77cm \quad (5)$$

Center distance of two wheels:

$$D = 1.5d = 21.3 \times 1.5 = 31.95\text{cm} \quad (6)$$

Equivalent area of two wheels:

$$A^* = L^* \times D = 22.77 \times 31.95 = 727.50cm^2 \quad (7)$$

2.3.1 Material parameter setting

Following the rules of < Specifications of asphalt pavement design > (JTG D50-2006), related data of asphalt pavement design and measured value of 28d modulus of compression resilience of cold regenerative mixture, the basic parameters such as elastic modulus and poisson's ratio of each layer of pavement structure are designed. Table 2. is design parameters of pavement structure layer.

Table 2. Design parameters of pavement structure layer

The structure layer	Thickness of structural layer /cm	Rebound modulus /MPa	Poisson's ratio
New asphalt pavement	8	1400	0.25
Stabilized soil regeneration base	10/20/30	1000	0.30
No stabilized soil regeneration base	30/20/10	1200	0.30
Original roadbed	-	50	0.35

2.4 Finite element model

When calculating pavement structure with finite element, its convergence is not only related to the reasonable division of the unit, but also related to the size of the spatial region. When the calculation range is large enough and the density degree of the element is appropriate to the field gradient, the calculation results converge to the exact solution. In the actual pavement structure, the foundation is an elastic half-space body. In order to make its infinite characteristic be reflected in the calculation, it is necessary to consider the space size range of the foundation.

Combined with the research practice in this paper, the foundation size is set as 9m×9m×9m. The model boundary condition is set as that the ground floor is fixed and the top surface is free. When the vehicle is driving in the middle of the road, there is basically no response from both sides of the road and both ends of the road. Therefore, the displacement of the corresponding direction is restrained. Fig. 6 and Fig. 7 are three-dimensional finite element mesh of the model.

Fig. 6. Equivalent diagram of double wheel load

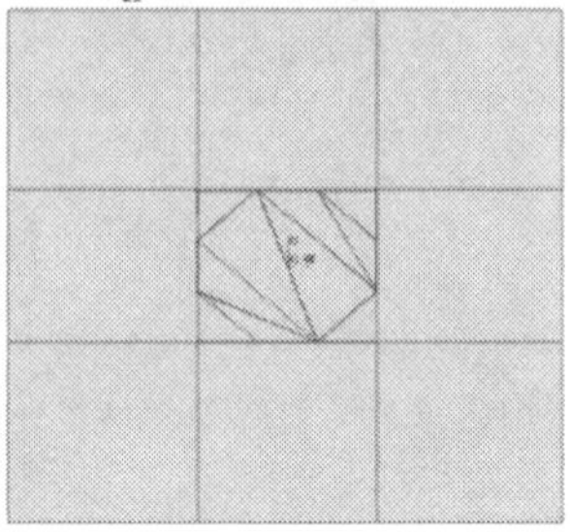

Fig. 7. Load model

4 Finite element analysis results

Stratified finite element simulation of XY surface shear stress, Z compressive stress and Z deformation of three types of cold regenerative pavement structures were carried out. The corresponding cloud maps are obtained. In the cloud map, the unit of stress is Pa and the unit of strain is m.

4.1 Finite element analysis of XY surface shear stress

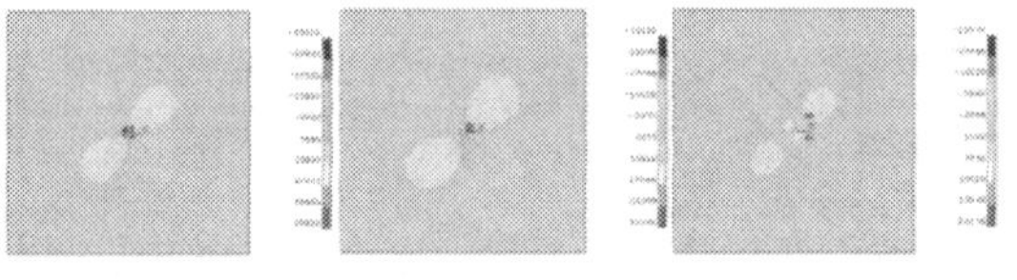

Fig. 8. The cloud map of the XY surface shear stress of the three structural layers of type 1

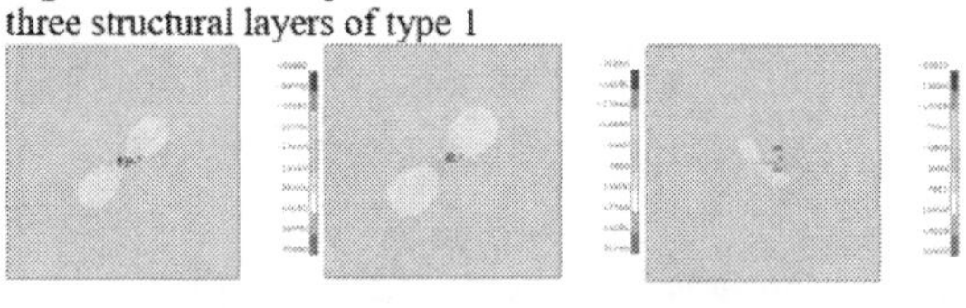

Fig. 9. The cloud map of the XY surfac eshear stress of the three structural layers of type 2

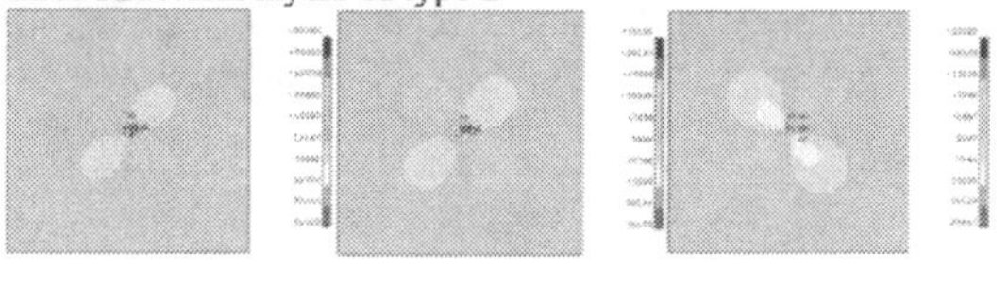

Fig. 10. The cloud map of the XY surface shear stress of the three structural layers of type 3

Under standard axle load, Table 3 is maximum shear stress on XY surface of each structure layer. For the new asphalt surface layer and the stabilized soil regeneration base, the maximum shear stress of XY surface is in the order of type 1 < type 2 < type 3. For the no stabilized soil regeneration base, the order of maximum shear stress on XY surface is type 1, >, type 2, >, type 3. However, when the structure type of cold regenerative road surface changes, the maximum XY surface corresponding to each structure layer is not much different. It shows that the change of pavement structure has little effect on the maximum shear stress of XY surface of each structure layer. The resistance to shear stress on XY surface of three types of cold regenerative pavement structures is comparable.

Table 3. Maximum XY surface shear stress of each structure layer

Structural Layer	Type 1	Type 2	Type 3
	Maximum XY surface shear stress /Pa		
New asphalt pavement	89822	90992	91325
No stabilized soil regeneration base	31255	31219	31105
Stabilized soil regeneration base	23076	23433	23828

4.2 Finite element analysis of Z compressive stress

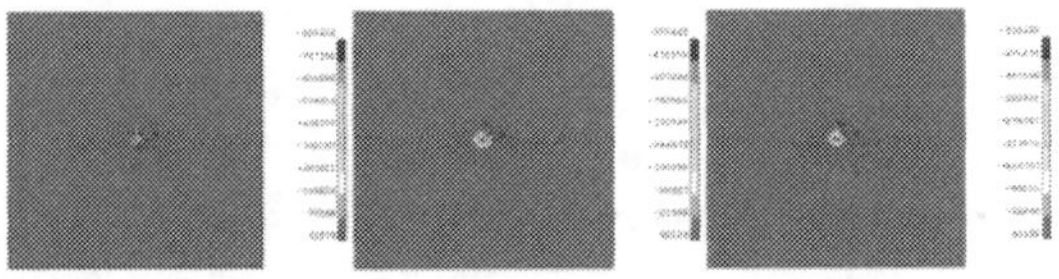

Fig. 11. The cloud map of the Z compressive stress of the three structural layers of type 1

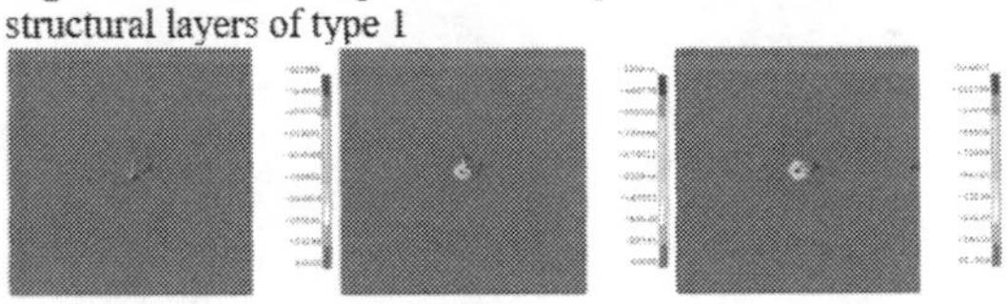

Fig. 12. The cloud map of the Z compressive stress of the three structural layers of type 2

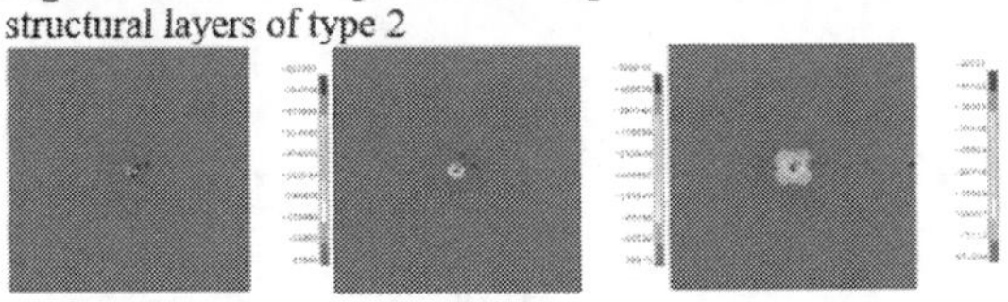

Fig. 13. The cloud map of the Z compressive stress of the three structural layers of type 3

Under standard axle load, Table 4 is maximum shear Z compressive stress of each structure layer. It can be seen that the Z maximum compressive stress corresponding to the three structural layers is arranged in the order of type 1 > type 2 > type 3. For the new asphalt surface layer and the no stabilized soil base, the Z maximum compressive stress of the three types is not much different. For the soil with stable soil base, the Z maximum compressive stress of type 1 is 2.15 times of that of type 2 and 5.84 times of that of type 3. So, when the structure type of the cold regenerative pavement changes, the Z maximum compressive stress of the soil with stable base is greatly affected. The order of the anti-z compressive stress capacity of the three types of cold regenerative pavement structure is type 3 > type 2 > type 1.

Table 4. Maximum Z compressive stress of each structure layer

Structural Layer	Type 1	Type 2	Type 3
	Maximum Z compressive stress /Pa		
New asphalt pavement	823418	822986	822767
No stabilized soil regeneration base	535445	530231	522270
Stabilized soil regeneration base	272609	126897	46715

4.3 Finite element analysis of Z deformation

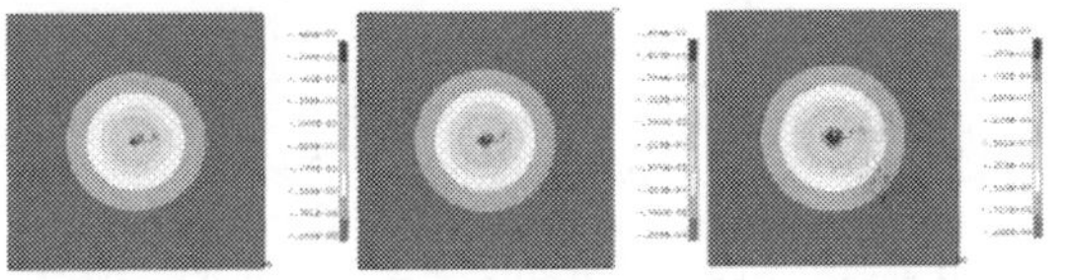

Fig. 14. The cloud map of the Z deformation of the three structural layers of type 1

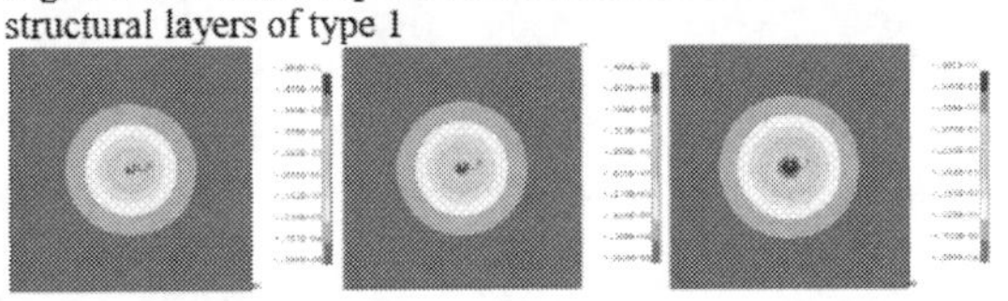

Fig. 15. The cloud map of the Z deformation of the three structural layers of type 2

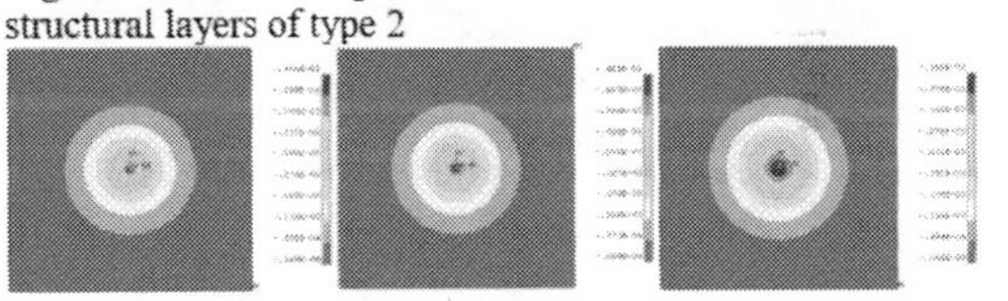

Fig. 16. The cloud map of the Z deformation of the three structural layers of type 3

Under standard axle load, Table 5 is maximum Z deformation of each structure layer. It can be seen that the for the new asphalt surface layer and unstable soil base, the size sequence of maximum Z deformation is type 1 < type 2 < type 3. However, for the soil with stable soil base, the order of maximum Z deformation is type 1 > type 2 > type 3.

Table 5. Maximum Z deformation of each structure layer

Structural Layer	Type 1	Type 2	Type 3
	Maximum Z deformation /mm		
New asphalt pavement	0.48	0.486	0.491
No stabilized soil regeneration base	0.454	0.460	0.465
Stabilized soil regeneration base	0.416	0.401	0.393

Taking into account the two factors of maximum Z deformation of each structure layer and z-deformation of special point with the change of pavement depth, it can be known that the size order of anti-z deformation ability of three types of cold-regenerative pavement structure types is type 1> type 2 > type 3.

5 Conclusion

Based on the above analysis, it can be seen that the anti-XY surface shear stress of the three types of cold regenerative pavement structure is equivalent. For the ability of anti-Z compressive stress and anti-z deformation, the Z compressive stress resistance of the cold-regenerative pavement structure type 1 is the worst, but the z-deformation resistance is the best. Type 3 has the best z-stress resistance, but the worst z-deformation resistance. The z-stress resistance and z-deformation resistance of type 2 are centered.

In order to make the cold-regenerative pavement structure anti-XY surface shear stress, anti-Z compressive stress and anti-z deformation three aspects are kept in a reasonable range. It is considered that the cold regenerative pavement structure type 2 is the best. Type 2 is specifically composed of 8 cm new asphalt surface layer +20 cm no stabilized soil regeneration base +20 cm Stabilized soil regeneration base + original road soil base. The results of this paper can provide some reference for the actual design of emulsified asphalt cold regeneration construction process.

References

1. Wang Hong, Hao Peiwen, Nan Bingzhang, Zhang Hang, Study on High Temperature Stability of Emulsified Asphalt Regenerated Mixture, Highway Engineering, **38** (2013)
2. Tood Thomas, Arlis Karmas. Performace cerelated tests and specifications for cold in-place recycling lab and field experience, TRB, (2002)
3. WEI Lian-yu, LIU Zhao-wei, ZHANG Hai-rong, WANG Jing, Experimental Analyse of Foamed Asphalt Stabilization Cold-recycled Mixture in Old Asphalt Pavement, Journal of Hebei Unicersity of Technology, **37** (2008)
4. Tood Thomas Arlis Karmas , John Huffman. Cold in-place recycling on Kansas US-283, TRB, (2000)
5. LING Tian-qing, HE Liang, MA Yu, A Cold Recycling Technique for Foamed Bitumen Mix, Journal of Civil Architectural & Environmental Engineering, **31** (2009)
6. Xiaobo Yin, Performance of Emulsified Asphalt Full Thickness Cold Recycling Subgrade Materials, Advanced Materials Research, (2014)

AUTHOR INDEX

AUTHOR INDEX

EDP Sciences – Web of Conferences
17, Avenue du Hoggar
Parc d'Activité de Courtabouf
BP 112
F-91944 Les Ulis Cedex A
France

ISBN 978-1-5108-7563-0

CURRAN ASSOCIATES INC.
proceedings.com

Seminary on Geomatics, Civil and Environmental Engineering (2018 BGC)

E3S Web of Conferences Volume 63 (2018)

Olsztyn, Poland
21-23 June 2018

edp sciences